SCHAUM'S OUTLINE OF

# THEORY AND PROBLEMS

OF

# APPLIED PHYSICS

## Third Edition

•

ARTHUR BEISER, Ph.D.

## SCHAUM'S OUTLINE SERIES
McGRAW-HILL, INC.

New York   San Francisco   Washington, D.C.   Auckland   Bogotá   Caracus   Lisbon
London   Madrid   Mexico City   Milan   Montreal   New Dehli
San Juan   Singapore   Sydney   Tokyo   Toronto

**ARTHUR BEISER** received his Ph.D. from New York University, where he subsequently served as Assistant and Associate Professor of Physics. He has been a consultant to various industrial firms and government agencies and is the author of more than a dozen textbooks of physics and mathematics.

Schaum's Outline of Theory and Problems of

APPLIED PHYSICS

5  6  7  8  9  10  PRS  PRS  4  3  2  1  0

ISBN 0-07-005201-8

Sponsoring Editor, Arthur Biderman
Production Supervisor, Leroy A. Young
Editing Supervisor, Patty Andrews


**Library of Congress Cataloging-in-Publication Data**


Beiser, Arthur.
    Schaum's outline of theory and problems of applied physics /
Arthur Beiser. — 3rd ed.
      p.   cm. — (Schaum's outline series)
    Includes index.
    ISBN 0–07–005201–8
    1. Physics.   I. Title.   II. Title: Theory and problems of applied
physics.
QC21.2.B45  1994
530′. 076—dc20                                              94–24928
                                                               CIP

*McGraw-Hill*

*A Division of The McGraw·Hill Companies*

# Preface

This book is intended to provide students of applied physics with help in mastering those physical principles that underlie modern technology. A wide spectrum of topics is covered, so that the reader may select those which correspond to his or her particular needs. SI (metric) units are mainly employed, although a few problems are worked out with British units to illustrate their use.

Each section of a chapter begins with an outline of its subject. The solved problems that follow are of two kinds: those that show how numerical answers are found to typical questions, and those that review important facts and ideas. A new feature of this third edition is a collection of multiple choice exercises. These exercises serve as a gauge of the reader's grasp of the chapter contents. The supplementary problems then give the reader a chance to practice what has been learned.

I wish to thank my McGraw-Hill editor, Arthur Biderman, for the many improvements which he suggested.

*Arthur Beiser*

# Contents

# Chapter 1

# Useful Math

## ALGEBRA

Algebra is the arithmetic of symbols that represent numbers. Not being restricted to relationships among specific numbers, algebra can express more general relationships among quantities whose numerical values need not be known.

The operations of addition, subtraction, multiplication, and division have the same meaning in algebra as in arithmetic. Thus the formula

$$x = \frac{(a + b)c}{d} - e$$

means that to find the value of $x$, we must first add $a$ and $b$, next multiply by $c$, then divide by $d$, and finally subtract $e$.

The rules for multiplying and dividing positive and negative quantities are as follows: If the quantities are both positive or both negative, the result is positive; if one is positive and the other negative, the result is negative. In symbolic form,

$$(+a)(+b) = (-a)(-b) = +ab \qquad (-a)(+b) = (+a)(-b) = -ab$$

$$\frac{+a}{+b} = \frac{-a}{-b} = +\frac{a}{b} \qquad\qquad \frac{-a}{+b} = \frac{+a}{-b} = -\frac{a}{b}$$

## SOLVED PROBLEM 1.1

Remove the parentheses from $2a - 3(a + b) + 4(2a - b)$.

Since $3(a + b) = 3a + 3b$ and $4(2a - b) = 8a - 4b$,

$$2a - 3(a + b) + 4(2a - b) = 2a - 3a - 3b + 8a - 4b$$
$$= (2 - 3 + 8)(a) - (3 + 4)(b) = 7a - 7b = 7(a - b)$$

## SOLVED PROBLEM 1.2

Find the value of

$$v = 5\left(\frac{x - y}{z}\right) + w$$

when $x = 15$, $y = 3$, $z = 4$, and $w = 10$.

We proceed as follows:

1. Subtract $y$ from $x$ to give

$$x - y = 15 - 3 = 12$$

2. Divide $x - y$ by $z$ to give

$$\frac{x - y}{z} = \frac{12}{4} = 3$$

1

3. Multiply $(x - y)/z$ by 5 to give

$$5\left(\frac{x - y}{z}\right) = (5)(3) = 15$$

4. Add $w$ to $5(x - y)/z$ to give

$$v = 5\left(\frac{x - y}{z}\right) + w = 15 + 10 = 25$$

## SOLVED PROBLEM 1.3

Examples of multiplication and division:

$$(-3)(-5) = 15 \qquad \frac{-16}{-4} = 4$$

$$(2)(-4) = -8 \qquad \frac{10}{-5} = -2$$

$$(-12)(6) = -72 \qquad \frac{-24}{4} = -6$$

## SOLVED PROBLEM 1.4

Find the value of

$$w = \frac{xy}{x + y}$$

when $x = 5$ and $y = -6$.

Here $xy = (5)(-6) = -30$ and $x + y = 5 + (-6) = 5 - 6 = -1$. Hence

$$w = \frac{xy}{x + y} = \frac{-30}{-1} = 30$$

## EQUATIONS

An equation is a statement of equality: Whatever is on the left-hand side of any equation is equal to whatever is on the right-hand side. The symbols in an algebraic equation usually cannot have arbitrary values if the equality is to hold. To *solve* an equation is to find the possible values of these symbols. The solution of the equation $5x - 10 = 20$ is $x = 6$ because only when $x$ is 6 is this equation a true statement.

The algebraic procedures that can be used to solve an equation are all based on the principle that any operation performed on one side of an equation must be performed on the other side as well. Thus an equation remains valid when the same quantity is added to or subtracted from both sides or is used to multiply or divide both sides. Other operations, such as raising to a power or taking a root, also do not change an equality if the same thing is done to both sides.

Two helpful rules follow from the above principle. First, any term on one side of an equation may be transposed to the other side by changing its sign. Thus if $a + b = c$, then $a = c - b$; and if $a - d = e$, then $a = e + d$. Second, a quantity that multiplies one side of an equation may be transposed so as to divide the other side, and vice versa. Thus if $ab = c$, then $a = c/b$; and if $a/d = e$, then $a = de$.

When each side of an equation consists of a fraction, all we need to do to remove the fractions is to *cross multiply*:

$$\frac{a}{b} = \frac{c}{d} \rightarrow ad = bc$$

What was originally the denominator (lower part) of each fraction now multiplies the numerator (upper part) of the other side of the equation.

## SOLVED PROBLEM 1.5

Solve $5x - 10 = 20$ for $x$.

We want to have just $x$ on the left-hand side of the equation. The first step is to transpose the $-10$ to the right-hand side, where it becomes $+10$:

$$5x - 10 = 20$$

$$5x = 20 + 10 = 30$$

Now we transpose the 5 so that it divides the right-hand side:

$$5x = 30$$

$$x = \frac{30}{5} = 6$$

The solution is $x = 6$.

## SOLVED PROBLEM 1.6

Examples of cross multiplication:

$$\frac{x}{2} = \frac{y}{7} \qquad\qquad \frac{y}{8} = \frac{5}{x}$$
$$7x = 2y \qquad\qquad xy = (5)(8) = 40$$

$$\frac{3x}{5} = \frac{4y}{3} \qquad\qquad \frac{y}{5} = \frac{3x + 2}{4}$$
$$3(3x) = 5(4y) \qquad\qquad 4y = 5(3x + 2) = 15x + 10$$
$$9x = 20y$$

## SOLVED PROBLEM 1.7

Solve the equation

$$\frac{5}{a + 2} = \frac{3}{a - 2}$$

for the value of $a$.

We proceed as follows:

| | |
|---|---|
| Cross multiply | $5(a - 2) = 3(a + 2)$ |
| Multiply both sides | $5a - 10 = 3a + 6$ |
| Transpose $-10$ and $3a$ | $5a - 3a = 6 + 10$ |
| Carry out indicated addition and subtraction | $2a = 16$ |
| Divide both sides by 2 | $a = 8$ |

**SOLVED PROBLEM 1.8**

Solve the equation

$$\frac{16x - 2}{8} = 3x$$

for the value of $x$.

Cross multiply                     $16x - 2 = 8(3x) = 24x$

Transpose $-2$ and $24x$       $16x - 24x = 2$

Combine like terms                $-8x = 2$

Divide both sides by $-8$         $x = -\frac{1}{4} = -0.25$

**SOLVED PROBLEM 1.9**

Solve the equation

$$\frac{4x - 35}{3} = 9(1 - x)$$

for the value of $x$.

Cross multiply                     $4x - 35 = (3)[9(1 - x)] = 27(1 - x)$

Multiply the right-hand side        $4x - 35 = 27 - 27x$

Transpose the $-35$ and $-27x$     $4x + 27x = 27 + 35$

Combine like terms                  $31x = 62$

Divide both sides by 31             $x = \frac{62}{31} = 2$

## EXPONENTS

There is a special shorthand way of expressing a quantity that is to be multiplied by itself one or more times. In this scheme a superscript number, called an *exponent*, indicates how many times the self-multiplication is to be carried out, as follows:

$$a = a^1 \qquad (a)(a) = a^2 \qquad (a)(a)(a) = a^3 \qquad \text{and so on}$$

The quantity $a^2$ is read as "$a$ squared" because it is equal to the area of a square whose sides are $a$ long, and $a^3$ is read as "$a$ cubed" because it is equal to the volume of a cube whose edges are $a$ long. For an exponent $n$ greater than 3, we read $a^n$ as "$a$ to the $n$th power" so that $a^5$ is "$a$ to the fifth power." The product of two powers of the same quantity, say $a^n$ and $a^m$, is that quantity raised to the sum of the two exponents: $(a^n)(a^m) = a^{n+m}$. Thus $(a^2)(a^5) = a^7$.

Reciprocal quantities are expressed according to the above scheme but with negative exponents:

$$\frac{1}{a} = a^{-1} \qquad \frac{1}{a^2} = a^{-2} \qquad \frac{1}{a^3} = a^{-3} \qquad \text{and so on}$$

In general, $1/a^n = (1/a)^n = a^{-n}$. A quantity raised to the zeroth power, $a^0$ for instance, is always equal to 1: $a^0 = 1$. To see why, we note that $a/a = 1$ can also be written $a/a = (a^1)(a^{-1}) = a^{1-1} = a^0$.

It is not necessary that an exponent be a whole number. A fractional exponent signifies a *root* of a quantity. The "square root of $a$," customarily written $\sqrt{a}$, is that quantity which, multiplied by itself once, is equal to $a$: $(\sqrt{a})(\sqrt{a}) = a$. Using exponents, we write the square root of $a$ as $\sqrt{a} = a^{1/2}$, because $(a^{1/2})(a^{1/2}) = a^{1/2+1/2} = a^1 = a$. In general, the $n$th root of any quantity is indicated by the exponent $1/n$.

Some rules concerning exponents and roots are the following:

$$(ab)^n = a^n b^n \qquad (a^n)^m = a^{nm} \qquad \left(\frac{a}{b}\right)^n = \frac{a^n}{b^n}$$

$$(ab)^{1/m} = a^{1/m} b^{1/m} \qquad \left(\frac{a}{b}\right)^{1/m} = \frac{a^{1/m}}{b^{1/m}} \qquad (a^n)^{1/m} = a^{n/m}$$

## SOLVED PROBLEM 1.10

Examples of exponents:

$$a^2 a^{-3} = a^{2-3} = a^{-1} \qquad (a^2)^{-4} = a^{(2)(-4)} = a^{-8} \qquad (a^3)^{-1/3} = a^{(-1/3)(3)} = a^{-1}$$

$$a^{-1} a^{-4} = a^{-1-4} = a^{-5} \qquad (a^{-3})^{-2} = a^{(-3)(-2)} = a^6 \qquad a^6 a^{1/2} = a^{6+1/2} = a^{13/2}$$

$$(a^{-2})^4 = a^{(-2)(4)} = a^{-8} \qquad (a^{1/2})^6 = a^{(6)(1/2)} = a^3 \qquad (a^6)^{1/2} = a^{(1/2)(6)} = a^3$$

## SOLVED PROBLEM 1.11

The area $A$ of a circle whose radius is $r$ is given by $A = \pi r^2$. Find the area of a circle whose radius is 4.00 meters (m).

The value of $\pi$ is $3.14159\cdots$. Many calculators have a key for this quantity. We have here

$$A = \pi r^2 = \pi(4.00 \text{ m})^2 = \pi(4.00 \text{ m})(4.00 \text{ m}) = 50.3 \text{ m}^2$$

The symbol "$m^2$" stands for "square meters."

## SOLVED PROBLEM 1.12

Find the radius of a cylindrical tank whose cross-sectional area is 5.00 m$^2$.

We begin by solving $A = \pi r^2$ for $r$ and then substitute $A = 5.00$ m$^2$ to find the value of $r$:

$$A = \pi r^2$$

$$r^2 = \frac{A}{\pi}$$

$$r = \sqrt{\frac{A}{\pi}} = \sqrt{\frac{5.00 \text{ m}^2}{\pi}} = \sqrt{1.59 \text{ m}^2} = 1.26 \text{ m}$$

We note that $\sqrt{m^2} = m$.

## POWERS OF 10

Very small and very large numbers are common in science and engineering and are best expressed with the help of powers of 10. A number in decimal form can be written as a number between 1 and 10 multiplied by a power of 10:

$$834 = 8.34 \times 10^2 \qquad 0.00072 = 7.2 \times 10^{-4}$$

The powers of 10 from $10^{-6}$ to $10^6$ are as follows:

$10^0 = 1$        = 1 with decimal point moved 0 places

$10^{-1} = 0.1$        = 1 with decimal point moved 1 place to the left

$10^{-2} = 0.01$        = 1 with decimal point moved 2 places to the left

$10^{-3} = 0.001$        = 1 with decimal point moved 3 places to the left

$10^{-4} = 0.0001$        = 1 with decimal point moved 4 places to the left

$10^{-5} = 0.00001$        = 1 with decimal point moved 5 places to the left

$10^{-6} = 0.000001$        = 1 with decimal point moved 6 places to the left

$10^0 = 1$        = 1 with decimal point moved 0 places

$10^1 = 10$        = 1 with decimal point moved 1 place to the right

$10^2 = 100$        = 1 with decimal point moved 2 places to the right

$10^3 = 1000$        = 1 with decimal point moved 3 places to the right

$10^4 = 10,000$        = 1 with decimal point moved 4 places to the right

$10^5 = 100,000$        = 1 with decimal point moved 5 places to the right

$10^6 = 1,000,000$        = 1 with decimal point moved 6 places to the right

When numbers written in powers-of-10 notation are to be added or subtracted, they must all be expressed in terms of the *same* power of 10:

$$3 \times 10^2 + 4 \times 10^3 = 0.3 \times 10^3 + 4 \times 10^3 = 4.3 \times 10^3$$

It does not matter which power of 10 is used, as long as it is the same for all the numbers. Thus we get the same answer as above if we proceed instead as follows:

$$3 \times 10^2 + 4 \times 10^3 = 3 \times 10^2 + 40 \times 10^2 = 43 \times 10^2 = 4.3 \times 10^3$$

Since a step is saved if the power of 10 used is that of the larger number, it makes sense to do this.

To subtract one number from another, the same procedure is followed. If the number being subtracted is the larger of the two, the answer will be negative in sign, just as $3 - 5 = -2$.

To multiply two powers of 10, add their exponents; to divide one power of 10 by another, subtract the exponent of the latter from that of the former:

$$(10^n)(10^m) = 10^{n+m} \qquad \frac{10^n}{10^m} = 10^{n-m}$$

Reciprocals follow the pattern

$$\frac{1}{10^n} = 10^{-n}$$

The rules of finding powers and roots of powers of 10 are

$$(10^n)^m = 10^{nm} \qquad \sqrt[m]{10^n} = (10^n)^{1/m} = 10^{n/m}$$

In taking the *m*th root, the power of 10 should be chosen to be a multiple of *m*. Thus

$$\sqrt{10^{15}} = (\sqrt{10})(\sqrt{10^{14}}) = \sqrt{10} \times 10^7 = 3.16 \times 10^7$$

### SOLVED PROBLEM 1.13

Examples of powers-of-10 notation.

$$20 = 2 \times 10 = 2 \times 10^1$$
$$3043 = 3.043 \times 1000 = 3.043 \times 10^3$$
$$8,700,000 = 8.7 \times 1,000,000 = 8.7 \times 10^6$$
$$0.22 = 2.2 \times 0.1 = 2.2 \times 10^{-1}$$
$$0.000035 = 3.5 \times 0.00001 = 3.5 \times 10^{-5}$$

### SOLVED PROBLEM 1.14

Examples of addition and subtraction. Note that $10^{-2}$ is larger than $10^{-3}$ and that $10^{-4}$ is larger than $10^{-5}$.

$$6 \times 10^2 + 5 \times 10^4 = 0.06 \times 10^4 + 5 \times 10^4 = 5.06 \times 10^4$$
$$2 \times 10^{-2} + 3 \times 10^{-3} = 2 \times 10^{-2} + 0.3 \times 10^{-2} = 2.3 \times 10^{-2}$$
$$7 + 2 \times 10^{-2} = 7 + 0.02 = 7.02$$
$$6 \times 10^4 - 4 \times 10^2 = 6 \times 10^4 - 0.04 \times 10^4 = 5.96 \times 10^4$$
$$3 \times 10^{-2} - 5 \times 10^{-3} = 3 \times 10^{-2} - 0.5 \times 10^{-2} = 2.5 \times 10^{-2}$$
$$7 \times 10^{-5} - 2 \times 10^{-4} = 0.7 \times 10^{-4} - 2 \times 10^{-4} = -1.3 \times 10^{-4}$$
$$6.23 \times 10^{-3} - 6.28 \times 10^{-3} = -0.05 \times 10^{-3} = -5 \times 10^{-5}$$

### SOLVED PROBLEM 1.15

Examples of multiplication and division.

$$(10^5)(10^{-2}) = 10^{5-2} = 10^3 \qquad\qquad \frac{10^4}{10^{-3}} = 10^{4-(-3)} = 10^{4+3} = 10^7$$
$$\frac{10^3}{10^6} = 10^{3-6} = 10^{-3} \qquad\qquad \frac{(10^5)(10^{-7})}{10^2} = 10^{5-7-2} = 10^{-4}$$

**SOLVED PROBLEM 1.16**

A sample calculation.

$$\frac{(460)(0.00003)(100,000)}{(9000)(0.0062)} = \frac{(4.6 \times 10^2)(3 \times 10^{-5})(10^5)}{(9 \times 10^3)(6.2 \times 10^{-3})} = \left(\frac{4.6 \times 3}{9 \times 6.2}\right)\left(\frac{10^2 \times 10^{-5} \times 10^5}{10^3 \times 10^{-3}}\right)$$

$$= (0.25)\left(\frac{10^{2-5+5}}{10^{3-3}}\right) = (0.25)\left(\frac{10^2}{10^0}\right) = 25$$

**SOLVED PROBLEM 1.17**

Examples of powers of numbers.

$$(10^2)^4 = 10^{(2)(4)} = 10^8$$

$$(10^{-3})^5 = 10^{(-3)(5)} = 10^{-15}$$

$$(10^{-4})^{-3} = 10^{(-4)(-3)} = 10^{12}$$

$$(3 \times 10^3)^2 = 3^2 \times (10^3)^2 = 9 \times 10^6$$

$$(4 \times 10^{-5})^3 = 4^3 \times (10^{-5})^3 = 64 \times 10^{-15} = 6.4 \times 10^{-14}$$

$$(2 \times 10^{-2})^{-4} = \frac{1}{2^4} \times (10^{-2})^{-4} = \frac{1}{16} \times 10^8 = 0.0625 \times 10^8 = 6.25 \times 10^6$$

**SOLVED PROBLEM 1.18**

Examples of square roots involving even exponents.

$$\sqrt{10^6} = 10^{6/2} = 10^3$$

$$\sqrt{10^{-4}} = 10^{-4/2} = 10^{-2}$$

$$\sqrt{5 \times 10^4} = \sqrt{5} \times \sqrt{10^4} = 2.24 \times 10^2$$

$$\sqrt{1.3 \times 10^{-8}} = \sqrt{1.3} \times \sqrt{10^{-8}} = 1.14 \times 10^{-4}$$

$$\sqrt{9 \times 10^{-2}} = \sqrt{9} \times \sqrt{10^{-2}} = 3 \times 10^{-1} = 0.3$$

**SOLVED PROBLEM 1.19**

Examples of square roots involving odd exponents.

$$\sqrt{10^5} = \sqrt{1 \times 10^5} = \sqrt{10 \times 10^4} = \sqrt{10} \times \sqrt{10^4} = 3.16 \times 10^2$$

$$\sqrt{10^{-3}} = \sqrt{1 \times 10^{-3}} = \sqrt{10 \times 10^{-4}} = \sqrt{10} \times \sqrt{10^{-4}} = 3.16 \times 10^{-2}$$

$$\sqrt{3 \times 10^5} = \sqrt{30 \times 10^4} = \sqrt{30} \times \sqrt{10^4} = 5.48 \times 10^2$$

$$\sqrt{6 \times 10^{-7}} = \sqrt{60 \times 10^{-8}} = \sqrt{60} \times \sqrt{10^{-8}} = 7.75 \times 10^{-4}$$

$$\sqrt{0.000025} = \sqrt{2.5 \times 10^{-5}} = \sqrt{25 \times 10^{-6}} = \sqrt{25} \times \sqrt{10^{-6}} = 5 \times 10^{-3}$$

**SOLVED PROBLEM 1.20**

Examples of cube roots.

$$\sqrt[3]{10^9} = 10^{9/3} = 10^3$$

$$\sqrt[3]{10^8} = \sqrt[3]{10^2 \times 10^6} = \sqrt[3]{100} \times \sqrt[3]{10^6} = 4.64 \times 10^2$$

$$\sqrt[3]{3.8 \times 10^{19}} = \sqrt[3]{38 \times 10^{18}} = \sqrt[3]{38} \times \sqrt[3]{10^{18}} = 3.36 \times 10^6$$

$$\sqrt[3]{2.7 \times 10^{-5}} = \sqrt[3]{27 \times 10^{-6}} = \sqrt[3]{27} \times \sqrt[3]{10^{-6}} = 3 \times 10^{-2}$$

## UNITS

Units are algebraic quantities and may be multiplied and divided by one another. To convert a quantity expressed in a certain unit to its equivalent in a different unit of the same kind, we use the fact that multiplying or dividing anything by 1 does not affect its value. For instance, 12 in. = 1 ft, so 12 in./ft = 1, and we can convert a length $s$ expressed in feet to its value in inches by multiplying $s$ by 12 in./ft:

$$4 \text{ ft} = (4 \text{ ft})\left(12\frac{\text{in.}}{\text{ft}}\right) = 48 \text{ in.}$$

Conversion factors for the most common British and SI (metric) units are given in Appendix B.

Subdivisions and multiples of metric units are designated by prefixes that indicate an appropriate power of 10. Standard prefixes are shown in Table 1.1.

**SOLVED PROBLEM 1.21**

Rome is 1440 km by road from Paris. How far is this in miles?

Since 1 km = 0.621 mi,

$$1440 \text{ km} = (1440 \text{ km})\left(0.621 \frac{\text{mi}}{\text{km}}\right) = 894 \text{ mi}$$

**Table 1.1**

| Prefix | Power | Abbreviation | Pronunciation | Example |
|--------|-------|--------------|---------------|---------|
| pico- | $10^{-12}$ | p | PEE · koh | 1 pf = 1 picofarad = $10^{-12}$ farad |
| nano- | $10^{-9}$ | n | NAN · oh | 1 ns = 1 nanosecond = $10^{-9}$ second |
| micro- | $10^{-6}$ | $\mu$ | MY · crow | 1 $\mu$A = 1 microampere = $10^{-6}$ ampere |
| milli- | $10^{-3}$ | m | MILL ·ee | 1 mH = 1 millihenry = $10^{-3}$ henry |
| centi- | $10^{-2}$ | c | SEN· tee | 1 cm = 1 centimeter = $10^{-2}$ meter |
| hecto- | $10^{2}$ | h | HEK · toe | 1 hL = 1 hectoliter = $10^{2}$ liters |
| kilo- | $10^{3}$ | k | KILL · oh | 1 kg = 1 kilogram = $10^{3}$ grams |
| mega- | $10^{6}$ | M | MEG · ah | 1 MW = 1 megawatt = $10^{6}$ watts |
| giga- | $10^{9}$ | G | GEE · gah | 1 GeV = 1 gigaelectronvolt = $10^{9}$ electronvolts |
| tera- | $10^{12}$ | T | TER · ah | 1 TJ = 1 terajoule = $10^{12}$ joules |

**SOLVED PROBLEM 1.22**

A man is 6 ft 2 in. tall. How many centimeters is this?
Since 1 ft = 12 in. and 1 in. = 2.54 cm,

$$6 \text{ ft } 2 \text{ in.} = [(6)(12) + 2] \text{ in.} = (74 \text{ in.})\left(2.54 \frac{\text{cm}}{\text{in.}}\right) = 188 \text{ cm}$$

**SOLVED PROBLEM 1.23**

How many square feet are there in 1 m$^2$?

Since 1 m = 3.28 ft,

$$1 \text{ m}^2 = (1 \text{ m}^2)\left(3.28 \frac{\text{ft}}{\text{m}}\right)^2 = 10.76 \text{ ft}^2$$

**SOLVED PROBLEM 1.24**

Express a velocity of 60 mi/h in feet per second.

There are 5280 ft in 1 mi and 3600 s in 1 h, so

$$60 \frac{\text{mi}}{\text{h}} = \left(60 \frac{\text{mi}}{\text{h}}\right)\left(5280 \frac{\text{ft}}{\text{mi}}\right)\left(\frac{1}{3600 \text{ s/h}}\right) = 88 \frac{\text{ft}}{\text{s}}$$

**SOLVED PROBLEM 1.25**

The unit of electric charge, symbol $Q$, is the coulomb (abbreviated C); the unit of voltage, symbol $V$, is the volt (V); and the unit of capacitance, symbol $C$, is the farad (F). When the voltage across a capacitor of capacitance $C$ is $V$, the charge on the capacitor is $Q = CV$. Find the charge on a 200-pF capacitor when it is connected to a 3-kV source.

Here $C = 200 \text{ pF} = 200$ picofarads $= 200 \times 10^{-12} \text{ F} = 2 \times 10^{-10} \text{ F}$ and $V = 3 \text{ kV} = 3$ kilovolts $= 3 \times 10^3 \text{ V}$. Hence

$$Q = CV = (2 \times 10^{-10} \text{ F})(3 \times 10^3 \text{ V}) = 6 \times 10^{-7} \text{ C}$$

The answer can be expressed as $Q = 0.6 \text{ }\mu\text{C} = 0.6$ microcoulomb since $1 \text{ }\mu\text{C} = 10^{-6} \text{ C}$.

## SIGNIFICANT FIGURES

An advantage of powers-of-10 notation is that it provides us with a clear way to express the accuracy with which a quantity is known. For instance, the speed of light $c$ in empty space is often given as $2.998 \times 10^8$ m/s. If this value were written out as 299,800,000 m/s, we might think that this speed is precisely equal to this many meters per second, right down to the last zero. Actually, the speed of light is 299,792,458 m/s. If we do not need this much detail, we write $c = 2.998 \times 10^8$ m/s to indicate both how large the number is (the $10^8$ tells us this) and how precise the quoted figure is (the 2.998 tells us that $c$ is closer to $2.998 \times 10^8$ m/s than it is to either 2.997 or $2.999 \times 10^8$ m/s).

Thus, giving the speed of light as

$$c = 2.998 \times 10^8 \text{ m/s}$$

means that

$$c = (2.998 \pm 0.0005) \times 10^8 \text{ m/s}$$

The accurately known digits, plus one uncertain digit, are called *significant figures*. Here $c$ has four significant figures; 2, 9, 9, and 8. The first three digits are accurately known, whereas the last (the 8) is uncertain; it could be a 7 or a 9.

To be sure, sometimes one or more final zeros in a number are meaningful in their own right. In the case of the speed of light, we can legitimately say that, to three-digit accuracy,

$$c = 3.00 \times 10^8 \text{ m/s}$$

because $c$ is closer to this figure than to either 2.99 or 3.01 $\times 10^8$ m/s.

When two or more quantities are combined arithmetically, the result is no more accurate than the quantity with the largest uncertainty. For instance, a 72 kg person might pick up a 0.33-kg apple. The total mass of person + apple is still considered to be 72 kg because we have only two significant figures. All we know of the person's mass is that it is somewhere between 71.5 and 72.5 kg, which means an uncertainty of more than the apple's mass. If the person's mass is given instead as 72.0 kg, the mass of person + apple is 72.3 kg; if it is given as 72.00 kg, the mass of person + apple is 72.33 kg. Thus

$$72 \text{ kg} + 0.33 \text{ kg} = 72 \text{ kg} \qquad \text{(2 significant figures)}$$
$$72.0 \text{ kg} + 0.33 \text{ kg} = 72.3 \text{ kg} \qquad \text{(3 significant figures)}$$
$$72.00 \text{ kg} + 0.33 \text{ kg} = 72.33 \text{ kg} \qquad \text{(4 significant figures)}$$

Significant figures must be taken into account in other arithmetical operations also. If we divide 7.9 by 3.24, we are not justified in writing

$$\frac{7.9}{3.24} = 2.43827\ldots$$

The 7.9 has only two significant figures, and the answer cannot have more than this. Hence the correct answer is 7.9/3.24 = 2.4.

When a calculation has several steps, it is a good idea to keep an extra digit in the intermediate steps and wait until the end to round off the final result to the correct number of significant figures. As an example,

$$\frac{7.9}{3.24} + \frac{1.8}{0.35} = 2.44 + 5.14 = 7.58 = 7.6$$

If the intermediate results had been rounded off to two digits, however, the result would have been the following, which is incorrect:

$$\frac{7.9}{3.24} + \frac{1.8}{0.35} = 2.4 + 5.1 = 7.5 \qquad \text{(incorrect)}$$

For simplicity in the text and clarity in the illustrations, in this book zeros after the decimal point have usually been omitted from values given in the problems. It should be assumed that, for example, when a resistance of 5 $\Omega$ is quoted, what is really meant is 5.00 $\Omega$.

## *Multiple-Choice Questions*

**1.1.** In which of the following equations is $x$ not equal to 2?

(a)  $x^2 + 4x = 12$      (c)  $\sqrt{6x^2 + 25} = 7(5 - x^2)$

(b)  $\sqrt{2x^3 + 5x} = 14$      (d)  $(x^2 + 6)^2 = 50x$

**1.2.** The value of $(5.0 \times 10^{-4})(6.0 \times 10^{-4})/(1.5 \times 10^3)$ is

(a)  $2.0 \times 10^{-11}$      (c)  $2.0 \times 10^{-10}$
(b)  $9.5 \times 10^{-11}$      (d)  $2.0 \times 10^{-4}$

**1.3.** The value of $1/(2.0 \times 10^{-6})^2$ is

(a)  $2.5 \times 10^5$      (c)  $5.0 \times 10^{11}$
(b)  $2.5 \times 10^{11}$      (d)  $2.5 \times 10^{12}$

**1.4.** The prefix *kilo-* represents

(a)  10      (c)  1000
(b)  100      (d)  1,000,000

**1.5.** A millimeter is

(a)  0.001 m      (c)  0.1 m
(b)  0.01 m      (d)  10 m

**1.6.** The longest of the following is

(a)  1 mm      (c)  0.01 in.
(b)  0.00001 km      (d)  0.001 ft

**1.7.** The shortest of the following is

(a)  $10^4$ in.      (c)  $10^3$ ft
(b)  $10^4$ m      (d)  0.1 mi

**1.8.** A person is 180 cm tall. This is equivalent to approximately

(a)  4 ft 6 in.      (c)  5 ft 11 in.
(b)  5 ft 9 in.      (d)  7 ft 1 in.

**1.9.** A square foot contains approximately

(a)  144 cm$^2$      (c)  929 cm$^2$
(b)  366 cm$^2$      (d)  1000 cm$^2$

**1.10.** A week contains approximately

(a)  $1.0 \times 10^4$ s      (c)  $6.05 \times 10^5$ s
(b)  $6.05 \times 10^4$ s      (d)  $2.6 \times 10^6$ s

# Supplementary Problems

**1.1.** Remove the parentheses from the following, and combine like terms:

(a) $a + (b + c - d)$      (f) $b - 3(b + 3)$
(b) $a - (b + c - d)$      (g) $2a - 3b - 4(a - 2b)$
(c) $a + 2(b - 4)$          (h) $3(a + b) - 3(a + 2b)$
(d) $a - 2(b - 4)$          (i) $2(a + b) - 3(a - b) + 4(a + 2b - c)$
(e) $b + 3(b + 3)$          (j) $5(a + b + c) - 5(a - b - c)$

**1.2.** Evaluate the following:

(a) $\dfrac{3(x + y)}{2}$ when $x = 5$ and $y = -2$

(b) $\dfrac{1}{x - y} - \dfrac{1}{x + y}$ when $x = 3$ and $y = 2$

(c) $\dfrac{4xy}{y + 3x} + 5$ when $x = 1$ and $y = -2$

(d) $\dfrac{x + y}{2z} + \dfrac{z}{x - y}$ when $x = -2$, $y = 2$, and $z = 4$

(e) $\dfrac{x + z}{y} - \dfrac{xy}{2}$ when $x = 2$, $y = 8$, and $z = 10$

(f) $\dfrac{3(x + 7)}{y + 2}$ when $x = 3$ and $y = -6$

(g) $\dfrac{5(3 - x)}{2(x + y)}$ when $x = -5$ and $y = 7$

(h) $\dfrac{x^2 + y^2}{2z} + \dfrac{z}{x - y}$ when $x = -2$, $y = 2$, and $z = 4$

(i) $\dfrac{(x + z)^2}{y} - \dfrac{xy}{2}$ when $x = 2$, $y = 8$, and $z = 10$

**1.3.** Solve each of the following equations for $x$:

(a) $\dfrac{4x - 35}{3} = 9(1 - x)$

(b) $\dfrac{3x - 42}{9} = 2(7 - x)$

(c) $x^3 + 27 = 0$

(d) $2x^4 - 32 = 0$

(e) $3x^2 = 6x$

(f) $(x + 3)(x - y) = z^2 + x^2$

(g) $y\sqrt{2x} = 12$

(h) $z = \dfrac{x + y}{x - y}$

(i) $\dfrac{x}{y} = \dfrac{4}{z}$

(j) $\dfrac{y}{x} = \dfrac{x}{5}$

(k) $\dfrac{1}{x + 1} = \dfrac{1}{2x - 1}$

(l) $\dfrac{3}{x - 1} = \dfrac{5}{x + 1}$

(m) $\dfrac{1}{3x + 4} = \dfrac{2}{x + 8}$

(n) $\dfrac{8}{x} = \dfrac{1}{4 - x}$

(o) $\dfrac{x}{2x - 1} = \dfrac{5}{7}$

(p) $\dfrac{y}{x^2} = \dfrac{z}{x}$

(q) $x^2 = \dfrac{8}{x}$

(r) $\dfrac{2}{x} = \dfrac{18}{x^3}$

**1.4.** Evaluate the following:

(a) $a^2 a^5$

(b) $a^4 a^{-2}$

(c) $a^{14} a^{-14}$

(d) $\dfrac{a^6}{a^2}$

(e) $\dfrac{a^2}{a^6}$

(f) $\dfrac{a^5 a^{-2}}{a^8}$

(g) $\dfrac{a^3 a^{-2}}{a^{-5}}$

(h) $a^2 + a^2$

(i) $a^2 - a^2$

(j) $a^2 + a^5$

(k) $a^{3/2} a^{1/2}$

(l) $a^{1/2} a^{1/2} a^{1/2}$

(m) $a a^{1/3}$

(n) $a^6 a^{-1/2}$

(o) $\dfrac{a}{a^{1/2}}$

(p) $\dfrac{a}{a^{-1/2}}$

(q) $(a^4)^{1/2}$

(r) $(a^4)^{-1/2}$

(s) $(a^{-4})^{1/3}$

(t) $a^5 + a^{1/2}$

(u) $a^4 (ab)^{-2}$

(v) $(ab)^3 (a^2 b)^{-2}$

(w) $7b^2 \left(\dfrac{a^2}{b}\right)^3$

(x) $(a^2 b^4)^{1/2} \left(\dfrac{b^2}{a^3}\right)^3$

(y) $2a \left(\dfrac{3a}{b}\right)^2 \left(\dfrac{b^2}{a}\right)^3$

(z) $(ab)^3 \left(\dfrac{a^2}{b^8}\right)^{1/4}$

**1.5.** Express the following numbers in powers-of-10 notation:

(a) 720

(b) 890,000

(c) 0.02

(d) 0.000062

(e) 3.6

(f) 0.4

(g) 49,527

(h) 0.002943

(i) 0.0014

(j) 49,000,000,000

(k) 0.000000011

(l) 1.4763

**1.6.** Express the following numbers in decimal notation:

(a) $3 \times 10^{-4}$

(b) $7.5 \times 10^3$

(c) $8.126 \times 10^{-5}$

(d) $1.01 \times 10^8$

(e) $5 \times 10^2$

(f) $3.2 \times 10^{-2}$

(g) $4.32145 \times 10^3$

(h) $6 \times 10^6$

(i) $5.7 \times 10^0$

(j) $6.9 \times 10^{-5}$

**1.7.** Perform the following additions and subtractions:

(a) $3 \times 10^2 + 4 \times 10^3$

(b) $2 \times 10^4 + 5 \times 10^6$

(c) $7 \times 10^{-2} + 2 \times 10^{-3}$

(d) $4 \times 10^{-5} + 5 \times 10^{-3}$

(e) $2 \times 10^1 + 2 \times 10^{-1}$

(f) $6.32 \times 10^2 + 5$

(g) $4 \times 10^3 - 3 \times 10^2$

(h) $5 \times 10^7 - 9 \times 10^4$

(i) $3.2 \times 10^{-4} - 5 \times 10^{-5}$

(j) $7 \times 10^4 - 2 \times 10^5$

(k) $4.6 \times 10^5 - 3.2 \times 10^7$

(l) $3 \times 10^5 - 2.98 \times 10^5$

(m) $4.76 \times 10^{-3} - 4.81 \times 10^{-3}$

(n) $7 \times 10^3 + 5 \times 10^2 - 9 \times 10^2$

(o) $3 \times 10^{-4} + 6 \times 10^{-5} - 7 \times 10^{-3}$

**1.8.** Perform the following multiplications and divisions, using powers-of-10 notation:

(a) $(5000)(0.005)$

(b) $\dfrac{5000}{0.005}$

(c) $\dfrac{(500,000)(18,000)}{9,000,000}$

(d) $\dfrac{(30)(80,000,000,000)}{0.0004}$

(e) $\dfrac{(30,000)(0.0000006)}{(1000)(0.02)}$

(f) $\dfrac{0.0001}{(60,000)(200)}$

(g) $\dfrac{(200)(0.00004)}{400,000}$

(h) $\dfrac{(0.002)(0.00000005)}{0.000004}$

(i) $\dfrac{(400)(0.00006)}{(0.2)(20,000)}$

(j) $\dfrac{(0.06)(0.0001)}{(0.00003)(40,000)}$

**1.9.** Evaluate the following and express the results in powers-of-10 notation:

(a)  $(4 \times 10^9)^3$        (e)  $(3 \times 10^{-8})^2$

(b)  $(2 \times 10^7)^2$        (f)  $(5 \times 10^{11})^{-2}$

(c)  $(2 \times 10^7)^{-2}$     (g)  $(3 \times 10^{-4})^{-3}$

(d)  $(2 \times 10^{-2})^5$

**1.10.** Evaluate the following and express the results in powers-of-10 notation. Note that $\sqrt{4} = 2$, $\sqrt{40} = 6.3$, $\sqrt[3]{4} = 1.6$, $\sqrt[3]{40} = 3.4$, and $\sqrt[3]{400} = 7.4$.

(a)  $\sqrt{4 \times 10^6}$           (f)  $\sqrt[3]{4 \times 10^{12}}$          (j)  $\sqrt[3]{4 \times 10^{-6}}$

(b)  $\sqrt{4 \times 10^7}$           (g)  $\sqrt[3]{4 \times 10^{13}}$          (k)  $\sqrt[3]{4 \times 10^{-7}}$

(c)  $\sqrt{4 \times 10^8}$           (h)  $\sqrt[3]{4 \times 10^{14}}$          (l)  $\sqrt[3]{4 \times 10^{-8}}$

(d)  $\sqrt{4 \times 10^{-4}}$        (i)  $\sqrt[3]{4 \times 10^{15}}$          (m)  $\sqrt[3]{4 \times 10^{-9}}$

(e)  $\sqrt{4 \times 10^{-5}}$

**1.11.** The earth is an average of $9.3 \times 10^7$ mi from the sun. How far is this in kilometers? In meters?

**1.12.** How many cubic feet are there in a cubic meter?

**1.13.** The speed limit in many European towns is 60 km/h. How many miles per hour is this?

**1.14.** The speed of light is $3.00 \times 10^8$ m/s. What is this speed in feet per second? In miles per second? In miles per hour?

**1.15.** A nautical mile is 6076 ft, and 1 knot (kn) is a unit of speed equal to 1 nautical mile/h. How fast is an 8-kn boat going in miles per hour? In feet per second?

**1.16.** Use the proper number of significant figures to express the results of the following calculations:

(a)  $93.2 + 8.56 - 12$

(b)  $(18.5)(2.32)/0.4163$

(c)  $(4.6 \times 10^5)(8.75 \times 10^3)$

(d)  $\dfrac{3.24 \times 10^3}{5.11 \times 10^{-2}} + 2.58 \times 10^2$

(e)  $\sqrt{43}$

## *Answers to Multiple-Choice Questions*

**1.1.** (b)      **1.6.** (b)

**1.2.** (c)      **1.7.** (d)

**1.3.** (b)      **1.8.** (c)

**1.4.** (c)      **1.9.** (c)

**1.5.** (a)      **1.10.** (c)

## Answers to Supplementary Problems

**1.1.** (a)  $a + b + c - d$        (e)  $4b + 9$          (h)  $-3b$

(b)  $a - b - c + d$        (f)  $-2b - 9$        (i)  $3a + 13b - 4c$

(c)  $a + 2b - 8$          (g)  $-2a + 5b$       (j)  $10b + 10c$

(d)  $a - 2b + 8$

**1.2.** (*a*) 4.5 (*b*) 0.8 (*c*) −3 (*d*) −1 (*e*) −6.5 (*f*) −7.5 (*g*) 10
(*h*) 0 (*i*) 10

**1.3.** (*a*) 2 (*g*) $72/y^2$ (*m*) 0
(*b*) 8 (*h*) $-y(1 + z)/(1 - z)$ (*n*) 32/9
(*c*) −3 (*i*) $4y/z$ (*o*) $\frac{5}{3}$
(*d*) 2 (*j*) $\sqrt{5y}$ (*p*) $y/z$
(*e*) 2 (*k*) 2 (*q*) 2
(*f*) $(z^2 + 3y)/(3 - y)$ (*l*) 4 (*r*) 3

**1.4.** (*a*) $a^7$ (*h*) $2a^2$ (*o*) $a^{1/2}$ (*v*) $b/a$
(*b*) $a^2$ (*i*) 0 (*p*) $a^{3/2}$ (*w*) $7a^6/b$
(*c*) 1 (*j*) $a^2 + a^5$ (*q*) $a^2$ (*x*) $(b/a)^8$
(*d*) $a^4$ (*k*) $a^2$ (*r*) $a^{-2}$ (*y*) $18b^4$
(*e*) $a^{-4}$ (*l*) $a^{3/2}$ (*s*) $a^{-4/3}$ (*z*) $a^{7/2}b$
(*f*) $a^{-5}$ (*m*) $a^{4/3}$ (*t*) $a^5 + a^{1/2}$
(*g*) $a^6$ (*n*) $a^{11/2}$ (*u*) $a^2/b^2$

**1.5.** (*a*) $7.2 \times 10^2$ (*d*) $6.2 \times 10^{-5}$ (*g*) $4.9527 \times 10^4$ (*j*) $4.9 \times 10^{10}$
(*b*) $8.9 \times 10^5$ (*e*) $3.6 \times 10^0$ (*h*) $2.943 \times 10^{-3}$ (*k*) $1.1 \times 10^{-8}$
(*c*) $2 \times 10^{-2}$ (*f*) $4 \times 10^{-1}$ (*i*) $1.4 \times 10^{-3}$ (*l*) $1.4763 \times 10^0$

**1.6.** (*a*) 0.0003 (*d*) 101,000,000 (*g*) 4321.45 (*j*) 0.000069
(*b*) 7500 (*e*) 500 (*h*) 6,000,000
(*c*) 0.00008126 (*f*) 0.032 (*i*) 5.7

**1.7.** (*a*) $4.3 \times 10^3$ (*e*) $2.02 \times 10^1$ (*i*) $2.7 \times 10^{-4}$ (*m*) $-5 \times 10^{-5}$
(*b*) $5.02 \times 10^6$ (*f*) $6.37 \times 10^2$ (*j*) $-1.3 \times 10^5$ (*n*) $6.6 \times 10^3$
(*c*) $7.2 \times 10^{-2}$ (*g*) $3.7 \times 10^3$ (*k*) $-3.154 \times 10^7$ (*o*) $-6.64 \times 10^{-3}$
(*d*) $5.04 \times 10^{-3}$ (*h*) $4.991 \times 10^7$ (*l*) $2 \times 10^3$

**1.8.** (*a*) $2.5 \times 10^1$ (*d*) $6 \times 10^{15}$ (*g*) $2 \times 10^{-8}$ (*j*) $5 \times 10^{-6}$
(*b*) $10^6$ (*e*) $9 \times 10^{-4}$ (*h*) $2.5 \times 10^{-5}$
(*c*) $10^3$ (*f*) $8.3 \times 10^{-12}$ (*i*) $6 \times 10^{-6}$

**1.9.** (*a*) $6.4 \times 10^{28}$ (*c*) $2.5 \times 10^{-15}$ (*e*) $9 \times 10^{-16}$ (*g*) $3.7 \times 10^{10}$
(*b*) $4 \times 10^{14}$ (*d*) $3.2 \times 10^{-9}$ (*f*) $4 \times 10^{-24}$

**1.10.** (*a*) $2 \times 10^3$ (*e*) $6.3 \times 10^{-3}$ (*i*) $1.6 \times 10^5$ (*m*) $1.6 \times 10^{-3}$
(*b*) $6.3 \times 10^3$ (*f*) $1.6 \times 10^4$ (*j*) $1.6 \times 10^{-2}$
(*c*) $2 \times 10^4$ (*g*) $3.4 \times 10^4$ (*k*) $7.4 \times 10^{-3}$
(*d*) $2 \times 10^{-2}$ (*h*) $7.4 \times 10^4$ (*l*) $3.4 \times 10^{-3}$

**1.11.** $1.5 \times 10^8$ km; $1.5 \times 10^{11}$ m

**1.12.** 35.3 ft$^3$

**1.13.** 37 mi/h

**1.14.** $9.84 \times 10^8$ ft/s; $1.86 \times 10^5$ mi/s; $6.72 \times 10^8$ mi/h

**1.15.** 9.2 mi/h; 13.5 ft/s

**1.16.** (*a*) 90 (*b*) 103 (*c*) $4.0 \times 10^9$ (*d*) $6.37 \times 10^4$ (*e*) 6.6

# Chapter 2

# Vectors

## SCALAR AND VECTOR QUANTITIES

A *scalar quantity* has only magnitude and is completely specified by a number and a unit. Examples are mass (a stone has a mass of 2 kg), volume (a bottle has a volume of 1.5 liters), and frequency (house current has a frequency of 60 Hz). Symbols of scalar quantities are printed in italic type ($m$ = mass, $V$ = volume). Scalar quantities of the same kind are added by using ordinary arithmetic.

A *vector quantity* has both magnitude and direction. Examples are displacement (an airplane has flown 200 km to the southwest), velocity (a car is moving at 60 km/h to the north), and force (a person applies an upward force of 25 newtons to a package). Symbols of vector quantities are printed in boldface type (**v** = velocity, **F** = force) and expressed in handwriting by arrows over the letters ($\vec{v}$, $\vec{F}$). The magnitude of a vector quantity is printed in italic type ($F$ is the magnitude of the force **F**). When vector quantities are added, their directions must be taken into account.

## VECTOR ADDITION: GRAPHICAL METHOD

A *vector* is an arrowed line whose length is proportional to a certain vector quantity and whose direction indicates the direction of the quantity.

To add vector **B** to vector **A**, draw **B** so that its tail is at the head of **A**. The vector sum **A** + **B** is the vector **R** that joins the tail of **A** and the head of **B** (Fig. 2-1). Usually **R** is called the *resultant* of **A** and **B**.

The order in which **A** and **B** are added is not significant, so that **A** + **B** = **B** + **A** (Figs. 2-1 and 2-2).

Exactly the same procedure is followed when more than two vectors of the same kind are to be added. The vectors are strung together head to tail (being careful to preserve their correct lengths and directions), and the resultant **R** is the vector drawn from the tail of the first vector to the head of the last. The order in which the vectors are added does not matter (Fig. 2-3).

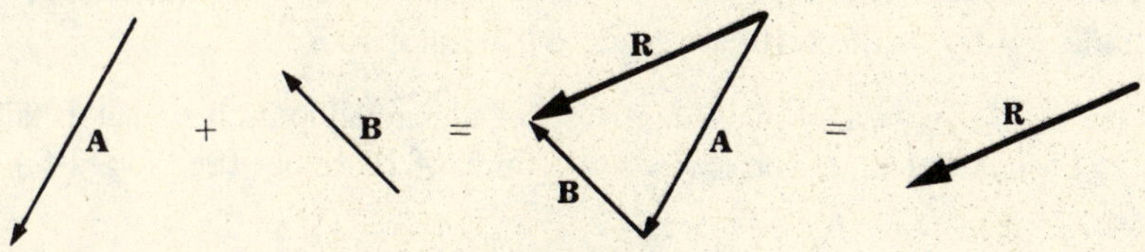

Fig. 2-1

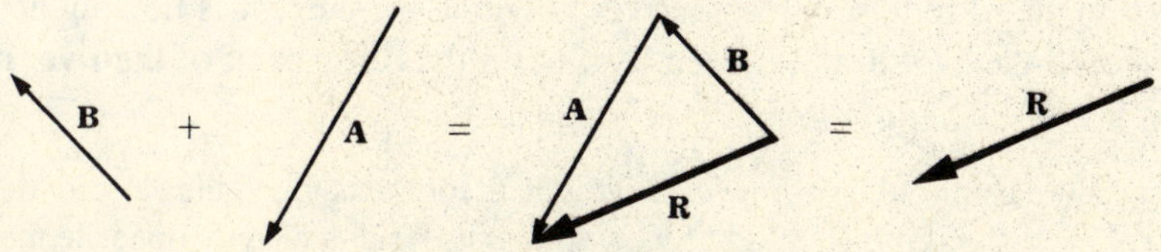

Fig. 2-2

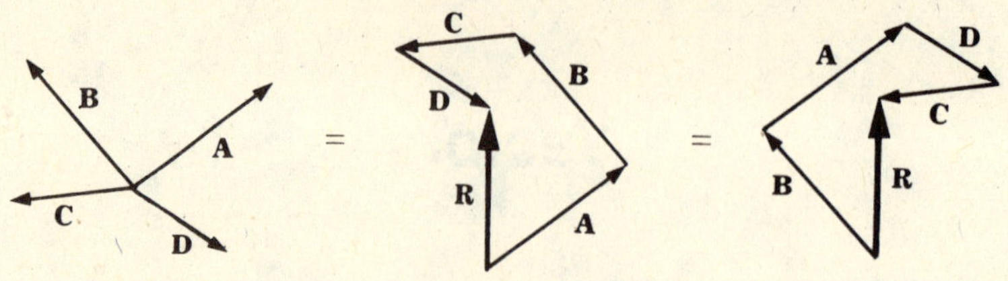

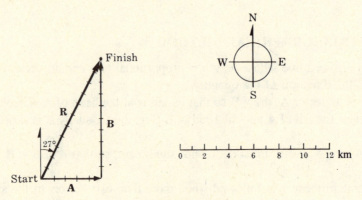

**Fig. 2-3**

## SOLVED PROBLEM 2.1

A woman walks eastward for 5 km and then northward for 10 km. How far is she from her starting point? If she had walked directly to her destination, in what direction would she have headed?

From Fig. 2-4, the length of the resultant vector **R** corresponds to a distance of 11.2 km, and a protractor shows that its direction is 27° east of north.

**Fig. 2-4**

## SOLVED PROBLEM 2.2

Two tugboats are towing a ship. Each exerts a force of 6 tons, and the angle between the towropes is 60°. What is the resultant force on the ship?

To add the force vectors **A** and **B**, vector **B** is shifted parallel to itself so that its tail is at the head of **A**. The length of the resultant **R** corresponds to a force of 10.4 tons. (See Fig. 2-5.)

## SOLVED PROBLEM 2.3

A boat moving at 5 km/h is to cross a river in which the current is flowing at 3 km/h. In what direction should the boat head to reach a point on the other bank of the river directly opposite its starting point?

The procedure here is to first draw the vector that represents $\mathbf{v}_{river}$, the velocity of the current. Then the vector $\mathbf{v}_{boat}$ is drawn from the head of $\mathbf{v}_{river}$ so that its head is directly opposite the tail of $\mathbf{v}_{river}$ (Fig. 2-6). A protractor shows the angle between $\mathbf{v}_{river}$ and $\mathbf{v}_{boat}$ to be 53°.

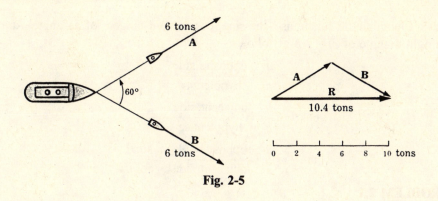

Fig. 2-5

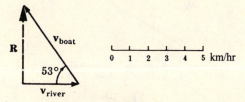

Fig. 2-6

## SOLVED PROBLEM 2.4

In going from one city to another, a car whose driver tends to get lost goes 30 km north, 50 km west, and 20 km southeast. Approximately how far apart are the cities?

The vectors representing the displacements are strung together head to tail, and their resultant is found to be 39 km (Fig. 2-7).

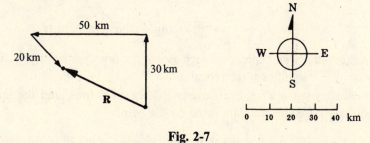

Fig. 2-7

## TRIGONOMETRY

Although it is possible to determine the magnitude and direction of the resultant of two or more vectors of the same kind graphically with ruler and protractor, this procedure is not very exact. For accurate results it is necessary to use trigonometry.

A *right triangle* is a triangle two of whose sides are perpendicular. The *hypotenuse* of a right triangle is the side opposite the right angle, as in Fig. 2-8; the hypotenuse is always the longest side.

The three basic trigonometric functions—the sine, cosine, and tangent of an angle—are defined in terms of the right triangle of Fig. 2-8 as follows:

$$\sin\theta = \frac{a}{c} = \frac{\text{opposite side}}{\text{hypotenuse}}$$

$$\cos\theta = \frac{b}{c} = \frac{\text{adjacent side}}{\text{hypotenuse}}$$

$$\tan\theta = \frac{a}{b} = \frac{\text{opposite side}}{\text{adjacent side}} = \frac{\sin\theta}{\cos\theta}$$

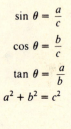

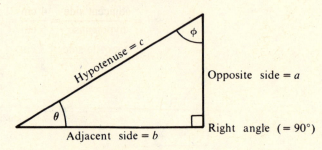

Fig. 2-8

The *inverse* of a trigonometric function is the angle whose function is given. Thus the inverse of $\sin\theta$ is the angle $\theta$. The names and abbreviations of the inverse trigonometric functions are as follows:

$$\sin\theta = x$$

$$\theta = \arcsin x = \sin^{-1} x = \text{angle whose sine is } x$$

$$\cos\theta = y$$

$$\theta = \arccos y = \cos^{-1} y = \text{angle whose cosine is } y$$

$$\tan\theta = z$$

$$\theta = \arctan z = \tan^{-1} z = \text{angle whose tangent is } z$$

Remember that in trigonometry an expression such as $\sin^{-1} x$ does *not* signify $1/(\sin x)$, even though in algebra the exponent $-1$ signifies a reciprocal.

## PYTHAGOREAN THEOREM

The *Pythagorean theorem* states that the sum of the squares of the short sides of a right triangle is equal to the squares of its hypotenuse. For the triangle of Fig. 2-8,

$$a^2 + b^2 = c^2$$

Hence we can always express the length of any of the sides of a right triangle in terms of the lengths of the other sides:

$$a = \sqrt{c^2 - b^2} \qquad b = \sqrt{c^2 - a^2} \qquad c = \sqrt{a^2 + b^2}$$

Another useful relationship is that the sum of the interior angles of any triangle is 180°. Since one of the angles in a right triangle is 90°, the sum of the other two must be 90°. Thus in Fig. 2-8, $\theta + \phi = 90°$.

Of the six quantities that characterize a triangle—three sides and three angles—we must know the values of at least three, including one of the sides, in order to calculate the others. In a right triangle, one of the angles is always 90°, so all we need are the lengths of any two sides or the length of one side plus the value of one of the other angles to find the other sides and angles.

## SOLVED PROBLEM 2.5

Find the values of the sine, cosine, and tangent of angle $\theta$ in Fig. 2-9.

$$\sin \theta = \frac{\text{opposite side}}{\text{hypotenuse}} = \frac{3 \text{ cm}}{5 \text{ cm}} = 0.6$$

$$\cos \theta = \frac{\text{adjacent side}}{\text{hypotenuse}} = \frac{4 \text{ cm}}{5 \text{ cm}} = 0.8$$

$$\tan \theta = \frac{\text{opposite side}}{\text{adjacent side}} = \frac{3 \text{ cm}}{4 \text{ cm}} = 0.75$$

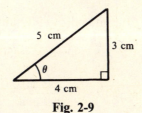

**Fig. 2-9**

## SOLVED PROBLEM 2.6

Find the angle whose cosine is 0.952.

With some calculators, the procedure is to enter 0.952 and then press the key marked $\cos^{-1}$ (or arccos). The result will appear as 17.8242, which can be rounded to 18°. In a table of cosines, such as the one in Appendix C, we look for the value nearest to 0.952 and then read across to find the corresponding angle. What we would find is that cos 17° = 0.956 and cos 18° = 0.951. Since 0.952 is closer to 0.951 than to 0.956, to the nearest degree we have $\cos^{-1} 0.952 = 18°$.

## SOLVED PROBLEM 2.7

In the triangle of Fig. 2-8, $c = 80$ cm and $\theta = 30°$. Find the values of $a$, $b$, and $\phi$.

We start with the length of the side $a$:

$$\sin \theta = \frac{a}{c}$$

$$a = c \sin \theta = (80 \text{ cm})(\sin 30°) = 40 \text{ cm}$$

To find $b$, we can proceed in two ways. In the first way we note that

$$\cos \theta = \frac{b}{c}$$

$$b = c \cos \theta = (80 \text{ cm})(\cos 30°) = 69 \text{ cm}$$

The other way is to use the Pythagorean theorem:

$$b = \sqrt{c^2 - a^2} = \sqrt{(80 \text{ cm})^2 - (40 \text{ cm})^2} = 69 \text{ cm}$$

## SOLVED PROBLEM 2.8

In the triangle of Fig. 2-8, $a = 70$ m and $b = 100$ m. Find the values of $c$, $\theta$, and $\phi$.

To find $c$, we use the Pythagorean theorem:

$$c = \sqrt{a^2 + b^2} = \sqrt{(70 \text{ m})^2 + (100 \text{ m})^2} = 122 \text{ m}$$

To find $\theta$, we proceed in the following way:

$$\tan \theta = \frac{a}{b} = \frac{70 \text{ m}}{100 \text{ m}} = 0.70$$

$$\theta = \tan^{-1} 0.70 = 35°$$

Since $\theta = 35°$,

$$\phi = 90° - \theta = 90° - 35° = 55°$$

## SOLVED PROBLEM 2.9

In the triangle of Fig. 2-8, $a = 30$ mm and $c = 90$ mm. Find the values of $b$, $\theta$ and $\phi$.

$$b = \sqrt{c^2 - a^2} = \sqrt{(90 \text{ mm})^2 - (30 \text{ mm})^2} = 85 \text{ mm}$$

$$\sin \theta = \frac{a}{c} = \frac{30 \text{ mm}}{90 \text{ mm}} = 0.33$$

$$\theta = \sin^{-1} 0.33 = 19°$$

$$\phi = 90° - \theta = 90° - 19° = 71°$$

## VECTOR ADDITION: TRIGONOMETRIC METHOD

It is easy to apply trigonometry to find the resultant **R** of two vectors **A** and **B** that are perpendicular to each other. The magnitude of the resultant is given by the Pythagorean theorem as

$$R = \sqrt{A^2 + B^2}$$

and the angle $\theta$ between **R** and **A** (Fig. 2-10) may be found from

$$\tan \theta = \frac{B}{A}$$

by examining a table of tangents or by using a calculator to determine $\tan^{-1} B/A$.

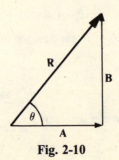

Fig. 2-10

## SOLVED PROBLEM 2.10

Use trigonometry to solve Prob. 2.1.

From the vector diagram of Fig. 2-11 we see that **A** and **B** are the sides of a right triangle and **R** is its hypotenuse. According to the Pythagorean theorem, the magnitudes $A$, $B$, and $R$ are related by $R^2 = A^2 + B^2$. Hence the magnitude $R$ is equal to

$$R = \sqrt{A^2 + B^2} = \sqrt{(5 \text{ km})^2 + (10 \text{ km})^2} = \sqrt{25 \text{ km}^2 + 100 \text{ km}^2}$$
$$= \sqrt{125 \text{ km}^2} = 11.2 \text{ km}$$

To find the direction of **R**, we note that

$$\tan \theta = \frac{B}{A} = \frac{10 \text{ km}}{5 \text{ km}} = 2$$

From a trigonometric table or with a calculator, we find that the angle whose tangent is closest to 2 is $\theta = 63°$. To express the direction of **R** in terms of north, we see from Fig. 2-11 that the angle $\phi$ between north and **R**, plus the angle $\theta$ between **R** and east, is equal to 90°. Since $\phi = \theta = 90°$,

$$\phi = 90° - \theta = 90° - 63° = 27°$$

The resultant **R** has a magnitude of 11.2 km, and its direction is 27° east of north.

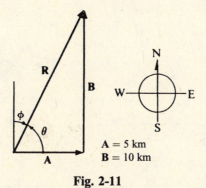

A = 5 km
B = 10 km

**Fig. 2-11**

## RESOLVING A VECTOR

Just as two or more vectors can be added to yield a single resultant vector, so it is possible to break up a single vector into two or more other vectors. If vectors **A** and **B** are together equivalent to vector **C**, then vector **C** is equivalent to the two vectors **A** and **B** (Fig. 2-12). When a vector is replaced by

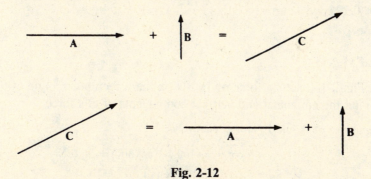

**Fig. 2-12**

two or more others, the process is called *resolving* the vector, and the new vectors are known as the *components* of the initial vector.

The components into which a vector is resolved are nearly always chosen to be perpendicular to one another. Figure 2-13 shows a wagon being pulled by a man with force **F**. Because the wagon moves horizontally, the entire force is not effective in influencing its motion. The force **F** may be resolved into two component vectors $\mathbf{F}_x$ and $\mathbf{F}_y$, where

$$\mathbf{F}_x = \text{horizontal component of } \mathbf{F}$$
$$\mathbf{F}_y = \text{vertical component of } \mathbf{F}$$

The magnitudes of these components are

$$F_x = F \cos \theta \qquad F_y = F \sin \theta$$

Evidently the component $\mathbf{F}_x$ is responsible for the wagon's motion, and if we were interested in working out the details of this motion, we would need to consider only $\mathbf{F}_x$.

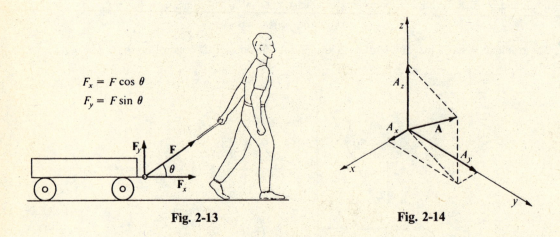

$$F_x = F \cos \theta$$
$$F_y = F \sin \theta$$

**Fig. 2-13**                                              **Fig. 2-14**

In Fig. 2-13 the force **F** lies in a vertical plane, and the two components $\mathbf{F}_x$ and $\mathbf{F}_y$ are enough to describe it. In general, however, three mutually perpendicular components are required to completely describe the magnitude and direction of a vector quantity. It is customary to call the directions of these components the $x$, $y$, and $z$ axes, as in Fig. 2-14. The components of some vector **A** in these directions are accordingly denoted $\mathbf{A}_x$, $\mathbf{A}_y$, and $\mathbf{A}_z$. If a component falls on the negative part of an axis, its magnitude is considered negative. Thus if $\mathbf{A}_z$ were downward in Fig. 2-14 instead of upward and its length were equivalent to, say, 12 N, we would write $A_z = -12$ N. [The newton (N) is the SI unit of force; it is equal to 0.225 lb.]

## SOLVED PROBLEM 2.11

The man in Fig. 2-13 exerts a force of 100 N on the wagon at an angle of $\theta = 30°$ above the horizontal. Find the horizontal and vertical components of this force.

The magnitudes of $\mathbf{F}_x$ and $\mathbf{F}_y$ are, respectively,

$$F_x = F \cos \theta = (100 \text{ N})(\cos 30°) = 86.6 \text{ N}$$
$$F_y = F \sin \theta = (100 \text{ N})(\sin 30°) = 50.0 \text{ N}$$

We note that $F_x + F_y = 136.6$ N although **F** itself has the magnitude $F = 100$ N. What is wrong? The answer is that nothing is wrong; because $F_x$ and $F_y$ are just the *magnitudes* of the vectors $\mathbf{F}_x$ and $\mathbf{F}_y$, it is meaningless to add them. However, we can certainly add the *vectors* $\mathbf{F}_x$ and $\mathbf{F}_y$ to find the magnitude of their resultant **F**. Because $\mathbf{F}_x$ and $\mathbf{F}_y$ are perpendicular,

$$F = \sqrt{F_x^2 + F_y^2} = \sqrt{(86.6 \text{ N})^2 + (50.0 \text{ N})^2} = 100 \text{ N}$$

as we expect.

## SOLVED PROBLEM 2.12

A woman in a car on a level road sees an airplane traveling in the same direction climbing at an angle of 30° above the horizontal. By driving at 110 km/h she is able to stay directly below the airplane. Find the airplane's velocity.

The car's velocity is equal to the horizontal component $\mathbf{v}_x$ of the airplane's velocity **v**, as in Fig. 2-15. Since

$$v_x = v \cos \theta$$

the airplane's velocity $v$ is

$$v = \frac{v_x}{\cos \theta} = \frac{110 \text{ km/h}}{\cos 30°} = 127 \text{ km/h}$$

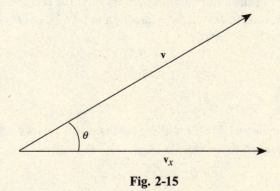

**Fig. 2-15**

## SOLVED PROBLEM 2.13

A car weighing 12.0 kN is on a hill that makes an angle of 20° with the horizontal. Find the components of the car's weight parallel and perpendicular to the road.

The weight of an object is the gravitational force with which the earth attracts it, and this force always acts vertically downward (Fig. 2-16). Because **w** is vertical and $\mathbf{F}_2$ is perpendicular to the road, the angle $\theta$ between **w** and $\mathbf{F}_2$ is the same as the angle $\theta$ between the road and the horizontal. Hence

$$F_1 = w \sin \theta = (12.0 \text{ kN})(\sin 20°) = 4.1 \text{ kN}$$
$$F_2 = w \cos \theta = (12.0 \text{ kN})(\cos 20°) = 11.3 \text{ kN}$$

## VECTOR ADDITION: COMPONENT METHOD

When vectors to be added are not perpendicular, the method of addition by components described below can be used. There do exist trigonometric procedures for dealing with oblique triangles (the *law*

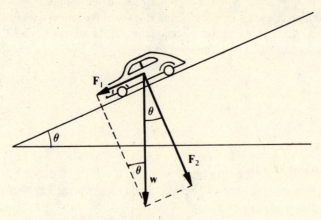

Fig. 2-16

*of sines* and the *law of cosines*), but these are not necessary since the component method is entirely general in its application.

To add two or more vectors **A**, **B**, **C**, ... by the component method, follow this procedure:

1. Resolve the initial vectors into components in the $x$, $y$, and $z$ directions.

2. Add the components in the $x$ direction to give $\mathbf{R}_x$, add the components in the $y$ direction to give $\mathbf{R}_y$, and add the components in the $z$ direction to give $\mathbf{R}_z$. That is, the magnitudes of $\mathbf{R}_x$, $\mathbf{R}_y$, and $\mathbf{R}_z$ are given by, respectively,

$$R_x = A_x + B_x + C_x + \cdots$$
$$R_y = A_y + B_y + C_y + \cdots$$
$$R_z = A_z + B_z + C_z + \cdots$$

3. Calculate the magnitude and direction of the resultant **R** from its components $\mathbf{R}_x$, $\mathbf{R}_y$, and $\mathbf{R}_z$ by using the Pythagorean theorem:

$$R = \sqrt{R_x^2 + R_y^2 + R_z^2}$$

If the vectors being added all lie in the same plane, only two components need to be considered.

## SOLVED PROBLEM 2.14

Use the component method of vector addition to solve Prob. 2.2.

The angle between the force each tugboat exerts on the ship and the direction of the ship's motion is 30°. Hence each force has a component in the direction of the ship's motion of

$$F_x = F \cos \theta = (6 \text{ tons})(\cos 30°) = 5.2 \text{ tons}$$

Since there are two tugboats, the resultant force on the ship is

$$R = 2F_x = (2)(5.2 \text{ tons}) = 10.4 \text{ tons}$$

## SOLVED PROBLEM 2.15

A boat is headed north at a velocity of 8.0 km/h. A strong wind is blowing whose pressure on the boat's superstructure causes it to move sideways to the west at a velocity of 2.0 km/h. There is also a tidal current present that flows in a direction 30° south of east at a velocity of 5.0 km/h. What is the boat's velocity relative to the earth's surface?

The first step is to establish a suitable set of coordinate axes, such as shown in Fig. 2-17($a$). Next we draw the three velocity vectors **A**, **B**, and **C** and calculate the magnitudes of their $x$ and $y$ components. We find these values:

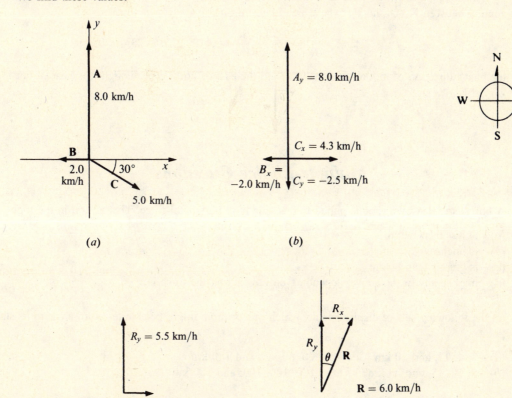

Fig. 2-17

| $x$ components | $y$ components |
|---|---|
| $A_x = 0$ | $A_y = 8.0$ km/h |
| $B_x = -2.0$ km/h | $B_y = 0$ |
| $C_x = C\cos 30°$ | $C_y = -C\sin 30°$ |
| $\quad = (5.0 \text{ km/h})(0.866)$ | $\quad = -(5.0 \text{ km/h})(0.500)$ |
| $\quad = 4.3$ km/h | $\quad = -2.5$ km/h |

These components are shown in Fig. 2-17($b$).

Now we add the values of the $x$ components to get $R_x$ and add the values of the $y$ components to get $R_y$:

$$R_x = A_x + B_x + C_x = 0 - 2.0 \text{ km/h} + 4.3 \text{ km/h} = 2.3 \text{ km/h}$$
$$R_y = A_y + B_y + C_y = 8.0 \text{ km/h} + 0 - 2.5 \text{ km/h} = 5.5 \text{ km/h}$$

The magnitude of the resultant **R** is therefore

$$R = \sqrt{R_x^2 + R_y^2} = \sqrt{(2.3 \text{ km/h})^2 + (5.5 \text{ km/h})^2} = \sqrt{35.54} \text{ km/h} = 6.0 \text{ km/h}$$

The boat's velocity relative to the earth's surface is 6.0 km/h.

The direction in which the boat is moving relative to the earth's surface can be given in terms of the angle $\theta$ between **R** and the $+y$ axis, which is north. Since

$$\tan \theta = \frac{R_x}{R_y} = \frac{2.3 \text{ km/h}}{5.5 \text{ km/h}} = 0.418 \qquad \theta = 23°$$

the boat's direction of motion is actually 23° to the east of north even though it is headed north [Fig. 2-17(d)].

# *Multiple-Choice Questions*

**2.1.**   A box suspended by a rope is pulled to one side by a horizontal force. The tension in the rope

   (a)   is less than before
   (b)   is unchanged
   (c)   is greater than before
   (d)   may be any of the above, depending on how strong the force is

**2.2.**   Of the following sets of displacements, which one or more might be able to return a car to its starting point?

   (a)   3, 4, 12, and 20 km          (c)   20, 60, 80, and 180 km
   (b)   5, 10, 15, and 20 km         (d)   100, 100, 100, and 100 km

**2.3.**   A boat whose velocity through the water is 14 km/h is moving in a river whose current is 6 km/h relative to the riverbed. The velocity of the boat relative to the riverbed must be between

   (a)   6 and 14 km/h          (c)   8 and 14 km/h
   (b)   6 and 20 km/h          (d)   8 and 20 km/h

**2.4.**   A ship travels 200 km to the south and then 400 km to the west. The ship's displacement from its starting point is

   (a)   200 km          (c)   450 km
   (b)   400 km          (d)   600 km

**2.5.**   At what angle west of south should the ship in Question 2.4 have headed to arrive at the same place in a straight path?

   (a)   22°          (c)   50°
   (b)   45°          (d)   63°

**2.6.**   A conveyor belt has a velocity of 4.00 m/s at an angle of 40° above the horizontal. The vertical component of its velocity is

   (a)   2.00 m/s          (c)   3.06 m/s
   (b)   2.57 m/s          (d)   3.36 m/s

**2.7.** An object is acted on by two forces of 20 N each. The angle between the forces is 120°. The resultant force on the object has the magnitude

  (a)  20 N    (c)  34 N
  (b)  28 N    (d)  40 N

**2.8.** A force **F** has the components **F$_x$** and **F$_y$**. The magnitude $F_x$ of the force component in the x direction is given by

  (a)  $F - F_y$          (c)  $\sqrt{F - F_y}$
  (b)  $\sqrt{F} - \sqrt{F_y}$     (d)  $\sqrt{F^2 - F_y^2}$

**2.9.** Forces of 20 N to the south, 40 N to the northeast, and 10 N to the east act on an object. The magnitude of the resultant force on the object is

  (a)  10 N    (c)  39 N
  (b)  20 N    (d)  46 N

**2.10.** The resultant for the forces in Question 2.9 points

  (a)  24° east of north    (c)  52° east of north
  (b)  45° east of north    (d)  78° east of north

# Supplementary Problems

**2.1.** Find the length of the hypotenuse of a right triangle whose legs are 483 and 620 m long.

**2.2.** The hypotenuse of a right triangle is 28 cm long and the length of one of the legs is 23 cm. Find the length of the other leg.

**2.3.** Find the values of the unknown sides and angles in the right triangles for which the following data are known (see Fig. 2-8).

  (a)  $\theta = 45°$, $a = 10$    (d)  $a = 3$, $b = 4$, $c = 5$
  (b)  $\theta = 15°$, $b = 4$     (e)  $a = 5$, $b = 12$, $c = 13$
  (c)  $\theta = 25°$, $c = 5$

**2.4.** Two forces, one of 10 N and the other of 6 N, act on a body. The directions of the forces are not known. (a) What is the minimum magnitude of the resultant of these forces? (b) What is the maximum magnitude?

**2.5.** A man drives 10 km to the north and then 20 km to the east. What are the magnitude and direction of his displacement from the starting point?

**2.6.** Find the magnitude and direction of the resultant force produced by a vertically upward force of 40 N and a horizontal force of 30 N.

**2.7.**   Find the vertical and horizontal components of a 50-N force that is directed 50° above the horizonal.

**2.8.**   A woman pushes a lawn mower with a force of 80 N. If the handle of the lawn mower is 40° above the horizontal, how much downward force is being exerted on the ground?

**2.9.**   An airplane is heading northeast at a velocity of 550 km/h. What is the northward component of its velocity? The eastward component?

**2.10.**  A boat heads northwest at 20 km/h in a river that flows east at 6 km/h. What are the magnitude and direction of the boat's velocity relative to the earth's surface?

**2.11.**  An airplane whose velocity is 120 km/h has just taken off from a runway. A car driving at 100 km/h on the runway is able to remain just below the airplane. At what angle is the airplane climbing?

**2.12.**  A boat moving at 36 km/h is crossing a river in which the current is flowing at 12 km/h. In what direction should the boat head if it is to reach a point on the other side of the river directly opposite its starting point?

**2.13.**  An airplane flies 400 km west from city A to city B, then 300 km northwest to city C, and finally 100 km north to city D. How far it is from city A to city D? In what direction must the airplane head to return directly to city A from city D?

**2.14.**  Find the magnitude and direction of the resultant of a 5-N force that acts at an angle of 37° clockwise from the +x axis, a 3-N force that acts at an angle of 180° clockwise from the +x axis, and a 7-N force that acts at an angle of 225° clockwise from the +x axis.

**2.15.**  Find the magnitude and direction of the resultant of a 60-N force that acts at an angle of 45° clockwise from the +y axis, a 20-N force that acts at an angle of 90° clockwise from the +y axis, and a 40-N force that acts at an angle of 300° clockwise from the +y axis.

## *Answers to Multiple-Choice Questions*

**2.1.**  (*c*)        **2.6.**   (*b*)

**2.2.**  (*b*), (*d*)  **2.7.**   (*a*)

**2.3.**  (*d*)        **2.8.**   (*d*)

**2.4.**  (*c*)        **2.9.**   (*c*)

**2.5.**  (*d*)        **2.10.**  (*d*)

## **Answers to Supplementary Problems**

**2.1.**   786 m

**2.2.**   16 cm

**2.3.**  (a)  $b = 10$, $c = 14.1$, $\phi = 45°$     (d)  $\theta = 37°$, $\phi = 53°$
         (b)  $a = 1.07$, $c = 4.14$, $\phi = 75°$     (e)  $\theta = 23°$, $\phi = 67°$
         (c)  $a = 2.11$, $b = 4.53$, $\phi = 65°$

**2.4.**  (a)  4 N     (b)  16 N

**2.5.**  22 km at 63° east of north

**2.6.**  50 N at 53° above the horizontal

**2.7.**  38.3 N; 32.2 N

**2.8.**  51.4 N

**2.9.**  389 km/h; 389 km/h

**2.10.**  16.3 km/h at 30° west of north

**2.11.**  34°

**2.12.**  19° upstream of directly across the river

**2.13.**  687 km at 63° east of south

**2.14.**  4.4 N at 206° clockwise from the $+x$ axis

**2.15.**  68 N at 24° clockwise from the $+y$ axis

# Chapter 3

## Motion in a Straight Line

### VELOCITY

The *velocity* of a body is a vector quantity that describes both how fast it is moving and the direction in which it is headed.

In the case of a body traveling in a straight line, its velocity is simply the rate at which it covers distance. The *average velocity* $\bar{v}$ of such a body when it covers the distance $s$ in the time $t$ is

$$\bar{v} = \frac{s}{t}$$

$$\text{Average velocity} = \frac{\text{distance}}{\text{time}}$$

The average velocity of a body during the time $t$ does not completely describe its motion, however, because during the time $t$ it may sometimes have gone faster than $\bar{v}$ and sometimes slower. The velocity of a body at any given moment is called its *instantaneous velocity* and is given by

$$v_{\text{inst}} = \frac{\Delta s}{\Delta t}$$

Here $\Delta s$ is the distance the body has gone in the very short time interval $\Delta t$ at the specified moment. ($\Delta$ is the capital Greek letter *delta*.) Instantaneous velocity is what a car's speedometer indicates.

When the instantaneous velocity of a body does not change, it is moving at *constant velocity*. For the case of constant velocity, the basic formula $v = s/t$ can be rewritten to give the distance covered in a given period of time:

$$s = vt$$

$$\text{Distance} = (\text{constant velocity})(\text{time})$$

Another way to write $v = s/t$ gives the time needed to cover a given distance at the constant velocity $v$:

$$t = \frac{s}{v}$$

$$\text{Time} = \frac{\text{distance}}{\text{constant velocity}}$$

### SOLVED PROBLEM 3.1

A ship travels 9 km in 45 min. What is its speed in kilometers per hour?

Since 45 min $= \frac{3}{4}$ h,

$$v = \frac{s}{t} = \frac{9 \text{ km}}{\frac{3}{4} \text{ h}} = 12 \text{ km/h}$$

### SOLVED PROBLEM 3.2

The velocity of sound in air at sea level is about 343 m/s. If a person hears a clap of thunder 3.00 s after seeing a lightning flash, how far away was the lightning?

The velocity of light is so great compared with the velocity of sound that the time needed for the light of the flash to reach the person can be neglected. Hence

$$s = vt = (343 \text{ m/s})(3.00 \text{ s}) = 1029 \text{ m} = 1.03 \text{ km}$$

## SOLVED PROBLEM 3.3

The velocity of light is $3.0 \times 10^8$ m/s. How long does it take light to reach the earth from the sun, which is $1.5 \times 10^{11}$ m away?

$$t = \frac{s}{v} = \frac{1.5 \times 10^{11} \text{ m}}{3.0 \times 10^8 \text{ m/s}} = 500 \text{ s} = 8.3 \text{ min}$$

## SOLVED PROBLEM 3.4

A car travels 270 km in 4.5 h. (*a*) What is its average velocity? (*b*) How far will it go in 7.0 h at this average velocity? (*c*) How long will it take to travel 300 km at this average velocity?

(*a*)
$$\bar{v} = \frac{s}{t} = \frac{270 \text{ km}}{4.5 \text{ h}} = 60 \text{ km/h}$$

(*b*)
$$s = \bar{v}t = (60 \text{ km/h})(7.0 \text{ h}) = 420 \text{ km}$$

(*c*)
$$t = \frac{s}{\bar{v}} = \frac{300 \text{ km}}{60 \text{ km/h}} = 5.0 \text{ h}$$

## SOLVED PROBLEM 3.5

An airplane whose velocity relative to the air is a constant 800 km/h has a constant tailwind of 240 km/h. How long will it take the airplane to cover 2000 km relative to the ground?

The ground velocity of the airplane is

$$v = 800 \text{ km/h} + 240 \text{ km/h} = 1040 \text{ km/h}$$

Hence the time needed to cover 2000 km over the ground is

$$t = \frac{s}{v} = \frac{2000 \text{ km}}{1040 \text{ km/h}} = 1.92 \text{ h} = 1 \text{ h } 55 \text{ min}$$

## SOLVED PROBLEM 3.6

A car travels at 100 km/h for 2 h, at 60 km/h for the next 2 h, and finally at 80 km/h for 1 h. What is the car's average velocity for the entire journey?

The car's average velocity equals the total distance it covers divided by the total time. Hence

$$\bar{v} = \frac{s_1 + s_2 + s_3}{t_1 + t_2 + t_3} = \frac{v_1 t_1 + v_2 t_2 + v_3 t_3}{t_1 + t_2 + t_3}$$
$$= \frac{(100 \text{ km/h})(2 \text{ h}) + (60 \text{ km/h})(2 \text{ h}) + (80 \text{ km/h})(1 \text{ h})}{2 \text{ h} + 2 \text{ h} + 1 \text{ h}}$$
$$= \frac{400 \text{ km}}{5 \text{ h}} = 80 \text{ km/h}$$

## ACCELERATION

A body whose velocity is changing is accelerated. A body is accelerated when its velocity is increasing, decreasing, or changing in direction. Accelerations that involve a change in direction are discussed in Chapter 9.

The *acceleration* of a body is the rate at which its velocity is changing. If a body moving in a straight line has a velocity of $v_0$ at the start of a certain time interval $t$ and of $v$ at the end, its acceleration is

$$a = \frac{v - v_0}{t}$$

$$\text{Acceleration} = \frac{\text{velocity change}}{\text{time}}$$

A positive acceleration means an increase in velocity; a negative acceleration (sometimes called a *deceleration*) means a decrease in velocity. Only constant accelerations are considered here.

The defining formula for acceleration can be rewritten to give the final velocity $v$ of an accelerated body:

$$v = v_0 + at$$

$$\text{Final velocity} = \text{initial velocity} + (\text{acceleration})(\text{time})$$

We can also solve for the time $t$ in terms of $v_0$, $v$, and $a$:

$$t = \frac{v - v_0}{a}$$

$$\text{Time} = \frac{\text{velocity change}}{\text{acceleration}}$$

Velocity has the dimensions of distance/time. Acceleration has the dimensions of velocity/time or distance/time$^2$. A typical acceleration unit is the meter/second$^2$ (meter per second squared). Sometimes two different time units are convenient; for instance, the acceleration of a car that goes from rest to 90 km/h in 10 s might be expressed as $a = 9$ (km/h)/s.

## SOLVED PROBLEM 3.7

A car starts from rest and reaches a velocity of 40 m/s in 10 s. (*a*) What is its acceleration? (*b*) If its acceleration remains the same, what will its velocity be 5 s later?

(*a*)   Here $v_0 = 0$. Hence

$$a = \frac{v}{t} = \frac{40 \text{ m/s}}{10 \text{ s}} = 4 \text{ m/s}^2$$

(*b*)   Now $v_0 = 40$ m/s, so

$$v = v_0 + at = 40 \text{ m/s} + (4 \text{ m/s}^2)(5 \text{ s}) = 40 \text{ m/s} + 20 \text{ m/s} = 60 \text{ m/s}$$

## SOLVED PROBLEM 3.8

A baseball is moving at 25 m/s when it is struck by a bat and moves off in the opposite direction at 35 m/s. If the impact lasted 0.010 s, find the baseball's acceleration during the impact.

The baseball's initial velocity is $v_0 = 25$ m/s and its final velocity is $-35$ m/s; the minus sign is needed because the baseball has reversed its direction of motion. Hence

$$a = \frac{v - v_0}{t} = \frac{(-35 \text{ m/s}) - 25 \text{ m/s}}{0.010 \text{ s}} = -\frac{60 \text{ m/s}}{0.010 \text{ s}} = -6000 \text{ m/s}^2 = -6.0 \text{ km/s}$$

The acceleration is negative because it is opposite in direction to the original direction of the baseball.

## SOLVED PROBLEM 3.9

(a) What is the acceleration of a car that goes from 20 to 30 km/h in 1.5 s? (b) At the same acceleration, how long will it take the car to go from 30 to 36 km/h?

(a)
$$a = \frac{v - v_0}{t} = \frac{30 \text{ km/h} - 20 \text{ km/h}}{1.5 \text{ s}} = 6.7 \text{ (km/h)/s}$$

(b)
$$t = \frac{v - v_0}{a} = \frac{36 \text{ km/h} - 30 \text{ km/h}}{6.7 \text{ (km/h)/s}} = 0.9 \text{ s}$$

## DISTANCE, VELOCITY, AND ACCELERATION

Let us consider a body whose velocity is $v_0$ when it starts to be accelerated at a constant rate. After time $t$ the final velocity of the body will be

$$v = v_0 + at$$

How far does the body go during the time interval $t$? The average velocity $\bar{v}$ of the body is

$$\bar{v} = \frac{v_0 + v}{2}$$

and so
$$s = \bar{v}t = \left(\frac{v_0 + v}{2}\right)t$$

Since $v = v_0 + at$, another way to specify the distance covered during $t$ is

$$s = \left(\frac{v_0 + v_0 + at}{2}\right)t = v_0 t + \tfrac{1}{2}at^2$$

If the body is accelerated starting from rest, $v_0 = 0$ and

$$s = \tfrac{1}{2}at^2$$

Another useful formula gives the final velocity of a body in terms of its initial velocity, its acceleration, and the distance it has traveled during the acceleration:

$$v^2 = v_0^2 + 2as$$

This can be solved for the distance $s$ to give

$$s = \frac{v^2 - v_0^2}{2a}$$

In the case of a body that starts from rest, $v_0 = 0$ and

$$v = \sqrt{2as} \qquad s = \frac{v^2}{2a}$$

Table 3.1 summarizes the formulas for motion under constant acceleration.

**Table 3.1.   Formulas for Motion
under Constant Acceleration**

| Distance | Final Velocity |
|---|---|
| $s = \left(\dfrac{v_0 + v}{2}\right)t$ | $v = v_0 + at$ |
| $s = v_0 t + \frac{1}{2}at^2$ | $v^2 = v_0^2 + 2as$ |

## SOLVED PROBLEM 3.10

A car has an acceleration of 8 m/s². (*a*) How much time is needed for it to reach a velocity of 24 m/s if it starts from rest? (*b*) How far does it go during this period?

(*a*)
$$t = \frac{v}{a} = \frac{24 \text{ m/s}}{8 \text{ m/s}^2} = 3 \text{ s}$$

(*b*)   Since the car starts from rest, $v_0 = 0$ and

$$s = \tfrac{1}{2}at^2 = (\tfrac{1}{2})(8 \text{ m/s}^2)(3 \text{ s})^2 = 36 \text{ m}$$

## SOLVED PROBLEM 3.11

The brakes of a certain car can produce an acceleration of 6 m/s². (*a*) How long does it take the car to come to a stop from a velocity of 30 m/s? (*b*) How far does the car travel during the time the brakes are applied?

(*a*)
$$t = \frac{v}{a} = \frac{30 \text{ m/s}}{6 \text{ m/s}^2} = 5 \text{ s}$$

(*b*)   Here the signs of $v_0$ and $a$ are important. The initial velocity of the car is $v_0 = +30$ m/s and its acceleration is $-6$ m/s², so

$$s = v_0 t + \tfrac{1}{2}at^2 = (30 \text{ m/s})(5 \text{ s}) - (\tfrac{1}{2})(6 \text{ m/s}^2)(5 \text{ s})^2 = 75 \text{ m}$$

## SOLVED PROBLEM 3.12

A car starts from rest with an acceleration of 2 m/s². What is its velocity after it has gone 200 m?

$$v = \sqrt{2as} = \sqrt{(2)(2 \text{ m/s}^2)(200 \text{ m})} = 28 \text{ m/s}$$

## SOLVED PROBLEM 3.13

An airplane must have a velocity of 50 m/s in order to take off. What must the airplane's acceleration be if it is to take off from a runway 500 m, long?

Since $v^2 = 2as$,

$$a = \frac{v^2}{2s} = \frac{(50 \text{ m/s})^2}{(2)(500 \text{ m})} = 2.5 \text{ m/s}^2$$

**SOLVED PROBLEM 3.14**

The brakes of a car whose initial velocity is 30 m/s are applied, and the car receives an acceleration of $-2$ m/s$^2$. How far will it have gone (a) when its velocity has decreased to 15 m/s and (b) when it has come to a stop?

(a)  Since $v^2 = v_0^2 + 2as$,

$$s = \frac{v^2 - v_0^2}{2a} = \frac{(15 \text{ m/s})^2 - (30 \text{ m/s})^2}{(2)(-2 \text{ m/s}^2)} = 169 \text{ m}$$

(b)  Here $v = 0$, and so

$$s = \frac{0 - (30 \text{ m/s})^2}{(2)(-2 \text{ m/s}^2)} = 225 \text{ m}$$

## *Multiple-Choice Questions*

**3.1.**  A snail travels 45 cm in 20 min. Its average velocity is

(a)  2.25 cm/h    (c)  90 cm/h
(b)  15 cm/h      (d)  135 cm/h

**3.2.**  A car travels at 20 km/h for 1.5 h, at 30 km/h for 2.0 h, and at 40 km/h for 0.75 h. The car's average velocity is approximately

(a)  19 km/h    (c)  28 km/h
(b)  21 km/h    (d)  30 km/h

**3.3.**  How long does a bicycle with an acceleration of 0.8 m/s$^2$ take to go from 4 to 12 m/s?

(a)  6.4 s    (c)  15 s
(b)  10 s     (d)  26 s

**3.4.**  A car that starts from rest has a constant acceleration of 4 m/s$^2$. In the first 3 s the car travels

(a)  6 m    (c)  18 m
(b)  12 m    (d)  72 m

**3.5.**  An airplane starting from rest takes 25 s and 500 m of runway at constant acceleration to leave the ground. Its velocity when it becomes airborne is

(a)  20 m/s    (c)  40 m/s
(b)  32 m/s    (d)  80 m/s

**3.6.**  A car has an acceleration of 1.2 m/s$^2$. If its initial velocity is 10 m/s, the distance the car covers in the first 5 s after the acceleration begins is

(a)  15 m    (c)  53 m
(b)  25 m    (d)  65 m

**3.7.**   A car with its brakes applied has an acceleration of $-1.2$ m/s². If its initial velocity is 10 m/s, the distance the car covers in the first 5 s after the acceleration begins is

(*a*)   15 m      (*c*)   35 m
(*b*)   32 m      (*d*)   47 m

**3.8.**   The distance the car in Question 3.7 travels before it comes to a stop is

(*a*)   6.5 m     (*c*)   21 m
(*b*)   8.3 m     (*d*)   42 m

# Supplementary Problems

**3.1.**   A ship steams at a constant velocity of 30 km/h. (*a*) How far does it travel in a day? (*b*) How long does it take to travel 500 km?

**3.2.**   A car moves at 50 km/h for $\frac{1}{2}$ h and then at 60 km/h for 2 h. (*a*) How far did it go? (*b*) What was its average velocity for the entire trip?

**3.3.**   An airplane whose airspeed is 500 km/h covers a distance of 1000 km in $2\frac{1}{2}$ h. How strong was the headwind against it?

**3.4.**   How long does it take an echo to return to a woman standing 300 m from a cliff? The velocity of sound in air is about 343 m/s.

**3.5.**   A pitcher takes 0.1 s to throw a baseball, which leaves his hand at a velocity of 30 m/s. What is the acceleration of the baseball while it is being thrown?

**3.6.**   A car comes to a stop in 6 s from a velocity of 30 m/s. (*a*) What is its acceleration? (*b*) At the same acceleration, how long would it take the car to come to a stop from a velocity of 40 m/s?

**3.7.**   The brakes of a car can slow it down from 60 to 40 km/h in 2 s. How long will it take to bring the car to a stop from an initial velocity of 25 km/h at the same acceleration?

**3.8.**   An object starts from rest with an acceleration of 10 m/s². (*a*) How far does it go in 0.5 s? (*b*) What is its velocity after 0.5 s?

**3.9.**   A car has an initial velocity of 10 m/s when it begins to be accelerated at 2.5 m/s². (*a*) How long does it take to reach a velocity of 25 m/s? (*b*) How far does it go during this period?

**3.10.**  The brakes of a car moving at 14 m/s are applied, and the car comes to a stop in 4 s. How far does the car go while it is slowing down from 14 to 10 m/s?

**3.11.**  A car whose velocity is 20 km/h is given an acceleration of 5 (km/h)/s. What is its velocity after it has gone 0.25 km?

**3.12.**  A sports car has an acceleration of 3 m/s². How much distance does it cover while its velocity is increased from 0 to 10 m/s? From 10 to 30 m/s?

**3.13.**  Find the acceleration of a car that comes to a stop from a velocity of 60 m/s in a distance of 360 m.

## *Answers to Multiple-Choice Questions*

**3.1.** (*d*)    **3.5.** (*c*)

**3.2.** (*c*)    **3.6.** (*d*)

**3.3.** (*b*)    **3.7.** (*c*)

**3.4.** (*c*)    **3.8.** (*d*)

## Answers to Supplementary Problems

**3.1.**   (*a*)  720 km     (*b*)  17.7 h

**3.2.**   (*a*)  145 km     (*b*)  58 km/h

**3.3.**   100 km

**3.4.**   1.75 s

**3.5.**   300 m/s$^2$

**3.6.**   (*a*)  5 m/s$^2$     (*b*)  8 s

**3.7.**   2.5 s

**3.8.**   (*a*)  1.25 m     (*b*)  5 m/s

**3.9.**   (*a*)  6 s     (*b*)  105 m

**3.10.**   13.7 m

**3.11.**   27 m/s = 97 km/h

**3.12.**   17 m; 133 m

**3.13.**   −5 m/s$^2$

# Chapter 4

## Motion in a Vertical Plane

### ACCELERATION OF GRAVITY

All bodies in free fall near the earth's surface have the same downward acceleration of

$$g = 9.8 \text{ m/s}^2 = 32 \text{ ft/s}^2$$

A body falling from rest in a vacuum thus has a velocity of 32 ft/s at the end of the first second, 64 ft/s at the end of the next second, and so forth. The farther the body falls, the faster it moves.

A body in free fall has the same downward acceleration whether it starts from rest or has an initial velocity in some direction.

The presence of air affects the motion of falling bodies partly through buoyancy and partly through air resistance. Thus two different objects falling in air from the same height will not, in general, reach the ground at exactly the same time. Because air resistance increases with velocity, eventually a falling body reaches a *terminal velocity* that depends on its mass, size, and shape, and it cannot fall any faster than that.

### FALLING BODIES

When buoyancy and air resistance can be neglected, a falling body has the constant acceleration $g$, and the formulas for uniformly accelerated motion apply. Thus a body dropped from rest has the velocity

$$v = gt$$

after time $t$, and it has fallen through a vertical distance of

$$h = \tfrac{1}{2}gt^2$$

From the latter formula we see that

$$t = \sqrt{\frac{2h}{g}}$$

and so the velocity of the body is related to the distance it has fallen by $v = gt$, or

$$v = \sqrt{2gh}$$

To reach a certain height $h$, a body thrown upward must have the same initial velocity as the final velocity of a body falling from that height, namely, $v = \sqrt{2gh}$.

*(Air resistance is assumed to be negligible in the problems of this chapter.)*

### SOLVED PROBLEM 4.1

A stone dropped from a bridge strikes the water 2.5 s later. (*a*) What is its final velocity in meters per second? (*b*) How high is the bridge?

(*a*) $$v = gt = (9.8 \text{ m/s}^2)(2.5 \text{ s}) = 24.5 \text{ m/s}$$

(*b*) $$h = \tfrac{1}{2}gt^2 = (\tfrac{1}{2})(9.8 \text{ m/s}^2)(2.5 \text{ s})^2 = 30.6 \text{ m}$$

## SOLVED PROBLEM 4.2

A ball is dropped from a window 64 ft above the ground. (*a*) How long does it take the ball to reach the ground? (*b*) What is its final velocity?

(*a*)    Since $h = \frac{1}{2}gt^2$,

$$t = \sqrt{\frac{2h}{g}} = \sqrt{\frac{(2)(64 \text{ ft})}{32 \text{ ft/s}^2}} = \sqrt{4 \text{ s}^2} = 2 \text{ s}$$

(*b*)                                   $v = gt = (32 \text{ ft/s}^2)(2 \text{ s}) = 64 \text{ ft/s}$

## SOLVED PROBLEM 4.3

What velocity must a ball have when thrown upward if it is to reach a height of 15 m?

The upward velocity the ball must have is the same as the downward velocity the ball would have if dropped from that height. Hence

$$v = \sqrt{2gh} = \sqrt{(2)(9.8 \text{ m/s}^2)(15 \text{ m})} = \sqrt{294 \text{ m}^2/\text{s}^2} = 17 \text{ m/s}$$

## SOLVED PROBLEM 4.4

A ball is thrown downward from the edge of a cliff with an initial velocity of 6 m/s. (*a*) How fast is it moving 2 s later? (*b*) How far does it fall in these 2 s?

(*a*)            $v = v_0 + gt = 6 \text{ m/s} + (9.8 \text{ m/s}^2)(2 \text{ s}) = 6 \text{ m/s} + 19.6 \text{ m/s} = 25.6 \text{ m/s}$

(*b*)        $h = v_0 t + \frac{1}{2}gt^2 = (6 \text{ m/s})(2 \text{ s}) + (\frac{1}{2})(9.8 \text{ m/s}^2)(2 \text{ s})^2 = 12 \text{ m} + 19.6 \text{ m} = 31.6 \text{ m}$

## SOLVED PROBLEM 4.5

A ball is thrown upward from the edge of a cliff with an initial velocity of 6 m/s. (*a*) How fast is it moving $\frac{1}{2}$ s later? In what direction? (*b*) How fast is it moving 2 s later? In what direction?

(*a*)    We consider upward as + and downward as −. Then $v_0 = +6$ m/s and $g = -9.8$ m/s$^2$, so

$$v = v_0 + gt = 6 \text{ m/s} + (-9.8 \text{ m/s}^2)(\tfrac{1}{2} \text{ s}) = 6 \text{ m/s} - 4.9 \text{ m/s} = 1.1 \text{ m/s}$$

which is positive and hence upward.

(*b*)    After 2 s

$$v = v_0 + gt = 6 \text{ m/s} + (-9.8 \text{ m/s}^2)(2 \text{ s}) = 6 \text{ m/s} - 19.6 \text{ m/s} = -13.6 \text{ m/s}$$

which is negative and hence downward.

## SOLVED PROBLEM 4.6

A person in a closed elevator with no floor indicator does not know whether it is stationary, moving upward at constant velocity, or moving downward at constant velocity. To try to find out, the person drops a coin from a height of 2 m and times its fall with a stopwatch. What would be noted in each case?

Since the coin has exactly the same initial velocity when it is dropped as the elevator, this experiment would give the same time of fall in each case. This time of fall is

$$t = \sqrt{\frac{2h}{g}} = \sqrt{\frac{(2)(2 \text{ m})}{9.8 \text{ m/s}^2}} = \sqrt{0.408 \text{ s}^2} = 0.64 \text{ s}$$

However, if the elevator were *accelerated* upward or downward, the time of fall would be, respectively, less or more than this.

## PROJECTILE MOTION

The formulas for straight-line motion can be used to analyze the horizontal and vertical aspects of a projectile's flight separately because these are independent of each other. If air resistance is neglected, the horizontal velocity component $v_x$ remains constant during the flight. The effect of gravity on the vertical component $v_y$ is to provide a downward acceleration. If $v_y$ is initially upward, $v_y$ first decreases to 0 and then increases in the downward direction.

The range of a projectile launched at an angle $\theta$ above the horizontal with initial velocity $v_0$ is

$$R = \frac{v_0^2}{g} \sin 2\theta$$

The time of flight is

$$T = \frac{2v_0 \sin \theta}{g}$$

If $\theta_1$ is an angle other than 45° that corresponds to a range $R$, then a second angle $\theta_2$ for the same range is given by

$$\theta_2 = 90° - \theta_1$$

as shown in Fig. 4-1.

## SOLVED PROBLEM 4.7

A ball is thrown horizontally at 8 m/s. (*a*) How fast is it moving 2 s later? (*b*) In what direction is it moving?

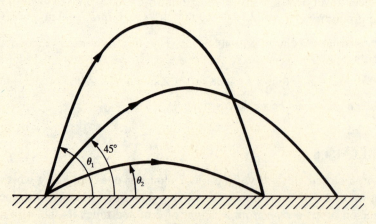

**Fig. 4-1**

(a)   The ball's velocity after 2 s has the horizontal component $v_x = 8$ m/s and the vertical component

$$v_y = gt = (9.8 \text{ m/s}^2)(2 \text{ s}) = 19.6 \text{ m/s}$$

Hence the magnitude of its final velocity is

$$v = \sqrt{v_x^2 + v_y^2} = \sqrt{(8 \text{ m/s})^2 + (19.6 \text{ m/s})^2} = 21.2 \text{ m/s}$$

(b)   From Fig. 4-2 we see that

$$\tan \theta = \frac{v_y}{v_x}$$

where $\theta$ is the angle of the ball's velocity below the horizontal. The angle $\theta$ is accordingly

$$\theta = \tan^{-1} \frac{v_y}{v_x} = \tan^{-1} \frac{19.6 \text{ m/s}}{8 \text{ m/s}} = 68°$$

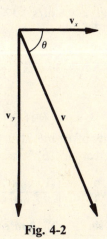

Fig. 4-2

## SOLVED PROBLEM 4.8

An airplane is in level flight at a velocity of 500 km/h and an altitude of 1500 m when a wheel falls off. (a) How long does the wheel take to reach the ground? (b) What horizontal distance will the wheel travel before it strikes the ground? (c) What will the wheel's velocity be when it strikes the ground?

(a)   The horizontal velocity of the wheel does not affect its vertical motion (Fig. 4-3). The wheel therefore reaches the ground at the same time as a wheel dropped from rest at an altitude of 1500 m, which is

$$t = \sqrt{\frac{2h}{g}} = \sqrt{\frac{(2)(1500 \text{ m})}{9.8 \text{ m/s}^2}} = 17.5 \text{ s}$$

(b)   The horizontal component of velocity of the wheel is

$$v_x = \left(500 \frac{\text{km}}{\text{h}}\right)\left(\frac{1000 \text{ m/km}}{3600 \text{ s/h}}\right) = 139 \text{ m/s}$$

In $t = 17.5$ s the wheel will travel a horizontal distance of

$$x = v_x t = (139 \text{ m/s})(17.5 \text{ s}) = 2433 \text{ m} = 2.43 \text{ km}$$

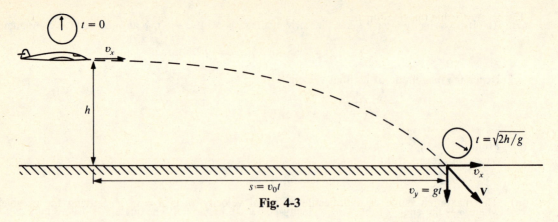

**Fig. 4-3**

(c)  The final velocity of the wheel has the horizontal component $v_x$ = 139 m/s and the vertical component

$$v_y = gt = (9.8 \text{ m/s}^2)(17.5 \text{ s}) = 172 \text{ m/s}$$

Hence the final velocity is

$$v = \sqrt{v_x^2 + v_y^2} = \sqrt{(139 \text{ m/s})^2 + (172 \text{ m/s})^2} = 221 \text{ m/s}$$

## SOLVED PROBLEM 4.9

A football is thrown with a velocity of 10 m/s at an angle of 30° above the horizontal. (a) How far away should its intended receiver be? (b) What will the time of flight be?

(a) $$R = \frac{v_0^2}{g} \sin 2\theta = \left[ \frac{(10 \text{ m/s})^2}{9.8 \text{ m/s}} \right] (\sin 60°) = 8.8 \text{ m}$$

(b) $$T = \frac{2v_0 \sin \theta}{g} = \frac{(2)(10 \text{ m/s})(\sin 30°)}{9.8 \text{ m/s}^2} = 1.02 \text{ s}$$

## SOLVED PROBLEM 4.10

An air rifle is fired at an angle of 60° above the horizontal. (a) If the pellet's initial velocity is 40 ft/s, how far does it go? (b) What is its time of flight?

(a)  Here $2\theta$ = 120°. Hence

$$R = \frac{v_0^2}{g} \sin 2\theta = \left[ \frac{(40 \text{ ft/s})^2}{32 \text{ ft/s}^2} \right] (\sin 120°) = 43 \text{ ft}$$

(b) $$T = \frac{2v_0 \sin \theta}{g} = \frac{(2)(40 \text{ ft/s})(\sin 60°)}{32 \text{ ft/s}^2} = 2.17 \text{ s}$$

## SOLVED PROBLEM 4.11

At what angle should a projectile be fired in order for its range to be a maximum?

The range of a projectile is given by

$$R = \frac{v_0^2}{g} \sin 2\theta$$

The greatest value the sine function can have is 1. Since $\sin 90° = 1$, the maximum range occurs when

$$2\theta = 90° \qquad \theta = 45°$$

Larger and smaller angles than 45° give shorter ranges. When $\theta = 45°$,

$$R_{max} = \frac{v_0^2}{g}$$

## SOLVED PROBLEM 4.12

(a) What minimum initial velocity must a projectile have to reach a target 90 km away?

(b) What would the time of flight be?

(a)   The maximum range of a projectile of initial velocity $v_0$ is $R = v_0^2/g$. Solving for $v_0$ gives $v_0 = \sqrt{Rg}$. Since here

$$R = (90\text{ km})(1000\text{ m/km}) = 9 \times 10^4\text{ m}$$

we have       $v_0 = \sqrt{Rg} = \sqrt{(9 \times 10^4\text{ m})(9.8\text{ m/s}^2)} = \sqrt{8.82 \times 10^5\text{ m}^2/\text{s}^2}$

To find a square root, the power of 10 must be an even number. We therefore write $8.82 \times 10^5$ as $88.2 \times 10^4$ and find that

$$v_0 = \sqrt{88.2 \times 10^4\text{ m}^2/\text{s}^2} = 9.39 \times 10^2\text{ m/s} = 939\text{ m/s}$$

(b)   Here $\theta = 45°$, corresponding to maximum range, so

$$T = \frac{2v_0 \sin\theta}{g} = \frac{(2)(939\text{ m/s})(\sin 45°)}{9.8\text{ m/s}^2} = 136\text{ s} = 2\text{ min } 16\text{ s}$$

## SOLVED PROBLEM 4.13

At what two angles above the horizontal can a projectile be fired in order to reach a distance equal to half its maximum range?

The maximum range is $R_{max} = v_0^2/g$, and so here $R = R_{max}/2 = v_0^2/(2g)$. The smaller angle $\theta_1$ is found as follows:

$$R = \frac{v_0^2}{g}\sin 2\theta_1 = \frac{v_0^2}{2g}$$
$$\sin 2\theta_1 = \tfrac{1}{2} = 0.5$$
$$2\theta_1 = \sin^{-1} 0.5$$
$$\theta_1 = \tfrac{1}{2}\sin^{-1} 0.5 = 15°$$

The larger angle $\theta_2$ is given by

$$\theta_2 = 90° - \theta_1 = 90° - 15° = 75°$$

# *Multiple Choice Questions*

4.1.   A stone thrown upward has an acceleration that is

(a)   smaller than that of a stone thrown downward

(b)   the same as that of a stone thrown downward

(c)  greater than that of a stone thrown downward

(d)  zero until it reaches the highest point in its path

**4.2.**  A ball is dropped from a roof at the same time as another ball is thrown upward from the roof. The two balls

(a)  have the same velocity when they reach the ground

(b)  have the same acceleration when they reach the ground

(c)  reach the ground at the same time

(d)  none of the above

**4.3.**  Ball A is dropped from a roof at the same time as ball B is thrown horizontally from the roof. Ball A

(a)  reaches the ground first

(b)  reaches the ground last

(c)  has the greater velocity when it reaches the ground

(d)  has the smaller velocity when it reaches the ground

**4.4.**  A bottle falls from a blimp whose altitude is 1200 m. The bottle reaches the ground in approximately

(a)  5 s        (c)  16 s

(b)  11 s       (d)  245 s

**4.5.**  After a stone dropped from a cliff has fallen 20 m, the stone's velocity is

(a)  10 m/s     (c)  196 m/s

(b)  20 m/s     (d)  392 m/s

**4.6.**  A ball is thrown vertically upward at 20 m/s. The ball comes to a momentary stop in approximately

(a)  0.5 s      (c)  1.5 s

(b)  1.0 s      (d)  2.0 s

**4.7.**  At what velocity will the ball of Question 4.6 reach the ground?

(a)  0.8 m/s    (c)  20 m/s

(b)  19.6 m/s   (d)  40 m/s

**4.8.**  A ball is thrown downward from a window 12.0 m above the ground. When it reaches the ground, the ball's velocity will be

(a)  9.3 m/s    (c)  21.5 m/s

(b)  15.3 m/s   (d)  52.8 m/s

**4.9.**  A ball is thrown horizontally from a roof at 12.0 m/s. The ball's velocity 2 s later will be

(a)  12.0 m/s   (c)  23.0 m/s

(b)  19.6 m/s   (d)  31.6 m/s

**4.10.**  A ball is thrown 40° above the horizontal at 4.0 m/s. After 0.50 s the horizontal component of the ball's velocity will be

(a)  2.6 m/s    (c)  3.4 m/s

(b)  3.1 m/s    (d)  5.5 m/s

**4.11.** A ball is thrown 40° below the horizontal at 4.0 m/s. After 0.50 s the horizontal component of the ball's velocity will be

(*a*)   2.6 m/s        (*c*)   3.4 m/s
(*b*)   3.1 m/s        (*d*)   5.5 m/s

**4.12.** After 0.50 s the vertical component of the velocity of the ball of Question 4.11 will be

(*a*)   2.6 m/s        (*c*)   4.9 m/s
(*b*)   3.1 m/s        (*d*)   7.5 m/s

# Supplementary Problems

**4.1.** A stone is dropped from the edge of a cliff. (*a*) What is its velocity in meters per second 3 s later? (*b*) How far does it fall in this time? (*c*) How far will it fall in the next second?

**4.2.** A girl throws a ball 20 m vertically into the air. (*a*) How long does she have to wait to catch it on the way down? (*b*) What was its initial velocity? (*c*) What will be its final velocity?

**4.3.** The Sears Tower in Chicago is 1454 ft high. (*a*) How long would it take an object dropped from the top of the building to reach the ground? (*b*) What would the object's final velocity be?

**4.4.** A body in free fall reaches the ground in 5 s. (*a*) From what height in meters was it dropped? (*b*) What is its final velocity? (*c*) How far did it fall in the last second of its descent?

**4.5.** The acceleration of gravity on the surface of Mars is 3.7 m/s$^2$. If a ball is thrown vertically downward at 10 m/s from a cliff on Mars, (*a*) what will its speed be after 1 s? (*b*) After 3 s?

**4.6.** A ball is thrown vertically upward with a velocity of 12 m/s. (*a*) At what height is the ball 1 s later? (*b*) 2*s* later? (*c*) What is the maximum height the ball reaches?

**4.7.** A ball is thrown horizontally with a velocity of 12 m/s. (*a*) How far has the ball fallen 1 s later? (*b*) 2 s later?

**4.8.** A ball is thrown vertically downward with a velocity of 12 m/s. (*a*) How far has the ball fallen 1 s later? (*b*) 2 s later?

**4.9.** A rifle with a muzzle velocity of 200 m/s is fired with its barrel horizontally at a height of 1.5 m above the ground. (*a*) How long is the bullet in the air? (*b*) How far away from the rifle does the bullet strike the ground? (*c*) If the muzzle velocity were 150 m/s, would there be any difference in these answers?

**4.10.** A ball is rolled off the edge of a table 3 ft high with a horizontal velocity of 4 ft/s. With what velocity does it strike the floor?

**4.11.** If the initial velocity of a projectile were doubled, how would its maximum range be affected?

**4.12.** A shell is fired at an angle of 40° above the horizontal at a velocity of 300 m/s. (*a*) What is its range? (*b*) What is its time of flight?

**4.13.** A golf ball leaves the club at 40 m/s at an angle of 55° above the horizontal. (*a*) What is its range? (*b*) What is its time of flight?

**4.14.** A rifle bullet has a muzzle velocity of 200 m/s. (*a*) At what angle should the rifle be pointed to achieve the maximum range? (*b*) What is that range? (*c*) What is the time of flight at the maximum range?

**4.15.** An arrow leaves a bow at 25 m/s. (*a*) What is its maximum range? (*b*) At what two angles can it be sent above the horizontal to reach a target 50 m away?

**4.16.** A soccer ball that was kicked at an angle of 40° above the horizontal strikes the ground 25 m away. What was its initial velocity?

## *Answers to Multiple-Choice Questions*

**4.1.** (*b*)    **4.7.** (*c*)

**4.2.** (*b*)    **4.8.** (*c*)

**4.3.** (*d*)    **4.9.** (*c*)

**4.4.** (*c*)    **4.10.** (*b*)

**4.5.** (*b*)    **4.11.** (*b*)

**4.6.** (*d*)    **4.12.** (*d*)

## Answers to Supplementary Problems

**4.1.** (*a*)  29.4 m/s    (*b*)  44.1 m    (*c*)  34.3 m

**4.2.** (*a*)  4.04 s    (*b*)  19.8 m/s    (*c*)  19.8 m/s

**4.3.** (*a*)  9.5 s    (*b*)  305 ft/s

**4.4.** (*a*)  123 m    (*b*)  49 m/s   (*c*) 44 m

**4.5.** (*a*)  13.7 m/s    (*b*)  43.3 m/s

**4.6.** (*a*)  7.1 m    (*b*)  4.4 m    (*c*)  7.3 m

**4.7.** (*a*)  4.9 m    (*b*)  19.6 m

**4.8.** (*a*)  16.9 m    (*b*)  43.6 m

**4.9.** (*a*)  0.56 s    (*b*)  112 m
(*c*)   The time would be unchanged since it depends only on the height of the rifle, but the distance would be reduced to 84 m.

**4.10.**   14.4 ft/s

**4.11.**   The maximum range would be four times greater.

**4.12.**   (*a*)   9.044 km       (*b*)   39 s

**4.13.**   (*a*)   153 m       (*b*)   6.7 s

**4.14.**   (*a*)   45°       (*b*)   4.08 km   (*c*)   29 s

**4.15.**   (*a*)   64 m       (*b*)   26°, 64°

**4.16.**   16 m/s

# Chapter 5

## Laws of Motion

### FIRST LAW OF MOTION

According to Newton's *first law of motion*, if no net force acts on it, a body at rest remains at rest and a body in motion remains in motion at constant velocity (that is, at constant speed in a straight line).

This law provides a definition of *force*: A force is any influence that can change the velocity of a body.

Two or more forces act on a body without affecting its velocity if the forces cancel one another out. What is needed for a velocity change is a *net force*, or *unbalanced force*. To accelerate something, a net force must be applied to it. Conversely, every acceleration is due to the action of a net force.

### MASS

The property a body has of resisting any change in its state of rest or of uniform motion is called *inertia*. The inertia of a body is related to what we can think of as the amount of matter it contains. A quantitative measure of inertia is *mass*: The more mass a body has, the less its acceleration when a given net force acts on it. The SI unit of mass is the *kilogram* (kg). A liter of water, which is 1.057 quarts, has a mass of almost exactly 1 kg.

### SECOND LAW OF MOTION

According to Newton's *second law of motion*, the net force acting on a body equals the product of the mass and the acceleration of the body. The direction of the force is the same as that of the acceleration.

In equation form,

$$\mathbf{F} = m\mathbf{a}$$

Net force = (mass) (acceleration)

Net force is sometimes designated $\Sigma \mathbf{F}$, where $\Sigma$ (Greek capital letter *sigma*) means "sum of." The second law of motion is the key to understanding the behavior of moving bodies since it links cause (force) and effect (acceleration) in a definite way.

In the SI system, the unit for force is the *newton* (N): A newton is that net force which, when applied to a 1-kg mass, gives it an acceleration of 1 m/s$^2$. The newton is not a basic unit, and in calculations it may have to be replaced with its equivalent in terms of the meter, the second, and the kilogram. This equivalent can be found from $F = ma$: 1 N = (1 kg) (1 m/s$^2$) = 1 kg·m/s$^2$. A newton is about 0.2248 lb, a little less than $\frac{1}{4}$ lb.

### SOLVED PROBLEM 5.1

A 10-kg body has an acceleration of 5 m/s$^2$. What is the net force acting on it?

$$F = ma = (10 \text{ kg})(5 \text{ m/s}^2) = 50 \text{ N}$$

## SOLVED PROBLEM 5.2

A force of 80 N gives an object of unknown mass an acceleration of 20 m/s$^2$. What is its mass?

$$m = \frac{F}{a} = \frac{80 \text{ N}}{20 \text{ m/s}^2} = 4 \text{ kg}$$

## SOLVED PROBLEM 5.3

A force of 3000 N is applied to a 1500-kg car at rest. (*a*) What is its acceleration? (*b*) What will its velocity be 5 s later?

(*a*)
$$a = \frac{F}{m} = \frac{3000 \text{ N}}{1500 \text{ kg}} = 2 \text{ m/s}^2$$

(*b*)
$$v = at = (2 \text{ m/s}^2)(5 \text{ s}) = 110 \text{ m/s}$$

## SOLVED PROBLEM 5.4

An empty truck whose mass is 2000 kg has a maximum acceleration of 1 m/s$^2$. What will its maximum acceleration be when it carries a load of 1000 kg?

The maximum force available is

$$F = ma = (2000 \text{ kg})(1 \text{ m/s}^2) = 2000 \text{ N}$$

When this force is applied to the total mass of the loaded truck, which is 3000 kg, the resulting acceleration will be

$$a = \frac{F}{m} = \frac{2000 \text{ N}}{3000 \text{ kg}} = \frac{2}{3} \text{ m/s}^2$$

## SOLVED PROBLEM 5.5

A 1000-kg car goes from 10 to 20 m/s in 5 s. What force is acting on it?

$$a = \frac{v - v_0}{t} = \frac{20 \text{ m/s} - 10 \text{ m/s}}{5 \text{ s}} = 2 \text{ m/s}^2$$

$$F = ma = (1000 \text{ kg})(2 \text{ m/s}^2) = 2000 \text{ N}$$

## SOLVED PROBLEM 5.6

A 60-g tennis ball approaches a racket at 15 m/s, is in contact with the racket for 0.005 s, and then rebounds at 20 m/s. Find the average force that the racket exerted on the ball.

The final velocity of the tennis ball was $v = -20$ m/s since it reversed direction when it was struck by the racket. Hence the ball's acceleration was

$$a = \frac{v - v_0}{t} = \frac{-20 \text{ m/s} - 15 \text{ m/s}}{0.005 \text{ s}} = \frac{-35 \text{ m/s}}{0.005 \text{ s}} = -7 \times 10^3 \text{ m/s}$$

Since the ball's mass is $m = 60$ g $= 0.06$ kg, the average force on it was

$$F = ma = (0.06 \text{ kg})(-7 \times 10^3 \text{ m/s}^2) = -420 \text{ N}$$

The minus sign means that the force was in the opposite direction to that in which the ball approached the racket.

**SOLVED PROBLEM 5.7**

The brakes of a 1000-kg car exert 3000 N. (*a*) How long will it take the car to come to a stop from a velocity of 30 m/s? (*b*) How far will the car travel during this time?

(*a*)   The acceleration the brakes can produce is

$$a = \frac{F}{m} = \frac{-3000 \text{ N}}{1000 \text{ kg}} = -3 \text{ m/s}^2$$

(The force is considered negative because it acts opposite to the car's direction of motion.) Here the initial velocity is $v_0 = 30$ m/s, the final velocity is $v = 0$, and the acceleration is $-3$ m/s$^2$. Hence

$$v = v_0 + at$$
$$0 = 30 \text{ m/s} - (3 \text{ m/s}^2)(t)$$
$$t = \frac{30 \text{ m/s}}{3 \text{ m/s}^2} = 10 \text{ s}$$

(*b*)        $s = v_0 t + \frac{1}{2}at^2 = (30 \text{ m/s})(10 \text{ s}) - (\frac{1}{2})(3 \text{ m/s}^2)(10 \text{ s})^2 = 300 \text{ m} - 150 \text{ m} = 150 \text{ m}$

## WEIGHT

The *weight* of a body is the gravitational force with which the earth attracts the body. If a person weighs 600 N (135 lb), this means the earth pulls that person down with a force of 600 N. Weight (a vector quantity) is different from mass (a scalar quantity), which is a measure of the response of a body to an applied force. The weight of a body varies with its location near the earth (or other astronomical body), whereas its mass is the same everywhere in the universe.

The weight of a body is the force that causes it to be accelerated downward with the acceleration of gravity *g*. Hence, from the second law of motion, with $F = w$ and $a = g$,

$$w = mg$$

Weight = (mass)(acceleration of gravity)

Because *g* is constant near the earth's surface, the weight of a body there is proportional to its mass—a large mass is heavier than a small one.

**SOLVED PROBLEM 5.8**

A 100-kg man slides down a rope at constant speed. (*a*) What minimum breaking strength must the rope have? (*b*) If the rope has precisely this strength, will it support the man if he tries to climb back up?

(*a*)   Since the man slides down at constant speed, the rope must support his weight of $mg = 980$ N.

(*b*)   For the man to climb up, an additional force must be exerted on the rope, so it will break.

**SOLVED PROBLEM 5.9**

(*a*) What is the weight of an object whose mass is 5 kg? (*b*) What is its acceleration when a net force of 100 N acts on it?

(*a*)                          $w = mg = (5 \text{ kg})(9.8 \text{ m/s}^2) = 49 \text{ N}$

(*b*)   From the second law of motion $F = ma$, we have

$$a = \frac{F}{m} = \frac{100 \text{ N}}{5 \text{ kg}} = 20 \text{ m/s}^2$$

## SOLVED PROBLEM 5.10

A force of 1 N acts on (*a*) a body whose mass is 1 kg and (*b*) a body whose weight is 1 N. Find their respective accelerations.

(*a*)
$$a = \frac{F}{m} = \frac{1 \text{ N}}{1 \text{ kg}} = 1 \text{ m/s}^2$$

(*b*)  The mass of a body whose weight is *w* is *m* = *w/g*. Hence, in general,

$$a = \frac{F}{m} = \frac{F}{w/g} = \frac{F}{w} g$$

Here *F* = *w* = 1 N, so the acceleration is *a* = *g* = 9.8 m/s$^2$.

## SOLVED PROBLEM 5.11

A net horizontal force of 4000 N is applied to a car at rest whose weight is 10,000 N. What will the car's speed be after 8 s?

To find the car's acceleration from the second law, we need its mass, not its weight. The mass is

$$m = \frac{w}{g} = \frac{10,000 \text{ N}}{9.8 \text{ m/s}^2} = 1020 \text{ kg}$$

The car's acceleration when a 4000-N net force is applied is

$$a = \frac{F}{m} = \frac{4000 \text{ N}}{1020 \text{ kg}} = 3.92 \text{ m/s}^2$$

The car's speed 8 s after the acceleration began is

$$v_f = at = (3.92 \text{ m/s}^2)(8 \text{ s}) = 31.4 \text{ m/s}$$

## BRITISH SYSTEM OF UNITS

In the British system, the unit of mass is the *slug* and the unit of force is the *pound* (lb). A net force of 1 lb acting on a mass of 1 slug gives it an acceleration of 1 ft/s$^2$. Here is how units of mass and force in the SI and British systems are related:

1 kg = 10$^3$ g = 0.0685 slug       (1 kg corresponds to 2.21 lb in the sense
                                                     that the *weight* of 1 kg is 2.21 lb)

1 slug = 14.6 kg       (1 slug corresponds to 32 lb in the sense
                                     that the *weight* of 1 slug is 32 lb)

1 N = 0.225 lb

1 lb = 4.45 N

Table 5.1 relates the SI and British systems to mass and weight.

## SOLVED PROBLEM 5.12

(*a*) What is the weight of an object whose mass is 50 slugs? (*b*) What is the mass of an object whose weight is 50 lb?

**Table 5.1   Mass and Weight in the SI and British Systems.**

| System of Units | To find mass $m$ given weight $w$ | To find weight $w$ given mass $m$ |
|---|---|---|
| SI | $m \text{ kg} = \dfrac{w \text{ N}}{9.8 \text{ m/s}^2}$ | $w \text{ N} = (m \text{ kg})(9.8 \text{ m/s}^2)$ |
| British | $m \text{ slugs} = \dfrac{w \text{ lb}}{32 \text{ ft/s}^2}$ | $w \text{ lb} = (m \text{ slugs})(32 \text{ ft/s}^2)$ |

($a$) $$w = mg = (50 \text{ slugs})(32 \text{ ft/s}^2) = 1600 \text{ lb}$$

($b$) $$m = \frac{w}{g} = \frac{50 \text{ lb}}{32 \text{ ft/s}^2} = 1.56 \text{ slugs}$$

## SOLVED PROBLEM 5.13

($a$) What is the mass of a 160-lb man? ($b$) With what force is he attracted to the earth? ($c$) If he jumps from a diving board, what will his downward acceleration be?

($a$) $$m = \frac{w}{g} = \frac{160 \text{ lb}}{32 \text{ ft/s}^2} = 5 \text{ slugs}$$

($b$)   The attractive force exerted on the man by the earth is his weight of 160 lb.

($c$)   His downward acceleration is the acceleration of gravity $g = 32 \text{ ft/s}^2$.

## SOLVED PROBLEM 5.14

A net force of 75 lb acts on a body of mass 25 slugs which is initially at rest. ($a$) Find its acceleration. ($b$) How fast will the body be moving 12 s later?

($a$) $$a = \frac{F}{m} = \frac{75 \text{ lb}}{25 \text{ slugs}} = 3 \text{ ft/s}^2$$

($b$) $$v = at = (3 \text{ ft/s}^2)(12 \text{ s}) = 36 \text{ ft/s}$$

## SOLVED PROBLEM 5.15

A net force of 150 lb acts on a body whose weight is 96 lb. What is its acceleration?

The mass of the body is

$$m = \frac{w}{g} = \frac{96 \text{ lb}}{32 \text{ ft/s}^2} = 3 \text{ slugs}$$

Its acceleration is therefore

$$a = \frac{F}{m} = \frac{150 \text{ lb}}{3 \text{ slugs}} = 50 \text{ ft/s}^2$$

## SOLVED PROBLEM 5.16

How much force is needed to bring a 3200-lb car from rest to a velocity of 44 ft/s (30 mi/h) in 8 s?

The car's mass is

$$m = \frac{w}{g} = \frac{3200 \text{ lb}}{32 \text{ ft/s}^2} = 100 \text{ slugs}$$

and its acceleration is

$$a = \frac{v}{t} = \frac{44 \text{ ft/s}}{8 \text{ s}} = 5.5 \text{ ft/s}^2$$

Hence

$$F = ma = (100 \text{ slugs})(5.5 \text{ ft/s}^2) = 550 \text{ lb}$$

## FREE-BODY DIAGRAMS AND TENSION

In all but the simplest problems that involve the second law of motion, it is helpful to draw a *free-body diagram* of the situation. This is a vector diagram that shows all the forces which act *on* the body whose motion is being studied. Forces that the body exerts on anything else should not be included, since such forces do not affect the body's motion.

Forces are often transmitted by *cables*, a general term that includes strings, ropes, and chains. Cables can change the direction of a force with the help of a pulley while leaving the magnitude of the force unchanged. The *tension T* in a cable is the magnitude of the force that any part of the cable exerts on the adjoining part (Fig. 5-1). The tension is the same in both directions in the cable and $T$ is the same along the entire cable if the cable's mass is small. Only cables of negligible mass will be considered here, so $T$ can be thought of as the magnitude of the force that either end of a cable exerts on whatever it is attached to.

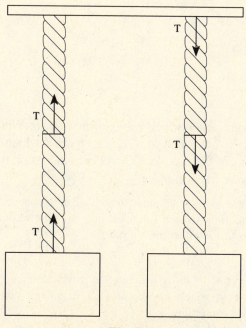

Fig. 5-1

**SOLVED PROBLEM 5.17**

A 1200-kg elevator is supported by a cable in which the maximum safe tension is 14,000 N.
(*a*) What is the greatest upward acceleration of the elevator? (*b*) The greatest downward
acceleration?

(*a*)  The elevator's weight is

$$w = mg = (1200 \text{ kg})(9.8 \text{ m/s}^2) = 11,760 \text{ N}$$

Figure 5-2 is a free-body diagram of the situation. Because the maximum tension in the cable is
$T = 14,000$ N, the net force $F$ available to accelerate the elevator upward is a maximum of

$$F = T - w = 14,000 \text{ N} - 11,760 \text{ N} = 2240 \text{ N}$$

The elevator's acceleration when this net force is applied to it is

$$a = \frac{F}{m} = \frac{2240 \text{ N}}{1200 \text{ kg}} = 1.87 \text{ m/s}^2$$

(*b*)  The cable is flexible and so cannot push down on the elevator. The maximum downward
acceleration therefore corresponds to free fall with an acceleration of $g = 9.8$ m/s².

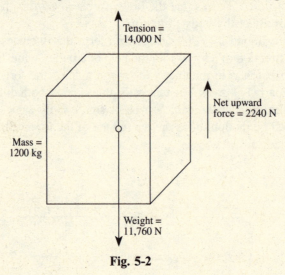

**Fig. 5-2**

**SOLVED PROBLEM 5.18**

Figure 5-3 shows a 5-kg block *A* which hangs from a string which passes over a frictionless
pulley and is joined at its other end to a 12-kg block *B* which lies on a frictionless table.
(*a*) Find the accelerations of the two blocks. (*b*) Find the tension in the string. What would the
tension be if block *B* were fixed in place?

(*a*)  The blocks have accelerations of the same magnitude *a* because they are joined by the string. The
net force $F_B$ on *B* equals the tension $T$ in the string. From the second law of motion, taking the left
as the + direction so that *a* will come out positive,

$$F_B = T = m_B a$$

The net force $F_A$ on *A* is the difference between its weight $m_A g$, which acts downward, and the
tension $T$ in the string, which acts upward on it. Taking downward as + so that the two
accelerations have the same sign,

$$F_A = m_A g - T = m_A a$$

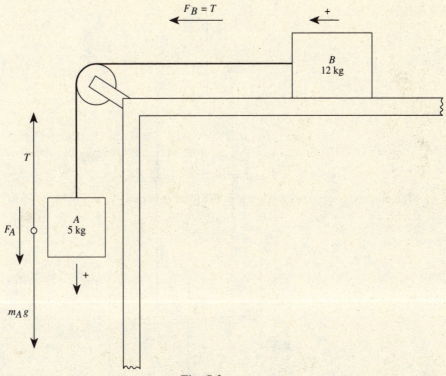

**Fig. 5-3**

We now have two equations in the two unknowns, $a$ and $T$. The easiest way to solve them is to start by substituting $T = m_B a$ from the first equation into the second. This gives

$$m_A g - T = m_A g - m_B a = m_A a$$
$$m_A g = (m_A + m_B)a$$
$$a = \frac{m_A g}{m_A + m_B} = \frac{(5 \text{ kg})(9.8 \text{ m/s}^2)}{5 \text{ kg} + 12 \text{ kg}} = 2.9 \text{ m/s}^2$$

(b)  We can use either of the original equations to find the tension $T$. From the first,

$$T = m_B a = (12 \text{ kg})(2.9 \text{ m/s}^2) = 35 \text{ N}$$

If $B$ were fixed in place, the tension would equal the weight of $A$, which is $m_A g = 49$ N. Here the tension is less because $B$ moves in response to the pull of $A$'s weight, but it is not zero because of $B$'s inertia.

## SOLVED PROBLEM 5.19

Figure 5-4 shows the same two blocks, $A$ and $B$, from Fig. 5-3 but now suspended by a string on either side of a frictionless pulley. (a) Find the accelerations of the two blocks. (b) Find the tension in the string.

(a)  Here $A$ moves upward and $B$ moves downward, so we let upward be $+$ for $A$ and downward be $-$ for $B$. Both blocks have accelerations of the same magnitude $a$. Applying the second law of motion to the two blocks with the help of the free-body diagrams in Fig. 5-4 gives

$$F_A = T - m_A g = m_A a$$
$$F_B = m_B g - T = m_B a$$

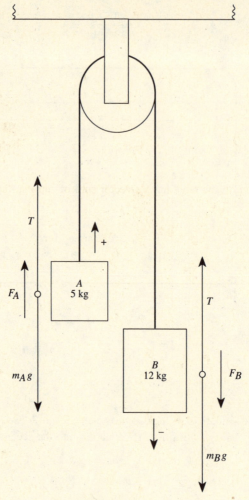

**Fig. 5-4**

From the first equation we have for the tension $T$

$$T = m_A a + m_A g$$

Substituting this value of $T$ in the second equation gives

$$m_B g - m_A a - m_A g = m_B a$$

$$(m_B - m_A)g = (m_A + m_B)a$$

$$a = \frac{(m_B - m_A)g}{m_A + m_B} = \frac{(12 \text{ kg} - 5 \text{ kg})(9.8 \text{ m/s}^2)}{5 \text{ kg} + 12 \text{ kg}} = 4.0 \text{ m/s}^2$$

(b)   We can use either of the two equations to find the tension $T$. From the first equation,

$$T = m_A a + m_A g = m_A(a + g) = (5 \text{ kg})(4.0 + 9.8) \text{ m/s}^2 = 69 \text{ N}$$

## THIRD LAW OF MOTION

According to Newton's *third law of motion*, when one body exerts a force on another body, the second body exerts on the first an equal force in the opposite direction.

The third law of motion applies to two different forces on two different bodies: the *action force* one body exerts on the other, and the equal but opposite *reaction force* the second body exerts on the first. Action and reaction forces never balance out because they act on different bodies.

### SOLVED PROBLEM 5.20

A book rests on a table. (*a*) Show the forces acting on the table and the corresponding reaction forces. (*b*) Why do the forces acting on the table not cause it to move?

(*a*)   See Fig. 5-5.
(*b*)   The forces that act on the table have a vector sum of zero, so there is no net force acting on it.

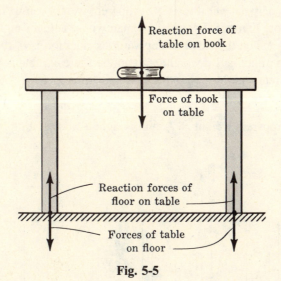

**Fig. 5-5**

### SOLVED PROBLEM 5.21

In the process of walking, what force makes a person move forward?

The person's foot exerts a backward force on the ground; the forward reaction force of the ground on the foot produces the forward motion.

### SOLVED PROBLEM 5.22

A 2-kg block *A* and a 3-kg block *B* are in contact on a frictionless table, as in Fig. 5-6. A horizontal force of 10 N is applied to *A*. Find the force with which *B* resists the pressure of *A* on it.

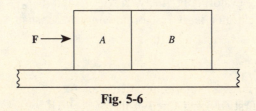

**Fig. 5-6**

The blocks have the same acceleration of

$$a = \frac{F}{m_A + m_B} = \frac{10 \text{ N}}{2 \text{ kg} + 3 \text{ kg}} = 2 \text{ m/s}^2$$

To give $B$ this acceleration, the force that $A$ exerts on $B$ must be $F_{AB} = m_B a$. The reaction force $F_{BA}$ has the same magnitude but the opposite direction, so

$$F_{BA} = -F_{AB} = -m_B a = -(3 \text{ kg})(2 \text{ m/s}^2) = -6 \text{ N}$$

## APPARENT WEIGHT

The *actual weight* of a body is the gravitational force that acts on it. The body's *apparent weight* is the force the body exerts on whatever it rests on. Apparent weight can be thought of as the reading on a scale a body is placed on. Figure 5-7 shows a woman whose actual weight is 700 N who is standing on a scale in an elevator. When the elevator's upward acceleration is $a$, the magnitude of the upward force $F$ on her is the sum of her actual weight $mg$ and the force $ma$ that is accelerating her upward, so $F = mg + ma$. Her apparent weight $w_{app}$ is the reaction force to $F$:

$$w_{app} = mg \qquad + ma$$

Apparent weight = Actual weight + accelerating force

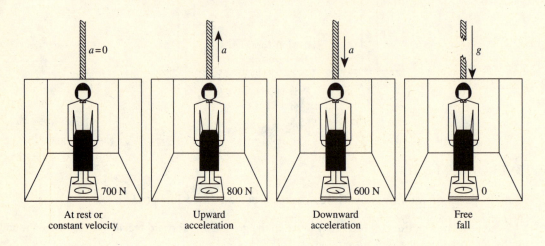

Fig. 5-7

If $a$ is positive, corresponding to an upward acceleration, $w_{app} > mg$, and the scale reading will be greater than the person's 700-N actual weight. If $a$ is negative, corresponding to a downward acceleration, $w_{app} < mg$, and the scale reading will be less than 700 N. If the cable that supports the elevator breaks and the elevator falls freely, $a = -g$, and $w_{app} = mg - mg = 0$; the woman is "weightless." When the elevator is at rest or moving at constant speed up or down, $a = 0$ and $w_{app} = mg$.

## SOLVED PROBLEM 5.23

The scale in the elevator of Fig. 5-7 reads 600 N. Find the elevator's acceleration.

Here $w_{app} < mg$, so the elevator's acceleration is downward. The person's mass is

$$m = \frac{mg}{g} = \frac{700\ \text{N}}{9.8\ \text{m/s}^2} = 71.4\ \text{kg}$$

From $w_{app} = mg + ma$ we have

$$a = \frac{w_{app}}{m} - g = \frac{600\ \text{N}}{71.4\ \text{kg}} - 9.8\ \text{m/s}^2 = (8.4 - 9.8)\ \text{m/s}^2 = -1.4\ \text{m/s}^2$$

## TWO AND THREE DIMENSIONS

When the forces that act on a body do not all lie along the same straight line, vector methods are needed to analyze the situation. In component form, the second law of motion is

$$F_x = ma_x$$
$$F_y = ma_y$$
$$F_z = ma_z$$

where $\mathbf{F} = \mathbf{F}_x + \mathbf{F}_y + \mathbf{F}_z$ is the net force on a body and $\mathbf{a} = \mathbf{a}_x + \mathbf{a}_y + \mathbf{a}_z$ is its acceleration. The best way to orient the coordinate axes depends on the problem. Often the forces involved lie in a plane, in which case only two force components and two acceleration components are needed.

## SOLVED PROBLEM 5.24

A 60-kg sprinter presses down on the ground with a force of 1000 N at an angle of 30° with the horizontal at the start of a race. What is his forward acceleration as his legs straighten out?

Figure 5-8 is a free-body diagram of the sprinter. The upward reaction force $\mathbf{F}_N$ of the ground on the sprinter is equal and opposite to the downward force $\mathbf{w}$ of the earth on him (which is his weight), so these forces cancel out. The net force on the sprinter is the reaction force $\mathbf{F}$ that the ground exerts on him in response to the pressure of his feet. The horizontal component $\mathbf{F}_x$ of this reaction force is what gives the sprinter his forward acceleration, where

$$F_x = F\ \cos\ \theta = (1000\ \text{N})(\cos\ 30°) = 866\ \text{N}$$

The horizontal component of the sprinter's acceleration is therefore

$$a_x = \frac{F_x}{m} = \frac{866\ \text{N}}{60\ \text{kg}} = 14\ \text{m/s}^2$$

(The net force $\mathbf{F}$ has a vertical component $\mathbf{F}_y$ that gives the sprinter a vertical acceleration. As a result he rises to a more erect posture as he accelerates away from the starting line.)

## SOLVED PROBLEM 5.25

A box is sliding down a frictionless ramp at an angle of 40° with the horizontal. What is the acceleration of the box?

The net force on the box is the vector sum of the two forces shown in Fig. 5-9:

1. The box's weight of $\mathbf{w} = m\mathbf{g}$ acting vertically downward.
2. The reaction force $\mathbf{F}_N$ of the ramp on the box. This force acts perpendicular to the ramp because there is no friction present that might produce a force parallel to the ramp. The reaction force is equal in magnitude to $\mathbf{w}_y$.

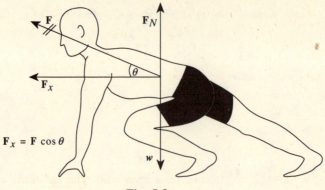

Fig. 5-8

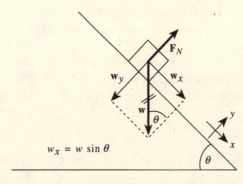

Fig. 5-9

Since all we are concerned with here is motion along the ramp, which is the $x$ axis in the figure, we can ignore the reaction force $\mathbf{F}_N$. The component of the box's weight in the $x$ direction is

$$F = w_x = w \sin \theta = mg \sin \theta$$

The acceleration of the box along the ramp is therefore

$$a_x = \frac{F}{m} = \frac{mg \sin \theta}{m} = g \sin \theta = (9.8 \text{ m/s}^2)(\sin 40°) = 6.3 \text{ m/s}^2$$

## SOLVED PROBLEM 5.26

A force parallel to the ramp of what magnitude is needed to pull a 50-kg box up a frictionless ramp at an angle of 25° with the horizontal so that the box has an acceleration of 3 m/s²?

To give the box an upward acceleration of $a_x$ along the ramp, the applied force $F$ must exceed the component $w_x$ parallel to the plane of the box's weight $\mathbf{w}$ by the amount $ma_x$. From Fig. 5-9, $w_x = w \sin \theta = mg \sin \theta$, so

$$F = mg \sin \theta + ma_x = m(g \sin \theta + a_x) = (50 \text{ kg})[(9.8 \text{ m/s}^2)(\sin 25°) + 3 \text{ m/s}^2] = 357 \text{ N}$$

# *Multiple-Choice Questions*

**5.1.** Compared with her mass and weight on Earth, an astronaut on Venus, where the acceleration of gravity is 8.8 m/s$^2$, has

    (*a*)   less mass and less weight
    (*b*)   less mass and the same weight
    (*c*)   less mass and more weight
    (*d*)   the same mass and less weight

**5.2.** A car towing a trailer is accelerating on a level road. The car exerts a force on the trailer whose magnitude is

    (*a*)   the same as that of the force the trailer exerts on the car
    (*b*)   the same as that of the force the trailer exerts on the road
    (*c*)   the same as that of the force the road exerts on the trailer
    (*d*)   greater than that of the force the trailer exerts on the car

**5.3.** In Newton's third law of motion the action and reaction forces

    (*a*)   act on the same object
    (*b*)   act on different objects
    (*c*)   do not necessarily have the same magnitude and do not necessarily have the same line of action
    (*d*)   have the same magnitude but do not necessarily have the same line of action

**5.4.** A jumper of weight *w* presses down on the floor with a force of magnitude *F*, and the jumper leaves the floor as a result. The force the floor exerted on the jumper must have had a magnitude

    (*a*)   equal to *w* and less than *F*
    (*b*)   equal to *w* and equal to *F*
    (*c*)   more than *w* and equal to *F*
    (*d*)   more than *w* and more than *F*

**5.5.** A force of 1.0 N acts on a 1.0-kg object that can move freely. The object's acceleration is

    (*a*)   0.102 m/s$^2$    (*c*)   1.0 m/s$^2$
    (*b*)   0.5 m/s$^2$     (*d*)   9.8 m/s$^2$

**5.6.** A force of 1.0 N acts on a 1.0-N object that can move freely. The object's acceleration is

    (*a*)   0.102 m/s$^2$    (*c*)   1.0 m/s$^2$
    (*b*)   0.5 m/s$^2$     (*d*)   9.8 m/s$^2$

**5.7.** Four hundred grams of salami weighs

    (*a*)   0.041 N    (*c*)   400 N
    (*b*)   3.9 N      (*d*)   3.9 kN

**5.8.** A force that gives a 2.0-kg object an acceleration of 1.6 m/s$^2$ would give an 8.0-kg object an acceleration of

    (*a*)   0.2 m/s$^2$    (*c*)   1.6 m/s$^2$
    (*b*)   0.4 m/s$^2$    (*d*)   6.4 m/s$^2$

**5.9.**   A 20-kg crate is lifted by an upward force of 200 N. The crate's upward acceleration is

   (*a*)   0.20 m/s$^2$      (*c*)   10 m/s$^2$
   (*b*)   9 m/s$^2$         (*d*)   98 m/s$^2$

**5.10.**  A vehicle slows down from 50 to 15 ft/s in 10 s when its brakes exert a force of 200 lb on it. The vehicle weighs approximately

   (*a*)   57 lb       (*c*)   1830 lb
   (*b*)   560 lb      (*d*)   22,400 lb

**5.11.**  A 5.0-kg object, initially at rest, is acted on by a net force of 3.0 N for 3.0 s. During that time the object moves

   (*a*)   0.3 m       (*c*)   1.8 m
   (*b*)   0.9 m       (*d*)   2.7 m

**5.12.**  A 300-g ball at rest is struck with a bat with a force of 150 N. If the bat was in contact with the ball for 0.020 s, the ball's velocity is

   (*a*)   0.01 m/s    (*c*)   2.5 m/s
   (*b*)   0.1 m/s     (*d*)   10 m/s

# Supplementary Problems

**5.1.**   Since action and reaction forces are always equal in magnitude and opposite in direction, how can anything ever be accelerated?

**5.2.**   A horse is pulling a cart. (*a*) What is the force that causes the horse to move forward? (*b*) What is the force that causes the cart to move forward?

**5.3.**   Is it possible for something to have a downward acceleration greater than *g*? If so, how can this be accomplished?

**5.4.**   (*a*) When a horizontal force equal to its weight is applied to an object on a frictionless surface, what is its acceleration? (*b*) What is its acceleration when the force is applied vertically upward?

**5.5.**   Compare the tension in the coupling between the first two cars in a train with the tension in the coupling between the last two cars when (*a*) the train's speed is constant and (*b*) the train is accelerating.

**5.6.**   (*a*) What is the weight of 6 kg of potatoes? (*b*) What is the mass of 6 N of potatoes?

**5.7.**   A force of 10 N is applied to (*a*) a body of mass 5 kg and (*b*) a body of weight 5 N. Find their accelerations.

**5.8.**   (*a*) What is the weight of 2 slugs of salami? (*b*) What is the mass of 2 lb of salami?

**5.9.**   (*a*) How much upward force is needed to support a 20-kg object at rest? (*b*) To give it an upward acceleration of 2 m/s$^2$? (*c*) To give it a downward acceleration of 2 m/s$^2$?

**5.10.** What is the acceleration of a 5-kg object suspended by a string when an upward force of (*a*) 39 N, (*b*) 49 N, and (*c*) 59 N is applied to the string?

**5.11.** How much applied force is needed to given an 8-N object (*a*) an upward acceleration of 2 m/s$^2$ and (*b*) a downward acceleration of 2 m/s$^2$? (*c*) In what direction must the latter force act?

**5.12.** A net force of 12 N gives an object an acceleration of 4 m/s$^2$. (*a*) What net force is needed to give it an acceleration of 1 m/s$^2$? (*b*) An acceleration of 10 m/s$^2$?

**5.13.** A certain net force gives a 2-kg object an acceleration of 0.5 m/s$^2$. What acceleration would the same force give a 10-kg object?

**5.14.** A 12,000-kg airplane launched by a catapult from an aircraft carrier is accelerated from 0 to 200 km/h in 3 s. (*a*) How many times the acceleration of gravity is the airplane's acceleration? (*b*) What is the average force that the catapult exerts on the airplane?

**5.15.** When a 5-kg rifle is fired, the 9-g bullet receives an acceleration of $3 \times 10^4$ m/s$^2$ while it is in the barrel. (*a*) How much force acts on the bullet? (*b*) Does any force act on the rifle? If so, how much and in what direction? (*c*) The bullet is accelerated for 0.007 s. How fast does it leave the barrel of the rifle?

**5.16.** How much force is needed to accelerate a train whose mass is 1000 metric tons (1 metric ton = 1000 kg) from rest to a velocity of 6 m/s in 2 min?

**5.17.** (*a*) How much force is needed to increase the velocity of a 6400-lb truck from 20 to 30 ft/s in 5 s? (*b*) How far does the truck travel in this time?

**5.18.** (*a*) How much force is needed to decrease the velocity of a 6400-lb truck from 30 to 20 ft/s in 5 s? (*b*) How far does the truck travel in this time?

**5.19.** A car strikes a stone wall at a velocity of 12 m/s. (*a*) The car is rigidly built, and the 60-kg driver comes to a stop in a time of 0.05 s. How much force acts on her? (*b*) The car is built so that its front end collapses gradually, and the driver comes to a stop in 0.2 s. How much force acts on her in this case?

**5.20.** A 0.05-kg snail goes from rest to a velocity of 0.01 m/s in 5 s. (*a*) how much force does it exert? (*b*) How far does it go during this time?

**5.21.** A 430-g soccer ball moving toward a player at 8 m/s is kicked and flies off in the opposite direction at 12 m/s. If the ball is in contact with the player's foot for 0.01 s, find the average force on the ball.

**5.22.** The cable supporting a 2000-kg elevator can safely withstand a tension of 25 kN. What is the maximum upward acceleration the elevator can have if the tension in the cable is not to exceed this figure?

**5.23.** An 800-N man stands on a scale in an elevator. What does the scale read when the elevator is (*a*) ascending at a constant velocity of 3 m/s, (*b*) ascending at a constant acceleration of 0.8 m/s$^2$, (*c*) descending at a constant velocity of 3 m/s, (*d*) descending at a constant acceleration of 0.8 m/s$^2$, and (*e*) in free fall because the cable has broken?

**5.24.** An 80-kg woman stands on a scale in an elevator. When it starts to move, the scale reads 700 N. (*a*) Is the elevator moving upward or downward? (*b*) Is its velocity constant? If so, what is it? If not, what is the elevator's acceleration?

**5.25.** Two boxes, one of mass 20 kg and the other of mass 30 kg, are sliding down a frictionless inclined plane that makes an angle of 25° with the horizontal. Find their respective accelerations.

**5.26.** A force of 50 lb is used to pull a 50-lb crate up a frictionless plane that is inclined at 30° with the horizontal. Find the acceleration of the crate.

**5.27.** An 800-kg car is towed up an 8° hill by a rope attached to a truck. The tension in the rope is 2000 N, and there is no frictional resistance to the car's motion. How much time is needed to tow the car for 50 m starting from rest?

## *Answers to Multiple-Choice Questions*

**5.1.**  (*d*)     **5.7.**  (*b*)

**5.2.**  (*a*)     **5.8.**  (*b*)

**5.3.**  (*b*)     **5.9.**  (*a*)

**5.4.**  (*c*)     **5.10.**  (*c*)

**5.5.**  (*c*)     **5.11.**  (*d*)

**5.6.**  (*d*)     **5.12.**  (*d*)

## **Answers to Supplementary Problems**

**5.1.** The action and reaction forces always act on different bodies.

**5.2.**  (*a*)   The reaction force the ground exerts on its feet.     (*b*)   The force the horse exerts on it.

**5.3.** Yes, by an applied downward force in addition to the downward force of gravity.

**5.4.**  (*a*)  g    (*b*)  0

**5.5.**  (*a*)   The tensions are the same.
      (*b*)   The front coupling is under the greater tension because it has a larger mass behind it to accelerate.

**5.6.**  (*a*)  59 N     (*b*)  0.61 kg

**5.7.**  (*a*)  2m/s$^2$     (*b*)  19.6 m/s$^2$

**5.8.**  (*a*)  64 lb     (*b*)  0.0625 slug

**5.9.**  (*a*)  196 N     (*b*)  236 N     (*c*)  156 N

**5.10.**   (*a*)   2 m/s$^2$ downward        (*b*)   0        (*c*)   2 m/s$^2$ upward

**5.11.**   (*a*)   9.63 N        (*b*)   6.37 N        (*c*)   Upward

**5.12.**   (*a*)   3 N        (*b*)   30 N

**5.13.**   0.1 m/s$^2$

**5.14.**   (*a*)   1.89 g        (*b*)   222 kN

**5.15.**   (*a*)   270 N        (*b*)   270 N backward        (*c*)   210 m/s

**5.16.**   5 × 10$^4$ N

**5.17.**   (*a*)   400 lb        (*b*)   125 ft

**5.18.**   (*a*)   400 lb        (*b*)   125 ft

**5.19.**   (*a*)   14.4 kN        (*b*)   3.6 kN

**5.20.**   (*a*)   10$^{-4}$ N        (*b*)   0.025 m

**5.21.**   860 N

**5.22.**   2.7 m/s$^2$

**5.23.**   (*a*)   800 N        (*b*)   865 N        (*c*)   800 N        (*d*)   735 N        (*e*)   0

**5.24.**   (*a*)   Downward        (*b*)   No; 1.05 m/s$^2$

**5.25.**   4.14 m/s$^2$; 4.14 m/s$^2$

**5.26.**   16 ft/s$^2$

**5.27.**   9.4 s

# Chapter 6

# Friction

## STATIC AND KINETIC FRICTION

Frictional forces act to oppose relative motion between surfaces that are in contact. Such forces act parallel to the surfaces.

*Static friction* occurs between surfaces at rest relative to each other. When an increasing force is applied to a book resting on a table, for instance, the force of static friction at first increases as well to prevent motion. In a given situation, static friction has a certain maximum value called *starting friction*. When the force applied to the book is greater than the starting friction, the book begins to move across the table. The *kinetic friction* (or *sliding friction*) that occurs afterward is usually less than the starting friction, so less force is needed to keep the book moving than to start it moving (Fig. 6-1).

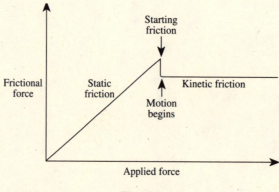

**Fig. 6-1**

## COEFFICIENT OF FRICTION

The frictional force between two surfaces depends on the normal (perpendicular) force $N$ pressing them together and on the natures of the surfaces. The latter factor is expressed quantitatively in the *coefficient of friction* $\mu$ (Greek letter *mu*) whose value depends on the materials in contact. The frictional force is experimentally found to be

$$F_f \leq \mu_s N \qquad \text{\textit{Static friction}}$$
$$F_f = \mu_k N \qquad \text{\textit{Kinetic friction}}$$

where the symbol $\leq$ means "less than or equal to." In the case of static friction, $F_f$ increases as the applied force increases until the limiting value of $\mu_s N$ is reached. Thus when there is no motion, $\mu_s N$ gives the starting frictional force, not the actual frictional force. Up to $\mu_s N$, the actual frictional force $F_f$ has the same magnitude as the applied force but is in the opposite direction.

When the applied force exceeds the starting frictional force $\mu_s N$, motion begins and now the coefficient of kinetic friction $\mu_k$ governs the frictional force. In this case $\mu_k N$ gives the actual amount of $F_f$, which no longer depends on the applied force and is constant over a fairly wide range of relative velocities.

## ROLLING FRICTION

Ideally there should be no rolling friction, because there should be no relative motion between the surfaces in contact. In reality, a wheel or ball is slightly flattened when it rests on a surface, which itself is slightly dented. A resistive force arises when the wheel or ball rolls, partly because it and the surface must be continually deformed and partly because there is some relative motion between them owing to the deformation. Coefficients of rolling friction $\mu_r$ are nevertheless much smaller than those of kinetic friction: For a rubber tire rolling on a concrete road $\mu_r$ is about 0.04, for instance, whereas it is 0.7 for the same tire sliding on the road.

### SOLVED PROBLEM 6.1

How much force is needed to keep a 1200-kg car moving at constant velocity on a level concrete road? Assume that the car is moving too slowly for air resistance to be important, and use $\mu_k = 0.04$ for the coefficient of rolling friction.

The normal force is the car's weight $mg$. Hence

$$F = F_f = \mu_k N = \mu_k mg = (0.04)(1200 \text{ kg})(9.8 \text{ m/s}^2) = 470 \text{ N}$$

### SOLVED PROBLEM 6.2

A force of 200 N is just sufficient to start a 50-kg steel trunk moving across a wooden floor. Find the coefficient of static friction.

The normal force is the trunk's weight $mg$. Hence

$$\mu_s = \frac{F}{N} = \frac{F}{mg} = \frac{200 \text{ N}}{(50 \text{ kg})(9.8 \text{ m/s}^2)} = 0.41$$

### SOLVED PROBLEM 6.3

A 40-kg wooden crate is being pushed across a wooden floor with a force of 160 N. If $\mu_k = 0.3$, find the acceleration of the crate.

The applied force $F_A = 160$ N is opposed by the frictional force

$$F_f = \mu_k N = \mu_k mg = (0.3)(40 \text{ kg})(9.8 \text{ m/s}^2) = 118 \text{ N}$$

The net force on the crate is therefore

$$F = F_A - F_f = 160 \text{ N} - 118 \text{ N} = 42 \text{ N}$$

and its acceleration is

$$a = \frac{F}{m} = \frac{42 \text{ N}}{40 \text{ kg}} = 1.05 \text{ m/s}^2$$

### SOLVED PROBLEM 6.4

A bowling ball with an initial velocity of 3 m/s rolls along a level floor for 50 m before coming to a stop. What is the coefficient of rolling friction?

We begin by finding the ball's acceleration. From Chapter 3, $v^2 = v_0^2 + 2as$. Here $v = 0$, $v_0 = 3$ m/s, and $s = 50$ m, and so

$$a = -\frac{v_0^2}{2s} = -\frac{(3 \text{ m/s})^2}{(2)(50 \text{ m})} = -0.09 \text{ m/s}^2$$

The force that corresponds to this acceleration is $ma$, which is equal and opposite to the frictional force $\mu_r N = \mu_r mg$. Hence

$$ma = -\mu_r mg \qquad \mu_r = -\frac{a}{g} = -\frac{-0.09 \text{ m/s}^2}{9.8 \text{ m/s}^2} = 0.0092$$

## SOLVED PROBLEM 6.5

The coefficient of kinetic friction between a rubber tire and a wet concrete road is 0.5. (a) Find the minimum time in which a car whose initial velocity is 50 km/h can come to a stop on such a road. (b) What distance will the car cover in this time?

(a)   The maximum available frictional force is $\mu_k N = \mu_k mg$, and so

$$F_f = \mu_k mg = -ma$$

$$a = -\mu_k g = -(0.5)(9.8 \text{ m/s}^2) = -4.9 \text{ m/s}^2$$

Here
$$v_0 = \left(50 \frac{\text{km}}{\text{h}}\right)\left(\frac{1000 \text{ m/km}}{3600 \text{ s/h}}\right) = 13.9 \text{ m/s}$$

and the final velocity of the car is $v = 0$. Hence

$$v = v_0 + at = 0 \qquad v_0 = -at$$

and the time required for the car to come to a stop is

$$t = -\frac{v_0}{a} = -\frac{13.9 \text{ m/s}}{-4.9 \text{ m/s}^2} = 2.84 \text{ s}$$

(b)   The distance covered by the car in coming to a stop is

$$s = v_0 t + \tfrac{1}{2}at^2 = (13.9 \text{ m/s})(2.84 \text{ s}) + (\tfrac{1}{2})(-4.9 \text{ m/s}^2)(2.84 \text{ s})^2 = 19.7 \text{ m}$$

Another way to obtain this result is to use the formula $v^2 = v_0^2 + 2as$. here $v = 0$, so

$$s = -\frac{v_0^2}{2a} = -\frac{13.9 \text{ m/s}^2}{(2)(-4.9 \text{ m/s}^2)} = 19.7 \text{ m}$$

## SOLVED PROBLEM 6.6

A block at rest on an adjustable inclined plane begins to move when the angle between the plane and the horizontal reaches a certain value $\theta$, which is known as the *angle of repose*. How is this angle related to the coefficient of static friction between the block and the plane?

The weight **w** of the block can be resolved into a component $\mathbf{F}_w$ parallel to the plane and another component **N** perpendicular to the plane. From Fig. 6-2 the magnitudes of $\mathbf{F}_w$ *and* **N** are

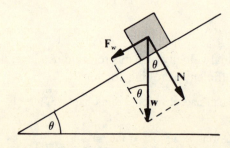

**Fig. 6-2**

$$F_w = w \sin \theta \qquad N = w \cos \theta$$

When the block just begins to move, the downward force along the plane $\mathbf{F}_w$ must be equal to the maximum force $\mu_s N$ of static friction, so

$$F_w = \mu_s N$$
$$w \sin \theta = \mu_s w \cos \theta$$
$$\mu_s = \frac{\sin \theta}{\cos \theta} = \tan \theta$$

Thus the coefficient of static friction equals the tangent of the angle of repose. The weight of the block has no significance here.

## SOLVED PROBLEM 6.7

A lathe mounted on wooden skids is to be slid down a pair of planks placed against the back of a truck. (*a*) If the coefficient of friction is 0.28, what angle should the planks make with the ground in order for the lathe to slide down at constant velocity? (*b*) When the planks are at this angle, will the lathe start to slide down of its own accord?

(*a*)   When the lathe moves at constant velocity, there is no net force on it, according to Newton's first law of motion. The downward force due to the lathe's weight therefore exactly balances the retarding force of sliding friction, and by the reasoning of Prob. 6.6,

$$\tan \theta = \mu_k = 0.28 \qquad \theta = 16°$$

(*b*)   Since the coefficient of static friction exceeds that of sliding friction, the lathe will have to be given a push to start it moving.

## SOLVED PROBLEM 6.8

A woman whose shoes have leather soles and heels is able to stand without slipping on a wooden surface that makes an angle of 25° with the horizontal. What is the minimum coefficient of static friction for leather on wood?

Since the angle of repose is at least 25°,

$$\mu_s \geq \tan 25° \geq 0.47$$

where the symbol $\geq$ means "greater than or equal to."

## SOLVED PROBLEM 6.9

A 5000-ton ship rests on launching ways that slope down to the water at an angle of 10°. If the coefficient of sliding friction is 0.18, how much force is needed to winch the ship down the ways into the water?

The frictional force to be overcome is

$$F_f = \mu N = \mu w \cos \theta = (0.18)(5000 \text{ tons})(\cos 10°) = 886 \text{ tons}$$

The component of the ship's weight parallel to the ways is

$$F_w = w \sin \theta = (5000 \text{ tons})(\sin 10°) = 868 \text{ tons}$$

Therefore an additional force of

$$F_f - F_w = 886 \text{ tons} - 868 \text{ tons} = 18 \text{ tons}$$

is required.

## *Multiple-Choice Questions*

**6.1.** Relative to the force needed to keep a box moving at constant velocity across a floor, to start the box moving usually needs

    (*a*)   less force
    (*b*)   the same force
    (*c*)   more force
    (*d*)   any of the above, depending on the natures of the surfaces in contact

**6.2.** When two surfaces are in contact, the frictional force between them depends on which one or more of the following?

    (*a*)   the normal force pressing one surface against the other
    (*b*)   the areas of the surfaces
    (*c*)   whether the surfaces are stationary or in relative motion
    (*d*)   whether a lubricant is used or not

**6.3.** The coefficient of kinetic friction between two oiled steel surfaces is

    (*a*)   0.03      (*c*)   0.03 N/kg
    (*b*)   0.03 N    (*d*)   0.03 kg/N

**6.4.** A 60-N force is needed to start a 60-kg skater moving across a frozen lake. The coefficient of static friction for steel on ice is approximately

    (*a*)   0.06    (*c*)   0.6
    (*b*)   0.1     (*d*)   1

**6.5.** The coefficient of static friction for wood on concrete is 0.6. The force needed to set a 40-kg crate in motion on a concrete floor is

    (*a*)   2.4 N    (*c*)   235 N
    (*b*)   24 N     (*d*)   653 N

**6.6.** A 153-kg engine on wooden skids is resting on a level floor. The coefficients of static and kinetic friction are, respectively, 0.5 and 0.4. When two men push on the engine with a total horizontal force of 500 N, the frictional force that acts on the skids is

    (*a*)   500 N    (*c*)   750 N
    (*b*)   600 N    (*d*)   1500 N

**6.7.** The coefficient of static friction between a car's tires and a level road is 0.80. If the car is to be stopped in a maximum time of 3.0 s, its speed cannot exceed

    (*a*)   2.4 m/s    (*c*)   7.8 m/s
    (*b*)   2.6 m/s    (*d*)   23.5 m/s

**6.8.** A force applied to a 50-kg box on a level floor is just enough to start it moving. The coefficients of static and kinetic friction are, respectively, 0.5 and 0.3. If the same force continues to be applied to the box, it will have an acceleration of approximately

(a) $2 \text{ m/s}^2$       (c) $4 \text{ m/s}^2$

(b) $3 \text{ m/s}^2$       (d) $5 \text{ m/s}^2$

**6.9.** After descending a slope, a skier coasts on level snow for 20 m before coming to a stop. If the coefficient of friction between skis and snow is 0.05, the skier's speed at the foot of the slope was

(a) 3.1 m/s       (c) 6.3 m/s

(b) 4.4 m/s       (d) 19.6 m/s

# Supplementary Problems

**6.1.** An 80-lb wooden crate rests on a horizontal wooden floor. If the coefficient of static friction is 0.5, how much force is needed to set the crate in motion?

**6.2.** A force of 300 N is sufficient to keep a 100-kg wooden crate moving at constant velocity across a wooden floor. What is the coefficient of kinetic friction?

**6.3.** A force of 1000 N is applied to a 1200-kg car. If the coefficient of rolling friction is 0.04, what is the car's acceleration?

**6.4.** The coefficients of static and kinetic friction for stone on wood are, respectively, 0.5 and 0.4. If a 150-kg stone statue is pushed with just enough force to start it moving across a wooden floor and the same force continues to act afterward, find the statue's acceleration.

**6.5.** A car whose brakes are locked skids to a stop in 70 m from an initial velocity of 80 km/h. Find the coefficient of kinetic friction.

**6.6.** A driver sees a horse on the road and applies the brakes so hard that they lock and the car skids to a stop in 24 m. The road is level, and the coefficient of kinetic friction between tires and road is 0.7. How fast was the car going when the brakes were applied?

**6.7.** A truck moving at 100 km/h carries a steel girder that rests on its wooden floor. What is the minimum time in which the truck can come to a stop without the girder moving forward? The coefficient of static friction between steel and wood is 0.5.

**6.8.** A car with its brakes locked will remain stationary on an inclined plane of dry concrete when the plane is at an angle of less than 45° with the horizontal. What is the coefficient of static friction of rubber tires on dry concrete?

**6.9.** A steel ramp is to be built for sliding blocks of ice from a refrigeration plant down to ground level. If $\mu = 0.05$, find the angle with the horizontal at which the ice will slide at constant velocity.

**6.10.** A box slides down a plane 8 m long that is inclined at an angle of 30° with the horizontal. If the box starts from rest and $\mu = 0.25$, find (a) the acceleration of the box, (b) its velocity at the bottom of the plane, and (c) the time required for it to reach the bottom.

**6.11.** If the box of Prob. 6.10 has a mass of 60 kg, how much force is needed to move it up the plane (*a*) at constant velocity and (*b*) with an acceleration of 2 m/s$^2$?

**6.12.** A skier stands on a 5° slope. If the coefficient of static friction is 0.1, does the skier start to slide down?

# *Answers to Multiple-Choice Questions*

| | | | |
|---|---|---|---|
| **6.1** | (*c*) | **6.6.** | (*a*) |
| **6.2.** | (*a*), (*c*), (*d*) | **6.7.** | (*d*) |
| **6.3.** | (*a*) | **6.8.** | (*a*) |
| **6.4.** | (*b*) | **6.9.** | (*b*) |
| **6.5.** | (*c*) | | |

# **Answers to Supplementary Problems**

**6.1.**   40 lb

**6.2.**   0.306

**6.3.**   0.44 m/s$^2$

**6.4.**   0.98 m/s$^2$

**6.5.**   0.36

**6.6.**   18 m/s = 65 km/h

**6.7.**   5.67 s

**6.8.**   1.0

**6.9.**   3°

**6.10.**   (*a*)   2.78 m/s$^2$       (*b*)   6.67 m/s       (*c*)   2.40 s

**6.11.**   (*a*)   421 N       (*b*)   541 N

**6.12.**   No

# Chapter 7

# Energy

## WORK

*Work* is a measure of the amount of change (in a general sense) that a force produces when it acts on a body. The change may be in the velocity of the body, in its position, in its size or shape, and so forth.

By definition, the work done by a force acting on a body is equal to the product of the force and the distance through which the force acts, provided that **F** and **s** are in the same direction. Thus

$$W = Fs$$
$$\text{Work} = (\text{force})(\text{distance})$$

Work is a scalar quantity; no direction is associated with it.

If **F** and **s** are not parallel but **F** is at the angle $\theta$ with respect to **s**, then

$$W = Fs \cos \theta$$

Since $\cos 0 = 1$, this formula becomes $W = Fs$ when **F** is parallel to **s**. When **F** is perpendicular to **s**, $\theta = 90°$ and $\cos 90° = 0$. No work is done in this case (Fig. 7-1).

The unit of work is the product of a force unit and a length unit. In SI units, the unit of work is the *joule* (J).

SI units :                1 joule (J) = 1 newton-meter = 0.738 ft·lb

British units :        1 foot-pound (ft·lb) = 1.36 J

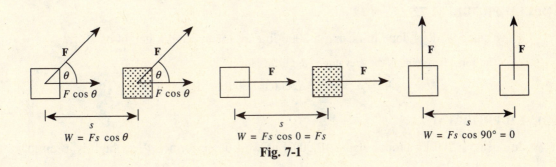

**Fig. 7-1**

## SOLVED PROBLEM 7.1

A horizontal force of 420 N is used to push a 100-kg crate for 5 m across a level warehouse floor. How much work is done?

The mass of the crate does not matter here. Since the force is parallel to the displacement,

$$W = Fs = (420 \text{ N})(5 \text{ m}) = 2100 \text{ J} = 2.1 \text{ kJ}$$

75

## SOLVED PROBLEM 7.2

The 420-N force of Prob. 7.1 is instead exerted on the crate at an angle of 35° above the horizontal. How much work is done now?

$$W = Fs \cos \theta = (420 \text{ N})(5 \text{ m})(\cos 35°) = 1920 \text{ J} = 1.92 \text{ kJ}$$

## SOLVED PROBLEM 7.3

A 60-kg box is pushed 12 m across a horizontal floor by a horizontal force of 200 N. The coefficient of kinetic friction is 0.3. (*a*) How much work went into overcoming friction and (*b*) how much into accelerating the box?

(*a*)   The frictional force to be overcome is

$$F_f = \mu_k N = \mu_k mg = (0.3)(60 \text{ kg})(9.8 \text{ m/s}^2) = 176 \text{ N}$$

The work done opposing friction is therefore

$$W_f = F_f s = (176 \text{ N})(12 \text{ m}) = 2112 \text{ J}$$

(*b*)   The remainder $\Delta F$ of the applied force goes into accelerating the box. Since

$$\Delta F = F - F_f = 200 \text{ N} - 176 \text{ N} = 24 \text{ N}$$

the work done in accelerating the box is

$$W_a = (\Delta F)s = (24 \text{ N})(12 \text{ m}) = 288 \text{ J}$$

## SOLVED PROBLEM 7.4

How much work is done in raising a 2-kg book from the ground to a height of 1.8 m?

The force needed to raise the book is equal to its weight $mg$. This force acts parallel to the displacement of the book, where $s = h$, so

$$W = Fs = mgh = (2 \text{ kg})(9.8 \text{ m/s}^2)(1.8 \text{ m}) = 35 \text{ J}$$

## SOLVED PROBLEM 7.5

How much work is done in raising a 2000-lb elevator through a height of 80 ft?

The force here equals the elevator's weight $w$. Hence

$$W = Fs = wh = (2000 \text{ lb})(80 \text{ ft}) = 1.6 \times 10^5 \text{ ft·lb}$$

## SOLVED PROBLEM 7.6

In raising a 200-kg bronze statue 10,000 J of work is performed. How high is it raised?

Since $W = Fs = mgh$,

$$h = \frac{W}{mg} = \frac{1 \times 10^4 \text{ J}}{(200 \text{ kg})(9.8 \text{ m/s}^2)} = 5.1 \text{ m}$$

## SOLVED PROBLEM 7.7

The identical 65-kg twins Alpha and Beta climb a mountain 3800 m high. Alpha follows a trail that averages 20° above the horizontal, while Beta follows a different trail that averages 30° above the horizontal. How much work does each twin do?

In each case the height through which the twins raise their bodies is $h = 3800$ m. The exact route they follow is irrelevant here. Hence each twin does the work

$$W = Fs = mgh = (65 \text{ kg})(9.8 \text{ m/s}^2)(3800 \text{ m}) = 2.42 \times 10^6 \text{ J} = 2.42 \text{ MJ}$$

## POWER

*Power* is the rate at which work is done by a force. Thus

$$P = \frac{W}{t}$$

$$\text{Power} = \frac{\text{work done}}{\text{time}}$$

The more power something has, the more work it can perform in a given time.

Two special units of power are in wide use, the *watt* and the *horsepower*, where

$$1 \text{ watt (W)} = 1 \text{ J/s} = 1.34 \times 10^{-3} \text{ hp}$$

$$1 \text{ horsepower (hp)} = 550 \text{ ft·lb/s} = 746 \text{ W}$$

A *kilowatt* (kW) is equal to $10^3$ W, or 1.34 hp. A *kilowatthour* is the work done in 1 h by an agency whose power output is 1 kW; hence $1 \text{ kWh} = (1000 \text{ W})(3600 \text{ s}) = 3.6 \times 10^6$ J.

When a constant force **F** does work on a body that is moving at the constant velocity **v**, if **F** is parallel to **v** the power involved is

$$P = \frac{W}{t} = \frac{Fs}{t} = Fv$$

because $s/t = v$; that is,

$$P = Fv$$

$$\text{Power} = (\text{force})(\text{velocity})$$

## EFFICIENCY

The *efficiency*, abbreviated Eff, of a system of some kind that takes power from a source and transmits it elsewhere is given by the ratio between the power output of the system $P_{\text{out}}$ and its power input $P_{\text{in}}$:

$$\text{Eff} = \frac{P_{\text{out}}}{P_{\text{in}}}$$

An example of such a system is the winding drum of a hoist, which takes power from the rotating shaft of a motor and delivers it through a linear pull on an object being raised.

## SOLVED PROBLEM 7.8

A 40-kg woman runs up a staircase 4 m high in 5 s. Find her minimum power output.

The minimum downward force the woman's legs must exert is equal to her weight $mg$. Hence

$$P_{\text{min}} = \frac{W}{t} = \frac{Fs}{t} = \frac{mgh}{t} = \frac{(40 \text{ kg})(9.8 \text{ m/s}^2)(4 \text{ m})}{5 \text{ s}} = 314 \text{ W}$$

## SOLVED PROBLEM 7.9

A man uses a horizontal force of 200 N to push a crate up a ramp 8 m long that is 20° above the horizontal. (*a*) How much work does the man perform? (*b*) If the man takes 12 s to push the crate up the ramp, what is his power output in watts and in horsepower?

(*a*)                          $W = Fs \cos \theta = (200 \text{ N})(8 \text{ m})(\cos 20°) = 1504 \text{ J}$

(*b*)                          $P = \dfrac{W}{t} = \dfrac{1504 \text{ J}}{12 \text{ s}} = 125 \text{ W}$

Since 1 hp = 746 W,

$$P = \frac{125 \text{ W}}{746 \text{ W/hp}} = 0.17 \text{ hp}$$

## SOLVED PROBLEM 7.10

A hoist powered by a 10-kW motor is used to raise a bucket filled with concrete and having a total mass of 500 kg to a height of 80 m. If the efficiency of the hoist is 80 percent, find the time needed.

The upward force required is equal to the bucket's weight *mg*. The power available is

$$P_{out} = (\text{Eff})P_{in} = (0.8)(10 \text{ kW}) = 8 \text{ kW} = 8000 \text{ W}$$

Since $P = W/t = Fs/t$, here

$$t = \frac{Fs}{P} = \frac{mgh}{P} = \frac{(500 \text{ kg})(9.8 \text{ m/s}^2)(80 \text{ m})}{8000 \text{ W}} = 49 \text{ s}$$

## SOLVED PROBLEM 7.11

A horse has a power output of 1 kW when it pulls a wagon with a force of 400 N. What is the wagon's velocity?

Since $P = Fv$,

$$v = \frac{P}{F} = \frac{1000 \text{ W}}{400 \text{ N}} = 2.5 \text{ m/s}$$

## SOLVED PROBLEM 7.12

A motorboat requires 60 kW to move at the constant velocity of 4.5 m/s. How much resistive force does the water exert on the boat at this velocity?

Since $P = Fv$,

$$F = \frac{P}{v} = \frac{6 \times 10^4 \text{ W}}{4.5 \text{ m/s}} = 1.33 \times 10^4 \text{ N} = 13.3 \text{ kN}$$

## SOLVED PROBLEM 7.13

A 2400-lb car ascends a 10° slope at 30 mi/h. If the overall efficiency is 70 percent, what is the power output of the car's engine?

From Fig. 7-2 the weight $w$ of the car has a component $F$ along the slope of magnitude

$$F = w \sin \theta = (2400 \text{ lb})(\sin 10°) = 417 \text{ lb}$$

The force supplied by the car's engine must equal this amount. The car's velocity is

$$v = (30 \text{ mi/h})\left(1.47 \frac{\text{ft/s}}{\text{mi/h}}\right) = 44 \text{ ft/s}$$

and so, since $P = Fv$, the required power at 100 percent efficiency is

$$P = Fv = (417 \text{ lb})(44 \text{ ft/s}) = 1.835 \times 10^4 \text{ ft·lb/s}$$

Since 1 hp = 550 ft·lb/s,

$$P = \frac{1.835 \times 10^4 \text{ ft·lb/s}}{550 \text{ (ft·lb/s)/hp}} = 33.4 \text{ hp}$$

At 70 percent efficiency,

$$P = \frac{33.4 \text{ hp}}{0.70} = 47.7 \text{ hp}$$

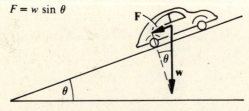

**Fig. 7-2**

## KINETIC ENERGY

*Energy* is that property something has which enables it to do work. The more energy something has, the more work it can perform. Every kind of energy falls into one of three general categories: kinetic energy, potential energy, and rest energy.

The units of energy are the same as those of work, namely, the joule and the foot-pound.

The energy a body has by virtue of its motion is called *kinetic energy*. If the body's mass is $m$ and its velocity is $v$, its kinetic energy is

$$\text{Kinetic energy} = \text{KE} = \tfrac{1}{2}mv^2$$

## SOLVED PROBLEM 7.14

Fin ic energy of a 1000-kg car whose velocity is 20 m/s.

$$\text{KE} = \tfrac{1}{2}mv^2 = (\tfrac{1}{2})(1000 \text{ kg})(20 \text{ m/s})^2 = 2 \times 10^5 \text{ J}$$

## SOLVED PROBLEM 7.15

What velocity does a 1-kg object have when its kinetic energy is 1 J?

Since $\text{KE} = \tfrac{1}{2}mv^2$,

$$v = \sqrt{\frac{2(\text{KE})}{m}} = \sqrt{\frac{2(1 \text{ J})}{1 \text{ kg}}} = \sqrt{2} \text{ m/s} = 1.4 \text{ m/s}$$

**SOLVED PROBLEM 7.16**

A 128-lb woman skates at a velocity of 15 ft/s. What is her kinetic energy?

The woman's mass is $m = w/g = 128$ lb/32 ft/s$^2 = 4$ slugs, and so her kinetic energy is

$$\text{KE} = \tfrac{1}{2}mv^2 = (\tfrac{1}{2})(4 \text{ slugs})(15 \text{ ft/s})^2 = 450 \text{ ft·lb}$$

**SOLVED PROBLEM 7.17**

A hammer with a 1.5-kg head is used to drive a nail into a wooden board. If the hammerhead is moving at 3 m/s when it strikes the nail and the nail moves 10 mm into the board, find the average force the hammerhead exerts on the nail.

The kinetic energy of the hammerhead is entirely converted into work done in driving the nail through the distance $s = 10$ mm $= 0.01$ m. To find the force $F$ exerted on the nail, we proceed as follows:

$$\text{KE of hammerhead} = \text{work done on nail}$$
$$\tfrac{1}{2}mv^2 = Fs$$

$$F = \frac{mv^2}{2s} = \frac{(1.5 \text{ kg})(3 \text{ m/s})^2}{(2)(0.01 \text{ m})} = 1350 \text{ N} = 1.35 \text{ kN}$$

**SOLVED PROBLEM 7.18**

Find the power output of the engine of a 1200-kg car while the car accelerates from 30 to 100 km/h in 10 s.

Here

$$v_1 = \left(30\frac{\text{km}}{\text{h}}\right)\left(\frac{1000 \text{ m/km}}{3600 \text{ s/h}}\right) = 8.33 \text{ m/s}$$

$$v_2 = \left(100\frac{\text{km}}{\text{h}}\right)\left(\frac{1000 \text{ m/km}}{3600 \text{ s/h}}\right) = 27.78 \text{ m/s}$$

The work the engine must do to accelerate the car from $v_1$ to $v_2$ is

$$W = \text{KE}_2 - \text{KE}_1 = \tfrac{1}{2}mv_2^2 - \tfrac{1}{2}mv_1^2 = \tfrac{1}{2}m(v_2^2 - v_1^2)$$
$$= \frac{1200 \text{ kg}}{2}\left[(27.78 \text{ m/s})^2 - (8.33 \text{ m/s})^2\right] = 4.21 \times 10^5 \text{ J}$$

The power needed to provide this amount of work in time $t = 10$ s is

$$P = \frac{W}{t} = \frac{4.21 \times 10^5 \text{ J}}{10 \text{ s}} = 4.1 \times 10^4 \text{ J} = 42.1 \text{ kW}$$

which is equivalent to

$$P = (42.1 \text{ kW})\left(1.34 \frac{\text{hp}}{\text{kW}}\right) = 56.3 \text{ hp}$$

## POTENTIAL ENERGY

The energy a body has by virtue of its position is called *potential energy*. A book held above the floor has gravitational potential energy because the book can do work on something else as it falls; a nail held near a magnet has magnetic potential energy because the nail can do work as it moves toward the magnet; the wound spring in a watch has elastic potential energy because the spring can do work as it unwinds.

The gravitational potential energy of a body of mass $m$ at a height $h$ above a given reference level is

$$\text{Gravitational potential energy} = \text{PE} = mgh$$

where $g$ is the acceleration of gravity. In terms of the weight $w$ of the body,

$$\text{PE} = wh$$

## SOLVED PROBLEM 7.19

A 1.5-kg book is held 60 cm above a desk whose top is 70 cm above the floor. Find the potential energy of the book (*a*) with respect to the desk, and (*b*) with respect to the floor.

(*a*)   Here $h = 60$ cm $= 0.6$ m, so

$$\text{PE} = mgh = (1.5 \text{ kg})(9.8 \text{ m/s}^2)(0.6 \text{ m}) = 8.8 \text{ J}$$

(*b*)   The book is $h = 60$ cm $+ 70$ cm $= 130$ cm $= 1.3$ m above the floor, so its PE with respect to the floor is

$$\text{PE} = mgh = (1.5 \text{ kg})(9.8 \text{ m/s}^2)(1.3 \text{ m}) = 19.1 \text{ J}$$

## SOLVED PROBLEM 7.20

Compare the potential energy of a 1200-kg car at the top of a hill 30 m high with its kinetic energy when its velocity is 100 km/h (27.8 m/s).

$$\text{PE} = mgh = (1200 \text{ kg})(9.8 \text{ m/s}^2)(30 \text{ m}) = 3.5 \times 10^5 \text{ J}$$

$$\text{KE} = \tfrac{1}{2}mv^2 = (\tfrac{1}{2})(1200 \text{ kg})(27.8 \text{ m/s})^2 = 4.6 \times 10^5 \text{ J}$$

The KE of the car at this velocity is greater than its PE at the top of the hill. This means that a crash at 100 km/h (62 mi/h) into a stationary obstacle will yield more work—that is, do more damage—than dropping the car from a height of 30 m (98 ft).

## SOLVED PROBLEM 7.21

(*a*) A 125-lb woman jumps off a wall 3 ft high and lands on a concrete road with her knees stiff. Her body is compressed by 3 in. at the moment she hits the road. What is the average force exerted on her by the road? (*b*) If the woman bends her knees on impact so that she comes to a stop over a distance of 12 in., what would the force on her be?

(*a*)   The woman's potential energy with respect to the road is converted to work done on her body. Here $s = 3$ in. $= 0.25$ ft. Setting her initial PE equal to the work done,

$$wh = Fs$$

$$F = \frac{wh}{s} = \frac{(125 \text{ lb})(3 \text{ ft})}{0.25 \text{ ft}} = 1500 \text{ lb}$$

(*b*)   Now $s = 12$ in. $= 1$ ft, and

$$F = \frac{wh}{s} = \frac{(125 \text{ lb})(3 \text{ ft})}{1 \text{ ft}} = 375 \text{ lb}$$

Clearly it is safer to land with bent knees rather than rigid knees after falling through even a short distance.

## REST ENERGY

According to *Einstein's theory of relativity*, matter can be converted to energy and energy can be converted to matter. The *rest energy* of a body is the energy it has by virtue of its mass alone. Thus mass can be regarded as a form of energy. The rest energy of a body is in addition to any KE or PE it might have.

If the mass of a body is $m_0$ when it is at rest, its rest energy is

$$\text{Rest energy} = E_0 = m_0 c^2$$

In this formula $c$ is the velocity of light, whose value is

$$c = 3.00 \times 10^8 \text{ m/s} = 9.83 \times 10^8 \text{ ft/s} = 186,000 \text{ mi/s}$$

The rest mass $m_0$ is specified here because the mass of a moving body increases with its velocity. The increase of mass with increasing velocity is significant only at velocities near the velocity of light. (Nothing can move faster than $c$.) At a velocity of $0.1c$, the mass increase is just 0.5 percent. At $0.9999c$, however, the mass increase is 7100 percent. Only subatomic particles such as protons and electrons can be given such high velocities.

## SOLVED PROBLEM 7.22

Approximately $4 \times 10^9$ kg of matter is converted to energy in the sun each second. What is the power output of the sun?

The energy produced by the sun per second is

$$E_0 = m_0 c^2 = (4 \times 10^9 \text{ kg})(3 \times 10^8 \text{ m/s})^2 = 3.6 \times 10^{26} \text{ J}$$

Hence the power output is

$$P = \frac{E_0}{t} = \frac{3.6 \times 10^{26} \text{ J}}{1 \text{ s}} = 3.6 \times 10^{26} \text{ W}$$

## SOLVED PROBLEM 7.23

How much mass is converted to energy per day in a nuclear power plant operated at a level of 100 MW ($100 \times 10^6$ W)?

There are $60 \times 60 \times 24 = 86,400$ s/day, so the energy liberated per day is

$$E_0 = Pt = (10^8 \text{ W})(8.64 \times 10^4 \text{ s}) = 8.64 \times 10^{12} \text{ J}$$

Since $E_0 = m_0 c^2$,

$$m_0 = \frac{E_0}{c^2} = \frac{8.64 \times 10^{12} \text{ J}}{(3 \times 10^8 \text{ m/s})^2} = 9.6 \times 10^{-5} \text{ kg}$$

## CONSERVATION OF ENERGY

According to the law of *conservation of energy*, energy cannot be created or destroyed, although it can be transformed from one kind to another. The total amount of energy in the universe is constant. A falling stone provides a simple example: More and more of its initial potential energy turns to kinetic energy as its velocity increases, until finally all its PE has become KE when it strikes the ground. The KE of the stone is then transferred to the ground as work by the impact.

In general,

Work done *on* an object = change in object's KE + change in object's PE + work done *by* object

Work done by an object against friction becomes heat, as discussed in later chapters.

### SOLVED PROBLEM 7.24

At her highest point, a girl on a swing is 7 ft above the ground, and at her lowest point she is 3 ft above the ground. What is her maximum velocity?

The girl's maximum velocity $v$ occurs at the lowest point. Her kinetic energy there equals her loss of potential energy in descending through a height of $h = 7$ ft $-$ 3 ft $= 4$ ft. Hence

$$KE = PE$$
$$\tfrac{1}{2}mv^2 = mgh$$
$$v = \sqrt{2gh} = \sqrt{(2)(32 \text{ ft/s}^2)(4 \text{ ft})} = \sqrt{256} \text{ ft/s} = 16 \text{ ft/s}$$

This result is independent of the girl's mass.

### SOLVED PROBLEM 7.25

A man skis down a slope 200 m high. If his velocity at the bottom of the slope is 20 m/s, what percentage of his initial potential energy was lost due to friction and air resistance?

$$\frac{\text{Final KE}}{\text{Initial PE}} = \frac{\tfrac{1}{2}mv^2}{mgh} = \frac{v^2}{2gh} = \frac{(20 \text{ m/s})^2}{(2)(9.8 \text{ m/s}^2)(200 \text{ m})} = 0.102 = 10.2\%$$

which means 89.8 percent of the initial PE was lost.

### SOLVED PROBLEM 7.26

A 30-kg crate is pulled up a ramp 15 m long and 2 m high by a constant force of 100 N. The crate starts from rest and has a velocity of 2 m/s when it reaches the top of the ramp. What is the frictional force between the crate and the ramp?

According to conservation of energy,

$$W = \Delta KE + \Delta PE + W_f$$

$$\begin{matrix}\text{Work done by applied} \\ \text{force on crate}\end{matrix} = \begin{matrix}\text{change in KE} \\ \text{of crate}\end{matrix} + \begin{matrix}\text{change in PE} \\ \text{of crate}\end{matrix} + \begin{matrix}\text{work done against} \\ \text{friction}\end{matrix}$$

Since the length of the ramp is $s = 15$ m and its height is $h = 2$ m,

$$W = Fs = (100 \text{ N})(15 \text{ m}) = 1500 \text{ J}$$

$$\Delta KE = KE_{final} - KE_{initial} = \tfrac{1}{2}mv^2 = (\tfrac{1}{2})(30 \text{ kg})(2 \text{ m/s})^2 = 60 \text{ J}$$

$$\Delta PE = PE_{final} - PE_{initial} = mgh = (30 \text{ kg})(9.8 \text{ m/s}^2)(2 \text{ m}) = 588 \text{ J}$$

The work done against friction is therefore

$$W_f = W - \Delta KE - \Delta PE = 1500 \text{ J} - 60 \text{ J} - 588 \text{ J} = 852 \text{ J}$$

If $F_f$ is the frictional force, then $W_f = F_f s$ and

$$F_f = \frac{W_f}{s} = \frac{852 \text{ J}}{15 \text{ m}} = 56.8 \text{ N}$$

# *Multiple-Choice Questions*

**7.1.** To raise a 200-kg steel beam to a height of 10 m on a bridge being built requires work of

(*a*)  2 kJ          (*c*)  19.6 kJ
(*b*)  10 kJ          (*d*)  98 kJ

**7.2.** A total of 15,000 ft·lb of work is used to lift a load of bricks to a height of 50 ft. The weight of the bricks is

(*a*)  30.6 lb          (*c*)  300 lb
(*b*)  94.0 lb          (*d*)  2940 lb

**7.3.** The work a 300-W electric grinder can do in 5.0 min is

(*a*)  1 kJ          (*c*)  25 kJ
(*b*)  1.5 kJ          (*d*)  90 kJ

**7.4.** An 80-lb girl climbs a 12-ft rope in 7.5 s. Her average power output is

(*a*)  0.13 hp          (*c*)  0.23 hp
(*b*)  0.19 hp          (*d*)  7.4 hp

**7.5.** A 150-kg yak has an average power output of 120 W. The yak can go up a mountain 1.2 km high in

(*a*)  25 min          (*c*)  13.3 h
(*b*)  4.1 h          (*d*)  14.7 h

**7.6.** A 700-kg horse is pulling a sled at a velocity of 3.5 m/s. If the horse's power output is 1.0 hp, the force the horse exerts on the sled is

(*a*)  0.16 kN          (*c*)  0.57 kN
(*b*)  0.21 kN          (*d*)  15.2 kN

**7.7.** The KE of a 900-kg car whose velocity is 60 km/h is

(*a*)  12.8 kJ          (*c*)  1.23 MJ
(*b*)  125 kJ          (*d*)  1.62 MJ

**7.8.** The velocity of a 5000-kg truck whose KE is 360 kJ is

(*a*)  12 km/h          (*c*)  43 km/h
(*b*)  31 km/h          (*d*)  144 km/h

**7.9.** The KE of a 2400-lb car whose velocity is 40 mi/h is

(*a*)  $1.9 \times 10^3$ ft·lb          (*c*)  $7.0 \times 10^4$ ft·lb
(*b*)  $6.0 \times 10^4$ ft·lb          (*d*)  $1.3 \times 10^5$ ft·lb

**7.10.** A 50-kg mass has a PE of 4.9 kJ relative to the ground. The height of the mass above the ground is

(*a*)  10 m          (*c*)  960 m
(*b*)  98 m          (*d*)  245 km

**7.11.** A 16-slug mass is raised by 10 ft. The PE of the mass increases by

    (*a*)   5 ft·lb     (*c*)   160 ft·lb
    (*b*)  20 ft·lb    (*d*)  5120 ft·lb

**7.12.** A girl on a swing varies in height from 60 cm above the ground to 180 cm. The girl's greatest velocity

    (*a*)  is 1.5 m/s    (*c*)  is 24 m/s
    (*b*)  is 4.8 m/s    (*d*)  depends on her mass

**7.13.** The 2.50-kg head of an ax exerts a force of 80 kN as it penetrates 18 mm into the trunk of a tree. The velocity of the ax head when it strikes the tree is

    (*a*)  1.2 m/s    (*c*)  34 m/s
    (*b*)  3.4 m/s    (*d*)  107 m/s

**7.14.** The mass equivalent of 6.0 MJ is

    (*a*)  $6.7 \times 10^{-11}$ kg    (*c*)  $6.7 \times 10^{-3}$ kg
    (*b*)  $5.4 \times 10^{-9}$ kg     (*d*)  $5.0 \times 10^{-2}$ kg

# Supplementary Problems

**7.1.** The earth exerts a gravitational force of $2 \times 10^{20}$ N on the moon, and the moon travels $2.4 \times 10^9$ m each time it orbits the earth. How much work does the earth do on the moon in each orbit?

**7.2.** (*a*) How much work must be done to raise a 1100-kg car 2 m above the ground? (*b*) What is the car's potential energy afterward?

**7.3.** A 20-lb object is raised to a height of 40 ft above the ground. (*a*) How much work was done? (*b*) What is the potential energy of the object? (*c*) If the object is dropped, what will its kinetic energy be just before it strikes the ground?

**7.4.** A boy pulls a wagon with a force of 45 N by means of a rope that makes an angle of 40° with the ground. How much work does he do in moving the wagon 50 m?

**7.5.** A horse exerts a force of 200 lb while pulling a sled for 3 mi. (*a*) How much work does the horse do? (*b*) If the trip takes 30 min, what is the power output of the horse in horsepower?

**7.6.** In 1970 the population of the world was about $3.5 \times 10^9$, and about $2 \times 10^{20}$ J of work was performed under human control. Find the average power consumption per person in watts and in horsepower (1 year $= 3.15 \times 10^7$ s).

**7.7.** A certain 80-kg mountain climber has an average power output of 0.1 hp. (*a*) How much work does she perform in climbing a mountain 2000 m high? (*b*) How long does she take to climb the mountain? (*c*) What is her potential energy at the top?

**7.8.** A man uses a rope and a system of pulleys to raise a 200-lb box to a height of 10 ft. He exerts a force of 60 lb on the rope and pulls a total of 40 ft of rope through the pulleys. (*a*) How much work does he perform? (*b*) By how much is the potential energy of the box increased? (*c*) If these answers are different, what do you think the reason is?

**7.9.** A total of $10^4$ kg of water per second flows over a waterfall 25 m high. If 50 percent of the power this flow represents could be converted into electricity, how many 100-W light bulbs could be supplied?

**7.10.** The four engines of a DC-8 airplane develops a total of 22 MW when its velocity is 240 m/s. How much force do the engines exert?

**7.11.** Neglecting friction and air resistance, is more work needed to accelerate a car from 10 to 20 km/h or from 20 to 30 km/h?

**7.12.** A 3000-lb car has an engine which can deliver 80 hp to the driving wheels. What is the maximum velocity at which the car can climb a 15° hill?

**7.13.** Find the kinetic energy of a 2-g (0.002-kg) insect when it is flying at 0.4 m/s.

**7.14.** The electrons in a television picture tube whose impacts on the screen produce the flashes of light that make up the image have masses of $9.1 \times 10^{-31}$ kg and typical velocities of $3 \times 10^7$ m/s. What is the kinetic energy of such an electron?

**7.15.** A 15-kg object initially at rest is raised to a height of 8 m by a force of 200 N. What is the velocity of the object at this height?

**7.16.** A 7-kg iron shot is thrown 18 m. What was its minimum initial kinetic energy?

**7.17.** An 800-kg car moving at 70 km/h is carrying two 75-kg people. If the power output of the car's engine is 30 kW, how much time is needed for the car to reach a velocity of 110 km/h? Neglect friction and air resistance.

**7.18.** (*a*) What velocity does a 1-slug object have when its kinetic energy is 1 ft·lb? (*b*) What velocity does a 1-lb object have when its kinetic energy is 1 ft·lb?

**7.19.** A stone is dropped from a height of 100 m. At what height is half of its energy potential and half kinetic?

**7.20.** A 10-g bullet has a velocity of 600 m/s when it leaves the barrel of a rifle. If the barrel is 60 cm long, find the average force on the bullet while it is in the barrel.

**7.21.** A 16-lb shell has a velocity of 2000 ft/s when it leaves the barrel of a cannon. If the barrel is 10 ft long, find the average force on the shell while it is in the barrel.

**7.22.** (*a*) An 8-N force pushes a 0.5-kg ball on a horizontal table for 3 m, starting from rest. If there is no friction, what is the final KE of the ball? (*b*) The same force is used to raise the ball a height of 3 m, starting from rest. What is its final KE now?

**7.23.** An 800-kg car moving at 6 m/s begins to coast down a hill 40 m high with its engine off. The driver applies the brakes so that the car's speed at the bottom of the hill is 20 m/s. How much energy was lost to friction?

**7.24.** This book weighs about 1.5 lb. What is its rest energy in foot-pounds? In joules?

**7.25.** Approximately 12 MJ of energy is liberated when 1 kg of dynamite explodes. How much matter is converted to energy in this process?

**7.26.** A sedentary person uses energy at an average rate of about 70 W. (*a*) How many joules of energy does this person use per day? (*b*) All this energy originates in the sun. How much matter is converted to energy per day to supply such a person?

# Answers to Multiple-Choice Questions

**7.1.** (*c*)  **7.8.**  (*c*)

**7.2.** (*c*)  **7.9.**  (*d*)

**7.3.** (*d*)  **7.10.** (*a*)

**7.4.** (*c*)  **7.11.** (*d*)

**7.5.** (*b*)  **7.12.** (*b*)

**7.6.** (*b*)  **7.13.** (*c*)

**7.7.** (*b*)  **7.14.** (*a*)

# Answers to Supplementary Problems

**7.1.** No work is done because the force on the moon is perpendicular to its direction of motion.

**7.2.** (*a*)  21.6 kJ      (*b*)  21.6 kJ

**7.3.** (*a*)  800 ft·lb     (*b*)  800 ft·lb     (*c*)  800 ft·lb

**7.4.** 1.72 kJ

**7.5.** (*a*)  $3.17 \times 10^6$ ft·lb     (*b*)  3.2 hp

**7.6.** 1.8 kW; 2.4 hp

**7.7.** (*a*)  1.57 MJ     (*b*)  5 h 50 min     (*c*)  1.57 MJ

**7.8.** (*a*)  2400 ft·lb   (*b*)  2000 ft·lb
      (*c*)  400 ft·lb was used in doing work against frictional forces in the pulleys

**7.9.** 12,250 bulbs

**7.10.** 92 kN

**7.11.** From 20 to 30 km/h

**7.12.** 56.7 ft/s = 38.6 mi/h

**7.13.** $1.6 \times 10^{-4}$ J

**7.14.** $4.1 \times 10^{-16}$ J

**7.15.** 7.5 m/s

**7.16.** 0.62 kJ

**7.17.** 8.8 s

**7.18.** (*a*)  1.4 ft/s      (*b*)  8 ft/s

**7.19.** 50 m

**7.20.** 3 kN

**7.21.** $10^5$ lb

**7.22.** (*a*)  24 J      (*b*)  9.3 J

**7.23.** 168 J

**7.24.** $4.5 \times 10^{16}$ ft·lb; $6.1 \times 10^{16}$ J

**7.25.** $1.3 \times 10^{-10}$ kg

**7.26.** (*a*)  $6.05 \times 10^6$ J      (*b*)  $6.72 \times 10^{-11}$ kg

# Chapter 8

# Momentum

## LINEAR MOMENTUM

Work and energy are scalar quantities that have no directions associated with them. When two or more bodies interact with one another, or a single body breaks up into two or more others, the various directions of motion cannot be related by energy considerations alone. The vector quantities called *linear momentum* and *impulse* are important in analyzing such events.

The linear momentum (usually called simply *momentum*) of a body of mass $m$ and velocity $\mathbf{v}$ is the product of $m$ and $\mathbf{v}$:

$$\text{Momentum} = m\mathbf{v}$$

The units of momentum are kilogram-meters per second and slug-feet per second. The direction of the momentum of a body is the same as the direction in which it is moving.

The greater the momentum of a body, the greater its tendency to continue in motion. Thus a baseball that is solidly struck by a bat ($v$ large) is harder to stop than a baseball thrown by hand ($v$ small), and an iron shot ($m$ large) is harder to stop than a baseball ($m$ small) of the same velocity.

## IMPULSE

A force $\mathbf{F}$ that acts on a body during time $t$ provides the body with an *impulse* of $\mathbf{F}t$:

$$\text{Impulse} = \mathbf{F}t = (\text{force})(\text{time interval})$$

The units of impulse are newton-seconds and pound-seconds.

When a force acts on a body to produce a change in its momentum, the momentum change $m(\mathbf{v}_2 - \mathbf{v}_1)$ is equal to the impulse provided by the force. Thus

$$\mathbf{F}t = m(\mathbf{v}_2 - \mathbf{v}_1)$$
$$\text{Impulse} = \text{momentum change}$$

## SOLVED PROBLEM 8.1

Find the momentum of a 50-kg boy running at 6 m/s.

$$mv = (50\ \text{kg})(6\ \text{m/s}) = 300\ \text{kg·m/s}$$

## SOLVED PROBLEM 8.2

A 160-lb woman runs 1 mi in 5 min. What is her average momentum?

The woman's mass $m$ and average velocity $\bar{v}$ are, respectively,

$$m = \frac{w}{g} = \frac{160\ \text{lb}}{32\ \text{ft/s}^2} = 5\ \text{slugs} \qquad \bar{v} = \frac{(1\ \text{mi})(5280\ \text{ft/mi})}{(5\ \text{min})(60\ \text{s/min})} = 17.6\ \text{ft/s}$$

89

Hence her average momentum is $m\bar{v} = (5 \text{ slugs})(17.6 \text{ ft/s}) = 88 \text{ slug·ft/s}$.

## SOLVED PROBLEM 8.3

A 46-g golf ball is struck by a club and flies off at 70 m/s. If the head of the club was in contact with the ball for 0.5 ms, what was the average force on the ball during the impact?

The ball started from rest, so $v_1 = 0$ and its momentum change is

$$m(v_2 - v_1) = mv_2 = (0.046 \text{ kg})(70 \text{ m/s}) = 3.22 \text{ kg·m/s}$$

Since 1 ms = 1 millisecond = $10^{-3}$ s, here $t = 0.5$ ms = $5 \times 10^{-4}$ s and

$$F = \frac{m(v_2 - v_1)}{t} = \frac{3.22 \text{ kg·m/s}}{5 \times 10^{-4} \text{ s}} = 6.4 \times 10^3 \text{ N} = 6.4 \text{ kN}$$

## SOLVED PROBLEM 8.4

A certain DC-9 airplane has a mass of 50,000 kg and a cruising velocity of 700 km/h. Its engines develop a total thrust of 70,000 N. If air resistance, change in altitude, and fuel consumption are ignored, how long does it take the airplane to reach its cruising velocity, starting from rest?

The airplane's initial velocity is $v_1 = 0$, and its final velocity is

$$v_2 = (700 \text{ km/h})\left(\frac{1000 \text{ m/km}}{3600 \text{ s/h}}\right) = 19.4 \text{ m/s}$$

Hence       Impulse = momentum change

$$Ft = m(v_2 - v_1) = mv_2$$

$$t = \frac{mv_2}{F} = \frac{(50,000 \text{ kg})(19.4 \text{ m/s})}{70,000 \text{ N}} = 139 \text{ s} = 2 \text{ min } 19 \text{ s}$$

## SOLVED PROBLEM 8.5

A 2400-lb car strikes a fence at 30 ft/s (about 20 mi/h) and comes to a stop in 1 s. What average force acted on the car?

The initial and final velocities of the car are 30 ft/s and 0, respectively. Hence

Impulse = momentum change

$$Ft = m(v_2 - v_1) = \frac{w}{g}(v_2 - v_1)$$

$$F = \frac{w(v_2 - v_1)}{gt} = \frac{(3200 \text{ lb})(0 - 30 \text{ ft/s})}{(32 \text{ ft/s}^2)(1 \text{ s})} = -3000 \text{ lb}$$

The minus sign means that the force which acted to stop the car is in the opposite direction to its initial velocity.

## SOLVED PROBLEM 8.6

The *thrust* of a rocket is the force developed by the expulsion of its exhaust gases. (*a*) Find the thrust of a rocket that uses 30 kg/s of fuel and whose exhaust gases leave the rocket at 3 km/s.

(b) If the initial mass of the rocket is 5000 kg of which 3500 kg is fuel, find its initial and final accelerations.

(a) When a mass $\Delta m$ of exhaust gas leaves a rocket in the time interval $\Delta t$, its momentum change $\Delta(mv)$ is

$$\Delta(mv) = v \, \Delta m$$

Since the impulse $Ft$ given to the rocket equals the momentum change of the exhaust gases,

$$F \, \Delta t = \Delta(mv)$$

$$F = \frac{\Delta(mv)}{\Delta t} = v \frac{\Delta m}{\Delta t}$$

The thrust of a rocket is the product of the exhaust speed and the rate at which fuel is consumed. Here $v = 3$ km/s $= 3000$ m/s and $\Delta m/\Delta t = 30$ kg/s, so

$$F = v \frac{\Delta m}{\Delta t} = (3000 \text{ m/s})(30 \text{ kg/s}) = 9 \times 10^4 \text{ N} = 90 \text{ kN}$$

(b) The initial mass of the rocket is $m_0 = 5000$ kg, and the initial force available for its acceleration is $F - m_0 g$. Hence

$$a_0 = \frac{F - m_0 g}{m_0} = \frac{9 \times 10^4 \text{ N} - (5000 \text{ kg})(9.8 \text{ m/s}^2)}{5000 \text{ kg}} = 8.2 \text{ m/s}^2$$

The final mass of the rocket is 1500 kg. Hence its final acceleration before it runs out of fuel is

$$a = \frac{F - mg}{m} = \frac{9 \times 10^4 \text{ N} - (1500 \text{ kg})(9.8 \text{ m/s}^2)}{1500 \text{ kg}} = 50.2 \text{ m/s}^2$$

## CONSERVATION OF LINEAR MOMENTUM

According to the *law of conservation of linear momentum*, when the vector sum of the external forces that act on a system of bodies equals zero, the total linear momentum of the system remains constant no matter what momentum changes occur within the system.

Although interactions within the system may change the *distribution* of the total momentum among the various bodies in the system, the total momentum does not change. Such interactions can give rise to two general classes of events: explosions, in which an original single body flies apart into separate bodies, and collisions, in which two or more bodies collide and either stick together or move apart, in each case with a redistribution of the original total momentum. Collisions will be examined in the next section.

## SOLVED PROBLEM 8.7

A rocket explodes in midair. How does this affect (a) its total momentum and (b) its total kinetic energy?

(a) The total momentum remains the same because no external forces acted on the rocket.

(b) The total kinetic energy increases because the rocket fragments received additional KE from the explosion.

**SOLVED PROBLEM 8.8**

An airplane's velocity is doubled. (*a*) What happens to its momentum? Is the law of conservation of momentum obeyed? (*b*) What happens to its kinetic energy? Is the law of conservation of energy obeyed?

(*a*)    The airplane's momentum $mv$ also doubles. Momentum is conserved because the increase in the airplane's velocity is accompanied by the backward motion of air through the action of its engines, so the total momentum of airplane plus air remains the same when their opposite directions are taken into account.

(*b*)    The airplane's kinetic energy $\frac{1}{2}mv^2$ increases fourfold. Energy is conserved because the additional KE comes from chemical potential energy released in the airplane's engines.

**SOLVED PROBLEM 8.9**

A 15-g (0.015-kg) bullet is fired from a 5-kg rifle at a muzzle velocity of 600 m/s. Find the recoil velocity of the rifle.

From conservation of momentum, $m_r v_r = m_b v_b$, and so

$$v_r = \left(\frac{m_b}{m_r}\right)(v_b) = \left(\frac{0.015 \text{ kg}}{5 \text{ kg}}\right)(600 \text{ m/s}) = 1.8 \text{ m/s}$$

**SOLVED PROBLEM 8.10**

An astronaut is in space at rest relative to an orbiting spacecraft. His total weight is 300 lb, and he throws away a 1-lb wrench at a velocity of 15 ft/s relative to the spacecraft. How fast does he move off in the opposite direction?

From conservation of momentum,

$$m_a v_a = m_w v_w$$

$$\left(\frac{w_a}{g}\right)(v_a) = \left(\frac{w_w}{g}\right)(v_w)$$

$$v_a = \left(\frac{w_w}{w_a}\right)(v_w) = \left(\frac{1 \text{ lb}}{300 \text{ lb}}\right)(15 \text{ ft/s}) = 0.05 \text{ ft/s}$$

We notice that the *g*'s have canceled, so it was not necessary to find the mass values first; the ratio of two masses is always the same as the ratio of the corresponding weights.

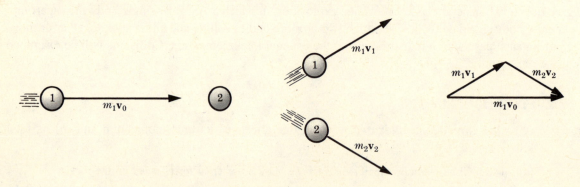

**Fig. 8-1**

## COLLISIONS

Momentum is also conserved in collisions. If a moving billiard ball strikes a stationary one, the two move off in such a way that the vector sum of their momenta is the same as the initial momentum of the first ball (Fig. 8-1). This is true even if the balls move off in different directions.

A perfectly *elastic* collision is one in which the bodies involved move apart in such a way that kinetic energy as well as momentum is conserved. In a perfectly *inelastic* collision, the bodies stick together and the kinetic energy loss is the maximum possible consistent with momentum conservation. Most collisions are intermediate between these two extremes.

### SOLVED PROBLEM 8.11

A 2000-lb car moving at 50 mi/h collides head-on with a 3000-lb car moving at 20 mi/h, and the two cars stick together. Which way does the wreckage move?

The 2000-lb car had the greater initial momentum, so the wreckage moves in the same direction it had.

### SOLVED PROBLEM 8.12

A 0.5-kg snowball moving at 20 m/s strikes and sticks to a 70-kg man standing on the frictionless surface of a frozen pond. What is the man's final velocity?

Let $v_1$ = snowball's velocity and $v_2$ = final velocity of man plus snowball. Then

Initial momentum of snowball = final momentum of man + snowball

$$m_s v_1 = (m_m + m_s)v_2$$

$$v_2 = \left(\frac{m_s}{m_m + m_s}\right)(v_1) = \left(\frac{0.5 \text{ kg}}{70.5 \text{ kg}}\right)(20 \text{ m/s}) = 0.14 \text{ m/s}$$

### SOLVED PROBLEM 8.13

A 40-kg skater traveling at 4 m/s overtakes a 60-kg skater traveling at 2 m/s in the same direction and collides with her. (*a*) If the two skaters remain in contact, what is their final velocity? (*b*) How much kinetic energy is lost?

(*a*)   Let $v_1$ = initial velocity of 40-kg skater, $v_2$ = initial velocity of 60-kg skater, and $v_3$ = final velocity of the two skaters. Then

Initial total momentum = final total momentum

$$m_1 v_1 + m_2 v_2 = (m_1 + m_2)v_3$$

$$v_3 = \frac{m_1 v_1 + m_2 v_2}{m_1 + m_2} = \frac{(40 \text{ kg})(4 \text{ m/s}) + (60 \text{ kg})(2 \text{ m/s})}{40 \text{ kg} + 60 \text{ kg}} = 2.8 \text{ m/s}$$

(*b*)                    Initial KE $= \frac{1}{2}m_1 v_1^2 + \frac{1}{2}m_2 v_2^2$

$$= (\tfrac{1}{2})(40 \text{ kg})(4 \text{ m/s})^2 + (\tfrac{1}{2})(60 \text{ kg})(2 \text{ m/s})^2 = 440 \text{ J}$$

Final KE $= \frac{1}{2}(m_1 + m_2)v_3^2 = (\tfrac{1}{2})(100 \text{ kg})(2.8 \text{ m/s})^2 = 392 \text{ J}$

Therefore 48 J of energy is lost, 11 percent of the original amount.

## SOLVED PROBLEM 8.14

The two skaters of Prob. 8.13 are moving in opposite directions and collide head-on. ($a$) If they remain in contact, what is their final velocity? ($b$) How much kinetic energy is lost?

($a$)   We take into account the opposite directions of motion by letting $v_1 = +4$ m/s and $v_2 = -2$ m/s. Then

$$v_3 = \frac{m_1 v_1 + m_2 v_2}{m_1 + m_2} = \frac{(40\text{ kg})(4\text{ m/s}) - (60\text{ kg})(2\text{ m/s})}{40\text{ kg} + 60\text{ kg}} = +0.4\text{ m/s}$$

Since $v_3$ is $+0.4$ m/s, the two skaters move off in the same direction as the 40-kg skater had originally, which is to be expected since she had the greater initial momentum.

($b$)                              Initial KE $= 440$ J        (as in Prob. 8.13)

Final KE $= \frac{1}{2}(m_1 + m_2)v_3^2 = (\frac{1}{2})(100\text{ kg})(0.4\text{ m/s})^2 = 8$ J

Therefore 432 J of energy is lost, 98 percent of the original amount. This is the reason why head-on collisions of automobiles produce much more damage than overtaking collisions.

## SOLVED PROBLEM 8.15

A 1000-kg car moving north at 20 m/s collides with a 1500-kg car moving west at 12 m/s. If the cars stick together after the collision, at what velocity and in what direction does the wreckage begin to move?

Linear momentum must be conserved separately in both the north-south and east-west directions, which we call the $y$ and $x$ axes, respectively, as in Fig. 8-2($a$). Thus we have

Momentum before crash = momentum after crash

$x$ direction                          $m_A v_{Ax} + m_B v_{Bx} = m_{AB} v_{ABx}$
$y$ direction                          $m_A v_{Ay} + m_B v_{By} = m_{AB} v_{ABy}$

Here

| | | |
|---|---|---|
| $m_A = 1000$ kg | $m_B = 1500$ kg | $m_{AB} = m_A + m_B = 2500$ kg |
| $v_{Ax} = 0$ | $v_{Bx} = -12$ m/s | $v_{ABx} = ?$ |
| $v_{Ay} = 20$ m/s | $v_{By} = 0$ | $v_{ABy} = ?$ |

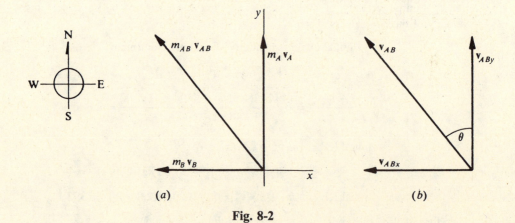

(a)                                                            (b)

Fig. 8-2

In the $x$ direction we have

$$v_{ABx} = \frac{m_A v_{Ax} + m_B v_{Bx}}{m_{AB}} = \frac{0 + (1500 \text{ kg})(-12 \text{ m/s})}{2500 \text{ kg}} = -7.2 \text{ m/s}$$

and in the $y$ direction we have

$$v_{ABy} = \frac{m_A v_{Ay} + m_B v_{By}}{m_{AB}} = \frac{(1000 \text{ kg})(20 \text{ m/s})}{2500 \text{ kg}} = 8 \text{ m/s}$$

The magnitude of the velocity $\mathbf{v}_{AB}$ is therefore

$$v_{AB} = \sqrt{v_{ABx}^2 + v_{ABy}^2} = \sqrt{(-7.2 \text{ m/s})^2 + (8 \text{ m/s})^2} = 10.8 \text{ m/s}$$

The direction of $\mathbf{v}_{AB}$ may be specified by the angle $\theta$ between the $+y$ direction, which is north, and $\mathbf{v}_{AB}$. From Fig. 8-2($b$), ignoring the sign of $v_{ABx}$, we get

$$\tan \theta = \frac{v_{ABx}}{v_{ABy}} = \frac{7.2 \text{ m/s}}{8 \text{ m/s}} = 0.9 \qquad \theta = 42°$$

The wreckage begins to move at 10.8 m/s in a direction 42° west of north.

## COEFFICIENT OF RESTITUTION

The *coefficient of restitution e* is defined as the ratio between the relative velocity of recession $v_2' - v_1'$ after a collision between two bodies and their relative velocity of approach $v_1 - v_2$:

$$\text{Coefficient of restitution} = e = \frac{v_2' - v_1'}{v_1 - v_2}$$

Values of $e$ range from 0 to 1. In a perfectly elastic collision, $e = 1$ and the relative velocity after the collision is the same as the relative velocity before it. In a perfectly inelastic collision, $e = 0$.

## SOLVED PROBLEM 8.16

An object of mass $m_1$ and initial velocity $v_1$ undergoes an elastic head-on collision with another object of mass $m_2$ that is at rest. How do the results of the collision depend on $m_1$ and $m_2$?

Let us call the final velocities of the objects $v_1'$ and $v_2'$. From conservation of momentum,

$$m_1 v_1 = m_1 v_1' + m_2 v_2'$$

The coefficient of restitution for an elastic collision is $e = 1$, so here, with $v_2 = 0$,

$$v_1 = v_2' - v_1' \qquad v_2' = v_1 + v_1'$$

Substituting for $v_2'$ in the first equation gives

$$m_1 v_1 = m_1 v_1' + m_2 v_1 + m_2 v_1'$$
$$(m_1 + m_2)(v_1') = (m_1 - m_2)(v_1)$$
$$v_1' = \left(\frac{m_1 - m_2}{m_1 + m_2}\right)(v_1)$$

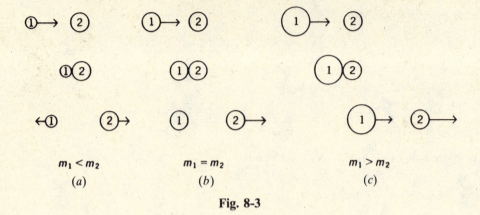

$$m_1 < m_2 \qquad\qquad m_1 = m_2 \qquad\qquad m_1 > m_2$$

$$(a) \qquad\qquad\qquad (b) \qquad\qquad\qquad (c)$$

**Fig. 8-3**

In a similar way we find that

$$v'_2 = \left(\frac{2m_1}{m_1 + m_2}\right)(v_1)$$

From these results we draw the following conclusions, which are illustrated in Fig. 8-3.

(a)   For $m_1 < m_2$: Here $v'_1$ is opposite in direction to $v_1$, so the incoming lighter object rebounds from the heavier one. If $m_1 \ll m_2$, for instance, in the case of a ball striking a wall, then $v'_1 \approx -v_1$.

(b)   For $m_1 = m_2$: Here $v'_1 = 0$ and $v'_2 = v_1$, so the incoming object $m_1$ comes to a stop, and the struck object $m_2$ moves away with the same velocity as $m_1$ had initially. An example is a moving billiard ball striking a stationary one head-on.

(c)   For $m_1 > m_2$: The incoming object $m_1$ continues with reduced velocity while the struck object $m_2$ moves away ahead of $m_1$ at a higher velocity. An example is a tennis serve. If $m_1 \gg m_2$, then $v'_2 \approx 2v_1$.

## SOLVED PROBLEM 8.17

A ball dropped on a floor from a height of 1.5 m bounces back to a height of 0.85 m. What is the coefficient of restitution?

The floor does not move, so $v_2 = v'_2 = 0$. We consider up as $+$ and down as $-$. The velocity of the ball after it has fallen from an initial height $h$ is $v_1 = -1\sqrt{2gh}$. The velocity that it needs to reach the height $h'$ after bouncing on the floor is $v'_1 = +\sqrt{2gh'}$. Hence

$$e = \frac{v'_2 - v'_1}{v_1 - v_2} = \frac{0 - v'_1}{v_1 - 0} = \frac{-\sqrt{2gh'}}{-\sqrt{2gh}} = \sqrt{\frac{h'}{h}}$$

and so here

$$e = \sqrt{\frac{h'}{h}} = \sqrt{\frac{0.85 \text{ m}}{1.5 \text{ m}}} = 0.75$$

## SOLVED PROBLEM 8.18

A 1-kg ball moving at 5 m/s collides with a 2-kg ball moving in the opposite direction at 4 m/s. If the coefficient of restitution is 0.7, find the velocities of the two balls after the impact.

Here

$$m_1 = 1 \text{ kg} \qquad v_1 = 5 \text{ m/s} \qquad v'_1 = ?$$
$$m_2 = 2 \text{ kg} \qquad v_2 = -4 \text{ m/s} \qquad v'_2 = ?$$

First we make use of the fact that $e = 0.7$:

$$e = \frac{v'_2 - v'_1}{v_1 - v_2}$$
$$v'_2 - v'_1 = e(v_1 - v_2) = 0.7[5 \text{ m/s} - (-4 \text{ m/s})] = 6.3 \text{ m/s}$$
$$v'_2 = v'_1 + 6.3 \text{ m/s}$$

From conservation of momentum we have

Momentum before = momentum after
$$m_1 v_1 + m_2 v_2 = m_1 v'_1 + m_2 v'_2$$
$$(1 \text{ kg})(5 \text{ m/s}) - (2 \text{ kg})(4 \text{ m/s}) = (1 \text{ kg})(v'_1) + (2 \text{ kg})(v'_2)$$
$$v'_2 = \frac{-3 \text{ kg} \cdot \text{m/s} - v'_1 \text{ kg}}{2 \text{ kg}}$$

We now set equal the two expressions for $v'_2$ and solve for $v'_1$:

$$v'_1 + 6.3 \text{ m/s} = \frac{-3 \text{ kg} \cdot \text{m/s} - v'_1 \text{ kg}}{2 \text{ kg}}$$
$$v'_1(1 + 0.5) = \frac{-3 \text{ kg} \cdot \text{m/s}}{2 \text{ kg}} - 6.3 \text{ m/s} = -7.8 \text{ m/s}$$
$$v'_1 = -5.2 \text{ m/s}$$

The minus sign means that the 1-kg ball moves off in the opposite direction from its initial one. To find $v'_2$, it is simplest to use $v'_2 = v'_1 + 6.3$ m/s:

$$v'_2 = -5.2 \text{ m/s} + 6.3 \text{ m/s} = 1.1 \text{ m/s}$$

The 2-kg ball also reverses its direction as a result of the collision.

## *Multiple-Choice Questions*

**8.1.** An elastic collision conserves

(a) momentum but not KE    (c) both momentum and KE
(b) KE but not momentum    (d) neither momentum nor KE

**8.2.** An inelastic collision conserves

(a) momentum but not KE    (c) both momentum and KE
(b) KE but not momentum    (d) neither momentum nor KE

**8.3.** A 60-kg skater pushes a 50-kg skater, who moves away at 2.0 m/s. As a result, the first skater moves backward at

(a) 0.6 m/s    (c) 2.0 m/s
(b) 1.7 m/s    (d) 2.4 m/s

**8.4.**   A 60-g tennis ball moving at 8.0 m/s strikes a stationary tennis racket perpendicularly and bounces off at 6.0 m/s. The impulse given to the racket is

(a)   0.12 N·s      (c)   0.48 N·s
(b)   0.36 N·s      (d)   0.84 N·s

**8.5.**   During a serve, a tennis racket exerts an average force of 250 N on a 60-g tennis ball, initially at rest, for 5.0 ms (0.0050 s). The ball's KE afterward is

(a)   0.78 J      (c)   13 J
(b)   1.25 J      (d)   127 J

**8.6.**   A 1500-kg truck whose velocity is 60 km/h overtakes a 4000-kg truck moving in the same direction at 35 km/h. The trucks collide and stick together, and the initial velocity of the wreckage is

(a)   9.1 km/h      (c)   48 km/h
(b)   42 km/h      (d)   53 km/h

**8.7.**   The trucks in Question 8.6 are headed in opposite directions when they collide. The trucks again stick together, and the wreckage now has an initial velocity of

(a)   9.1 km/h      (c)   48 km/h
(b)   42 km/h      (d)   53 km/h

**8.8.**   The trucks in Question 8.6 lost a total KE that is

(a)   0
(b)   less than the KE lost by the trucks in Question 8.7
(c)   the same as the KE lost by the trucks in Question 8.7
(d)   more than the KE lost by the trucks in Question 8.7

**8.9.**   An 800-kg car headed south at 40 km/h strikes a 1200-kg car headed west at 25 km/h. The cars stick together and the initial velocity of the wreckage is

(a)   22 km/h      (c)   33 km/h
(b)   31 km/h      (d)   47 km/h

**8.10.**   The wreckage of Question 8.9 moves off at

(a)   20° W of S      (c)   47° W of S
(b)   43° W of S      (d)   70° W of S

**8.11.**   A ball is dropped from a height of 300 cm. If the coefficient of restitution is 0.600, the ball rebounds to a height of

(a)   104 cm      (c)   134 cm
(b)   108 cm      (d)   180 cm

# Supplementary Problems

**8.1.**   Find the momentum of a 100-kg ostrich running at 15 m/s.

**8.2.**  Find the momentum of a 3200-lb car moving at 60 mi/h (88 ft/s).

**8.3.**  An object at rest explodes and breaks up into two parts that fly off. Must they move in opposite directions?

**8.4.**  A moving object strikes a stationary one. After the collision, must they move in the same direction?

**8.5.**  A 2500-kg truck crashes into a wall at 40 km/h and comes to a stop in 0.5 s. What is the average force on the truck?

**8.6.**  A 5-kg rifle and a 7-kg rifle fire identical bullets with the same muzzle velocities. Compare the recoil momenta and recoil velocities of the two rifles.

**8.7.**  An empty dump truck is coasting with its engine off along a level road when rain starts to fall. Neglecting friction, what (if anything) happens to the velocity of the truck?

**8.8.**  A 60-kg woman dives horizontally from a 250-kg boat with a velocity of 2 m/s. What is the recoil velocity of the boat?

**8.9.**  Four 50-kg girls simultaneously dive horizontally at 2.5 m/s from the same side of a boat, whose recoil velocity is 0.1 m/s. What is the mass of the boat?

**8.10.**  A spacecraft's motors provide a total thrust of 1.5 MN. If the exhaust speed is 2.5 km/s, at what rate is fuel being used?

**8.11.**  A rocket whose initial weight is 10,000 lb uses 50 lb/s of fuel that it exhausts at 8000 ft/s. Find the initial acceleration of the rocket.

**8.12.**  An unoccupied 1200-kg car has coasted down a hill and is moving along a level road at 10 m/s. A 6000-kg truck moving in the opposite direction collides head-on with it. What was the truck's velocity if both vehicles come to a stop after the collision?

**8.13.**  A 50-kg boy at rest on roller skates catches a 0.6-kg ball moving toward him at 30 m/s. How fast does he move backward as a result?

**8.14.**  A 1200-kg car traveling at 10 m/s overtakes a 1000-kg car traveling at 8 m/s and collides with it. (*a*) If the two cars stick together, what is their final velocity? (*b*) How much kinetic energy is lost? What percentage of the original KE is this?

**8.15.**  The cars of Prob. 8.14 are moving in opposite directions and collide head-on. (*a*) If they stick together, what is their final velocity? (*b*) How much kinetic energy is lost? What percentage of the original KE is this?

**8.16.**  A 1-lb stone moving south at 8 ft/s collides with a 5-lb lump of clay moving west at 3 ft/s and becomes embedded in the clay. Find the velocity (magnitude and direction) of the composite body.

**8.17.**  A 5-kg ball moving at 6 m/s strikes a 3-kg ball initially at rest. The 5-kg ball continues moving in the same direction at 2 m/s. (*a*) Find the velocity and direction of the 3-kg ball. (*b*) Find the coefficient of restitution.

**8.18.**  A 5-kg ball moving to the right at 3 m/s overtakes and collides with a 10-kg ball moving to the right at 2.4 m/s. Find the final velocities of the two balls if the coefficient of restitution is 0.8.

**8.19.** A rubber ball is dropped on the ground from a height of 5 m. If the coefficient of restitution is 0.7, find the height to which the ball rebounds.

**8.20.** A ball dropped from a height of 3 m bounces up to a height of 1 m. Find the coefficient of restitution.


## *Answers to Multiple-Choice Questions*

**8.1.** (*c*)      **8.7.**   (*a*)

**8.2.** (*a*)      **8.8.**   (*b*)

**8.3.** (*b*)      **8.9.**   (*a*)

**8.4.** (*d*)      **8.10.**  (*b*)

**8.5.** (*c*)      **8.11.**  (*b*)

**8.6.** (*b*)


## Answers to Supplementary Problems

**8.1.** 1500 kg·m/s

**8.2.** 8800 slug·ft/s

**8.3.** Yes

**8.4.** No; no; they can move in any direction provided that the vector sum of their momenta equals the initial momentum of the first object.

**8.5.** 18,500 N

**8.6.** The recoil momenta are the same, but the lighter rifle has a higher recoil velocity.

**8.7.** The truck's velocity decreases as rainwater accumulates in it, since the total momentum must remain constant despite the increase in mass.

**8.8.** 0.48 m/s

**8.9.** 5000 kg

**8.10.** 600 kg/s

**8.11.** 8 m/s$^2$

**8.12.** 2 m/s

**8.13.** 0.36 m/s

**8.14.** (*a*)  9.09 m/s      (*b*)   1100 J; 1.2%

**8.15.** (*a*)  1.82 m/s      (*b*)   88.4 kJ; 96%

**8.16.** 2.83 ft/s at 28° south of west

**8.17.** (*a*)  6.67 m/s in the same direction as the 5-kg ball      (*b*)   0.78

**8.18.** 2.3 m/s; 2.8 m/s

**8.19.** 2.45 m

**8.20.** 0.58

# Chapter 9

## Circular Motion and Gravitation

### CENTRIPETAL ACCELERATION

A body that moves in a circular path at a velocity whose magnitude is constant is said to undergo *uniform circular motion*.

Although the velocity of a body in uniform circular motion is constant in magnitude, its direction changes continually. The body is therefore accelerated. The direction of this *centripetal acceleration* is toward the center of the circle in which the body moves, and its magnitude is

$$a_c = \frac{v^2}{r}$$

$$\text{Centripetal acceleration} = \frac{(\text{velocity of body})^2}{\text{radius of circular path}}$$

Because the acceleration is perpendicular to the path followed by the body, its velocity changes only in direction, not in magnitude.

### CENTRIPETAL FORCE

The inward force that must be applied to keep a body moving in a circle is called *centripetal force*. Without centripetal force, circular motion cannot occur. Since $F = ma$, the magnitude of the centripetal force on a body in uniform motion is

$$\text{Centripetal force} = F_c = \frac{mv^2}{r}$$

### SOLVED PROBLEM 9.1

A ball is whirled at the end of a string in a horizontal circle 60 cm in radius at the rate of 1 revolution (rev) every 2 s. Find the ball's centripetal acceleration.

The distance the ball travels per revolution is

$$s = 2\pi r = (2\pi)(0.6 \text{ m}) = 3.77 \text{ m}$$

Since each revolution takes 2 s, the ball's velocity is

$$v = \frac{s}{t} = \frac{3.77 \text{ m}}{2 \text{ s}} = 1.88 \text{ m/s}$$

The ball's centripetal acceleration is therefore

$$a_c = \frac{v^2}{r} = \frac{(1.88 \text{ m/s})^2}{0.6 \text{ m}} = 5.9 \text{ m/s}^2$$

### SOLVED PROBLEM 9.2

A 1000-kg car rounds a turn of radius 30 m at a velocity of 9 m/s. (*a*) How much centripetal force is required? (*b*) Where does this force come from?

(a)
$$F_c = \frac{mv^2}{r} = \frac{(1000 \text{ kg})(9 \text{ m/s})^2}{30 \text{ m}} = 2700 \text{ N}$$

(b)  The centripetal force on a car making a turn on a level road is provided by the road acting via friction on the car's tires.

## SOLVED PROBLEM 9.3

How much centripetal force is needed to keep a 160-lb skater moving in a circle 20 ft in radius at a velocity of 10 ft/s?

The skater's mass is $m = w/g = 160 \text{ lb}/32 \text{ ft/s}^2 = 5$ slugs. Hence

$$F_c = \frac{mv^2}{r} = \frac{(5 \text{ slugs})(10 \text{ ft/s})^2}{20 \text{ ft}} = 25 \text{ lb}$$

## SOLVED PROBLEM 9.4

The maximum force a road can exert on the tires of a 1500-kg car is 8500 N. What is the maximum velocity at which the car can round a turn of radius 120 m?

Solving the formula $F_c = mv^2/r$ for $v$ gives for the maximum velocity

$$v = \sqrt{\frac{F_c r}{m}} = \sqrt{\frac{(8500 \text{ N})(120 \text{ m})}{1500 \text{ m}}} = 26 \text{ m/s}$$

which is

$$\left(26 \frac{\text{m}}{\text{s}}\right)\left(\frac{3600 \text{ s/h}}{1000 \text{ m/kh}}\right) = 94 \text{ km/h}$$

## SOLVED PROBLEM 9.5

A car is traveling at 20 mi/h on a level road where the coefficient of static friction between tires and road is 0.8. Find the minimum turning radius of the car.

The maximum centripetal force that friction can provide here is

$$F_f = \mu_s N = \mu_s w = \mu_s mg$$

Hence                          Centripetal force = frictional force

$$\frac{mv^2}{r} = \mu_s mg$$

$$r = \frac{v^2}{\mu_s g}$$

The car's velocity is

$$v = (20 \text{ mi/h})\left(1.47 \frac{\text{ft/s}}{\text{mi/h}}\right) = 29.4 \text{ ft/s}$$

and so the minimum turning radius is

$$r = \frac{v^2}{\mu_s g} = \frac{(29.4 \text{ ft/s})^2}{(0.8)(32 \text{ ft/s}^2)} = 34 \text{ ft}$$

**SOLVED PROBLEM 9.6**

Highway curves are usually *banked* (tilted inward) at an angle $\theta$ such that the horizontal component of the reaction force of the road on a car traveling at the design velocity equals the required centripetal force. Find the proper banking angle for a car making a turn of radius $r$ at velocity $v$.

In the absence of friction, the reaction force **F** of the road on the car is perpendicular to the road surface. The vertical component $\mathbf{F}_y$ of this force supports the weight **w** of the car, and its horizontal component $\mathbf{F}_x$ provides the centripetal force $mv^2/r$. From Fig. 9-1 we have

$$F_x = F \sin \theta = \frac{mv^2}{r}$$
$$F_y = F \cos \theta = w = mg$$

Dividing $F_x$ by $F_y$ gives

$$\frac{F \sin \theta}{F \cos \theta} = \frac{mv^2}{mgr} \qquad \tan \theta = \frac{v^2}{gr}$$

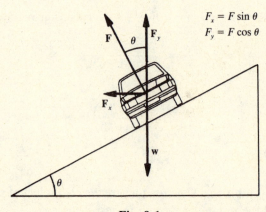

$$F_x = F \sin \theta$$
$$F_y = F \cos \theta$$

**Fig. 9-1**

The proper banking angle $\theta$ depends only on $v$ and $r$, not on the mass of the car. If a car goes around a banked turn more slowly than at the design velocity, friction tends to keep it from sliding down the inclined roadway; if the car goes faster, friction tends to keep it from skidding outward. If the car's speed is too great, however, friction may not be sufficient to keep the car on the road.

**SOLVED PROBLEM 9.7**

(*a*) Find the proper banking angle for cars moving at 50 mi/h to go around a curve 1000 ft in radius. (*b*) If the curve were not banked, what coefficient of friction would be required between the tires and the road?

(*a*)   Here $v = (50 \text{ mi/h})[1.47 \text{ (ft/s)(mi/h)}] = 73.5$ ft/s, and so, from the solution to Prob. 9.6,

$$\tan \theta = \frac{v^2}{gr} = \frac{(73.5 \text{ ft/s})^2}{(32 \text{ ft/s}^2)(1000 \text{ ft})} = 0.169 \qquad \theta = 9.6°$$

(b)   From the solution to Prob. 9.5, $r = v^2/(\mu_s g)$. Hence

$$\mu_s = \frac{v^2}{gr} = 0.169$$

## MOTION IN A VERTICAL CIRCLE

When a body moves in a vertical circle at the end of a string, the tension **T** in the string varies with the body's position. The centripetal force **F**$_c$ on the body at any point is the vector sum of **T** and the component of the body's weight **w** toward the center of the circle. At the top of the circle, as in Fig. 9-2(a), the weight **w** and the tension **T** both act toward the center of the circle, and so

$$T = F_c - w$$

At the bottom of the circle, **w** acts away from the center of the circle, and so

$$T = F_c + w$$

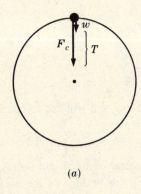

(a)

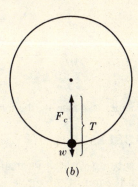

(b)

**Fig. 9-2**

## SOLVED PROBLEM 9.8

A string 0.5 m long is used to whirl a 1-kg stone in a verticle circle at a uniform velocity of 5 m/s. What is the tension in the string (a) when the stone is at the top of the circle and (b) when the stone is at the bottom of the circle?

(a)   The centripetal force needed to keep the stone moving at 5 m/s is

$$F_c = \frac{mv^2}{r} = \frac{(1 \text{ kg})(5 \text{ m/s})^2}{0.5 \text{ m}} = 50 \text{ N}$$

The weight of the stone is $w = mg = (1 \text{ kg})(9.8 \text{ m/s}^2) = 9.8$ N. At the top of the circle,

$$T = F_c - w = 50 \text{ N} - 9.8 \text{ N} = 40.2 \text{ N}$$

(b)   At the bottom of the circle,

$$T = F_c + w = 59.8 \text{ N}$$

## SOLVED PROBLEM 9.9

An airplane pulls out of a dive in a circular arc whose radius is 1200 m. The airplane's velocity is a constant 200 m/s. Find the force with which the 80-kg pilot presses down on his seat at the bottom of the arc.

The downward force $F$ the pilot exerts on his seat is the equal and opposite reaction to the upward force of the seat on him. This upward force both supports his weight $w$ and provides the centripetal force $F_c$ that keeps him in a circular path. The upward force of the seat on the pilot plays the same role here as the tension in the string of Fig. 9-2(b) does in the motion of a body moving in a vertical circle when the body is at the bottom of its path. Hence

$$F = F_c + w = \frac{mv^2}{r} + mg = \frac{(80 \text{ kg})(200 \text{ m/s})^2}{1200 \text{ m}} + (80 \text{ kg})(9.8 \text{ m/s}^2) = (2667 + 784) \text{ N} = 3451 \text{ N}$$

The pilot presses down on his seat with a force of 3451 N, which is 4.4 times his weight.

## SOLVED PROBLEM 9.10

A ball at the end of an 80-cm string is being whirled in a vertical circle. At what critical velocity $v_0$ will the string begin to go slack at the top of the ball's path?

The situation here corresponds to that shown in Fig. 9-2(a), with $T = F_c - w$. When the string goes slack, the tension in it is $T = 0$, and

$$F_c = w$$
$$\frac{mv_0^2}{r} = mg$$
$$v_0 = \sqrt{rg} = \sqrt{(0.8 \text{ m})(9.8 \text{ m/s}^2)} = 2.8 \text{ m/s}$$

For velocities less than $v_0$, the top of the ball's path will flatten out and will not be circular.

## SOLVED PROBLEM 9.11

A ball is being whirled vertically at constant energy at the end of an 80-cm string. If the ball's speed at the top of the circle is 3 m/s, what is its speed at the bottom of the circle?

At the top of the circle, the ball has a potential energy relative to the bottom of the circle of

$$PE_1 = mgh = mg(2r) = 2mgr$$

At the bottom of the circle, $PE_2 = 0$. Because the total energy of the ball is constant,

$$KE_2 + PE_2 = KE_1 + PE_1$$

where $KE_1$ is the ball's kinetic energy at the top and $KE_2$ is its kinetic energy at the bottom. Therefore

$$\tfrac{1}{2}mv_2^2 + 0 = \tfrac{1}{2}mv_1^2 + 2mgr$$
$$v_2^2 = v_1^2 + 4gr$$
$$v_2 = \sqrt{v_1^2 + 4gr} = \sqrt{(3 \text{ m/s})^2 + (4)(9.8 \text{ m/s}^2)(0.8 \text{ m})} = 6.35 \text{ m/s}$$

The ball's velocity at the bottom of its path is more than twice its velocity at the top. The difference is due to the conversion of the ball's potential energy at the top into additional kinetic energy at the bottom.

## GRAVITATION

According to Newton's *law of universal gravitation*, every body in the universe attracts every other body with a force that is directly proportional to each of their masses and inversely proportional to the square of the distance between them. In equation form,

$$\text{Gravitational force} = F = G\frac{m_1 m_2}{r^2}$$

where $m_1$ and $m_2$ are the masses of any two bodies, $r$ is the distance between them, and $G$ is a constant whose values in SI and British units are, respectively,

SI units :                                          $G = 6.67 \times 10^{-11}$ N·m$^2$/kg$^2$

British units :                                     $G = 3.44 \times 10^{-8}$ lb·ft$^2$/slug$^2$

A spherical body behaves gravitationally as though its entire mass were concentrated at its center.

## SOLVED PROBLEM 9.12

What gravitational force does a 1000-kg lead sphere exert on an identical sphere 3 m away?

$$F = G \frac{m_1 m_2}{r^2} = \frac{(6.67 \times 10^{-11} \text{N·m}^2/\text{kg}^2)(10^3 \text{ kg})(10^3 \text{ kg})}{(3 \text{ m})^2} = 7.4 \times 10^{-4} \text{ N}$$

This is less than the force that would result from blowing gently on one of the spheres. Gravitational forces are usually significant only when at least one of the bodies has a very large mass.

## SOLVED PROBLEM 9.13

A girl weighs 128 lb on the earth's surface. (*a*) What would she weigh at a height above the earth's surface of one earth radius? (*b*) What would her mass be there?

(*a*)   Since the gravitational force the earth exerts on an object a distance $r$ from its center varies as $1/r^2$, the gravitational force on the girl relative to its value $mg$ at the earth's surface (where $r = r_e$) is

$$F = \left(\frac{r_e}{r}\right)^2 mg$$

When the girl is a distance $r_e$ above the earth, $r = 2r_e$, and

$$F = \left(\frac{r_e}{r}\right)^2 mg = \left(\frac{r_e}{2r_e}\right)^2 w = (\tfrac{1}{4})(128 \text{ lb}) = 32 \text{ lb}$$

(*b*)   Her mass of

$$m = \frac{w}{g} = \frac{128 \text{ lb}}{32 \text{ ft/s}^2} = 4 \text{ slugs}$$

is the same everywhere.

## SOLVED PROBLEM 9.14

Find the acceleration of gravity at an altitude of 1000 km.

The gravitational force $Gm_e m/r^2$ of the earth on an object of mass $m$ at the distance $r$ from the earth's center is equal to the object's weight $mg$ at that distance, where $m_e$ is the earth's mass. Hence

$$\frac{Gm_e m}{r^2} = mg \qquad g = \frac{Gm_e}{r^2}$$

At the earth's surface, where $g = g_0 = 9.8$ m/s$^2$ and $r = r_e = 6400$ km,

$$g_0 = \frac{Gm_e}{r_e^2}$$

Hence at the distance $r$ from the earth's center, the acceleration of gravity is

$$g = \left(\frac{r_e}{r}\right)^2 g_0$$

When $r = r_e + h = 6400$ km $+ 1000$ km $= 7400$ km,

$$g = \left(\frac{6400 \text{ km}}{7400 \text{ km}}\right)^2 (9.8 \text{ m/s}^2) = 7.3 \text{ m/s}^2$$

## SATELLITE MOTION

Gravitation provides the centripetal forces that keep the planets in their orbits around the sun and the moon in its orbit around the earth. The same is true for artificial satellites put into orbit around the earth.

### SOLVED PROBLEM 9.15

Find the velocity an artificial satellite must have to pursue a circular orbit around the earth just above the surface.

In a stable orbit, the gravitational force $mg$ on the satellite must be equal to the centripetal force $mv^2/r$ required. Hence

$$\frac{mv^2}{r} = mg$$
$$v^2 = rg$$
$$v = \sqrt{rg}$$

The mass of the satellite is irrelevant. To find $v$, we use the radius of the earth $r_e$ for $r$ and the acceleration of gravity at the earth's surface for $g$. This gives

$$v = \sqrt{r_e g} = \sqrt{(6.4 \times 10^6 \text{ m})(9.8 \text{ m/s}^2)} = 7.9 \times 10^3 \text{ m/s}$$

which is about 18,000 mi/h. With a smaller velocity than this, a space vehicle projected horizontally above the earth will fall to the surface; with a larger velocity, it will have an elliptical rather than a circular orbit.

### SOLVED PROBLEM 9.16

A satellite in a *geostationary orbit* circles the earth above the equator with a period of exactly 1 day, so it stays above a particular place all the time. Most of the satellites in such orbits act as relays for telephone calls and television programs. Find the altitude of a geostationary orbit.

A satellite that moves in a circular orbit of radius $r$ covers a distance $2\pi r$ in period $T$. Hence its velocity is $v = 2\pi r/T$. From the solution to Prob. 9.15, another formula for the satellite speed is $v = \sqrt{rg}$. Setting these formulas equal allows us to eliminate $v$:

$$\frac{2\pi r}{T} = \sqrt{rg}$$

From the solution to Prob. 9.14, $g = (r_e/r)^2 g_0$, so

$$\frac{2\pi r}{T} = \sqrt{\frac{r_e^2 g_0}{r}}$$

Squaring both sides yields

$$\frac{4\pi^2 r^2}{T^2} = \frac{r_e^2 g_0}{r}$$

Multiplying both sides by $r$ gives

$$\frac{4\pi^2 r^3}{T^2} = r_e^2 g_0$$

Solving for $r$, we have

$$r = \left(\frac{r_e^2 g_0 T^2}{4\pi^2}\right)^{1/3}$$

Since here $T = 1$ day $= (24$ h$)(60$ min/h$)(60$ s/min$) = 8.64 \times 10^4$ s,

$$r = \left[\frac{(6.4 \times 10^6 \text{ m})^2 (9.8 \text{ m/s}^2)(8.64 \times 10^4 \text{ s})^2}{4\pi^2}\right]^{1/3} = 4.23 \times 10^7 \text{ m}$$

The corresponding altitude above the earth's surface is

$$h = r - r_e = (4.23 - 0.64) \times 10^7 \text{ m} = 3.59 \times 10^7 \text{ m} = 35{,}900 \text{ km}$$

# *Multiple-Choice Questions*

**9.1.** A 1200-kg car whose velocity is 6 m/s rounds a turn whose radius is 30 m. The centripetal force on the car is

(a)  48 N        (c)  240 N
(b)  147 N       (d)  1440 N

**9.2.** If the car of Question 9.1 rounds the same turn at 12 m/s, the required centripetal force is

(a)  halved      (c)  doubled
(b)  the same    (d)  quadrupled

**9.3.** The maximum centripetal force that friction can provide the car of Question 9.1 on a rainy day is 8000 N. The highest velocity at which the car can round the turn is

(a)  14 m/s      (c)  200 m/s
(b)  77 m/s      (d)  40 km/s

**9.4.** The rain stops, the road dries, and the coefficient of friction between the tires of the car of Question 9.1 and the road increases to 1.40 times its value in Question 9.3. The highest velocity at which the car can now round the turn is

(a)  unchanged              (c)  1.40 its former value
(b)  1.18 its former value  (d)  1.96 its former value

**9.5.** The coefficient of static friction between the tires of a truck whose velocity is 60 km/h and a road is 0.65. The smallest turning radius of the truck is

(a)  9.4 m       (c)  44 m
(b)  22 m        (d)  565 m

**9.6.** A 500-g ball is whirled in a vertical circle at the end of a string 60 cm long. If the velocity of the ball at the bottom of the circle is 4.0 m/s, the tension in the string there is

(a)  4.9 N       (c)  13.3 N
(b)  8.4 N       (d)  18.2 N

**9.7.** A car just leaves the road when it passes over a hump whose radius is 40 m. The car's velocity is

    (*a*)   20 m/s    (*c*)   392 m/s
    (*b*)   62 m/s    (*d*)   impossible to find without knowing the car's mass

**9.8.** An astronaut weighs 3200 N on a planet whose mass is the same as that of the earth but whose radius is half that of the earth. The astronaut's weight on the earth is

    (*a*)   400 N    (*c*)   1600 N
    (*b*)   800 N    (*d*)   3200 N

**9.9.** A satellite orbits the earth in a circle of radius 8000 km. At that distance from the earth $g = 6.2$ m/s$^2$. The velocity of the satellite is

    (*a*)   0.90 km/s    (*c*)   8.9 km/s
    (*b*)   7.0 km/s    (*d*)   impossible to find without knowing the satellite's mass

# Supplementary Problems

**9.1.** Can a body move in a curved path without being accelerated?

**9.2.** In what ways, if any, do $g$ and $G$ change with increasing height above the earth's surface?

**9.3.** The moon's mass is approximately 1 percent of the earth's mass. How does the gravitational pull of the earth on the moon compare with the gravitational pull of the moon on the earth?

**9.4.** A hole is drilled to the center of the earth, and a stone whose mass is 1 kg at the earth's surface is dropped into it. What is the mass of the stone when it is at the earth's center? What is its weight?

**9.5.** A lasso is whirled so that the loop at its end describes a horizontal circle 5 ft in radius. If each revolution of the lasso requires 4 s, find the centripetal acceleration of the loop.

**9.6.** A 0.02-kg ball is whirled in a horizontal circle at the end of a string 0.5 m long whose breaking strength is 1 N. Neglecting gravity, what is the maximum velocity the ball can have?

**9.7.** How much centripetal force is needed to keep a 4-lb iron ball moving in a horizontal circle of radius 5 ft at a velocity of 15 ft/s?

**9.8.** A 5000-kg airplane makes a horizontal turn 1 km in radius at a velocity of 50 m/s. How much centripetal force is required?

**9.9.** A motorcycle begins to skid when it makes a turn of 15-m radius at a velocity of 12 m/s. What is the highest velocity at which it can make a turn of 30-m radius?

**9.10.** What is the minimum coefficient of static friction required for a car traveling at 60 km/h to make a level turn 40 m in radius?

**9.11.** The coefficient of static friction between a car's tires and a certain concrete road is 1.0 when the road is dry and 0.7 when the road is wet. If the car can safely make a certain turn at 25 mi/h on a dry day, what is the maximum velocity on a rainy day?

**9.12.** An airplane whose velocity is 500 km/h banks at an angle of 40°. If the rudder is not used, what is the radius of the turn the airplane makes?

**9.13.** A highway curve 250 m in radius is banked at an angle of 14°. For what velocity is the curve designed?

**9.14.** Find the minimum velocity at which a car will leave the road when passing over a hump whose radius is 12 m.

**9.15.** A woman swings a pail of water at constant velocity in a vertical circle 100 cm in radius. (*a*) If the water is not to spill, what is the minimum velocity the pail can have? (*b*) How much time per revolution is this equivalent to?

**9.16.** What is the gravitational attraction between an 80-kg man and a 50-kg woman who are 2 m apart?

**9.17.** What is the gravitational attraction between two 3200-lb elephants when they are 20 ft apart?

**9.18.** The acceleration of gravity on the surface of Mars is 0.4*g*. How much will a person whose weight on the Earth's surface is 180 lb weigh on the surface of Mars?

**9.19.** A man weighs 180 lb on the earth's surface. (*a*) What would he weigh at a distance from the center of the earth of three times the radius of the earth? (*b*) What would his mass there be?

**9.20.** A 10-kg monkey is $10^6$ m above the earth's surface. (*a*) What is its weight there? (*b*) What is its mass there?

**9.21.** Find the weight of an 80-kg person at the earth's surface and at an altitude of 2000 km.

**9.22.** At what altitude will a person's weight be half what it is at the earth's surface?

**9.23.** Find the velocity in meters per second that an artificial earth satellite must have to pursue a circular orbit at an altitude of half an earth radius.

**9.24.** Find the radius of a satellite orbit whose period is 12 h.

# *Answers to Multiple-Choice Questions*

**9.1.** (*d*)    **9.6.** (*d*)

**9.2.** (*d*)    **9.7.** (*a*)

**9.3.** (*a*)    **9.8.** (*b*)

**9.4.** (*b*)    **9.9.** (*b*)

**9.5.** (*c*)

# Answers to Supplementary Problems

**9.1.** No.

**9.2.** The value of $g$ decreases; the value of $G$ is the same everywhere in the universe.

**9.3.** They are equal.

**9.4.** 1 kg; 0

**9.5.** 12.3 ft/s$^2$

**9.6.** 5 m/s

**9.7.** 5.6 lb

**9.8.** $1.25 \times 10^4$ N

**9.9.** 17 m/s

**9.10.** 0.71

**9.11.** 21 mi/h

**9.12.** 2.35 km

**9.13.** 2.47 m/s

**9.14.** 11 m/s

**9.15.** (a)  3.13 m/s     (b)  2.00 s

**9.16.** $6.67 \times 10^{-8}$ N

**9.17.** $8.6 \times 10^{-7}$ lb

**9.18.** 72 lb

**9.19.** (a)  20 lb     (b)  5.6 slugs

**9.20.** (a)  73 N     (b)  10 kg

**9.21.** 784 N; 455 N

**9.22.** 2650 km

**9.23.** $6.5 \times 10^3$ m/s

**9.24.** $2.67 \times 10^7$ m

# Chapter 10

# Rotational Motion

## ANGULAR MEASURE

In everyday life, angles are measured in degrees, where 360° equals a full turn. A more suitable unit for technical purposes is the *radian* (rad). If a circle is drawn whose center is at the vertex of a particular angle (Fig. 10-1), the angle $\theta$ (Greek letter *theta*) in radians is equal to the ratio between the arc $s$ cut by the angle and the radius $r$ of the circle:

$$\theta = \frac{s}{r}$$

$$\text{Angle in radians} = \frac{\text{arc length}}{\text{radius}}$$

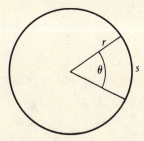

**Fig. 10-1**

Because the circumference of a circle of radius $r$ is $2\pi r$, there are $2\pi$ rad in a complete revolution (rev). Hence

$$1 \text{ rev} = 360° = 2\pi \text{ rad}$$

and so

$$1° = 0.01745 \text{ rad} \qquad 1 \text{ rad} = 57.30°$$

## ANGULAR VELOCITY

The *angular velocity* of a body describes how fast it is turning about an axis. If a body turns through the angle $\theta$ in the time $t$, its angular velocity $\omega$ (Greek letter *omega*) is

$$\omega = \frac{\theta}{t}$$

$$\text{Angular velocity} = \frac{\text{angular displacement}}{\text{time}}$$

Angular velocity is usually expressed in radians per second (rad/s), revolutions per second (rev/s or rps), and revolutions per minute (rev/min or rpm), where

$$1 \text{ rev/s} = 2\pi \text{ rad/s} = 6.28 \text{ rad/s}$$

$$1 \text{ rev/min} = \frac{2\pi}{60} \text{ rad/s} = 0.105 \text{ rad/s}$$

113

The linear velocity $v$ of a particle that moves in a circle of radius $r$ with the uniform angular velocity $\omega$ is given by

$$v = \omega r$$

Linear velocity = (angular velocity)(radius of circle)

This formula is valid only when $\omega$ is expressed in radian measure.

**SOLVED PROBLEM 10.1**

(a) Express 6 rev in radians. (b) How many revolutions are equivalent to 10 rad? (c) How many revolutions are equivalent to $\pi/2$ rad?

(a)                              $\theta = (6 \text{ rev})(2\pi \text{ rad/rev}) = 12\pi \text{ rad} = 37.7 \text{ rad}$

(b)                              $\theta = \dfrac{10 \text{ rad}}{2\pi \text{ rad/rev}} = 1.6 \text{ rev}$

(c)                              $\theta = \dfrac{\pi/2 \text{ rad}}{2\pi \text{ rad/rev}} = \dfrac{1}{4} \text{ rev}$

**SOLVED PROBLEM 10.2**

(a) Express $8°$ in radians. (b) Express 2.5 rad in degrees. (c) Express $\pi$ rad in degrees.

(a)                              $\theta = (8°)(0.01745 \text{ rad/}°) = 0.14 \text{ rad}$

(b)                              $\theta = (2.5 \text{ rad})(57.3°/\text{rad}) = 143 \text{ rad}$

(c)                              $\theta = (\pi \text{ rad})\left(\dfrac{360°}{2\pi \text{ rad}}\right) = 180°$

**SOLVED PROBLEM 10.3**

The shaft of a motor rotates at 1800 rev/min. Through how many radians does it turn in 18 s?

The angular velocity in radians per second that corresponds to 1800 rev/min is

$$\omega = (1800 \text{ rev/min})\left(0.105 \frac{\text{rad/s}}{\text{rev/min}}\right) = 189 \text{ rad/s}$$

Since $\omega = \theta/t$, in 18 s the shaft turns through

$$\theta = \omega t = (189 \text{ rad/s})(18 \text{ s}) = 3402 \text{ rad}$$

**SOLVED PROBLEM 10.4**

A phonograph record 30 cm in diameter turns through an angle of $200°$. How far does a point on the rim of the record travel?

The first step is to find the equivalent in radians of $200°$, which is

$$\theta = (200°)(0.01745 \text{ rad/}°) = 3.49 \text{ rad}$$

The radius of the record is 15 cm. Since $\theta = s/r$, a point on its rim travels

$$s = r\theta = (15 \text{ cm})(3.49 \text{ rad}) = 52 \text{ cm}$$

when it turns through $200°$.

## SOLVED PROBLEM 10.5

A wheel 80 cm in diameter turns at 120 rev/min. (*a*) What is the angular velocity of the wheel in radians per second? (*b*) What is the linear velocity of a point on the rim of the wheel in meters per second?

(*a*)   The angular velocity of the wheel is

$$\omega = (120 \text{ rev/min})\left(0.105 \ \frac{\text{rad/s}}{\text{rev/min}}\right) = 12.6 \text{ rad/s}$$

(*b*)   Since the radius of the wheel is $r = 40$ cm $= 0.4$ m, a point on its rim has the linear velocity

$$v = \omega r = (12.6 \text{ rad/s})(0.4 \text{ m}) = 5.04 \text{ m/s}$$

## SOLVED PROBLEM 10.6

A steel cylinder 60 mm in radius is to be machined in a lathe. If the desired linear velocity of the cylinder's surface is to be 0.7 m/s, at how many revolutions per minute should it rotate?

From the formula $v = \omega r$ we obtain, with $r = 60$ mm $= 0.06$ m,

$$\omega = \frac{v}{r} = \frac{0.7 \text{ m/s}}{0.06 \text{ m}} = 11.7 \text{ rad/s}$$

and so, since 1 rev/min = 0.105 rad/s,

$$\omega = \frac{11.7 \text{ rad/s}}{0.105 \text{ (rad/s)/(rev/min)}} = 111 \text{ rev/min}$$

## ANGULAR ACCELERATION

A rotating body whose angular velocity changes from $\omega_0$ to $\omega_f$ in the time interval $t$ has the *angular acceleration* $\alpha$ (Greek letter *alpha*) of

$$\alpha = \frac{\omega_f - \omega_0}{t}$$

$$\text{Angular acceleration} = \frac{\text{angular velocity change}}{\text{time}}$$

A positive value of $\alpha$ means that the angular velocity is increasing; a negative value means that it is decreasing. Only constant accelerations are considered here.

The formulas relating the angular displacement, velocity, and acceleration of a rotating body under constant acceleration are analogous to the formulas relating linear displacement, velocity, and acceleration given in Chapter 3. If a body has the initial angular velocity $\omega_0$, its angular velocity $\omega_f$ after a time $t$ during which its angular acceleration is $\alpha$ will be

$$\omega_f = \omega_0 + \alpha t$$

and in this time it will have turned through an angular displacement of

$$\theta = \omega_0 t + \tfrac{1}{2}\alpha t^2$$

A relationship that does not involve the time $t$ directly is sometimes useful:

$$\omega_t^2 = \omega_0^2 + 2\alpha\theta$$

**SOLVED PROBLEM 10.7**

An engine requires 5 s to go from its idling speed of 600 rev/min to 1200 rev/min. (*a*) What is its angular acceleration? (*b*) How many revolutions does it make in this period?

(*a*)   The initial and final angular velocities of the engine are, respectively,

$$\omega_0 = (600 \text{ rev/min})\left(0.105 \frac{\text{rad/s}}{\text{rev/min}}\right) = 63 \text{ rad/s}$$

$$\omega_f = (1200 \text{ rev/min})\left(0.105 \frac{\text{rad/s}}{\text{rev/min}}\right) = 126 \text{ rad/s}$$

and so its angular acceleration is

$$\alpha = \frac{\omega_f - \omega_0}{t} = \frac{126 \text{ rad/s} - 63 \text{ rad/s}}{5 \text{ s}} = 12.6 \text{ rad/s}^2$$

(*b*)   The angle through which the engine turns is

$$\theta = \omega_0 t + \tfrac{1}{2}\alpha t^2 = (63 \text{ rad/s})(5 \text{ s}) + (\tfrac{1}{2})(12.6 \text{ rad/s}^2)(5 \text{ s})^2$$
$$= 472.5 \text{ rad}$$

Since there are $2\pi$ rad in 1 rev,

$$\theta = \frac{472.5 \text{ rad}}{2\pi \text{ rad/rev}} = 75.2 \text{ rev}$$

**SOLVED PROBLEM 10.8**

A phonograph turntable initially rotating at 3.5 rad/s makes three complete turns before coming to a stop. (*a*) What is its angular acceleration? (*b*) How much time does it take to come to a stop?

(*a*)   The angle in radians that corresponds to 3 rev is

$$\theta = (3 \text{ rev})(2\pi \text{ rad/rev}) = 6\pi \text{ rad}$$

From the formula $\omega^2 = \omega_0^2 + 2\alpha\theta$ we find

$$\alpha = \frac{\omega_f^2 - \omega_0^2}{2\theta} = \frac{0 - (3.5 \text{ rad/s})^2}{(2)(6\pi \text{ rad})} = -0.325 \text{ rad/s}^2$$

(*b*)   Since $\omega_f = \omega_0 + \alpha t$, we have here

$$t = \frac{\omega_f - \omega_0}{\alpha} = \frac{0 - 3.5 \text{ rad/s}}{-0.325 \text{ rad/s}^2} = 10.8 \text{ s}$$

## MOMENT OF INERTIA

The rotational analog of mass is a quantity called *moment of inertia*. The greater the moment of inertia of a body, the greater its resistance to a change in its angular velocity. The value of the moment of inertia $I$ of a body about a particular axis of rotation depends not only upon the body's mass but also upon how the mass is distributed about the axis.

Let us imagine a rigid body divided into a great many small particles whose masses are $m_1$, $m_2$, $m_3$, ... and whose distances from the axis of rotation are respectively $r_1$, $r_2$, $r_3$, ... (Fig. 10-2). The moment of inertia of this body is given by

$$I = m_1 r_1^2 + m_2 r_2^2 + m_3 r_3^2 + \cdots = \Sigma \, mr^2$$

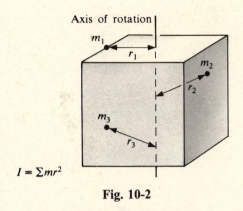

$$I = \Sigma mr^2$$

**Fig. 10-2**

where the symbol $\Sigma$ (Greek capital letter *sigma*) means "sum of" as before. The farther a particle is from the axis of rotation, the more it contributes to the moment of inertia. The units of $I$ are kg·m$^2$ and slug·ft$^2$. Some examples of moments of inertia of bodies of mass $M$ are shown in Fig. 10-3.

## TORQUE

The *torque* $\tau$ (Greek letter *tau*) exerted by a force on a body is a measure of its effectiveness in turning the body about a certain pivot point. The *moment arm* of a force **F** about a pivot point $O$ is the perpendicular distance $L$ between the line of action of the force and $O$ (Fig. 10-4). The torque $\tau$ exerted by the force about $O$ has the magnitude

$$\tau = FL$$

$$\text{Torque} = (\text{force})(\text{moment arm})$$

The torque exerted by a force is also known as the *moment* of the force. A force whose line of action passes through $O$ produces no torque about $O$ because its moment arm is zero.

Torque plays the same role in rotational motion that force plays in linear motion. A net force $F$ acting on a body of mass $m$ causes it to undergo the linear acceleration $a$ in accordance with Newton's

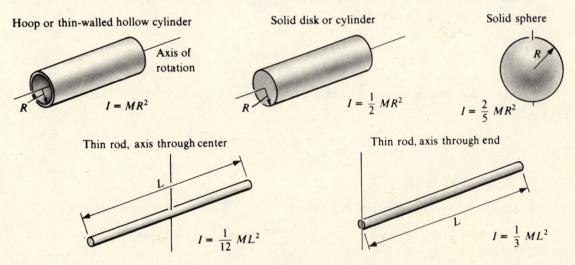

**Fig. 10-3**

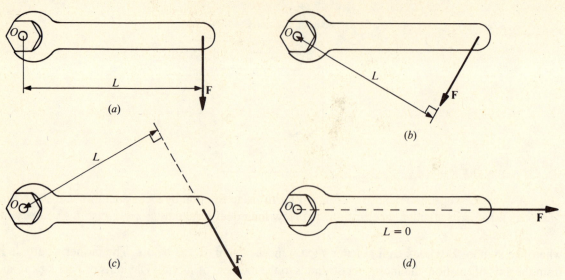

**Fig. 10-4** Four directions along which a force **F** can be applied to a wrench. In (*a*) the moment arm *L* is longest, hence the torque $\tau = FL$ is a maximum. In (*d*) the line of action of **F** passes through the pivot *O*, so $L = 0$ and $\tau = 0$.

second law of motion $F = ma$. Similarly a net torque $\tau$ acting on a body of moment of inertia $I$ causes it to undergo the angular acceleration $\alpha$ (in rad/s$^2$) in accordance with the formula

$$\tau = I\alpha$$

Torque = (moment of inertia)(angular acceleration)

In the SI system, the unit of torque is the newton-meter (N·m); in the British system, it is the pound-foot (lb·ft).

## SOLVED PROBLEM 10.9

The *radius of gyration* of a body about a particular axis is the distance from that axis to a point at which the body's entire mass may be considered to be concentrated. Thus the moment of inertia of a body of mass $M$ and radius of gyration $k$ is $I = Mk^2$. (*a*) The radius of gyration of a hollow sphere of radius $R$ and mass $M$ is $k = \sqrt{\frac{2}{3}}R$. What is its moment of inertia? (*b*) Find the radius of gyration of a solid sphere.

(*a*)
$$I = Mk^2 = M\left(\sqrt{\frac{2}{3}}R\right)^2 = \tfrac{2}{3}MR^2$$

(*b*)   From Fig. 10-3 the moment of inertia of a solid sphere is $I = \tfrac{2}{5}MR^2$. Hence its radius of gyration is

$$k = \sqrt{\frac{I}{M}} = \sqrt{\frac{2}{5}\frac{MR^2}{M}} = \sqrt{\frac{2}{5}}R$$

## SOLVED PROBLEM 10.10

The radius of gyration of a 200-lb flywheel is 1 ft. Find its moment of inertia.

The mass of the flywheel is $M = w/g = 200 \text{ lb}/(32 \text{ ft/s}^2) = 6.25$ slugs, and so its moment of inertia is

$$I = Mk^2 = (6.25 \text{ slugs})(1 \text{ ft})^2 = 6.25 \text{ slug·ft}^2$$

**SOLVED PROBLEM 10.11**

The starting cord of an outboard motor is wound around a pulley 18 cm in diameter that is attached to the motor's crankshaft. How much torque is applied to the crankshaft when the cord is pulled with a force of 50 N?

Here the moment arm of the force is the pulley's radius of 9 cm = 0.09 m. The torque is therefore

$$\tau = Fr = (50\text{ N})(0.09\text{ m}) = 4.5\text{ N·m} \quad.$$

**SOLVED PROBLEM 10.12**

A flywheel whose moment of inertia is 6 kg·m$^2$ is acted upon by a constant torque of 50 N·m. (a) What is its angular acceleration? (b) How long does it take to go from rest to a velocity of 90 rad/s?

(a)
$$\alpha = \frac{\tau}{I} = \frac{50\text{ N·m}}{6\text{ kg·m}^2} = 8.33\text{ rad/s}^2$$

(b)
$$t = \frac{\omega_f - \omega_0}{\alpha} = \frac{90\text{ rad/s}}{8.33\text{ rad/s}^2} = 10.8\text{ s}$$

**SOLVED PROBLEM 10.13**

The winding drum of an elevator is 4 ft in diameter. (a) At how many revolutions per minute should the drum rotate in order to raise the cab at 500 ft/min? (b) If the total load is 2 tons, how much torque is required?

(a)  Since

$$v = \frac{500\text{ ft/min}}{60\text{ s/min}} = 8.33\text{ ft/s}$$

the angular velocity of the drum must be

$$\omega = \frac{v}{r} = \frac{8.33\text{ ft/s}}{2\text{ ft}} = 4.17\text{ rad/s}$$

which is
$$\omega = \frac{4.17\text{ rad/s}}{(0.105\text{ rad/s})/(\text{rev/min})} = 39.7\text{ rev/min}$$

(b)  Since 1 ton = 2000 lb, the required torque is

$$\tau = Fr = (4000\text{ lb})(2\text{ ft}) = 8000\text{ lb·ft}$$

## ROTATIONAL ENERGY AND WORK

The kinetic energy of a body of moment of inertia $I$ whose angular velocity is $\omega$ (in rad/s) is

$$KE = \tfrac{1}{2}I\omega^2$$
$$\text{Kinetic energy} = (\tfrac{1}{2})(\text{moment of inertia})(\text{angular velocity})^2$$

The work done by a constant torque $\tau$ that acts on a body while it experiences the angular displacement $\theta$ rad is

$$W = \tau\theta$$
$$\text{Work} = (\text{torque})(\text{angular displacement})$$

The rate at which work is being done when a torque $\tau$ acts on a body that rotates at the constant angular velocity $\omega$ rad/s is

$$P = \tau\omega$$
$$\text{Power} = (\text{torque})(\text{angular velocity})$$

**SOLVED PROBLEM 10.14**

(a) Find the moment of inertia of a 1-kg phonograph turntable that is 34 cm in diameter. (b) What is its kinetic energy when it rotates at 45 rev/min?

(a)    The radius of the turntable is $R = 17$ cm $= 0.17$ m. Assuming that the turntable is a solid disk, we find for its moment of inertia

$$I = \tfrac{1}{2}MR^2 = (\tfrac{1}{2})(1 \text{ kg})(0.17 \text{ m})^2 = 0.0145 \text{ kg·m}^2$$

(b)    The angular velocity of the turntable is

$$\omega = (45 \text{ rev/min})\left(0.105 \, \frac{\text{rad/s}}{\text{rev/min}}\right) = 4.73 \text{ rad/s}$$

and so its kinetic energy is

$$\text{KE} = \tfrac{1}{2}I\omega^2 = (\tfrac{1}{2})(0.0145 \text{ kg·m}^2)(4.73 \text{ rad/s})^2 = 0.162 \text{ J}$$

(Note that 1 kg·m$^2$/s$^2$ = 1 N·m = 1 J.)

**SOLVED PROBLEM 10.15**

A flywheel of moment of inertia 10 slug·ft$^2$ is rotating at 90 rad/s. (a) What constant torque is required to slow it down to 40 rad/s in 20 s? (b) What is the angular displacement of the flywheel while it is being slowed down? (c) How much kinetic energy is lost by the flywheel?

(a)                        $$\alpha = \frac{\omega_f - \omega_0}{t} = \frac{(40 - 90) \text{ rad/s}}{20 \text{ s}} = -2.5 \text{ rad/s}^2$$

The minus sign means that $\omega$ is decreasing. The required torque is

$$\tau = I\alpha = (10 \text{ slug·ft}^2)(-2.5 \text{ rad/s}^2) = -25 \text{ lb·ft}$$

(b)    Either $\theta = \omega_0 t + \tfrac{1}{2}\alpha t^2$ or $\omega_f^2 = \omega_0^2 + 2\alpha\theta$ can be used to find $\theta$. From the first formula,

$$\theta = \omega_0 t + \tfrac{1}{2}\alpha t^2 = (90 \text{ rad/s})(20 \text{ s}) - \tfrac{1}{2}(2.5 \text{ rad/s}^2)(20 \text{ s})^2 = 1300 \text{ rad}$$

(c)    The KE lost by the flywheel appears as work done against the applied torque. Hence

$$\Delta \text{KE} = W = \tau\theta = -(25 \text{ lb·ft})(1300 \text{ rad}) = -3.25 \times 10^4 \text{ ft·lb}$$

As a check, we can find $\Delta$KE directly:

$$\Delta \text{KE} = \text{KE}_f - \text{KE}_0 = \tfrac{1}{2}I(\omega_f^2 - \omega_0^2)$$
$$= \tfrac{1}{2}(10 \text{ slug·ft}^2)[(40 \text{ rad/s})^2 - (90 \text{ rad/s})^2] = -3.25 \times 10^2 \text{ ft·lb}$$

**SOLVED PROBLEM 10.16**

A solid cylinder rolls from rest down an inclined plane 1.2 m high without slipping. What is its linear velocity at the foot of the plane?

The potential energy of the cylinder at the top of the plane is PE = $mgh$. At the foot of the plane it has both the kinetic energy of translation $\tfrac{1}{2}mv^2$ and the kinetic energy of rotation $\tfrac{1}{2}I\omega^2$, and its total KE then is equal to its initial PE:

$$\text{PE} = \text{KE}$$
$$mgh = \tfrac{1}{2}mv^2 + \tfrac{1}{2}I\omega^2$$

Because the cylinder rolls without slipping, its linear and angular velocities are related by $v = \omega R$, so $\omega = v/R$. The moment of inertia of a solid cylinder is $I = \frac{1}{2}mR^2$. Hence

$$\tfrac{1}{2}I\omega^2 = \tfrac{1}{2}(\tfrac{1}{2}mR^2)\left(\frac{v}{R}\right)^2 = \tfrac{1}{4}mv^2$$

and the energy equation becomes

$$mgh = \tfrac{1}{2}mv^2 + \tfrac{1}{4}mv^2 = \tfrac{3}{4}mv^2$$

Thus two-thirds of the cylinder's KE resides in its translational motion and one-third in its rotational motion. Solving for $v$ yields

$$v = \sqrt{\frac{4}{3}gh} = \sqrt{\frac{4}{3}(9.8 \text{ m/s}^2)(1.2 \text{ m})} = 3.96 \text{ m/s}$$

**SOLVED PROBLEM 10.17**

Find the minimum horsepower needed for the motor of the elevator of Prob. 10.13.

The required power can be calculated both from $P = Fv$ and from $P = \tau\omega$:

$$P = Fv = (4000 \text{ lb})(8.33 \text{ ft/s}) = 3.34 \times 10^4 \text{ ft·lb/s}$$
$$P = \tau\omega = (8000 \text{ lb·ft})(4.17 \text{ rad/s}) = 3.34 \times 10^4 \text{ ft·lb/s}$$

Since 1 hp = 550 ft·lb/s,

$$P = \frac{3.34 \times 10^4 \text{ ft·lb/s}}{(550 \text{ ft·lb/s})/\text{hp}} = 60.6 \text{ hp}$$

**SOLVED PROBLEM 10.18**

A marine diesel engine develops 90 kW at 2000 rev/min. (a) How much torque can it exert at this velocity? (b) The engine delivers its power through a spur gear 23 cm in radius. If two teeth of the gear transmit torque to another gear at a certain moment, find the force on each gear tooth.

(a)   The angular velocity here is

$$\omega = (2000 \text{ rev/min})\left(0.105 \ \frac{\text{rad/s}}{\text{rev/min}}\right) = 210 \text{ rad/s}$$

Hence the torque is

$$\tau = \frac{P}{\omega} = \frac{90,000 \text{ W}}{210 \text{ rad/s}} = 429 \text{ N·m}$$

(b)   Since $\tau = Fr$, the total force is

$$F = \frac{\tau}{r} = \frac{429 \text{ N·m}}{0.23 \text{ m}} = 1863 \text{ N}$$

Each of the two gear teeth in contact with the driven gear exerts a force of half this amount, or 932 N.

**SOLVED PROBLEM 10.19**

An air compressor is powered by a 200 rad/s electric motor using a V-belt drive. The motor pulley is 8 cm in radius, and the tension in the V-belt is 135 N on one side and 45 N on the other. Find the power of the motor.

The torque exerted by the motor on the belt is, since the net force on the belt is the difference between the two tensions,

$$\tau = Fr = (135 \text{ N} - 45 \text{ N})(0.08 \text{ m}) = 7.2 \text{ N·m}$$

The power output of the motor is therefore

$$P = \tau\omega = (7.2 \text{ N·m})(200 \text{ rad/s}) = 1440 \text{ W} = 1.44 \text{ kW}$$

## ANGULAR MOMENTUM

The equivalent of linear momentum in rotational motion is *angular momentum*. The angular momentum **L** of a rotating body has the magnitude

$$L = I\omega$$

Angular momentum = (moment of inertia)(angular velocity)

The greater the angular momentum of a spinning object, such as a top, the greater its tendency to continue to spin.

Like linear momentum, angular momentum is a vector quantity with direction as well as magnitude. The direction of the angular momentum of a rotating body is given by the right-hand rule (Fig. 10-5): When the fingers of the right hand are curled in the direction of rotation, the thumb points in the direction of **L**.

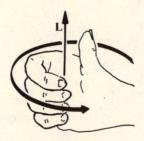

**Fig. 10-5**

According to the principle of *conservation of angular momentum*, the total angular momentum of a system of bodies remains constant in the absence of a net torque regardless of what happens within the system.

Skaters performing a spin make use of conservation of angular momentum. They begin to turn with arms and one leg outstretched and then bring them in close to the body to distribute the body's mass nearer to the axis of rotation, thereby decreasing the moment of inertia *I*. Since *L* must remain constant, the angular velocity $\omega$ increases.

Because angular momentum is a vector quantity, its conservation implies that the direction of the axis of rotation tends to remain unchanged. For this reason a spinning top stays upright, whereas a stationary one falls over immediately.

Table 10.1 compares linear and angular quantities.

## SOLVED PROBLEM 10.20

Why do all helicopters have two propellers?

If a single propeller were used, the helicopter itself would have to rotate in the opposite direction in order to conserve angular momentum.

**Table 10-1.  Comparison of Linear and Angular Quantities**

| Linear Quantity | | Angular Quantity | |
|---|---|---|---|
| Distance | $s = v_0 t + \frac{1}{2}at^2$ <br> $s = \left(\dfrac{v_0 + v_f}{2}\right)t$ | Angle | $\theta = \omega_0 t + \frac{1}{2}\alpha t^2$ <br> $\theta = \left(\dfrac{\omega_0 + \omega_f}{2}\right)t$ |
| Speed | $v = v_0 + at$ <br> $v^2 = v_0^2 + 2as$ | Angular speed | $\omega = \omega_0 + \alpha t$ <br> $\omega^2 = \omega_0^2 + 2\alpha\theta$ |
| Acceleration | $a = \Delta v/\Delta t$ | Angular acceleration | $\alpha = \Delta\omega/\Delta t$ |
| Mass | $m$ | Moment of inertia | $I$ |
| Force | $F = ma$ | Torque | $\tau = I\alpha$ |
| Momentum | $p = mv$ | Angular momentum | $L = I\omega$ |
| Work | $W = Fs$ | Work | $W = \tau\theta$ |
| Power | $P = Fv$ | Power | $P = \tau\omega$ |
| Kinetic energy | $\text{KE} = \frac{1}{2}mv^2$ | Kinetic energy | $\text{KE} = \frac{1}{2}I\omega^2$ |

**SOLVED PROBLEM 10.21**

A skater is spinning at 1 rev/s with her arms outstretched so that her moment of inertia is 2.4 kg·m². She then pulls her arms to her sides, which reduces her moment of inertia to 1.2 kg·m². (*a*) What is her new angular velocity? (*b*) How much work did she have to perform to pull her arms in?

(*a*)  From conservation of angular momentum,

$$I_1\omega_1 = I_2\omega_2$$

$$\omega_2 = \frac{I_1}{I_2}\,\omega_1 = \left(\frac{2.4 \text{ kg·m}^2}{1.2 \text{ kg·m}^2}\right)(1 \text{ rev/s}) = 2 \text{ rev/s}$$

The skater spins twice as fast as before.

(*b*)  The work done equals the difference between the initial and final kinetic energies of rotation. To find the KE values, the rad/s must be used as the unit of angular velocity. We have

$$\omega_1 = (1 \text{ rev/s})(2\pi \text{ rad/rev}) = 2\pi \text{ rad/s}$$
$$\omega_2 = (2 \text{ rev/s})(2\pi \text{ rad/rev}) = 4\pi \text{ rad/s}$$

and so

$$\text{KE}_1 = \tfrac{1}{2}I_1\omega_1^2 = \tfrac{1}{2}(2.4 \text{ kg·m}^2)(2\pi \text{ rad/s})^2 = 47 \text{ J}$$
$$\text{KE}_2 = \tfrac{1}{2}I_2\omega_2^2 = \tfrac{1}{2}(1.2 \text{ kg·m}^2)(4\pi \text{ rad/s})^2 = 95 \text{ J}$$

The work done is therefore

$$W = \text{KE}_2 - \text{KE}_1 = 95 \text{ J} - 47 \text{ J} = 48 \text{ J}$$

# *Multiple-Choice Questions*

**10.1.**   A quarter of a circle contains

    (*a*)  $\pi/4$ rad    (*c*)  $\pi$ rad
    (*b*)  $\pi/2$ rad    (*d*)  $2\pi$ rad

**10.2.** An angle of $\pi/18$ rad is equivalent to

(a)  10°        (c)  20°
(b)  18°        (d)  36°

**10.3.** The linear velocity of the rim of a wheel 80 cm in diameter when the wheel turns at 90 rev/min is

(a)  $0.6\pi$ m/s        (c)  $1.5\pi$ m/s
(b)  $1.2\pi$ m/s        (d)  $2.4\pi$ m/s

**10.4.** In 30 s the crankshaft of a truck engine operating at 2400 rev/min turns through

(a)  382 rad        (c)  3770 rad
(b)  1200 rad        (d)  7540 rad

**10.5.** A pulley is uniformly accelerated from rest to an angular velocity of 30 rad/s in 8.0 s. The total angle through which the pulley turned during the acceleration is

(a)  60 rad        (c)  240 rad
(b)  120 rad        (d)  3600 rad

**10.6.** A motor takes 6.0 s to go from 150 to 50 rad/s at constant angular acceleration. The total angle through which the motor's shaft turned during the acceleration is

(a)  300 rad        (c)  1200 rad
(b)  600 rad        (d)  3600 rad

**10.7.** A solid iron cylinder $A$ rolls down a ramp, and an identical iron cylinder $B$ slides down the same ramp without friction.

(a)  $A$ reaches the bottom first.
(b)  $B$ reaches the bottom first.
(c)  $A$ and $B$ reach the bottom together.
(d)  Any of the above, depending on the angle of the ramp.

**10.8.** When the cylinders of Question 10.7 reach the bottom of the ramp,

(a)  the KE of $A$ is more than the KE of $B$
(b)  the KE of $B$ is more than the KE of $A$
(c)  $A$ and $B$ have the same KE
(d)  any of the above, depending on the angle of the ramp

**10.9.** A solid wooden disk rolls down a ramp. The center of the disk is then cut out, and the resulting doughnut rolls down the same ramp. At the bottom of the ramp the doughnut's velocity is

(a)  less than that of the solid disk
(b)  the same as that of the solid disk
(c)  greater than that of the solid disk
(d)  any of the above, depending on the angle of the ramp

**10.10.** A flywheel whose moment of inertia is 4.0 kg·m$^2$ is acted on by a torque of 50 N·m. Six seconds after starting from an angular velocity of 40 rad/s the flywheel will have turned through

(a)  225 rad        (c)  465 rad
(b)  315 rad        (d)  3053 rad

**10.11.** The KE of the flywheel of Question 10.10 will have increased by

    (*a*)   8.05 kJ    (*c*)   23.25 kJ

    (*b*)   11.25 kJ   (*d*)   26.45 kJ

**10.12.** The torque needed to bring a turbine whose moment of inertia is 60 slug·ft$^2$ to rest in 12 s from an initial angular velocity of 80 rad/s is

    (*a*)   9 lb·ft    (*c*)   200 lb·ft

    (*b*)   16 lb·ft   (*d*)   400 lb·ft

**10.13.** A 5.0-N force acts tangentially on the rim of a wheel of radius 36 cm. Starting from rest, the wheel makes 10 rev in 4.0 s. The moment of inertia of the wheel is

    (*a*)   0.11 kg·m$^2$    (*c*)   0.46 kg·m$^2$

    (*b*)   0.23 kg·m$^2$    (*d*)   1.44 kg·m$^2$

**10.14.** An engine turning at 3000 rev/min develops 75 kW. The torque on the engine's shaft

    (*a*)   is 2.6 N·m    (*c*)   is 239 N·m

    (*b*)   is 25 N·m    (*d*)   depends on the shaft radius

# Supplementary Problems

**10.1.** (*a*) Express 12.5 rev in radians. (*b*) Express 0.2 rad in revolutions. (*c*) Express $3\pi$ rad in revolutions.

**10.2.** (*a*) Express 32° in radians. (*b*) Express 4.8 rad in degrees. (*c*) Express $\pi/8$ rad in degrees.

**10.3.** The minute hand of a clock is 40 cm long. How many centimeters does its tip move in 25 min?

**10.4.** A drill bit $\frac{1}{2}$ in. in diameter is turning at 400 rev/min. (*a*) What is its angular velocity in radians per second? (*b*) What is the linear velocity in feet per second of a point on its circumference?

**10.5.** The teeth of a certain circular saw blade are supposed to move at 15 m/s. If the blade is 60 cm in radius, at how many revolutions per minute should it turn?

**10.6.** A car whose tires have radii of 50 cm travels at 20 km/h. What is the angular velocity of the tires?

**10.7.** The earth's radius is $6.4 \times 10^6$ m. (*a*) Through how many radians does the earth turn in 1 year? (*b*) How far does a point on the equator move in 1 year owing to this rotation?

**10.8.** A grindstone that is rotating at 2000 rev/min requires 50 s to come to a stop when its motor is switched off. (*a*) Find the angular acceleration of the grindstone. (*b*) How many radians does it turn through before stopping? (*c*) Through how many revolutions?

**10.9.** A wheel rotating at 20 rev/s is brought to rest by a constant torque in 12 s. How many revolutions does it make in this time?

**10.10.** The propeller of a ship makes 300 rev while its speed increases from 200 to 500 rev/min. (*a*) What is its angular acceleration? (*b*) How much time did the increase in speed require?

**10.11.** The mass and radius of the earth are, respectively, $6 \times 10^{24}$ kg and $6.4 \times 10^6$ m. Assuming that it is a sphere of uniform density (which is not the case, since the earth's metallic core has a greater density than the mantle of rock around it), find the moment of inertia of the earth.

**10.12.** The moment of inertia of a 45-kg grindstone is 5 kg·m². Find its radius of gyration.

**10.13.** A solid cylinder and a hollow cylinder of the same mass and diameter, both initially at rest, roll down the same inclined plane without slipping. (*a*) Which reaches the bottom first? (*b*) How do their kinetic energies at the bottom compare?

**10.14.** A bowling ball rolling at 8 m/s begins to move up an inclined plane. What height does it reach?

**10.15.** A certain engine develops 40 hp at 1000 rev/min and 65 hp at 1600 rev/min. At which angular velocity does it produce the greatest torque?

**10.16.** The alternator on a truck engine produces 350 W of electric power when it rotates at 4000 rev/min. If it is 95 percent efficient, what should the difference (in newtons) between the tensions in the tight and slack parts of its V-belt drive be? The diameter of the alternator's pulley is 10 cm.

**10.17.** The anchor windlass of a boat is required to pull a load of 400 lb at 2 ft/s. (*a*) How many horsepower are required? (*b*) If the windlass drum is 8 in. in diameter, at how many revolutions per minute should it turn? (*c*) What torque is developed by the drum?

**10.18.** A constant frictional torque of 200 N·m is applied to a turbine initially rotating at 120 rad/s, and it comes to a stop in 80 s. What is the turbine's moment of inertia?

**10.19.** A wheel whose moment of inertia is 2 kg·m² has an initial angular velocity of 50 rad/s. (*a*) If a constant torque of 10 N·m acts on the wheel, how long does it take to be accelerated to 80 rad/s? (*b*) By how much does its kinetic energy increase?

**10.20.** A wheel whose moment of inertia is 0.4 slug·ft² is rotating at 1500 rev/min. (*a*) What constant torque is required to increase its angular velocity to 2000 rev/min? (*b*) How many turns does the wheel make while it is being accelerated? (*c*) How much work is done on the wheel?

**10.21.** A disk whose moment of inertia is 1 kg·m² is rotating at 100 rad/s. This disk is pressed against a similar disk that is able to rotate freely but is initially at rest. The two disks stick together and rotate as a unit. (*a*) Find the final angular velocity of the combination. (*b*) How much energy was lost to friction when the disks were brought together?

## Answers to Multiple-Choice Questions

**10.1.** (*b*)    **10.5.** (*b*)

**10.2.** (*a*)    **10.6.** (*b*)

**10.3.** (*b*)    **10.7.** (*b*)

**10.4.** (*d*)    **10.8.** (*c*)

**10.9.** (a)    **10.12.** (d)

**10.10.** (c)    **10.13.** (b)

**10.11.** (c)    **10.14.** (c)

# Answers to Supplementary Problems

**10.1.**    (a)  78.5 rad      (b)  0.032 rev      (c)  1.5 rev

**10.2.**    (a)  0.558 rad      (b)  275°      (c)  22.5°

**10.3.**    105 cm

**10.4**    (a)  42 rad/s      (b)  0.875 ft/s

**10.5.**    238 rev/min

**10.6.**    11.1 rad/s

**10.7.**    (a)  2293 rad      (b)  $1.47 \times 10^{10}$ m

**10.8.**    (a)  $-4.2$ rad/s$^2$      (b)  5250 rad      (c)  836 rev

**10.9.**    120 rev

**10.10.**    (a)  0.614 rad/s$^2$      (b)  51.3 s

**10.11.**    $9.8 \times 10^{37}$ kg·m$^2$

**10.12.**    33 cm

**10.13.**    (a) The solid cylinder reaches the bottom first because less of its total kinetic energy is KE of rotation. (b) The total kinetic energies of the cylinders are the same since they had the same potential energies at the top of the plane.

**10.14.**    4.57 m

**10.15.**    1600 rev/min

**10.16.**    17.5 N

**10.17.**    (a)  1.45 hp      (b)  57 rev/min      (c)  133 lb·ft

**10.18.**   133 kg·m$^2$

**10.19.**   (*a*)   6 s      (*b*)   3900 J

**10.20.**   (*a*)   2.63 lb·ft      (*b*)   234 turns      (*c*)   3.86 × 10$^3$ ft·lb

**10.21.**   (*a*)   50 rad/s      (*b*)   2500 J

# Chapter 11

# Equilibrium

## TRANSLATIONAL EQUILIBRIUM

A body is in *translational equilibrium* when no net force acts on it. Such a body is not accelerated, and it remains either at rest or in motion at constant velocity along a straight line, whichever its initial state was.

A body in translational equilibrium may have forces acting on it, but they must be such that their vector sum is zero. Thus the condition for the translational equilibrium of a body may be written

$$\Sigma \mathbf{F} = 0$$

where (as before) the symbol $\Sigma$ (Greek capital letter *sigma*) means "sum of" and $\mathbf{F}$ refers to the various forces that act on the body.

The procedure for working out a problem that involves translational equilibrium has three steps:

1. Draw a diagram of the forces that act *on* the body. As mentioned in Chapter 5, this is called a *free-body diagram*.

2. Choose a set of coordinate axes and resolve the various forces into their components along these axes.

3. Set the sum of the force components along each axis equal to zero so that

$$\text{Sum of } x \text{ force components} = \Sigma F_x = 0$$
$$\text{Sum of } y \text{ force components} = \Sigma F_y = 0$$
$$\text{Sum of } z \text{ force components} = \Sigma F_z = 0$$

In this way the vector equation $\Sigma \mathbf{F} = 0$ is replaced by three scalar equations. Then solve the resulting equations for the unknown quantities.

A proper choice of directions for the axes often simplifies the calculations. When all the forces lie in a plane, for instance, the coordinate system can be chosen so that the $x$ and $y$ axes lie in the plane; then the two equations $\Sigma F_x = 0$ and $\Sigma F_y = 0$ are enough to express the condition for translational equilibrium.

## SOLVED PROBLEM 11.1

A 100-N box is suspended from two ropes that each make an angle of 40° with the vertical. Find the tension in each rope.

The forces that act on the box are shown in the free-body diagram of Fig. 11-1($a$). They are

$$\mathbf{T}_1 = \text{tension in left-hand rope}$$
$$\mathbf{T}_2 = \text{tension in right-hand rope}$$
$$\mathbf{w} = \text{weight of box, which acts downward}$$

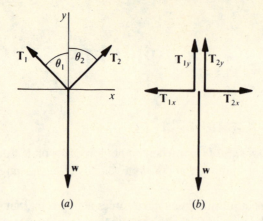

**Fig. 11-1**

Since the forces all lie in a plane, we need only $x$ and $y$ axes. In Fig. 11-1($b$) the forces are resolved into their $x$ and $y$ components, whose magnitudes are as follows:

$$T_{1x} = -T_1 \sin \theta_1 = -T_1 \sin 40° = -0.643T_1$$
$$T_{1y} = T_1 \cos \theta_1 = T_1 \cos 40° = 0.766T_1$$
$$T_{2x} = T_2 \sin \theta_2 = T_2 \sin 40° = 0.643T_2$$
$$T_{2y} = T_2 \cos \theta_2 = T_2 \cos 40° = 0.766T_2$$
$$w = -100 \text{ N}$$

Because $\mathbf{T}_{1x}$ and $\mathbf{w}$ are, respectively, in the $-x$ and $-y$ directions, both have negative magnitudes.

Now we are ready for step 3. First we add the $x$ components of the forces and set the sum equal to zero. This yields

$$\Sigma F_x = T_{1x} + T_{2x} = 0$$
$$-0.643T_1 + 0.643T_2 = 0$$
$$T_1 = T_2 = T$$

Evidently the tensions in the two ropes are equal. Next we do the same for the $y$ components:

$$\Sigma F_y = T_{1y} + T_{2y} + w = 0$$
$$0.766T_1 + 0.766T_2 - 100 \text{ N} = 0$$
$$0.766(T_1 + T_2) = 100 \text{ N}$$
$$T_1 + T_2 = \frac{100 \text{ N}}{0.766} = 130.5 \text{ N}$$

Since $T_1 = T_2 = T$,

$$T_1 + T_2 = 2T = 130.5 \text{ N}$$
$$T = 65 \text{ N}$$

The tension in each rope is 65 N.

## SOLVED PROBLEM 11.2

A 500-kg load is suspended from the end of a horizontal boom, as in Fig. 11-2($a$). The angle between the boom and the cable supporting its end is 45°. Assuming that the boom's mass can be neglected compared with that of the load, find ($a$) the tension in the cable and ($b$) the inward force the boom exerts on the wall.

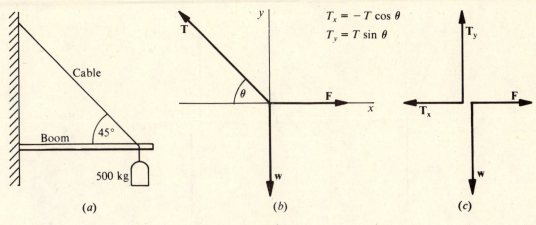

**Fig. 11-2**

(a)    The three forces that act on the end of the boom are the weight **w** of the load, the tension **T** in the cable, and the outward force **F** exerted by the boom. A free-body diagram of these forces is shown in Fig. 11-2(b). The $x$ and $y$ components of these forces have the magnitudes

$$T_x = -T \cos \theta = -T \cos 45° = -0.707T$$
$$T_y = T \sin \theta = T \sin 45° = 0.707T$$
$$w = -mg = -(500 \text{ kg})(9.8 \text{ m/s}^2) = -4900 \text{ N}$$
$$F = ?$$

The condition for translational equilibrium in the $y$ (vertical) direction yields

$$\Sigma F_y = T_y + w = 0$$
$$0.707T - 4900 \text{ N} = 0$$
$$T = \frac{4900 \text{ N}}{0.707} = 6930 \text{ N}$$

(b)    To find the inward force the boom exerts on the wall, we start with the condition for equilibrium in the $x$ (horizontal) direction:

$$\Sigma F_x = T_x + F = 0$$
$$-0.707T + F = 0$$
$$F = 0.707T = (0.707)(6930 \text{ N}) = 4900 \text{ N}$$

The inward force on the wall must have the same magnitude as the outward force on the load; hence, the inward force is also equal to 4900 N.

## SOLVED PROBLEM 11.3

A 50-lb box is suspended by a rope from the ceiling. If a horizontal force of 20 lb is applied to the box, what angle will the rope make with the vertical?

A free-body diagram of the forces acting on the box is shown in Fig. 11-3(a), and the forces are resolved into components in Fig. 11-3(b). The $x$ and $y$ components of the forces are

$$T_x = -T \sin \theta \qquad T_y = T \cos \theta$$
$$F = 20 \text{ lb} \qquad w = -50 \text{ lb}$$

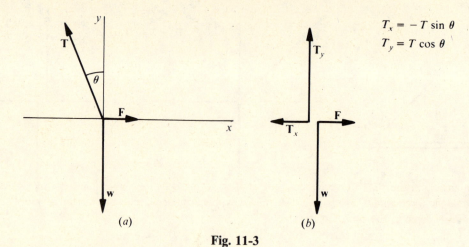

$$T_x = -T \sin \theta$$
$$T_y = T \cos \theta$$

(a)                                          (b)

**Fig. 11-3**

Applying the conditions for equilibrium yields

$$\Sigma F_x = T_x + F = 0 \qquad -T \sin \theta + 20 \text{ lb} = 0 \qquad \sin \theta = \frac{20 \text{ lb}}{T}$$

$$\Sigma F_y = T_y + w = 0 \qquad T \cos \theta - 50 \text{ lb} = 0 \qquad \cos \theta = \frac{50 \text{ lb}}{T}$$

If we divide the expression for sin $\theta$ by that for cos $\theta$, the $T$'s cancel to give

$$\frac{\sin \theta}{\cos \theta} = \tan \theta = \frac{20 \text{ lb}/T}{50 \text{ lb}/T} = \frac{20 \text{ lb}}{50 \text{ lb}} = 0.40$$

The angle whose tangent is nearest to 0.40 is 22°, and so $\theta = 22°$.

### SOLVED PROBLEM 11.4

To move a heavy crate across a floor, one end of a rope is tied to it and the other end is tied to a wall 10 m away. When a force of 400 N is applied to the midpoint of the rope, the rope stretches so that the midpoint moves to the side by 60 cm. What is the force on the crate?

The first step is to find the angle $\theta$ between either part of the rope and a straight line between the crate and the point of attachment of the rope to the wall. With the help of Fig. 11-4(a) we find that

$$\tan \theta = \frac{0.6 \text{ m}}{5 \text{ m}} = 0.12 \qquad \theta = 6.8°$$

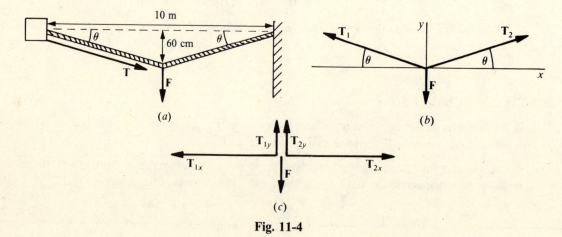

(a)                    (b)

(c)

**Fig. 11-4**

Figure 11-4($b$) is a free-body diagram of the forces acting on the midpoint of the rope; $\mathbf{T}_1$ and $\mathbf{T}_2$ are the tensions in the two parts of the rope, and $\mathbf{F} = -400$ N is the applied force. These forces are resolved in Fig. 11-4($c$). Since $T_1 = T_2 = T$,

$$T_{1y} = T_{2y} = T \sin \theta = T \sin 6.8° = 0.118T$$

At equilibrium

$$\Sigma F_y = T_{1y} + T_{2y} + F = 0$$
$$(2)(0.118T) - 400 \text{ N} = 0$$
$$T = 1695 \text{ N}$$

The tension in the rope provides the force applied to the crate. We note that the force on the crate exceeds the force applied to the rope, which is why this arrangement is used instead of simply applying the 400-N force to the crate directly.


## SOLVED PROBLEM 11.5

A boom hinged at the base of a vertical mast is used to lift a weight of 8000 N, as in Fig. 11-5($a$). Find the tension in the cable from the top of the mast to the top of the boom.

We begin by finding angles $\theta_1$ and $\theta_2$. Since the sum of the interior angles of a triangle is always 180°,

$$\theta_1 = 180° - 40° - 65° = 75°$$

Because the mast is vertical, it makes a 90° angle with the ground, and so

$$\theta_2 = 90° - 40° = 50°$$

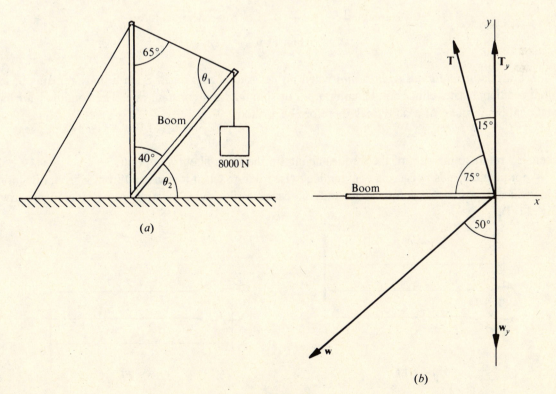

(a)

(b)

**Fig. 11-5**

We now let the $x$ axis be in the direction of the boom with the $y$ axis perpendicular to it, as in Fig. 11-5($b$). Since the boom is rigid, we need consider only the translational equilibrium of its upper end in the $y$ direction. The angle between **T** and the $y$ axis is $90° - 75° = 15°$, and so

$$T_y = T \cos 15° = 0.966T$$

The component of the weight **w** in the $y$ direction is

$$w_y = -(8000 \text{ N})(\cos 50°) = -5142 \text{ N}$$

At equilibrium

$$\Sigma F_y = T_y + w_y = 0 \qquad 0.966T - 5142 \text{ N} = 0 \qquad T = 5323 \text{ N}$$

## ROTATIONAL EQUILIBRIUM

When the lines of action of the forces that act on a body in translational equilibrium intersect at a common point, they have no tendency to turn the body. Such forces are said to be *concurrent*. When the lines of action do not intersect, the forces are *nonconcurrent* and exert a net torque that acts to turn the body even through the resultant of the forces is zero (Fig. 11-6).

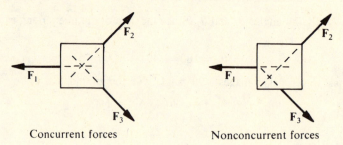

Concurrent forces                         Nonconcurrent forces

**Fig. 11-6**

A body is in *rotational equilibrium* when no net torque acts on it. Such a body remains in its initial rotational state, either not spinning at all or spinning at a constant rate. The condition for the rotational equilibrium of a body may therefore be written

$$\Sigma \tau = 0$$

where $\Sigma \tau$ refers to the sum of the torques acting on the body about any point.

A torque that tends to cause a counterclockwise rotation when it is viewed from a given direction is considered positive; a torque that tends to cause a clockwise rotation is considered negative (Fig. 11-7).

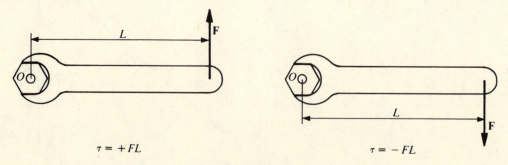

$\tau = +FL$                                      $\tau = -FL$

**Fig. 11-7**

To investigate the rotational equilibrium of a body, any convenient point may be used as the pivot point for calculating torques; if the sum of the torques on a body in translational equilibrium is zero about some point, it is zero about any other point.

## CENTER OF GRAVITY

The *center of gravity* of a body is that point at which the body's entire weight can be regarded as being concentrated. A body can be suspended in any orientation from its center of gravity without tending to rotate. In analyzing the equilibrium of a body, its weight can be considered as a downward force acting from its center of gravity.

### SOLVED PROBLEM 11.6

(*a*) Under what circumstances is it necessary to consider torques in analyzing an equilibrium situation? (*b*) About what point should torques be calculated when this is necessary?

(*a*)   Torques must be considered when the various forces that act on the body are nonconcurrent, that is, when their lines of action do not intersect at a common point.

(*b*)   Torques may be calculated about any point whatever for the purpose of determining the equilibrium of the body. Hence it makes sense to use a point that minimizes the labor involved, which usually is the point through which pass the maximum number of lines of action of the various forces; this is because a force whose line of action passes through a point exerts no torque about that point.

### SOLVED PROBLEM 11.7

A beam 3 m long has a weight of 200 N at one end and another weight of 80 N at the other end. The weight of the beam itself is negligible. Find the balance point of the beam.

When the beam is supported at its balance point, the torques of the two weights cancel, and the beam has no tendency to rotate. The supporting force **F** exerts no torque since it acts through the balance point. If the balance point is the distance $x$ from the 200-N weight, as in Fig. 11-8, it is the distance $3 \text{ m} - x$ from the 80-N weight. Since the beam is horizontal, the moment arms of the weights are, respectively, $x$ and $3 \text{ m} - x$, and the torques the weights exert about the balance point $O$ are

$$\tau_1 = w_1 L_1 = 200x \text{ N} \qquad \tau_2 = -w_2 L_2 = -80(3 \text{ m} - x) \text{ N}$$

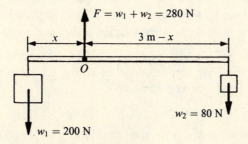

**Fig. 11-8**

The torque $\tau_1$ is positive because it tends to cause a counterclockwise rotation; the torque $\tau_2$ is negative because it tends to cause a clockwise rotation. The condition for rotational equilibrium yields

$$\Sigma\tau = \tau_1 + \tau_2 = 0$$
$$200x \text{ N} - 80(3 \text{ m} - x) \text{ N} = 0$$
$$200x \text{ N} = 240 \text{ N·m} - 80x \text{ N}$$
$$280x \text{ N} = 240 \text{ N·m}$$
$$x = 0.86 \text{ m} = 86 \text{ cm}$$

## SOLVED PROBLEM 11.8

A 4-m wooden platform weighing 160 N is suspended from the roof of a house by ropes attached to its ends. A painter weighing 640 N stands 1.2 m from the left-hand end of the platform. Find the tension in each of the ropes.

With the left-hand end of the platform as the pivot point (Fig. 11-9), the condition for rotational equilibrium yields

$$\Sigma\tau = -w_1L_1 - w_2L_2 + T_2L_3 = 0$$
$$-(640 \text{ N})(1.2 \text{ m}) - (160 \text{ N})(2 \text{ m}) + 4T_2 \text{ m} = 0$$
$$4T_2 \text{ m} = 768 \text{ N·m} + 320 \text{ N·m} = 1088 \text{ N·m}$$
$$T_2 = 272 \text{ N}$$

To find $T_1$ we proceed as follows:

$$T_1 + T_2 = w_1 + w_2$$
$$T_1 = 640 \text{ N} + 160 \text{ N} - 272 \text{ N} = 528 \text{ N}$$

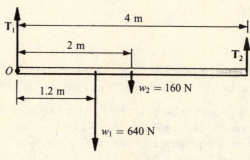

**Fig. 11-9**

## SOLVED PROBLEM 11.9

A *cantilever* is a beam that projects beyond its supports, such as a diving board. Figure 11-10 shows a 30-kg diving board 3.6 m long that has a 50-kg woman standing at its end. The board's supports are 1.0 m apart. Find the force that each support exerts.

First we calculate the downward force $F_1$ on the left-hand support. The easiest way to do this is to calculate torques about the other support, since we do not yet know the value of $F_2$. If the board is uniform, its center of gravity is at its center, and so

$$x_1 = 1.0 \text{ m} \qquad w_1 = m_1g = (30 \text{ kg})(9.8 \text{ m/s}^2) = 294 \text{ N}$$
$$x_2 = 0.8 \text{ m} \qquad w_2 = m_2g = (50 \text{ kg})(9.8 \text{ m/s}^2) = 490 \text{ N}$$
$$x_3 = 2.8 \text{ m}$$

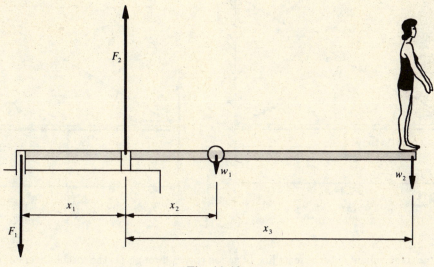

**Fig. 11-10**

The three torques that act about the right-hand support are

$$\tau_1 = +F_1 x_1 = +(F_1)(1.0 \text{ m}) = 1.0 F_1 \text{ m}$$
$$\tau_2 = -w_1 x_2 = -(294 \text{ N})(0.8 \text{ m}) = -235 \text{ N·m}$$
$$\tau_3 = -w_2 x_3 = -(490 \text{ N})(2.8 \text{ m}) = -1372 \text{ N·m}$$

The condition for rotational equilibrium then gives

$$\Sigma\tau = \tau_1 + \tau_2 + \tau_3 = 0$$
$$1.0 F_1 \text{ m} - 235 \text{ N·m} - 1372 \text{ N·m} = 0$$
$$F_1 = \frac{1607 \text{ N·m}}{1.0 \text{ m}} = 1607 \text{ N}$$

To find $F_2$ the upward force exerted by the right-hand support, we consider the translational equilibrium of the system of board plus woman:

$$\Sigma F = -F_1 + F_2 - w_1 - w_2$$
$$F_2 = F_1 + w_1 + w_2 = 1607 \text{ N} + 294 \text{ N} + 490 \text{ N} = 2391 \text{ N}$$

## SOLVED PROBLEM 11.10

A horizontal boom 2.4 m long is attached to a wall at its inner end and is supported at its outer end by a cable that makes an angle of 30° with the boom. The boom weighs 200 N, and a load of 1500 N is attached to its outer end. Find (*a*) the tension in the cable and (*b*) the compression force in the boom.

(*a*)  Four forces act on the boom, as in Fig. 11-11: the load $\mathbf{w}_1$ at its outer end, the boom's own weight $\mathbf{w}_2$ that acts from its center, the tension $\mathbf{T}$ in the cable, and the force $\mathbf{F}$ that the wall exerts on the inner end of the boom. We can disregard $\mathbf{F}$ by calculating torques about the inner end of the boom. It simplifies matters to use the vertical component $\mathbf{T}_y$ of the tension in the cable instead of $\mathbf{T}$, since the moment arm of $\mathbf{T}_y$ is the boom's length $L_1$. The horizontal component of the tension $\mathbf{T}_x$ exerts

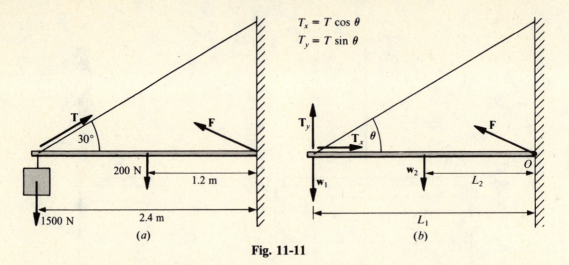

**Fig. 11-11**

no torque about $O$ because its line of action passes through $O$. The torques exerted by the load $\mathbf{w}_1$, the boom's weight $\mathbf{w}_2$, and the vertical component of tension $\mathbf{T}_y$ are, respectively,

$$\tau_1 = w_1 L_1 = (1500 \text{ N})(2.4 \text{ m}) = 3600 \text{ N·m}$$

$$\tau_2 = w_2 L_2 = (200 \text{ N})(1.2 \text{ m}) = 240 \text{ N·m}$$

$$\tau_3 = -T_y L_1 = -2.4T \sin 30° \text{ m} = -1.2T \text{ m}$$

For the boom to be in equilibrium,

$$\Sigma \tau = \tau_1 + \tau_2 + \tau_3 = 0$$

$$3600 \text{ N·m} + 240 \text{ N·m} - 1.2T \text{ m} = 0$$

$$T = \frac{3840 \text{ N·m}}{1.2 \text{ m}} = 3200 \text{ N}$$

(b)   The compression force $\mathbf{F}_x$ in the boom is equal in magnitude to the horizontal component of the cable tension $\mathbf{T}$. Thus

$$F_x = T \cos \theta = (3200 \text{ N})(\cos 30°) = 2771 \text{ N}$$

## SOLVED PROBLEM 11.11

A gate 1.8 m long and 1.2 m high has hinges at the top and bottom of one edge. (a) If the entire 300-N weight of the gate is supported by the lower hinge, find the force the gate exerts on the upper hinge. (b) Find the force the gate exerts on the lower hinge.

(a)   The weight of the gate acts from its center of gravity, which we assume is its geometric center, as in Fig. 11-12(a). The gate exerts a force that is downward and to the right on the lower hinge, which in response exerts a reaction force $\mathbf{F}_2$ on the gate that is upward and to the left. The gate exerts a force to the left on the upper hinge, which exerts a reaction force $\mathbf{F}_1$ to the right on the gate. To find the magnitude of $\mathbf{F}_1$, it is easiest to calculate torques about the lower hinge. The torques exerted on the gate by the upper hinge and by its own weight are, respectively,

$$\tau_1 = -F_1 L_1 = -1.2F_1 \text{ m}$$

$$\tau_3 = wL_3 = (300 \text{ N})(0.9 \text{ m}) = 270 \text{ N·m}$$

$$\Sigma \tau = \tau_1 + \tau_3 = 0$$

Hence

$$-1.2F_1 \text{ m} + 270 \text{ N·m} = 0$$

$$F_1 = 225 \text{ N}$$

The force exerted by the gate on the upper hinge has the same magnitude.

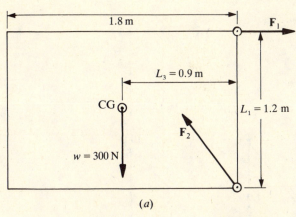

**Fig. 11-12**

(b) In order that the gate be in translational equilibrium, $\mathbf{F}_{2x}$ must be equal and opposite to $\mathbf{F}_1$, and $\mathbf{F}_{2y}$ must be equal and opposite to $\mathbf{w}$. Hence

$$F_{2x} = 225 \text{ N} \qquad F_{2y} = 300 \text{ N}$$

Since $\mathbf{F}_{2x}$ and $\mathbf{F}_{2y}$ are perpendicular, the magnitude of $F$ is

$$F = \sqrt{F_{2x}^2 + F_{2y}^2} = 375 \text{ N}$$

If $\theta$ is the angle between $\mathbf{F}_2$ and the vertical, as in Fig. 11-12(b),

$$\tan \theta = \frac{F_{2x}}{F_{2y}} = 0.75 \qquad \theta = 37°$$

The force the gate exerts on the lower hinge is equal and opposite to $\mathbf{F}_2$, hence it acts downward at an angle of 37° from the vertical and toward the right.

### SOLVED PROBLEM 11.12

A 12-ft ladder that weighs 50 lb rests against a frictionless wall at a point 10 ft above the ground. How much force does the ladder exert (a) on the ground and (b) on the wall?

(a) The forces that act on the ladder are its weight $\mathbf{w}$ acting downward from its center, the horizontal reaction force $\mathbf{F}_1$ of the wall (there is no vertical force component because the wall is frictionless), and the reaction force $\mathbf{F}_2$ of the ground, which has both vertical and horizontal components. Since $\mathbf{F}_{2y}$ and $\mathbf{w}$ are the only vertical forces,

$$F_{2y} = w = 50 \text{ lb}$$

To find $F_{2x}$, we begin by finding the value of the angle between the ladder and the ground. From Fig. 11-13

$$\sin \theta = \frac{10 \text{ ft}}{12 \text{ ft}} = 0.833 \qquad \theta = 56°$$

Now we calculate the torques produced by $\mathbf{F}_{2y}$, $\mathbf{F}_{2x}$, and $\mathbf{w}$, respectively, about the upper end of the ladder:

$$\tau_1 = -F_{2y}L_1 = -(50 \text{ lb})(12 \text{ ft})(\cos\ 56°) = -336 \text{ lb·ft}$$
$$\tau_2 = F_{2x}L_2 = (F_{2x})(10 \text{ ft}) = 10F_{2x} \text{ ft}$$
$$\tau_3 = wL_3 = (50 \text{ lb})(6 \text{ ft})(\cos\ 56°) = 168 \text{ lb·ft}$$

Applying the condition for rotational equilibrium yields

$$\Sigma \tau = \tau_1 + \tau_2 + \tau_3 = 0$$
$$-336 \text{ lb·ft} + 10F_{2x} \text{ ft} + 168 \text{ lb·ft} = 0$$
$$F_{2x} = 16.8 \text{ lb}$$

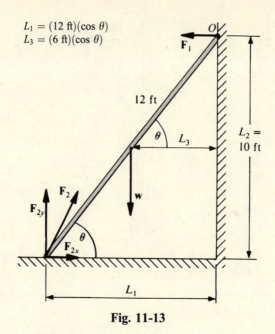

**Fig. 11-13**

The total force the ground exerts on the ladder is

$$F_2 = \sqrt{F_{2x}^2 + F_{2y}^2} = 53 \text{ lb}$$

The force the ladder exerts on the ground has the same magnitude.

(b)   The force the ladder exerts on the wall is equal in magnitude to $F_{2x}$, namely, 16.8 lb.

## SOLVED PROBLEM 11.13

The front wheels of a truck support 8 kN, and its rear wheels support 14 kN. The axles are 4 m apart. Where is the center of gravity of the truck located?

With $x$ the distance between the front axle and the center of gravity, as in Fig. 11-14, calculating torques about the center of gravity yields

$$\Sigma \tau = w_1 x - w_2(4 \text{ m} - x) = 0$$
$$8 \text{ kN} - 14(4 \text{ m} - x) \text{ kN} = 0$$
$$22x \text{ kN} = 56 \text{ kN·m}$$
$$x = 2.55 \text{ m}$$

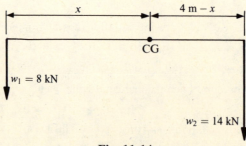

**Fig. 11-14**

## FINDING A CENTER OF GRAVITY

The center of gravity (CG) of an object of regular form and uniform composition is located at its geometric center. In the case of a complex object, the way to find its center of gravity is to consider it as a system of separate particles and then find the balance point of the system. An example is the massless rod of Fig. 11-15, which has three particles $m_1$, $m_2$, and $m_3$ attached to it. The CG of the system is at a distance $X$ from the end of the rod such that the torque exerted by a single particle of mass $M = m_1 + m_2 + m_3$ at $X$ equals the sum of the torques exerted by the particles at their locations $x_1$, $x_2$, and $x_3$. Thus

$$m_1 g x_1 + m_2 g x_2 + m_3 g x_3 = MgX = (m_1 + m_2 + m_3)gX$$
$$X = \frac{m_1 x_1 + m_2 x_2 + m_3 x_3}{m_1 + m_2 + m_3}$$

This formula can be extended to any number of particles. If the complex object involves two or three dimensions rather than just one, the same procedure is applied along two or three coordinate axes to find $X$ and $Y$ or $X$, $Y$, and $Z$, which are the coordinates of the center of gravity.

## SOLVED PROBLEM 11.14

Find the location of the center of gravity of the $L$-shaped piece of plywood shown in Fig. 11-16.

Let us consider the plywood piece to be made up of two sections, one a rectangle 3 m long and 1 m wide and the other a square 1 m on each side. The centers of gravity of these sections are at their geometric centers, as in Fig. 11-16($b$). If the plywood has a mass of $m_0$ per square meter, section 1 has a

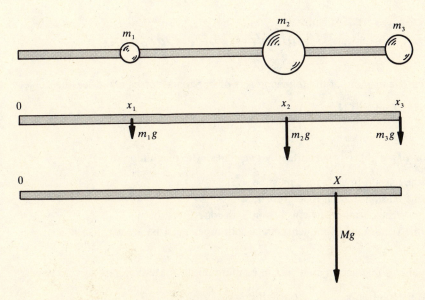

**Fig. 11-15**

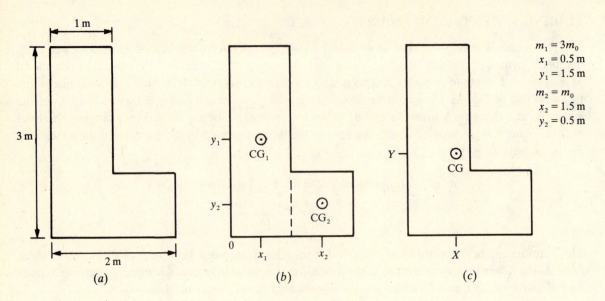

**Fig. 11-16**

mass of $3m_0$ and section 2 has a mass of $m_0$, since their respective areas are 3 and 1 m$^2$. The $x$ and $y$ coordinates of the center of gravity of the entire plywood piece are therefore

$$X = \frac{m_1 x_1 + m_2 x_2}{m_1 + m_2} = \frac{(3m_0)(0.5 \text{ m}) + (m_0)(1.5 \text{ m})}{3m_0 + m_0} = 0.75 \text{ m}$$

$$Y = \frac{m_1 y_1 + m_2 y_2}{m_1 + m_2} = \frac{(3m_0)(1.5 \text{ m}) + (m_0)(0.5 \text{ m})}{3m_0 + m_0} = 1.25 \text{ m}$$

$$m_1 = 3m_0 \qquad m_2 = m_0$$
$$x_1 = 0.5 \text{ m} \qquad x_2 = 1.5 \text{ m}$$
$$y_1 = 1.5 \text{ m} \qquad y_2 = 0.5 \text{ m}$$

# Multiple-Choice Questions

**11.1.** Which one or more of the following sets of horizontal forces could leave an object in equilibrium?

(a)  5, 10, and 20 N     (c)  8, 8, and 8 N
(b)  6, 12, and 18 N     (d)  2, 4, 8, and 16 N

**11.2.** A box of weight $w$ is supported by two ropes. The magnitude of

(a)  the tension in each rope must be $w/2$
(b)  the tension in each rope must be $w$
(c)  the vector sum of the tensions in both ropes must be $w$
(d)  the vector sum of the tensions in both ropes must be greater than $w$

**11.3.** The torques that act on an object in equilibrium have a vector sum of zero about

(a)  one point only         (c)  all points
(b)  one or more points     (d)  any of the above, depending on the situation

**11.4.** The point about which torques are calculated when studying the equilibrium of an object

  (*a*) must pass through the center of gravity of the object
  (*b*) must pass through one end of the object
  (*c*) must be located inside of the object
  (*d*) may be located anywhere

**11.5.** A 250-N box hangs from a rope. If the box is pushed with a horizontal force of 145 N, the angle between the rope and the vertical is

  (*a*) 30°    (*c*) 60°
  (*b*) 45°    (*d*) 75°

**11.6.** A picture hangs from two wires that go from its upper corners to a nail in the wall. If the picture weighs 8.0 N and each wire makes a 30° angle with the vertical, the tension in each wire is

  (*a*) 4.0 N    (*c*) 8.0 N
  (*b*) 4.6 N    (*d*) 9.2 N

**11.7.** An 0.80-kN load hangs from the end of a horizontal boom 2.0 m long hinged to a vertical mast. A rope 2.5 m long joins the end of the boom with a point on the mast 1.5 m above the hinge. The tension in the rope is

  (*a*) 0.48 kN    (*c*) 1.00 kN
  (*b*) 0.80 kN    (*d*) 1.33 kN

**11.8.** The boom of Question 11.7 exerts an inward force on the hinge of

  (*a*) 0.60 kN    (*c*) 1.07 kN
  (*b*) 0.80 kN    (*d*) 1.67 kN

**11.9.** A 50-kg barrel hangs from one end of a 30-kg beam 3.0 m long. The distance from the loaded end to the balance point is

  (*a*) 56 cm    (*c*) 94 cm
  (*b*) 75 cm    (*d*) 113 cm

**11.10.** A 50-kg steel pipe 4.0 m long is supported by a rope attached 1.7 m from one end. The downward force that must be applied to the end of the pipe closest to the rope to keep the pipe horizontal is

  (*a*) 8.8 N    (*c*) 227 N
  (*b*) 86 N    (*d*) 490 N

## Supplementary Problems

**11.1.** A weight is suspended from the middle of a rope whose ends are at the same height. Is it possible for the tension in the rope to be sufficiently great to prevent the rope from sagging at all?

**11.2.** A *couple* consists of two equal forces that act along parallel lines of action in opposite directions. If each of the forces in a couple has the magnitude $F$ and their lines of action are $d$ apart, find the torque exerted by the couple.

**11.3.** The wheels of a certain bus are 2 m apart, and the bus falls over when it is tilted sideways at a 45° angle. How high above the road is the center of gravity of the bus?

**11.4.** A 40-N box is suspended from two ropes which each make a 45° angle with the vertical. What is the tension in each rope?

**11.5.** A 100-kg box is suspended from two ropes; the left one makes an angle of 20° with the vertical, and the other makes an angle of 40°. What is the tension in each rope?

**11.6.** A load of unknown weight is suspended from the end of a horizontal boom whose own weight is negligible. The angle between the boom and the cable supporting its end is 30°, and the tension in the cable is 400 N. Find the weight of the load.

**11.7.** A 2000-N steel beam is raised by a crane, and a horizontal rope is used to pull it into position in a bridge under construction. What is the tension in the rope when the supporting cable is at an angle of 15° from the vertical?

**11.8.** A 1-kg pigeon sits on the middle of a clothesline whose supports are 10 m apart. The clothesline sags by 1 m. If the weight of the clothesline is negligible, find the tension in it.

**11.9.** A 30-lb weight is attached to one end of a 6-ft uniform beam whose own weight is 20 lb. Where is the balance point of the system?

**11.10.** A pail of water is to be carried by a man and a boy who each hold one end of a 180-cm pole thrust through the pail's handle. Where should the pail be located along the pole so that the man carries twice as much weight as the boy? Neglect the pole's weight.

**11.11.** A 20-kg child and a 30-kg child sit at opposite ends of a 4-m seesaw that is pivoted at its center. Where should another 20-kg child sit in order to balance the seesaw?

**11.12.** The axles of a 2400-lb car are 7 ft apart. If the center of gravity of the car is 3 ft behind the front axle, how much weight is supported by each of the car's wheels?

**11.13.** (*a*) A 50-kg horizontal boom 4 m long is hinged to a vertical mast, and its outer end is supported by a cable attached to the mast 3 m above the hinge pin. Find the tension in the cable. (*b*) A load of 200 kg is suspended from the outer end of the boom. Find the new tension in the cable.

**11.14.** Solve Prob. 11.13 by considering the rotational equilibrium of the boom, using the lower end of the boom as the pivot point.

**11.15.** A door 7 ft high and 3 ft wide has hinges at the top and bottom of one edge. Half the 40-lb weight of the door is supported by each hinge. Find the horizontal components of the force the door exerts on each hinge.

**11.16.** A ladder that weighs 240 N rests against a frictionless wall at an angle of 60° from the ground. A 600-N man stands on the ladder three-fourths of the distance from the top. How much force does the top of the ladder exert on the wall?

**11.17.** A 15-kg ladder 3 m long rests against a frictionless wall at a point 2.4 m from the ground. Find the vertical and horizontal components of the force that the ladder exerts on the floor.

**11.18.** A steel pipe 2 m long has another steel pipe of the same kind 1 m long welded across one end, so the result has the form of a T. Where is the center of gravity of this object located?

# *Answers to Multiple-Choice Questions*

**11.1.** (*b*), (*c*)          **11.6.**     (*b*)

**11.2.** (*c*)          **11.7.**     (*d*)

**11.3.** (*c*)          **11.8.**     (*c*)

**11.4.** (*d*)          **11.9.**     (*a*)

**11.5.** (*a*)          **11.10.**   (*b*)

# Answers to Supplementary Problems

**11.1.** No. The rope must sag in order for its tension to provide an upward component of force to support the weight. The greater the tension, the less the sag, but it is impossible for the rope to be perfectly horizontal.

**11.2.** *Fd*

**11.3.** 1 m

**11.4.** 28 N in each rope

**11.5.** The tension in the left-hand rope is 727 N and that in the right-hand rope is 387 N.

**11.6.** 200 N

**11.7.** 536 N

**11.8** 26 N

**11.9.** The balance point is 2.4 ft from the 30-lb weight.

**11.10.** The pail should be 60 cm from the man.

**11.11.** The third child should be 1 m from the 20-kg child.

**11.12.** The front wheels each support 686 lb, and the rear wheels each support 514 lb.

**11.13.** (*a*)  306 N     (*b*)  2756 N

**11.14.**  5323 N

**11.15.**  8.6 lb; 8.6 lb

**11.16.**  440 N

**11.17.**  147 N; 55 N

**11.18.**  67 cm from the crossbar.

# Chapter 12

# Simple Machines

## MACHINES

A *machine* is a device that changes the magnitude, direction, or mode of application of a force or torque while transmitting it for a particular purpose. There are only three basic machines, of which all others are developments: the lever, the inclined plane, and the hydraulic press. The hydraulic press is described in Chapter 16.

## MECHANICAL ADVANTAGE

The *actual mechanical advantage* (AMA) of a machine is the ratio between the output force $F_{out}$ it exerts and the input force $F_{in}$ that is applied to it:

$$AMA = \frac{F_{out}}{F_{in}}$$

$$\text{Actual mechanical advantage} = \frac{\text{output force}}{\text{input force}}$$

An AMA greater than 1 means that $F_{out}$ exceeds $F_{in}$; an AMA less than 1 means that $F_{out}$ is smaller than $F_{in}$.

The *ideal mechanical advantage* (IMA) of a machine is its mechanical advantage in the absence of friction. If the input force acts through the distance $s_{in}$ when the output force acts through the distance $s_{out}$, then according to the principle of conservation of energy

$$\text{Work input} = \text{work output}$$

$$F_{in}s_{in} = F_{out}s_{out}$$

and so

$$\frac{F_{out}}{F_{in}} = \frac{s_{in}}{s_{out}}$$

when there is no friction present. Because friction acts to decrease the ratio $F_{out}/F_{in}$ in an actual machine but does not change the ratio $s_{in}/s_{out}$, it is customary to define ideal mechanical advantage in terms of the latter:

$$IMA = \frac{s_{in}}{s_{out}}$$

$$\text{Ideal mechanical advantage} = \frac{\text{input distance}}{\text{output distance}}$$

## EFFICIENCY

The *efficiency* (Eff) of a machine equals the ratio between its actual and its ideal mechanical advantages:

$$\text{Eff} = \frac{AMA}{IMA} = \frac{\text{work output}}{\text{work input}} = \frac{\text{power output}}{\text{power input}}$$

If we know the ideal mechanical advantage and the efficiency of a machine,

$$F_{out} = (\text{Eff})(\text{IMA})(F_{in})$$

## COMPOUND MACHINES

When two or more machines are coupled together so that the output of one is the input of the next, the overall mechanical advantage (IMA or AMA) of the combination is the product of the mechanical advantages of each one:

$$MA = (MA_1)(MA_2)(MA_3) \cdots$$

The overall efficiency of two or more machines in combination is similarly

$$\text{Eff} = (\text{Eff}_1)(\text{Eff}_2)(\text{Eff}_3) \cdots$$

### SOLVED PROBLEM 12.1

Find the IMA of the pulley system shown in Fig. 12-1.

To raise the load through a height $h$, a length of rope equal to $4h$ must be pulled through the pulley system since the load is supported by four strands of rope. Hence

$$\text{IMA} = \frac{s_{in}}{s_{out}} = \frac{4h}{h} = 4$$

As a general rule, the IMA of a pulley system is equal to the number of strands of rope that support the load. The rope leaving the upper pulley in Fig. 12-1 does not help to support the load and hence is not counted in determining the IMA of the system.

### SOLVED PROBLEM 12.2

Find the IMA of the pulley system shown in Fig. 12-2.

This pulley system is the same as that in Fig. 12-1 but is inverted. As a result, five strands of rope now support the load, and the IMA is accordingly 5.

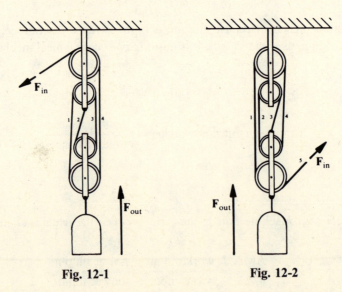

Fig. 12-1                    Fig. 12-2

**SOLVED PROBLEM 12.3**

Find the IMA of the pulley system shown in Fig. 12-3.

There are actually two pulley systems here. The load is initially supported by the two strands of ropes 1 and 2 that pass around pulley $A$, so $IMA_1 = 2$. The tension in rope 2 is then supported by strands 3 and 4 of another rope that passes around pulley $B$, so $IMA_2 = 2$. Pulley $C$ merely changes the direction of the rope, and strand 5 does not help to support the load. Hence

$$IMA = (IMA_1)(IMA_2) = (2)(2) = 4$$

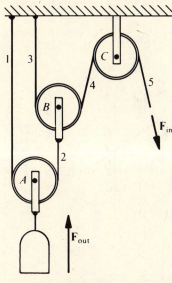

**Fig. 12-3**

**SOLVED PROBLEM 12.4**

A woman uses a force of 80 N to raise a load of 280 N by 1.5 ft with a pulley system. If she pulls a total of 6 m of rope through the system, what is the efficiency of the system?

The IMA and AMA of the system are, respectively,

$$IMA = \frac{s_{in}}{s_{out}} = \frac{6 \text{ m}}{1.5 \text{ m}} = 4$$

$$AMA = \frac{F_{out}}{F_{in}} = \frac{280 \text{ N}}{80 \text{ N}} = 3.5$$

Hence the efficiency of the system is

$$Eff = \frac{AMA}{IMA} = \frac{3.5}{4} = 0.875 = 87.5\%$$

**SOLVED PROBLEM 12.5**

How much force is needed to raise a load of 500 N with the pulley system of Prob. 12.4?

We must use the actual mechanical advantage to find $F_{in}$. Hence

$$F_{in} = \frac{F_{out}}{AMA} = \frac{500 \text{ N}}{3.5} = 143 \text{ N}$$

**SOLVED PROBLEM 12.6**

A *differential pulley* (often called a *chain hoist* because a chain running over notched pulleys is used to prevent slipping) is shown in Fig. 12-4. When the free loop of the chain is pulled as shown, strand 1 of the chain supporting the movable pulley is shortened while strand 2 is lengthened. Because strand 1 is shortened by more than strand 2 is lengthened, the load is raised. Pulling down on the other strand of the free loop lowers the load. (*a*) Find the IMA of the differential pulley. (*b*) If the upper pulleys of such a hoist are 14 and 15 cm in radius and the hoist is 75 percent efficient, how heavy a load can be raised by a force of 150 N? (*c*) If the load is to be raised by 1 m, how much chain must be pulled through the pulleys?

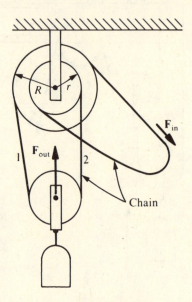

**Fig. 12-4**

(*a*) The radius of the large upper pulley is $R$ and that of the small one is $r$. In a complete rotation of the upper pulleys, the input force $F_{in}$ acts through a distance equal to the circumference of the large pulley, so $s_{in} = 2\pi R$. In this rotation, strand 1 has been shortened by $2\pi R$ while strand 2 has been lengthened by $2\pi r$, so the total decrease in length is $2\pi R - 2\pi r = 2\pi(R - r)$. Since the decrease in length is shared by both strands 1 and 2, the movable pulley is raised by half this amount and $s_{out} = \pi(R - r)$. The IMA of the system is accordingly

$$\text{IMA} = \frac{s_{in}}{s_{out}} = \frac{2\pi R}{\pi(R - r)} = \frac{2R}{R - r}$$

(*b*) Since $R = 15$ cm and $r = 14$ cm, the IMA of the hoist is

$$\text{IMA} = \frac{2R}{R - r} = \frac{(2)(15 \text{ cm})}{15 \text{ cm} - 14 \text{ cm}} = 30$$

At 75 percent efficiency,

$$\text{AMA} = (\text{Eff})(\text{IMA}) = (0.75)(30) = 22.5$$

An applied force of 150 N can therefore lift a load of

$$F_{out} = (\text{AMA})(F_{in}) = (22.5)(150 \text{ N}) = 3375 \text{ N}$$

(*c*) Given that $s_{out} = 1$ m and IMA = 30, we have

$$s_{in} = (\text{IMA})(s_{out}) = (30)(1 \text{ m}) = 30 \text{ m}$$

## THE LEVER

The IMA of a *lever* is equal to the ratio between its moment arms $L_{in}$ and $L_{out}$ (Fig. 12-5):

$$\text{IMA} = \frac{L_{in}}{L_{out}}$$

The wheel and axle, belt and gear drives, and pulley systems are all developments of the lever.

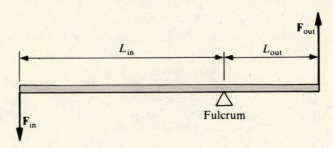

**Fig. 12-5**

## SOLVED PROBLEM 12.7

One end of a 200-kg crate is to be lifted from the ground by using a 3-m plank as a lever. If the maximum force that can be applied to the plank is 350 N, where should the fulcrum be placed? Assume that the crate's contents are uniformly distributed, so the mass to be raised is 100 kg.

The weight of a 100-kg mass is

$$w = mg = (100 \text{ kg})(9.8 \text{ m/s}^2) = 980 \text{ N}$$

Since the plank is 3 m long (Fig. 12-6),

$$L_{in} + L_{out} = 3 \text{ m}$$
$$L_{in} = 3 \text{ m} - L_{out}$$

In the absence of friction,

$$\frac{F_{out}}{F_{in}} = \frac{L_{in}}{L_{out}}$$
$$\frac{980 \text{ N}}{350 \text{ N}} = \frac{3 \text{ m} - L_{out}}{L_{out}}$$
$$3.8 L_{out} = 3 \text{ m}$$
$$L_{out} = 0.79 \text{ m}$$

The fulcrum should be located 79 cm from the end of the crate.

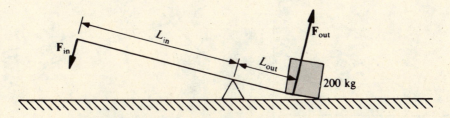

**Fig. 12-6**

### SOLVED PROBLEM 12.8

The *wheel and axle* (Fig. 12-7) is a development of the lever that permits continuous motion. (*a*) Where is the fulcrum? (*b*) What is the IMA of a wheel and axle?

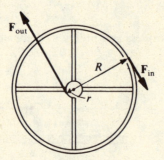

**Fig. 12-7**

(*a*)   The center of the axle acts as the fulcrum.

(*b*)   The input force acts tangentially on the wheel rim, and the output force acts tangentially on the axle rim. When the combination makes one complete turn, a point on the wheel moves through $s_{in} = 2\pi R$ and a point on the axle moves through $s_{out} = 2\pi r$. Hence

$$\text{IMA} = \frac{s_{in}}{s_{out}} = \frac{2\pi R}{2\pi r} = \frac{R}{r}$$

The larger the wheel radius relative to that of the axle, the greater the mechanical advantage.

### THE INCLINED PLANE

The IMA of an *inclined plane* is equal to the ratio between its length and its height (Fig. 12-8):

$$\text{IMA} = \frac{L}{h}$$

Wedges, cams, and screws are all developments of the inclined plane.

A *screw* is an inclined plane wrapped around a cylinder in the form of a helix. The *pitch p* of a screw is the distance from one thread to the next (Fig. 12-9). If the head of a screw is turned by a tangential force applied at a distance $L$ from its axis, the input force travels the distance $s_{in} = 2\pi L$ while the screw advances $s_{out} = p$. Hence the IMA of a screw is

$$\text{IMA} = \frac{s_{in}}{s_{out}} = \frac{2\pi L}{p}$$

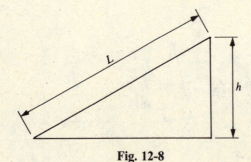

**Fig. 12-8**                                              **Fig. 12-9**

**SOLVED PROBLEM 12.9**

A ramp 25 m long slopes down 1.4 m to the edge of a lake. How much force is needed to pull out a 360-kg boat on a 90-kg trailer if friction in the trailer's wheels reduces the efficiency to 90 percent?

The IMA of the ramp is

$$\text{IMA} = \frac{L}{h} = \frac{25 \text{ m}}{1.4 \text{ m}} = 17.9$$

and its AMA is

$$\text{AMA} = (\text{Eff})(\text{IMA}) = (0.9)(17.9) = 16.1$$

The output force is the combined weight of boat and trailer, which is

$$F_{\text{out}} = w = (m_1 + m_2)g = (450 \text{ kg})(9.8 \text{ m/s}^2) = 4410 \text{ N}$$

The required input force is

$$F_{\text{in}} = \frac{F_{\text{out}}}{\text{AMA}} = \frac{4410 \text{ N}}{16.1} = 274 \text{ N}$$

**SOLVED PROBLEM 12.10**

The screw of a machinist's vise has a pitch of 5 mm and a handle 20 cm long. If the efficiency is 40 percent, how much force is developed between the jaws of the vise when a force of 40 N is applied to the end of the handle?

The IMA of the vise is

$$\text{IMA} = \frac{2\pi L}{p} = \frac{(2\pi)(200 \text{ mm})}{5 \text{ mm}} = 251$$

and its AMA is

$$\text{AMA} = (\text{Eff})(\text{IMA}) = (0.4)(251) = 100.4$$

The output force is therefore

$$F_{\text{out}} = (\text{AMA})(F_{\text{in}}) = (100.4)(40 \text{ N}) = 4016 \text{ N}$$

**SOLVED PROBLEM 12.11**

A screw jack has a pitch of 5 mm and a handle 60 cm long. If a force of 50 N is needed to raise a load of 700 kg, find the efficiency of the jack.

The IMA and AMA of the jack are, respectively,

$$\text{IMA} = \frac{2\pi L}{p} = \frac{(2\pi)(60 \text{ cm})}{0.5 \text{ cm}} = 754$$

$$\text{AMA} = \frac{F_{\text{out}}}{F_{\text{in}}} = \frac{mg}{F_{\text{in}}} = \frac{(700 \text{ kg})(9.8 \text{ m/s}^2)}{50 \text{ N}} = 137$$

The efficiency of the jack is

$$\text{Eff} = \frac{\text{AMA}}{\text{IMA}} = \frac{137}{754} = 0.18 = 18\%$$

## TORQUE TRANSMISSION

Belt and gear drives make possible the transmission of torques from one shaft to another (Fig. 12-10). The actual mechanical advantage of any such system is

$$AMA = \frac{\tau_{out}}{\tau_{in}} = \frac{\text{output torque}}{\text{input torque}}$$

The ideal mechanical advantage of a pulley system equals the ratio between the radius of the driven pulley $r_{out}$ and that of the driving pulley $r_{in}$, which is the same as the ratio of their diameters:

$$IMA = \frac{r_{out}}{r_{in}} = \frac{d_{out}}{d_{in}}$$

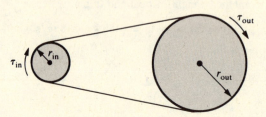

**Fig. 12-10**

In the case of a gear drive, since the number of teeth on a gear is proportional to its radius,

$$IMA = \frac{N_{out}}{N_{in}} = \frac{\text{number of teeth on driven gear}}{\text{number of teeth on driving gear}}$$

In any torque transmission system, the angular velocity ratio is the inverse of the IMA:

$$\frac{\text{Output angular velocity}}{\text{Input angular velocity}} = \frac{\omega_{out}}{\omega_{in}} = \frac{d_{in}}{d_{out}} = \frac{1}{IMA}$$

An increase in torque is accompanied by a decrease in speed of rotation, and vice versa.

## SOLVED PROBLEM 12.12

A 1200 rev/min motor is connected to a 12-in.-diameter circular saw blade with a V-belt and a pair of step pulleys (Fig. 12-11). (*a*) If the pulley diameters are 4, 5, and 6 in. in each set, find

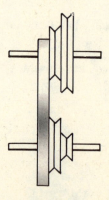

**Fig. 12-11**

the possible angular velocities of the saw blade in revolutions per minute. (*b*) Find the corresponding linear velocities of the saw's teeth in feet per minute.

(*a*)   The three possible ratios of pulley diameters are $\frac{2}{3}$, 1, and $\frac{3}{2}$. Since

$$\omega_{\text{out}} = \omega_{\text{in}} \frac{d_{\text{in}}}{d_{\text{out}}}$$

the three possible angular velocities of the saw blade are

$$\omega_1 = (1200 \text{ rev/min})(\tfrac{2}{3}) = 800 \text{ rev/min}$$
$$\omega_2 = (1200 \text{ rev/min})(1) = 1200 \text{ rev/min}$$
$$\omega_3 = (1200 \text{ rev/min})(\tfrac{3}{2}) = 1800 \text{ rev/min}$$

(*b*)   The linear velocity $v$ of the saw's teeth is related to the radius $r$ of the saw blade and its angular velocity $\omega$ by the formula $v = \omega r$, where $\omega$ must be expressed in radian measure. Since there are $2\pi$ rad in a revolution, the above angular velocities can also be expressed as

$$\omega_1 = \left(2\pi \, \frac{\text{rad}}{\text{rev}}\right)(800 \text{ rev/min}) = 5027 \text{ rad/min}$$

$$\omega_2 = \left(2\pi \, \frac{\text{rad}}{\text{rev}}\right)(1200 \text{ rev/min}) = 7540 \text{ rad/min}$$

$$\omega_3 = \left(2\pi \, \frac{\text{rad}}{\text{rev}}\right)(1800 \text{ rev/min}) = 11,310 \text{ rad/min}$$

The radius of the saw blade is $r = 6$ in. $= 0.5$ ft, and so the possible linear velocities of its teeth are

$$v_1 = \omega_1 r = 2514 \text{ ft/min} \qquad v_2 = \omega_2 r = 3770 \text{ ft/min} \qquad v_3 = \omega_3 r = 5655 \text{ ft/min}$$

### SOLVED PROBLEM 12.13

An engine develops 6 kW at 4000 rev/min. What gear ratio is needed if an output torque of 60 N·m is required?

Since

$$\omega = (4000 \text{ rev/min})\left(0.105 \, \frac{\text{rad/s}}{\text{rev/min}}\right) = 420 \text{ rad/s}$$

the engine itself develops a torque of

$$\tau = \frac{P}{\omega} = \frac{6000 \text{ W}}{420 \text{ rad/s}} = 14.3 \text{ N·m}$$

Hence the required gear ratio is

$$\frac{N_{\text{out}}}{N_{\text{in}}} = \frac{\tau_{\text{out}}}{\tau_{\text{in}}} = \frac{60 \text{ N·m}}{14.3 \text{ N·m}} = 4.2$$

## *Multiple-Choice Questions*

**12.1.**   A machine can increase which one or more of the following?

(*a*)   work         (*c*)   force
(*b*)   speed        (*d*)   torque

**12.2.** A machine has an IMA of 6.0 and an AMA of 5.0. The work input needed by the machine to raise a 40-kg load by 3.0 m is

(*a*)  0.98 kJ     (*c*)  1.4 kJ
(*b*)  1.2 kJ      (*d*)  5.9 kJ

**12.3.** A person uses a force of 300 N to pry up one end of a 120-kg box with a lever 2.4 m long. The distance from the end of the box to the fulcrum is

(*a*)  40 cm      (*c*)  81 cm
(*b*)  49 cm      (*d*)  171 cm

**12.4.** The smallest number of pulleys needed for an IMA of 5 is

(*a*)  3     (*c*)  5
(*b*)  4     (*d*)  6

**12.5.** A force of 50 N is needed to raise a 240-N load with a pulley system. The load goes up 1 m for every 5 m of rope pulled through the pulleys. The efficiency of the system is

(*a*)  48%     (*c*)  96%
(*b*)  50%     (*d*)  104%

**12.6.** Two sets of spur gears, each set 96 percent efficient, drive a worm gear that is 80 percent efficient. The overall efficiency of the system is

(*a*)  74%     (*c*)  80%
(*b*)  77%     (*d*)  96%

**12.7.** A screw jack has a pitch of 4.0 mm and a handle 50 cm long. Its efficiency is 20 percent. The force that must be applied to the handle in order to lift a load of 400 kg is

(*a*)  1 N       (*c*)  5 N
(*b*)  2.5 N     (*d*)  25 N

# Supplementary Problems

**12.1.** How great a force is needed to lift a 20-kg mass by using the lever shown in Fig. 12-12?

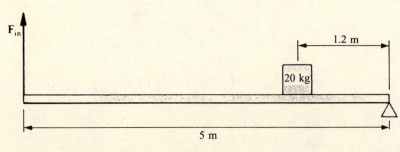

**Fig. 12-12**

**12.2.** A bolt is held in the jaws of a pair of pliers. If the bolt is 25 mm from the pivot and a force of 10 N is applied to each handle 150 mm from the pivot, find the total compressive force on the bolt.

**12.3.** Find the IMA of the pulley system shown in Fig. 12-13.

**12.4.** Find the IMA of the pulley system shown in Fig. 12-14.

**12.5.** Find the IMA of the pulley system shown in Fig. 12-15.

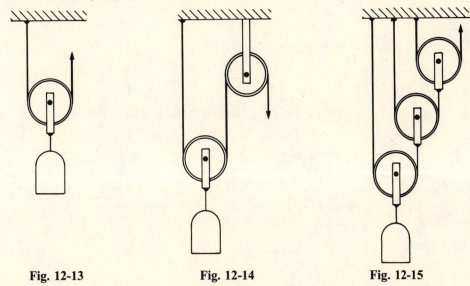

**Fig. 12-13**                          **Fig. 12-14**                          **Fig. 12-15**

**12.6.** What is the highest IMA that can be obtained with a system of two pulleys?

**12.7.** A chain hoist has upper pulleys 4 and 4.5 in. in radius (*a*) If the hoist is 80 percent efficient, how much force is needed to raise a load of 200 lb? (*b*) How much chain must be pulled through the pulleys to raise the load by 2 ft?

**12.8.** A chain hoist is used to lift a 180-kg engine block by 1.2 m. To do this, a force of 100 N must be applied to the chain and 30 m of chain pulled through the system. What is the efficiency of the hoist?

**12.9.** A 200-kg piano on frictionless casters is to be pushed up a 7-m ramp to a loading platform 1 m above the ground. How much force is needed?

**12.10.** (*a*) What kind of simple machine does a screwdriver represent? (*b*) The handle of a screwdriver is 3 cm in diameter, and its blade is 6 mm wide. What is its mechanical advantage?

**12.11.** A screwdriver whose handle is 1 in. in diameter is used to tighten a bolt which has 24 threads per inch. If the efficiency is 8 percent, find the force exerted on the nut when a force of 3 lb is applied to the handle.

**12.12.** A 600 rev/min motor is used to operate an air compressor at 200 rev/min through a V-belt drive. If the motor pulley is 15 cm in diameter, what should the diameter of the compressor pulley be?

**12.13.** The sprocket wheel on the rear axle of a certain bicycle is 10 cm in diameter, and the sprocket wheel to which the pedals are attached is 20 cm in diameter. The wheels of the bicycle are 65 cm in diameter. At how many revolutions per minute must the pedals be turned to travel at 5 m/s?

## *Answers to Multiple-Choice Questions*

**12.1.** (*b*), (*c*), (*d*)      **12.5.** (*c*)

**12.2.** (*c*)      **12.6.** (*a*)

**12.3.** (*c*)      **12.7.** (*d*)

**12.4.** (*b*)

## Answers to Supplementary Problems

**12.1.** 47 N      **12.8.** 71%

**12.2.** 120 N      **12.9.** 280 N

**12.3.** 2      **12.10.** (*a*) Wheel and axle   (*b*) 5

**12.4.** 2      **12.11.** 18 lb

**12.5.** 8      **12.12.** 45 cm

**12.6.** 3      **12.13.** 7.3 rev/min

**12.7.** (*a*) 13.9 lb      (*b*) 36 ft

# Chapter 13

## Elasticity

### STRESS AND STRAIN

The *stress* on a body acted on by a deforming force is equal to the magnitude of the force $F$ divided by the cross-sectional area $A$ over which it acts. The unit of stress in SI units is newtons per square meter, which is known as the *pascal* (Pa). In the British system it is customary to use pounds per square inch. The three categories of stress—*tension*, *compression*, and *shear*—are illustrated in Fig. 13-1. The unstressed shape is shown by the dashed lines, and the stressed shape by the solid lines.

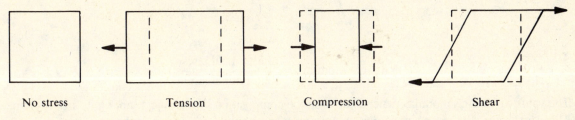

| No stress | Tension | Compression | Shear |

**Fig. 13-1**

The relative change in the size or shape of a body due to applied stress is called *strain*. Strain is a dimensionless quantity; for instance, the longitudinal strain that tension produces in a body is its change in length $\Delta L$ divided by its original length $L_0$, which is a pure number.

### ELASTICITY

The *elastic limit* of a material is the maximum stress that can be applied to a body without causing a permanent deformation. For stresses below the elastic limit, the material exhibits *elastic* behavior: when the stress is removed, the body returns to its original size and shape.

Below the elastic limit, strain is found to be proportional to stress. This relationship is known as *Hooke's law*. In the case of tension, for example, doubling the applied force on a body will double the amount by which the body stretches. The *modulus of elasticity* of a material subjected to a particular kind of stress below its elastic limit is defined by the relationship

$$\text{Modulus of elasticity} = \frac{\text{stress}}{\text{strain}}$$

The *ultimate strength* of a material is the greatest stress it can withstand without rupture. In many materials the ultimate strength considerably exceeds the elastic limit. When a stress greater than its elastic limit but less than its ultimate strength is applied to such a material, the result is a permanent deformation. Bending a piece of metal is an example.

### SOLVED PROBLEM 13.1

A nylon rope 24 mm in diameter has a breaking strength of 120 kN. Find the breaking strengths of similar ropes (*a*) 12 mm and (*b*) 48 mm in diameter.

Since the breaking stress $F/A$ is the same for all the ropes, their breaking strengths $F$ are in proportion to their cross-sectional areas $A$. The cross-sectional area of a cylinder of diameter $d$ is $A = \pi r^2 = \pi d^2/4$, and so in each case $F$ varies directly with $d^2$.

(a)   A rope 12 mm in diameter has an area $(\frac{1}{2})^2 = \frac{1}{4}$ that of a rope 24 mm in diameter. Hence its breaking strength is one-fourth as much, or 30 kN.

(b)   A rope 48 mm in diameter has an area $2^2 = 4$ times that of a rope 24 mm in diameter. Hence its breaking strength is four times as much, or 480 kN.

## YOUNG'S MODULUS

When a tension or compression force $F$ acts on an object of length $L_0$ and cross-sectional area $A$ (Fig. 13-2), the result is a change in length $\Delta L$. Below the elastic limit, the ratio between stress and strain in this situation is called *Young's modulus*:

$$Y = \frac{F/A}{\Delta L/L_0}$$

$$\text{Young's modulus} = \frac{\text{longitudinal stress}}{\text{longitudinal strain}}$$

The value of Young's modulus depends only on the composition of the object, not on its size or shape. The usual units of $Y$ are newtons per square meter and pounds per square inch.

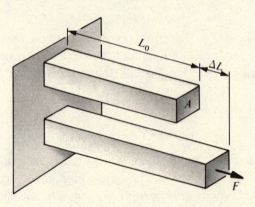

**Fig. 13-2**

## SOLVED PROBLEM 13.2

An aluminum wire 3 mm in diameter and 4 m long is used to support a mass of 50 kg. What is the elongation of the wire? Young's modulus for aluminum is $7 \times 10^{10}$ Pa.

The cross-sectional area of a wire of radius $r = 1.5$ mm $= 1.5 \times 10^{-3}$ m is

$$A = \pi r^2 = (\pi)(1.5 \times 10^{-3} \text{ m})^2 = 7.07 \times 10^{-6} \text{ m}^2$$

The applied force is

$$F = mg = (50 \text{ kg})(9.8 \text{ m/s}^2) = 490 \text{ N}$$

and so the elongation of the wire is

$$\Delta L = \frac{L_0}{Y}\frac{F}{A} = \frac{(4 \text{ m})(490 \text{ N})}{(7 \times 10^{10} \text{ N/m}^2)(7.07 \times 10^{-6} \text{ m}^2)} = 3.96 \times 10^{-3} \text{ m} = 3.96 \text{ mm}$$

## SOLVED PROBLEM 13.3

The elastic limit of aluminum is $1.3 \times 10^8$ Pa. What is the maximum mass that the wire of Prob. 13.2 can support without exceeding its elastic limit?

Since the cross-sectional area of the wire is $A = 7.07 \times 10^{-6}$ m$^2$ and

$$\left(\frac{F}{A}\right)_{\text{max}} = 1.3 \times 10^8 \text{ N/m}^2$$

the maximum force is

$$F_{\text{max}} = (1.3 \times 10^8 \text{ N/m}^2)(7.07 \times 10^{-6} \text{ m}^2) = 919 \text{ N}$$

which corresponds to a mass of

$$m = \frac{w}{g} = \frac{F_{\text{max}}}{g} = \frac{919 \text{ N}}{9.8 \text{ m/s}^2} = 94 \text{ kg}$$

## SOLVED PROBLEM 13.4

A wire 8 ft long with a cross-sectional area of 0.01 in.$^2$ stretches by 0.05 in. when a weight of 100 lb is suspended from it. Find the stress on the wire, the resulting strain, and the value of Young's modulus for the wire's material.

$$\text{Stress} = \frac{F}{A} = \frac{100 \text{ lb}}{0.01 \text{ in.}^2} = 10^4 \text{ lb/in.}^2$$

$$\text{Strain} = \frac{\Delta L}{L_0} = \frac{0.05 \text{ in.}}{96 \text{ in.}} = 5.2 \times 10^{-4}$$

$$Y = \frac{F/A}{\Delta L/L_0} = \frac{10^4 \text{ lb/in.}^2}{5.2 \times 10^{-4}} = 1.92 \times 10^7 \text{ lb/in.}^2$$

## SOLVED PROBLEM 13.5

A steel pipe 3.6 m long is placed vertically under a sagging floor to support it. The inside diameter of the pipe is 80 mm, its outside diameter is 100 mm, and $Y = 2 \times 10^{11}$ Pa. A sensitive strain gauge indicates that the pipe's length decreases by 0.1 mm. What is the magnitude of the load the pipe supports?

The cross-sectional area of the pipe (Fig. 13-3) is

$$A = \pi(R^2 - r^2) = \pi[(0.05 \text{ m})^2 - (0.04 \text{ m})^2] = 2.83 \times 10^{-3} \text{ m}^2$$

Here $\Delta L = 1 \times 10^{-4}$ m, so

$$F = YA\,\frac{\Delta L}{L_0} = (2 \times 10^{11} \text{ N/m}^2)(2.83 \times 10^{-3} \text{ m}^2)\left(\frac{1 \times 10^{-4} \text{ m}}{3.6 \text{ m}}\right) = 1.57 \times 10^4 \text{ N} = 15.7 \text{ kN}$$

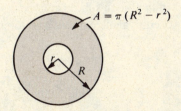

$$A = \pi(R^2 - r^2)$$

**Fig. 13-3**

**SOLVED PROBLEM 13.6**

By how much can a steel wire 3 m long and 2 mm in diameter be stretched before the elastic limit is exceeded? Young's modulus for the wire is $2 \times 10^{11}$ Pa, and its elastic limit is $2.5 \times 10^8$ Pa.

The cross-sectional area of a wire of radius $r = 1$ mm $= 10^{-3}$ m is

$$A = \pi r^2 = 3.14 \times 10^{-6} \text{ m}^2$$

The maximum force that can be applied without exceeding the elastic limit is therefore

$$F = \left(\frac{F}{A}\right)_{max} (A) = (2.5 \times 10^8 \text{ N/m}^2)(3.14 \times 10^{-6} \text{ m}^2) = 785 \text{ N}$$

When this force is applied, the wire will stretch by

$$\Delta L = \frac{L_0}{Y}\frac{F}{A} = \frac{(3 \text{ m})(785 \text{ N})}{(2 \times 10^{11} \text{ N/m}^2)(3.14 \times 10^{-6} \text{ m}^2)} = 3.75 \times 10^{-3} \text{ m} = 3.75 \text{ mm}$$

**SOLVED PROBLEM 13.7**

A steel cable whose cross-sectional area is 1 in.$^2$ is used to support an elevator cab weighing 5000 lb. If the stress in the cable is not to exceed 20 percent of the cable's elastic limit of 40,000 lb/in.$^2$, find the maximum permissible upward acceleration.

The force that corresponds to a maximum stress in the cable of $(0.20)(40,000 \text{ lb/in.}^2) = 8000$ lb/in.$^2$ is

$$F = \left(\frac{F}{A}\right)_{max} (A) = (8000 \text{ lb/in.}^2)(1 \text{ in.}^2) = 8000 \text{ lb}$$

This force is to equal the weight $w$ of the cab plus the force $ma$ that provides it with an upward acceleration so that

$$F = w + ma$$
$$a = \frac{F - w}{m}$$

Since $m = w/g$,

$$a = \frac{g(F - w)}{w} = \frac{(32 \text{ ft/s}^2)(8000 \text{ lb} - 5000 \text{ lb})}{5000 \text{ lb}} = 19.2 \text{ ft/s}^2$$

## SHEAR MODULUS

A shear stress changes the shape of an object, not its volume. Figure 13-4 shows a rectangular block acted on by shear forces $F$. The shearing stress is equal to $F/A$, and the shearing strain is equal to the *angle of shear* $\phi$, expressed in radians. Because $\phi$ is always small, it is very nearly the same as the ratio $s/d$ between the displacement $s$ of the block's faces and the distance $d$ between these faces. Below the elastic limit, then, there are two equivalent expressions for the *shear modulus* or (*modulus of rigidity*):

$$s = \frac{F/A}{\phi} = \frac{F/A}{s/d}$$

$$\text{Shear modulus} = \frac{\text{shear stress}}{\text{shear strain}}$$

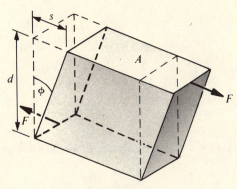

**Fig. 13-4**

### SOLVED PROBLEM 13.8

The shearing strength of a certain steel alloy is $2.5 \times 10^8$ Pa. Two 5 mm-diameter bolts of this alloy are used to fasten a bracket to a wall. What is the maximum load the bracket can support without shearing off the bolts?

The shearing stress here is exerted perpendicular to each bolt, so for each bolt $A = \pi r^2$ and, since $r = 0.0025$ m,

$$F = \left(\frac{F}{A}\right)_{max}(A) = (2.5 \times 10^8 \text{ N/m}^2)(\pi)(0.0025 \text{ m})^2 = 4.9 \text{ kN}$$

The combined load is twice this, or 9.8 kN. The corresponding mass is 1000 kg.

### SOLVED PROBLEM 13.9

How much force is required to punch a hole $\frac{1}{2}$ in. in diameter in a steel sheet $\frac{1}{8}$ in. thick whose shearing strength is $4 \times 10^4$ lb/in.$^2$?

The shear stress is exerted over the cylindrical surface that is the boundary of the hole (Fig. 13-5). The area of this surface is

$$A = 2\pi rh = (2\pi)(0.25 \text{ in.})(0.125 \text{ in.}) = 0.196 \text{ in.}^2$$

Since the minimum shear stress needed to rupture the steel is

$$\left(\frac{F}{A}\right)_{min} = 4 \times 10^4 \text{ lb/in.}^2$$

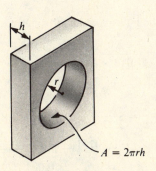

$A = 2\pi rh$

**Fig. 13-5**

the required force is

$$F = \left(\frac{F}{A}\right)_{min} (A) = \left(4 \times 10^4 \ \frac{lb}{in.^2}\right)(0.196 \ in.^2) = 7840 \ lb$$

## BULK MODULUS

When compressive forces act over the entire surface of a body, its volume decreases. If the compressive force per unit area $F/A$ is uniform, the *bulk modulus* is given by

$$B = -\frac{F/A}{\Delta V/V_0}$$

$$\text{Bulk modulus} = -\frac{\text{volume stress}}{\text{volume strain}}$$

The minus sign is included because an increase in the volume stress leads to a decrease in the volume.

Volume stresses occur when objects are immersed in liquids, since a liquid exerts a uniform force perpendicular to any surface in its interior. As will be discussed in Chapter 16, the stress $F/A$ exerted by a liquid is called *pressure p*, so we can also write

$$B = -\frac{p}{\Delta V/V_0}$$

## SOLVED PROBLEM 13.10

The pressure at a depth of 300 m in the ocean exceeds sea-level atmospheric pressure by 3.03 kPa. By how much does the volume of a 0.2-$m^3$ aluminum object contract when lowered to this depth in the ocean? The bulk modulus of aluminum is $7 \times 10^{10}$ Pa.

$$\Delta V = -\frac{pV_0}{B} = -\frac{(3.03 \times 10^3 \ Pa)(0.2 \ m^3)}{7 \times 10^{10} \ Pa} = -8.7 \times 10^{-9} \ m^3 = -8.7 \ mm^3$$

The minus sign means that the volume of the object decreases.

## SOLVED PROBLEM 13.11

The reciprocal of the bulk modulus $B$ of a liquid is called its *compressibility k*, so $k = 1/B$. The bulk modulus of water is $2.3 \times 10^9$ Pa. (*a*) Find its compressibility per atmosphere of pressure, where 1 atm = $1.013 \times 10^5$ Pa is the pressure exerted by the earth's atmosphere at sea level. (*b*) How much pressure in atmospheres is needed to compress a sample of water by 0.1 percent?

(*a*)   In terms of atmospheres, the bulk modulus of water is

$$B = \frac{2.3 \times 10^9 \ Pa}{1.013 \times 10^5 \ Pa/atm} = 2.27 \times 10^4 \ atm$$

and so its compressibility is

$$k = \frac{1}{B} = \frac{1}{2.27 \times 10^4 \ atm} = 4.4 \times 10^{-5} \ atm^{-1}$$

(*b*)   Here $\Delta V/V_0 = -0.1$ percent $= -0.001$. Hence the required pressure is

$$p = -\frac{1}{k}\frac{\Delta V}{V_0} = \frac{0.001}{4.4 \times 10^{-5} \ atm^{-1}} = 23 \ atm$$

# Multiple-Choice Questions

**13.1.** The stress on a wire that supports a load depends on which one or more of the following?

   (a) the wire's length    (c) acceleration of gravity
   (b) the wire's diameter    (d) the mass of the load

**13.2.** A load causes a wire to stretch by the amount $s$. If the same load is applied to another wire of the same material but twice as long and with twice the diameter, the second wire will stretch by the amount

   (a) $s/4$    (c) $s$
   (b) $s/2$    (d) $2s$

**13.3.** At its elastic limit, the first wire of Question 13.2 could support a load of $F$. At its elastic limit, the second wire could support a load of

   (a) $F/2$    (c) $2F$
   (b) $F$    (d) $4F$

**13.4.** Young's modulus for iron is $1.9 \times 10^{11}$ Pa. When an iron wire 1.0 m long with a cross-sectional area of 4.0 mm$^2$ supports a 100-kg load, the wire stretches by

   (a) 0.0027 mm    (c) 1.3 mm
   (b) 0.27 mm    (d) 3.7 mm

**13.5.** Young's modulus for brass is $1.3 \times 10^7$ lb/in.$^2$. When a brass rod 2 ft 4 in. long with a cross-sectional area of 0.50 in.$^2$ supports a 400-lb load, the rod stretches by

   (a) 0.00015 in.    (c) 0.0034 in.
   (b) 0.0017 in.    (d) 0.029 in.

**13.6.** Young's modulus for aluminum is $7.0 \times 10^{10}$ Pa. When an aluminum wire 1.5 mm in diameter and 50 cm long is stretched by 1.0 mm, the force applied to the wire is

   (a) 247 N    (c) 990 N
   (b) 315 N    (d) 2.42 kN

**13.7.** The elastic limit of aluminum is $1.8 \times 10^8$ Pa. The maximum mass the wire in Question 13.6 can support without going beyond its elastic limit is

   (a) 32 kg    (c) 623 kg
   (b) 65 kg    (d) 636 kg

**13.8.** The ultimate strength of concrete is $2.0 \times 10^7$ Pa in compression. A concrete cube 80 cm on each edge can support a maximum mass of

   (a) $1.3 \times 10^5$ kg    (c) $1.6 \times 10^6$ kg
   (b) $1.3 \times 10^6$ kg    (d) $1.3 \times 10^7$ kg

**13.9.** The shear strength of a steel sheet 2.5 mm thick is $3.0 \times 10^8$ Pa. The force needed to punch a hole 6.0 mm square in this sheet is

   (a) 7.2 kN    (c) 27 kN
   (b) 18 kN    (d) 18 MN

**13.10.** The bulk modulus of kerosene is 1.3 GPa. When a pressure of 1.8 MPa is applied to a liter of kerosene, its volume decreases by

(a)  1.4 mL      (c)  7.2 mL
(b)  2.3 mL      (d)  14 mL

# Supplementary Problems

**13.1.** A cable is shortened to half its original length. (a) How does this affect its elongation under a given load? (b) How does this affect the maximum load it can support without exceeding its elastic limit?

**13.2.** A cable is replaced by another one of the same length and material but of twice the diameter. (a) How does this affect its elongation under a given load? (b) How does this affect the maximum load it can support without exceeding its elastic limit?

**13.3.** A wire 5 m long and 4 mm in diameter supports a load of 80 kg. If the wire stretches by 2.6 mm, find the value of Young's modulus for its material.

**13.4.** A steel wire ($Y = 2.9 \times 10^7$ lb/in.$^2$) 6 ft long and 0.002 in.$^2$ in cross section supports a load of 15 lb. (a) What is its elongation? (b) What would its elongation be if the load were doubled to 30 lb?

**13.5.** A lead cube 20 mm on each edge is held in the jaws of a vise with a force of 3 kN. By how much is the cube compressed? Young's modulus for lead is $1.6 \times 10^{10}$ Pa.

**13.6.** The ultimate strength in tension of a certain type of steel is $5 \times 10^8$ Pa. What is the maximum tension a rod made of this steel and 25 mm in diameter can withstand?

**13.7.** An elevator cab weighing 3000 lb is designed for a maximum upward acceleration of 12 ft/s$^2$. If the stress in its cable is not to exceed 6000 lb/in.$^2$, what should the cable diameter be?

**13.8.** A steel cable of cross-sectional area 2.5 cm$^2$ supports a 1000-kg elevator. The elastic limit of the cable is $3 \times 10^8$ Pa. If the stress in the cable is not to exceed 20 percent of the elastic limit, find the maximum upward acceleration of the elevator.

**13.9.** Two steel beams are riveted together to form a single longer beam. Eight rivets 10 mm in diameter are used. If a tension force of 55 kN is applied to the new beam, what is the shearing stress on the rivets? How does this compare with their shearing strength of $3.5 \times 10^8$ Pa?

**13.10.** Find the force needed to punch a hole 1 in. square in a steel sheet 0.05 in. thick whose shearing strength is $5 \times 10^4$ lb/in.$^2$.

**13.11.** A punch press that exerts a force of 20 kN is used to punch holes 1 cm square in sheet aluminum. If the shear strength of aluminum is 70 MPa, what is the maximum thickness of aluminum sheet that can be used with the press?

**13.12.** When a pressure of 300 lb/in.$^2$ is applied to a mercury sample, it contracts by 0.008 percent. Find the bulk modulus of mercury.

## Answers to Multiple-Choice Questions

**13.1.** (*b*), (*c*), (*d*)     **13.6.**   (*a*)

**13.2.** (*b*)     **13.7.**   (*a*)

**13.3.** (*d*)     **13.8.**   (*b*)

**13.4.** (*c*)     **13.9.**   (*b*)

**13.5.** (*b*)     **13.10.**  (*a*)

## Answers to Supplementary Problems

**13.1.** (*a*) Its elongation will be half the former amount.   (*b*) No change.

**13.2.** (*a*) Its elongation will be one-quarter the former amount.

(*b*) The maximum load will be four times greater.

**13.3.** $1.2 \times 10^{11}$ Pa

**13.4.** (*a*) 0.0186 in.   (*b*) 0.0372 in.

**13.5.** $9.4 \times 10^{-3}$ mm

**13.6.** 245 kN

**13.7.** 0.936 in.$^2$

**13.8.** 5.2 m/s$^2$

**13.9.** $8.8 \times 10^7$ Pa; 25%

**13.10.** 10,000 lb

**13.11.** 7.1 mm

**13.12.** $3.8 \times 10^6$ lb/in.$^2$

# Chapter 14

## Simple Harmonic Motion

### RESTORING FORCE

When an elastic object such as a spring is stretched or compressed, a *restoring force* appears that tries to return the object to its normal length. It is this restoring force that must be overcome by the applied force in order to deform the object. From Hooke's law, the restoring force $F$ is proportional to the displacement $s$ provided the elastic limit is not exceeded. Hence

$$F_r = -ks$$

$$\text{Restoring force} = -(\text{force constant})(\text{displacement})$$

The minus sign is required because the restoring force acts in the opposite direction to the displacement. The greater the value of the *force constant $k$*, the greater the restoring force for a given displacement and the greater the applied force $F = ks$ needed to produce the displacement.

### ELASTIC POTENTIAL ENERGY

Because work must be done by an applied force to stretch or compress an object, the object has *elastic potential energy* as a result, where

$$\text{PE} = \tfrac{1}{2}ks^2$$

When a deformed elastic object is released, its elastic potential energy turns into kinetic energy or into work done on something else.

### SOLVED PROBLEM 14.1

A force of 5 N compresses a spring by 4 cm. (*a*) Find the force constant of the spring. (*b*) Find the elastic potential energy of the compressed spring.

(*a*) $$k = \frac{F}{s} = \frac{5 \text{ N}}{0.04 \text{ m}} = 125 \text{ N/m}$$

(*b*) $$\text{PE} = \tfrac{1}{2}ks^2 = (\tfrac{1}{2})(125 \text{ N/m})(0.04 \text{ m})^2 = 0.1 \text{ J}$$

### SOLVED PROBLEM 14.2

A spring with a force constant of 200 N/m is compressed by 8 cm. A 250-g ball is placed against the end of the spring, which is then released. What is the ball's velocity when it leaves the spring?

The kinetic energy of the ball equals the elastic potential energy of the compressed spring. Hence

$$\text{KE} = \text{PE}$$
$$\tfrac{1}{2}mv^2 = \tfrac{1}{2}ks^2$$

$$v = \sqrt{\frac{k}{m}}\, s = \sqrt{\frac{200 \text{ N/m}}{0.25 \text{ kg}}}(0.08 \text{ m}) = 2.26 \text{ m/s}$$

**SOLVED PROBLEM 14.3**

Two springs, one of force constant $k_1$ and the other of force constant $k_2$, are connected end-to-end. (a) Find the force constant $k$ of the combination. (b) If $k_1 = 5$ N/m and $k_2 = 10$ N/m, find $k$.

(a)  When a force $F$ is applied to the combination, each spring is acted on by this force. Hence the respective elongations of the springs are

$$s_1 = \frac{F}{k_1} \qquad s_2 = \frac{F}{k_2}$$

and the total elongation of the combination is

$$s = s_1 + s_2 = \frac{F}{k_1} + \frac{F}{k_2} = \frac{F(k_1 + k_2)}{k_1 k_2}$$

Since $F = ks$ for the combination,

$$k = \frac{F}{s} = \frac{k_1 k_2}{k_1 + k_2}$$

(b)
$$k = \frac{(5 \text{ N/m})(10 \text{ N/m})}{5 \text{ N/m} + 10 \text{ N/m}} = 3.33 \text{ N/m}$$

The force constant of the combination is smaller than either of the individual force constants.

## SIMPLE HARMONIC MOTION

In *periodic motion*, a body repeats a certain motion indefinitely, always returning to its starting point after a constant time interval and then starting a new cycle. *Simple harmonic motion* is periodic motion that occurs when the restoring force on a body displaced from an equilibrium position is proportional to the displacement and in the opposite direction. A mass $m$ attached to a spring executes simple harmonic motion when the spring is pulled out and released. The spring's PE becomes KE as the mass begins to move, and the KE of the mass becomes PE again as its momentum causes the spring to overshoot the equilibrium position and become compressed (Fig. 14-1).

The *amplitude A* of a body undergoing simple harmonic motion is the maximum value of its displacement on either side of the equilibrium position.

## PERIOD AND FREQUENCY

The *period T* of a body undergoing simple harmonic motion is the time needed for one complete cycle; $T$ is independent of the amplitude $A$. If the acceleration of the body is $a$ when its displacement is $s$,

$$T = 2\pi \sqrt{-\frac{s}{a}}$$

$$\text{Period} = 2\pi \sqrt{-\frac{\text{displacement}}{\text{acceleration}}}$$

In the case of a body of mass $m$ attached to a spring of force constant $k$, $F_r = -ks = ma$, and so $-s/a = m/k$. Hence

$$T = 2\pi \sqrt{\frac{m}{k}} \qquad \text{stretched spring}$$

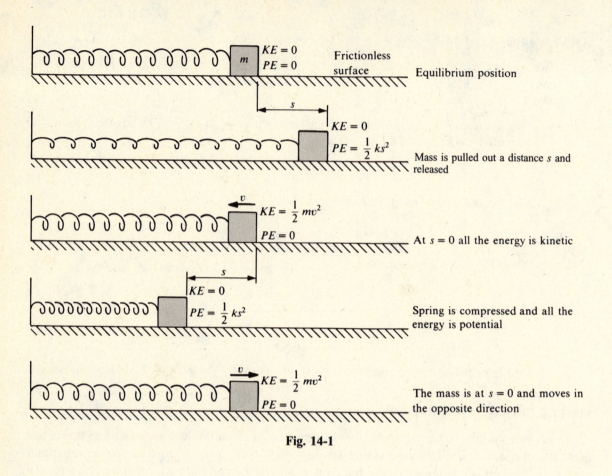

Fig. 14-1

The *frequency f* of a body undergoing simple harmonic motion is the number of cycles per second it executes, so that

$$f = \frac{1}{T}$$

$$\text{Frequency} = \frac{1}{\text{period}}$$

The unit of frequency is the *hertz* (Hz), where 1 Hz = 1 cycle/s.

### SOLVED PROBLEM 14.4

The amplitude of a simple harmonic oscillator is doubled. How does this affect (*a*) the period, (*b*) the total energy, and (*c*) the maximum velocity of the oscillator?

(*a*)   The period of such an oscillator does not depend on the amplitude of its motion; hence $T$ is unchanged.

(*b*)   The total energy of the oscillator is equal to the elastic potential energy $\frac{1}{2}kA^2$ at either extreme of its motion, when $v = 0$. Hence doubling $A$ means that the total energy increases fourfold.

(*c*)   The maximum velocity occurs at $s = 0$, the equilibrium position, when the entire energy of the oscillator is KE. Since KE $= \frac{1}{2}mv^2$ and the total energy is four times greater than originally, $v_{max}$ must double.

## SOLVED PROBLEM 14.5

A 100-g object is suspended from a spring whose force constant is 50 N/m. (*a*) By how much does the spring stretch? (*b*) What is the period of oscillation of the system? (*c*) What is its frequency?

(*a*)    Here $F = mg = (0.1 \text{ kg})(9.8 \text{ m/s}^2) = 0.98$ N, and so the spring stretches by

$$s = \frac{F}{k} = \frac{0.98 \text{ N}}{50 \text{ N/m}} = 0.0196 \text{ m} = 1.96 \text{ cm}$$

(*b*)

$$T = 2\pi\sqrt{\frac{m}{k}} = 2\pi\sqrt{\frac{0.1 \text{ kg}}{50 \text{ N/m}}} = 0.281 \text{ s}$$

(*c*)

$$f = \frac{1}{T} = \frac{1}{0.281 \text{ s}} = 3.56 \text{ s}^{-1} = 3.56 \text{ Hz}$$

## SOLVED PROBLEM 14.6

An object of unknown mass is suspended from a spring, which stretches by 10 cm as a result. If the system is set in oscillation, what will its frequency be?

The force $F$ that causes the spring to stretch by $s = 10 \text{ cm} = 0.1$ m is the weight $mg$ of the unknown mass, so the force constant of the spring is

$$k = \frac{F}{s} = \frac{mg}{0.1 \text{ m}} = 10mg \text{ m}^{-1}$$

The period of oscillation of the system is therefore

$$T = 2\pi\sqrt{\frac{m}{k}} = 2\pi\sqrt{\frac{m}{10mg \text{ m}^{-1}}} = 2\pi\sqrt{\frac{1}{(10 \text{ m}^{-1})(9.8 \text{ m/s}^2)}} = 0.635 \text{ s}$$

and the frequency is $f = 1/T = 1.58$ Hz.

## SOLVED PROBLEM 14.7

A spring whose force constant is 12 lb/ft oscillates up and down with a period of 0.5 s when a wrench is suspended from it. How much does the wrench weigh?

Since $m = w/g$,

$$T = 2\pi\sqrt{\frac{m}{k}} = 2\pi\sqrt{\frac{w}{gk}}$$

$$w = \frac{gkT^2}{4\pi^2} = \frac{(32 \text{ ft/s}^2)(12 \text{ lb/ft})(0.5 \text{ s})^2}{4\pi^2} = 2.4 \text{ lb}$$

## DISPLACEMENT, VELOCITY, AND ACCELERATION

If $t = 0$ when a body undergoing simple harmonic motion is in its equilibrium position of $s = 0$ and is moving in the direction of increasing $s$, then at any time $t$ thereafter its displacement is

$$s = A \sin 2\pi ft$$

Often this formula is written

$$s = A \sin \omega t$$

where $\omega = 2\pi f$ is the *angular frequency* of the motion in radians per second. (If instead the body is at $s = +A$ when $t = 0$, then $s = A \cos 2\pi ft = A \cos \omega t$.) Figure 14-2 is a graph of $s$ versus $t$.

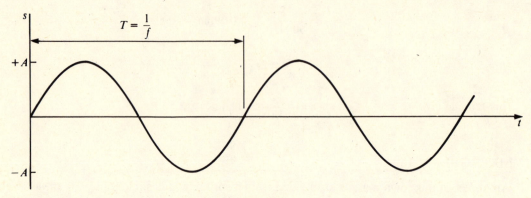

**Fig. 14-2**

The velocity of the body at the time $t$ is

$$v = 2\pi fA \cos 2\pi ft = \omega A \cos \omega t$$

When $v$ is $+$, the body is moving in the direction of increasing $s$; when $v$ is $-$, it is moving in the direction of decreasing $s$. In terms of the displacement $s$, the magnitude of the velocity is

$$v = 2\pi f \sqrt{A^2 - s^2}$$

The acceleration of the body at time $t$ is

$$a = -4\pi^2 f^2 A \sin 2\pi ft = -\omega^2 A \sin \omega t$$

In terms of the displacement $s$, the acceleration is

$$a = 4\pi^2 f^2 s$$

## SOLVED PROBLEM 14.8

An object is oscillating in simple harmonic motion with an amplitude of 10 cm and a period of 2 s. Find the magnitudes of its velocity and acceleration when its displacement $s$ from its equilibrium position is (*a*) 0, (*b*) +5 cm, and (*c*) −10 cm.

(*a*)  The frequency of the oscillations is $f = 1/T = 0.5$ s$^{-1}$. At $s = 0$,

$$v = 2\pi f \sqrt{A^2 - s^2} = 2\pi fA = (2\pi)(0.5 \text{ s}^{-1})(0.10 \text{ m}) = 0.314 \text{ m/s}$$

$$a = -4\pi^2 f^2 s = 0$$

(*b*)  At $s = +5$ cm,

$$v = 2\pi f \sqrt{A^2 - s^2} = (2\pi)(0.5 \text{ s}^{-1})\sqrt{(0.10 \text{ m})^2 - (0.05 \text{ m})^2} = 0.272 \text{ m/s}$$

$$a = -4\pi^2 f^2 s = -(4\pi^2)(0.5 \text{ s}^{-1})^2(0.05 \text{ m}) = -0.493 \text{ m/s}^2$$

(c)   At $s = -10$ cm $= -A$,

$$v = 2\pi f \sqrt{A^2 - s^2} = 0$$

$$a = -4\pi^2 f^2 s = -(4\pi^2)(0.5 \text{ s}^{-1})^2(0.10 \text{ m}) = 0.987 \text{ m/s}^2$$

## SOLVED PROBLEM 14.9

What is the maximum velocity of a body undergoing simple harmonic motion? At what displacement does this velocity occur?

For a body in simple harmonic motion, $v = 2\pi f \sqrt{A^2 - s^2}$, which is a maximum when $s = 0$. Hence

$$v_{max} = 2\pi f \sqrt{A^2 - 0} = 2\pi f A$$

This formula gives only the magnitude of $v_{max}$, not its direction.

## SOLVED PROBLEM 14.10

What is the maximum acceleration of a body undergoing simple harmonic motion? At what displacement does this acceleration occur?

For a body in simple harmonic motion, $a = -4\pi^2 f^2 s$, which is a maximum when $s = \pm A$, where $A$ is the amplitude. Hence

$$a_{max} = \pm 4\pi^2 f^2 A$$

The acceleration is negative when $s = +A$, and the acceleration is positive when $s = -A$.

## SOLVED PROBLEM 14.11

Each piston of a certain car engine has a mass of 0.5 kg and has a "stroke" (total travel distance) of 120 mm. When the engine is operating at 3000 rev/min, find (a) the maximum velocity of each piston, (b) its maximum acceleration, and (c) the maximum force on it. Assume that the pistons move up and down in simple harmonic motion, which is approximately true.

(a)   The frequency of oscillation of each piston is 3000 cycles/min, which is

$$f = \frac{3000 \text{ cycles/min}}{60 \text{ s/min}} = 50 \text{ Hz}$$

The amplitude of the motion is half the stroke, so $A = 60$ mm $= 0.6$ m. Hence the maximum piston velocity is

$$v_{max} = 2\pi f A = (2\pi)(50 \text{ s}^{-1})(0.06 \text{ m}) = 18.8 \text{ m/s}$$

(b)   The maximum acceleration is

$$a_{max} = 4\pi^2 f^2 A = (4\pi^2)(50 \text{ s}^{-1})^2(0.06 \text{ m}) = 5.92 \times 10^3 \text{ m/s}$$

(c)   The maximum force on each piston is

$$F_{max} = m a_{max} = (0.5 \text{ kg})(5.92 \times 10^3 \text{ m/s}^2) = 2.96 \times 10^3 \text{ N} = 2.96 \text{ kN}$$

## PENDULUMS

A *simple pendulum* has its entire mass concentrated at the end of a string, as in Fig. 14-3(a), and it undergoes simple harmonic motion provided that the arc through which it travels is only a few

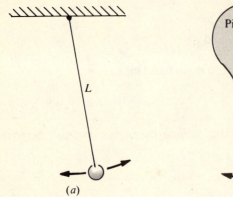
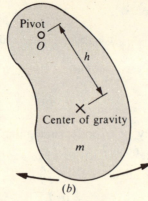
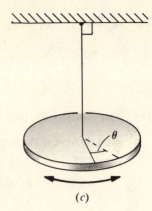

(a)                                    (b)                                    (c)

**Fig. 14-3**

degrees. The period of a simple pendulum of length $L$ is

$$T = 2\pi\sqrt{\frac{L}{g}} \quad \text{simple pendulum}$$

The *physical pendulum* of Fig. 14-3($b$) is an object of any kind which is pivoted so that it can oscillate freely. If the moment of inertia of the object about the pivot $O$ is $I$, its mass is $m$, and the distance from its center of gravity to the pivot is $h$, then its period is

$$T = 2\pi\sqrt{\frac{I}{mgh}} \quad \text{physical pendulum}$$

A *torsion pendulum* consists of an object suspended by a wire or thin rod, as in Fig. 14-3($c$), which undergoes rotational simple harmonic oscillations. From Hooke's law, the torque $\tau$ needed to twist the object through an angle $\theta$ is

$$\tau = K\theta$$

provided the elastic limit is not exceeded, where $K$ is a constant that depends on the material and dimensions of the wire. If $I$ is the moment of inertia of the object about its point of suspension, the period of the oscillations is

$$T = 2\pi\sqrt{\frac{I}{K}} \quad \text{torsion pendulum}$$

### SOLVED PROBLEM 14.12

($a$) A pendulum clock is in an elevator that descends at a constant velocity. Does it keep correct time? ($b$) The same clock is in an elevator in free fall. Does it keep correct time?

($a$)   The motion of the pendulum bob is not affected by motion of its support at constant velocity, so the clock keeps correct time.

($b$)   In free fall the pendulum's support has the same downward acceleration of $g$ as the bob, so no oscillations occur and the clock does not operate at all.

### SOLVED PROBLEM 14.13

A lamp is suspended from a high ceiling with a cord 12 ft long. Find its period of oscillation.

$$T = 2\pi\sqrt{\frac{L}{g}} = 2\pi\sqrt{\frac{12 \text{ ft}}{32 \text{ ft/s}^2}} = 3.85 \text{ s}$$

**SOLVED PROBLEM 14.14**

Find the length in meters of a simple pendulum whose period is 2 s.

The first step is to solve the formula $T = 2\pi\sqrt{L/g}$ for $L$. We proceed as follows:

$$T^2 = \frac{4\pi^2 L}{g}$$

$$L = \frac{gT^2}{4\pi^2}$$

Now we substitute $g = 9.8$ m/s$^2$ and $T = 2$ s, and we obtain

$$L = \frac{(9.8 \text{ m/s}^2)(2 \text{ s})^2}{4\pi^2} = 0.993 \text{ m}$$

**SOLVED PROBLEM 14.15**

A broomstick 1.5 m long is suspended from one end and set in oscillation. (a) What is the period of oscillation? (The moment of inertia of a thin rod pivoted at one end is $I = \frac{1}{3}mL^2$.) (b) What would be the length of a simple pendulum with the same period?

(a) The distance $h$ from the pivot to the center of gravity of the broomstick is $L/2$. Hence

$$T = 2\pi\sqrt{\frac{I}{mgh}} = 2\pi\sqrt{\frac{mL^2/3}{mgL/2}} = 2\pi\sqrt{\frac{2L}{3g}} = 2\pi\sqrt{\frac{(2)(1.5 \text{ m})}{(3)(9.8 \text{ m/s}^2)}} = 2.01 \text{ s}$$

(b) From the solution to Prob. 14.14,

$$L = \frac{gT^2}{4\pi^2} = \frac{(9.8 \text{ m/s}^2)(2.01 \text{ s})^2}{4\pi^2} = 1.0 \text{ m}$$

**SOLVED PROBLEM 14.16**

A certain torsion pendulum consists of a 2-kg horizontal aluminum disk 15 cm in radius that is suspended from its center by a wire. When a torque of 1 N·m is applied to the disk, it rotates through 15°. Find the frequency of oscillation of the disk.

Since 1° = 0.01745 rad, 15° = 0.262 rad, and the torque constant of the suspension wire is

$$K = \frac{\tau}{\theta} = \frac{1 \text{ N·m}}{0.262 \text{ rad}} = 3.82 \text{ N·m/rad}$$

From Fig. 10-3 the moment of inertia of the disk is

$$I = \frac{1}{2}MR^2 = (\frac{1}{2})(2 \text{ kg})(0.15 \text{ m})^2 = 0.0225 \text{ kg·m}^2$$

The period of oscillation is therefore

$$T = 2\pi\sqrt{\frac{I}{K}} = 2\pi\sqrt{\frac{0.0225 \text{ kg·m}^2}{3.82 \text{ N·m/rad}}} = 0.482 \text{ s}$$

The frequency is $f = 1/T = 1/0.482$ s = 2.07 Hz.

**SOLVED PROBLEM 14.17**

To determine its moment of inertia about a diameter, a brass hoop is suspended by a wire whose torsion constant is $K = 25$ N·m/rad. The hoop executes 6 oscillations per second. What is the moment of inertia?

From the formula $T = 2\pi\sqrt{I/K}$ for the period of a torsion pendulum we obtain $I = T^2K/(4\pi^2)$. Since $T = 1/f$, the moment of inertia of the hoop is

$$I = \frac{K}{4\pi^2 f^2} = \frac{25\ \text{N}\cdot\text{m/rad}}{(4\pi^2)(6\ \text{s}^{-1})^2} = 0.0176\ \text{kg}\cdot\text{m}^2$$

# *Multiple-Choice Questions*

**14.1.** The period of a simple harmonic oscillator does not depend on which one or more of the following?

    (*a*)   its mass         (*c*)   its force constant
    (*b*)   its frequency     (*d*)   its amplitude

**14.2.** When an object that is undergoing simple harmonic motion passes through its equilibrium position, its velocity is

    (*a*)   zero                (*c*)   its maximum value
    (*b*)   half its maximum value     (*d*)   none of the above

**14.3.** A spring is cut in three equal parts. If its original force constant was $k$, each new spring has a force constant of

    (*a*)   $k/3$     (*c*)   $3k$
    (*b*)   $k$       (*d*)   $9k$

**14.4.** A vertical spring 60 mm long resting on a table is compressed by 5.0 mm when a 200-g mass is placed on it. The force constant of the spring is

    (*a*)   1.6 N/m     (*c*)   196 N/m
    (*b*)   40 N/m      (*d*)   392 N/m

**14.5.** A spring is stretched by 30 mm when a force of 0.40 N is applied to it. The potential energy of the stretched spring is

    (*a*)   $2.8 \times 10^{-5}$ J     (*c*)   $6.0 \times 10^{-3}$ J
    (*b*)   $7.2 \times 10^{-5}$ J     (*d*)   $1.2 \times 10^{-2}$ J

**14.6.** If it is pressed down and released, the mass of Question 14.4 will oscillate up and down with a period of

    (*a*)   0.0032 s     (*c*)   0.14 s
    (*b*)   0.057 s      (*d*)   0.44 s

**14.7.** A particle that undergoes simple harmonic motion has a period of 0.40 s and an amplitude of 12 mm. The maximum velocity of the particle is

    (*a*)   3 cm/s     (*c*)   38 cm/s
    (*b*)   19 cm/s     (*d*)   43 cm/s

**14.8.** If a mass of 15 g is to oscillate at 12 Hz, it should hang from a spring whose force constant is

    (*a*)   1.1 N/m     (*c*)   7.1 N/m
    (*b*)   1.3 N/m     (*d*)   85 N/m

**14.9.**   The end of a diving board moves down by 30 cm when a girl stands on it. If she then bounces up and down, the frequency of the oscillations

   (*a*)   is 0.91 Hz      (*c*)   is 2.3 Hz
   (*b*)   is 1.1 Hz       (*d*)   depends on her mass

**14.10.**   To increase the period of a harmonic oscillator from 3 to 6 s, its original mass of 20 g should be changed to

   (*a*)   5 g       (*c*)   40 g
   (*b*)   10 g      (*d*)   80 g

**14.11.**   The 400-g piston in a compressor oscillates up and down through a total distance of 80 mm. The maximum force on the piston when it goes through 10 cycles/s is

   (*a*)   1.0 N      (*c*)   63 N
   (*b*)   6.3 N      (*d*)   126 N

**14.12.**   A boy holding on to the end of a rope swings back and forth once every 4.0 s. The length of the rope

   (*a*)   is 2.5 m      (*c*)   is 6.2 m
   (*b*)   is 4.0 m      (*d*)   cannot be found without knowing the boy's mass

# Supplementary Problems

**14.1.**   What change in mass is required to double the frequency of a harmonic oscillator?

**14.2.**   The total energy of a harmonic oscillator is doubled. What effect does this have (*a*) on the amplitude of the oscillations and (*b*) on their frequency?

**14.3.**   A 500-g object is dropped on a vertical spring from a height of 2 m. If the maximum compression of the spring is 10 cm, find its force constant.

**14.4.**   A toy gun uses a spring whose force constant is 100 N/m to propel an 8-g rubber pellet. If the spring is compressed by 5 cm when the trigger is pulled, what is the pellet's initial velocity?

**14.5.**   A 5-lb object is suspended from a spring whose force constant is 40 lb/ft. Find (*a*) the amount by which the spring is stretched, (*b*) the elastic potential energy of the stretched spring, and (*c*) the period of oscillation of the system.

**14.6.**   A 30-g mass is suspended from a spring whose force constant is 20 N/m. Find (*a*) the amount by which the spring is stretched, (*b*) the elastic potential energy of the stretched spring, and (*c*) the period of oscillation of the system.

**14.7.**   A 200-lb portable gasoline-powered generating set is placed on four springs, which are depressed by 0.5 in. What is the natural frequency of vibration of this system?

**14.8.**   A spring has a 0.3-s period of oscillation when a 30-N weight is suspended from it. By how much will the spring stretch when a 50-N weight replaces the 30-N weight?

**14.9.** A shaking table used to study the effects of accelerations on electronic equipment undergoes simple harmonic motion with an amplitude of 20 mm. At what frequency will the maximum acceleration equal 10g?

**14.10.** An object is oscillating in simple harmonic motion with an amplitude of 1 cm and a period of 0.2 s. Find the magnitudes of its velocity and acceleration when its displacement from the equilibrium position is 0.3 cm.

**14.11.** The piston of a refrigeration compressor weighs 1 lb and has a stroke of 3 in. When the compressor is operated at 600 rev/min, find the piston's maximum velocity and acceleration and the maximum force acting on it.

**14.12.** Find the frequency of a simple pendulum 20 cm long.

**14.13.** What is the length in inches of a simple pendulum whose period is 1 s?

**14.14.** A pendulum 1.00 m long oscillates 30.0 times per minute in a certain location. What is the value of $g$ there?

**14.15.** The prongs of a tuning fork vibrate at 440 Hz (the musical note A) with an amplitude of 0.5 mm at the tips. Find the maximum velocity of the tips.

**14.16.** A uniform steel girder 20 ft long is suspended from one end. What is the period of its oscillations?

**14.17.** A 24-kg wooden sphere whose diameter is 40 cm is suspended by a wire. A torque of 0.5 N·m is found to rotate the sphere through 10°. Find the period of oscillation of the sphere.

**14.18.** A disk whose moment of inertia is 0.5 kg·m$^2$ is suspended by a thin rod. When the disk is turned through a few degrees and then released, it oscillates back and forth twice per second. Find the torsion constant of the rod.

## *Answers to Multiple-Choice Questions*

| | | | |
|---|---|---|---|
| **14.1.** | (*d*) | **14.7.** | (*b*) |
| **14.2.** | (*c*) | **14.8.** | (*d*) |
| **14.3.** | (*c*) | **14.9.** | (*a*) |
| **14.4.** | (*d*) | **14.10.** | (*d*) |
| **14.5.** | (*c*) | **14.11.** | (*c*) |
| **14.6.** | (*c*) | **14.12.** | (*b*) |

# Answers to Supplementary Problems

**14.1.** The mass must be reduced to one-fourth its original value.

**14.2.** (*a*)  $\sqrt{2}$ greater    (*b*)  No change

**14.3.** 2.06 kN/m

**14.4.** 5.6 m/s

**14.5.** (*a*)  0.125 ft    (*b*)  0.3125 ft·lb    (*c*)  0.393 s

**14.6.** (*a*)  1.47 cm    (*b*)  $2.16 \times 10^{-3}$ J    (*c*)  0.24 s

**14.7.** 2.21 Hz

**14.8.** 37 mm

**14.9.** 11 Hz

**14.10.** 0.300 m/s; 2.96 m/s$^2$

**14.11.** 7.85 ft/s; 493 ft/s$^2$; 15.4 lb

**14.12.** 1.11 Hz

**14.13.** 9.73 in.

**14.14.** 9.87 m/s$^2$

**14.15.** 0.69 m/s

**14.16.** 7.33 s

**14.17.** 2.30 s

**14.18.** 79 N·m/rad

# Chapter 15

# Waves and Sound

## WAVES

A *wave* is, in general, a disturbance that moves through a medium. (An exception is an *electromagnetic wave*, which can travel through a vacuum. Examples are light and radio waves.) A wave carries energy, but there is no transport of matter. In a *periodic wave*, pulses of the same kind follow one another in regular succession.

In a *transverse wave*, the particles of the medium move back and forth perpendicular to the direction of the wave. Waves that travel down a stretched string when one end is shaken are transverse (Fig. 15-1).

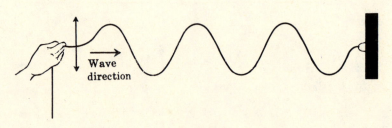

**Fig. 15-1**

In a *longitudinal wave*, the particles of the medium move back and forth in the same direction as the wave. Waves that travel down a coil spring when one end is pulled out and released are longitudinal (Fig. 15-2). Sound waves are also longitudinal.

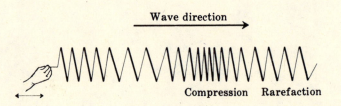

**Fig. 15-2**

Water waves are a combination of longitudinal and transverse waves. Each particle near the surface moves in a circular orbit, as shown in Fig. 15-3, so that a succession of crests and troughs occurs. At a crest, the surface water moves in the direction of the wave; at a trough, it moves in the opposite direction. As in all types of wave motion, there is no net movement of matter from one place to another.

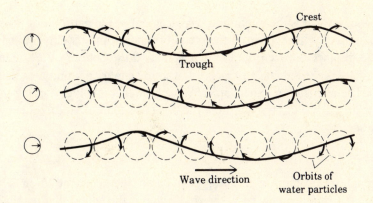

Fig. 15-3

## WAVE PROPERTIES

The *period T* of a wave is the time required for one complete wave to pass a given point. The *frequency f* is the number of waves that pass that point per second [Fig. 15-4(a)], so

$$f = \frac{1}{T}$$

$$\text{Frequency} = \frac{1}{\text{period}}$$

The *wavelength* $\lambda$ (Greek letter *lambda*) of a periodic wave is the distance between adjacent wave crests [Fig. 15-4(b)]. Frequency and wavelength are related to wave velocity by

$$v = f\lambda$$

$$\text{Wave velocity} = (\text{frequency})(\text{wavelength})$$

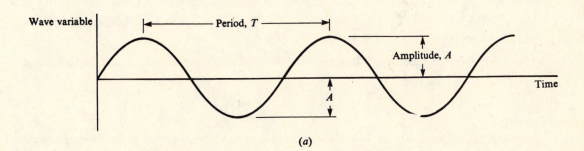

(a)

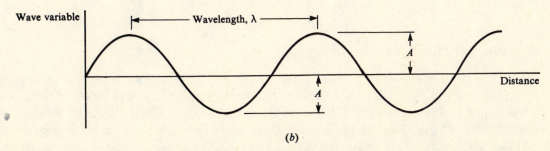

(b)

Fig. 15-4

The *amplitude A* of a wave is the maximum displacement of the particles of the medium through which the wave passes on either side of their equilibrium positions. In a transverse wave, the amplitude is half the distance between the top of a crest and the bottom of a trough (Fig. 15-4).

The *intensity I* of a wave is the rate at which it transports energy per unit area perpendicular to its direction of motion. The intensity of a mechanical wave (one that involves moving matter, in contrast to, say, an electromagnetic wave) is proportional to $f^2$, the square of its frequency, and to $A^2$, the square of its amplitude.

## SOLVED PROBLEM 15.1

As a phonograph record turns, a certain groove passes the needle at 25 cm/s. If the wiggles in the groove are 0.1 mm apart, what is the frequency of the sound that results?

Here the wavelength of the wiggles is $\lambda = 0.1$ mm $= 10^{-4}$, so they pass the needle at the rate of

$$f = \frac{v}{\lambda} = \frac{0.25 \text{ m/s}}{10^{-4} \text{ m}} = 2500 \text{ Hz}$$

This is therefore the frequency of the sound waves produced by the electronic system of the record player.

## SOLVED PROBLEM 15.2

The velocity of sound in seawater is 1531 m/s. Find the wavelength in seawater of a sound wave whose frequency is 256 Hz.

$$\lambda = \frac{v}{f} = \frac{1531 \text{ m/s}}{256 \text{ Hz}} = 5.98 \text{ m}$$

## SOLVED PROBLEM 15.3

A tuning fork vibrating at 300 Hz is placed in a tank of water. (*a*) Find the frequency and wavelength of the sound waves in the water. (*b*) Find the frequency and wavelength of the sound waves produced in the air above the tank by the vibrations of the water surface. The velocity of sound is 4913 ft/s in water and 1125 ft/s in air.

(*a*)   In the water, the frequency of the sound waves is the 300 Hz of their source, and their wavelength is

$$\lambda_1 = \frac{v_1}{f} = \frac{4913 \text{ ft/s}}{300 \text{ Hz}} = 16.4 \text{ ft}$$

(*b*)   In the air, the frequency of the sound waves is the same as the frequency of their source, which is the vibrating water surface. Hence $f = 300$ Hz. The wavelength is different, however:

$$\lambda_2 = \frac{v_2}{f} = \frac{1125 \text{ ft/s}}{300 \text{ Hz}} = 3.75 \text{ ft}$$

## SOLVED PROBLEM 15.4

A certain radar emits 9400-MHz radio waves in groups 0.08 $\mu$s in duration. (The time needed for these groups to reach a target, be reflected, and return to the radar indicates the distance of the target.) The velocity of these waves, like other electromagnetic waves, is $c = 3.00 \times 10^8$ m/s. Find (*a*) the wavelength of these waves, (*b*) the length of each wave group, which governs how precisely the radar can measure distances, and (*c*) the number of waves in each group.

(a)   Since 1 MHz = $10^6$ Hz, 9400 MHz = $9.4 \times 10^9$ Hz, and

$$\lambda = \frac{c}{f} = \frac{3.00 \times 10^8 \text{ m/s}}{9.4 \times 10^9 \text{ Hz}} = 3.19 \times 10^{-2} \text{ m} = 3.19 \text{ cm}$$

(b)   Since 1 $\mu s$ = $10^{-6}$ s, 0.08 $\mu s$ = $8 \times 10^{-8}$ s, and the length $s$ of each wave group is

$$s = ct = (3.00 \times 10^8 \text{ m/s})(8 \times 10^{-8} \text{ s}) = 24 \text{ m}$$

(c)   There are two ways to find the number $n$ of waves in each group:

$$n = ft = (9.4 \times 10^9 \text{ Hz})(8 \times 10^{-8} \text{ s}) = 752 \text{ waves}$$

$$n = \frac{s}{\lambda} = \frac{24 \text{ m}}{3.19 \times 10^{-2} \text{ m}} = 752 \text{ waves}$$

## SOLVED PROBLEM 15.5

An anchored boat is observed to rise and fall through a total range of 2 m once every 4 s as waves whose crests are 30 m apart pass it. Find (a) the frequency of the waves, (b) their velocity, (c) their amplitude, and (d) the velocity of an individual water particle at the surface.

(a)
$$f = \frac{1}{T} = \frac{1}{4 \text{ s}} = 0.25 \text{ Hz}$$

(b)
$$v = f\lambda = (0.25 \text{ Hz})(30 \text{ m}) = 7.5 \text{ m/s}$$

(c)   The amplitude is half the total range so $A = 1$ m.

(d)   As each wave passes, the water particles at the surface move in circular orbits of radius $r = A = 1$ m (see Fig. 15-3). The circumference of such an orbit is

$$s = 2\pi r = (2\pi)(1 \text{ m}) = 6.28 \text{ m}$$

The waves have the period 4 s, which means that each surface water particle must move through its 6.28-m orbit in 4 s. The velocity of such a water particle is therefore

$$V = \frac{s}{T} = \frac{6.28 \text{ m}}{4 \text{ s}} = 1.57 \text{ m/s}$$

Note that the *wave* velocity here is 7.5 m/s, nearly five times greater. This signifies that the motion of a wave can be much faster than the motions of the individual particles of the medium in which the wave travels.

## LOGARITHMS

Although logarithms have many other uses, their chief application in applied physics is in connection with the decibel, which is described in the next section. Logarithms are discussed here only to the extent required for this purpose.

The *logarithm* of a number $N$ is the power $n$ to which 10 must be raised in order that $10^n = N$. That is,

$$N = 10^n \qquad \text{therefore} \qquad \log N = n$$

(Logarithms are not limited to a base of 10, but base-10 logarithms are the most common and are all that are needed here.) For instance,

$$1000 = 10^3 \qquad \text{therefore} \qquad \log 1000 = 3$$
$$0.01 = 10^{-2} \qquad \text{therefore} \qquad \log 0.01 = -2$$

Logarithms are not limited to powers of 10 that are whole numbers. For instance,

$$5 = 10^{0.669} \qquad \text{therefore} \qquad \log 5 = 0.669$$
$$240 = 10^{2.380} \qquad \text{therefore} \qquad \log 240 = 2.380$$

To find the logarithm of a number with a calculator, enter the value of the number and press the LOG button.

Logarithms are defined only for positive numbers. The quantity $10^n$ is positive whether $n$ is negative, positive, or 0; and since $n$ is the logarithm of $10^n$, it can only describe a positive number.

The *antilogarithm* of a quantity $n$ is the number $N$ whose logarithm it is. That is,

$$\text{If} \qquad \log N = n \qquad \text{then} \qquad \text{antilog } n = N$$

To find an antilogarithm with a calculator, enter the value of the logarithm and press the $10^x$ (INV LOG on some calculators) button. For instance,

$$\text{If} \qquad \log 5 = 0.669 \qquad \text{then} \qquad \text{antilog } 0.669 = 5$$

Because of the way logarithms are defined, the logarithm of a product equals the sum of the logarithms of the factors:

$$\log xy = \log x + \log y$$

Other useful relations are

$$\log \frac{x}{y} = \log x - \log y$$
$$\log x^n = n \log x$$

## SOUND

*Sound waves* are longitudinal waves in which alternate regions of compression and rarefaction move away from a source. Sound waves can travel through solids, liquids, and gases. The velocity of sound is a constant for a given material at a given pressure and temperature; in air at 1-atm pressure and 20°C it is 343 m/s = 1125 ft/s.

When sound waves spread out uniformly in space, their intensity decreases inversely with the square of the distance $R$ from their source. Thus if the intensity of a certain sound is $I_1$ at the distance $R_1$, its intensity $I_2$ at the distance $R_2$ can be found from

$$\frac{I_2}{I_1} = \frac{R_1^2}{R_2^2}$$

The response of the human ear to sound intensity is not proportional to the intensity, so doubling the actual intensity of a certain sound does not lead to the sensation of a sound twice as loud but only of one that is slightly louder than the original. For this reason the *decibel* (dB) scale is used for sound intensity. An intensity of $10^{-12}$ W/m$^2$, which is just audible, is given the value 0 dB; a sound 10 times more intense is given the value 10 dB; a sound $10^2$ times more intense than 0 dB is given the value 20 dB; a sound $10^3$ times more intense than 0 dB is given the value 30 dB; and so forth. More formally, the intensity $I$ dB of a sound wave whose intensity is $I$ W/m$^2$ is given by

$$I \text{ dB} = 10 \log \frac{I}{I_0}$$

where $I_0 = 10^{-12}$ W/m². Normal conversation might be 60 dB, city traffic noise might be 90 dB, and a jet aircraft might produce as much as 140 dB (which produces damage to the ear) at a distance of 100 ft. Long-term exposure to intensity levels of over 85 dB usually leads to permanent hearing damage.

If the power input to an amplifier or other signal processing device is $P_{in}$ and the power output of the device is $P_{out}$, the *power gain G* of the system in decibels is defined as

$$G \text{ dB} = 10 \ \log \ \frac{P_{out}}{P_{in}}$$

A change in audio power output of 1 dB is about the minimum that can be detected by a person with good hearing; usually the change must be 2 or 3 dB to be apparent.

## SOLVED PROBLEM 15.6

How many times more intense is a 50-dB sound than a 40-dB sound? Than a 20-dB sound?

Each interval of 10 dB represents a change in sound intensity by a factor of 10. Hence a 50-dB sound is 10 times more intense than a 40-dB sound and $10 \times 10 \times 10 = 1000$ times more intense than a 20-dB sound.

## SOLVED PROBLEM 15.7

What is the intensity in watts per square meter of the 70-dB noise of a truck passing by?

An intensity of 0 dB is equivalent to $10^{-12}$ W/m². Since a sound of 70 dB is $10^7$ times more intense, it is equivalent to a rate of energy flow of

$$I = (10^7)(10^{-12} \text{ W/m}^2) = 10^{-5} \text{ W/m}^2$$

## SOLVED PROBLEM 15.8

A certain person speaking normally produces a sound intensity of 40 dB at a distance of 3 ft. If the threshold intensity for reasonable audibility is 20 dB, how far away can the person be heard clearly?

A change of 20 dB in sound intensity is equivalent to a ratio of $10 \times 10 = 100$. Hence

$$\frac{I_2}{I_1} = \frac{R_1^2}{R_2^2}$$

$$R_2 = R_1 \sqrt{\frac{I_1}{I_2}} = (3 \text{ ft})\left(\sqrt{100}\right) = 30 \text{ ft}$$

## SOLVED PROBLEM 15.9

Find the power gain of an amplifier whose power input is 0.2 W and whose power output is 80 W.

$$G \text{ dB} = 10 \ \log \ \frac{P_{out}}{P_{in}} = 10 \ \log \ \frac{80 \text{ W}}{0.2 \text{ W}} = 10 \ \log \ 400 = 10(2.60) = 26 \text{ dB}$$

### SOLVED PROBLEM 15.10

A record player pickup has an output of 0.002 $\mu$W. What is the output power when a 100-dB amplifier is used with it?

We start with the definition

$$G \text{ dB} = 10 \text{ log } \frac{P_{out}}{P_{in}}$$

and divide both sides by 10 to give

$$\frac{G \text{ dB}}{10} = \log \frac{P_{out}}{P_{in}}$$

Now we take the antilogarithm of both sides:

$$\text{antilog } \frac{G \text{ dB}}{10} = \frac{P_{out}}{P_{in}}$$

Hence

$$P_{out} = P_{in} \text{ antilog } \frac{G \text{ dB}}{10} = (2 \times 10^{-9} \text{ W})\left(\text{antilog } \frac{100}{10}\right)$$

$$= (2 \times 10^{-9} \text{ W})(\text{antilog } 10) = (2 \times 10^{-9} \text{ W})(10^{10}) = 2 \times 10^{1} \text{ W} = 20 \text{ W}$$

### SOLVED PROBLEM 15.11

An RG-58/U coaxial cable has a signal attenuation of 7 dB per 100 ft at a frequency of 160 mHz. A 50-ft length of this cable is used to couple a 25-W VHF radio transmitter that operates at 160 mHz to an antenna. How much power reaches the antenna?

*Attenuation* refers to a loss of power, so the power gain of this type of cable is −7 dB per 100 ft. Since the cable length here is 50 ft, the power gain is −7 dB/2 = −3.5 dB. From the solution to Prob. 15.10,

$$P_{out} = P_{in} \text{ antilog } \frac{G \text{ dB}}{10} = (25 \text{ W})\left(\text{antilog } \frac{-3.5}{10}\right)$$

$$= (25 \text{ W})(0.447) = 11.2 \text{ W}$$

Less than half the power reaches the antenna.

### SOLVED PROBLEM 15.12

An audio system is made up of components with the following power gains: preamplifier, +35 dB; attenuator, −10 dB; amplifier, +70 dB. What is the overall gain of the system?

Since power gains in decibels are logarithmic quantities, the overall gain in decibels of a system of several devices is equal to the sum of the separate gains in decibels of the devices:

$$G \text{ (overall)} = G_1 + G_2 + G_3 + \cdots$$
$$G \text{ (overall)} = +35 \text{ dB} - 10 \text{ dB} + 70 \text{ dB} = +95 \text{ dB}$$

### DOPPLER EFFECT

When there is relative motion between a source of waves and an observer, the apparent frequency of the waves is different from their frequency $f_S$ at the source. This change in frequency is called the

*Doppler effect.* When the source approaches the observer (or vice versa), the observed frequency is higher; when the source recedes from the observer (or vice versa), the observed frequency is lower. In the case of sound waves, the frequency $f$ that a listener hears is given by

$$f = f_S \left( \frac{v + v_L}{v - v_S} \right) \qquad \text{sound}$$

In this formula $v$ is the velocity of sound, $v_L$ is the velocity of the listener (considered positive for motion toward the source and negative for motion away from the source), and $v_S$ is the velocity of the source (considered positive for motion toward the listener and negative for motion away from the listener).

The Doppler effect in electromagnetic waves (light and radio waves are examples) obeys the formula

$$f = f_S \left[ \frac{1 + (v/c)}{1 - (v/c)} \right]^{1/2} \qquad \text{electromagnetic waves}$$

Here $c$ is the velocity of light ($3.00 \times 10^8$ m/s), and $v$ is the relative velocity between source and observer (considered positive if they are approaching and negative if they are receding). Astronomers use the Doppler effect in light to determine the motions of stars; police use the effect in radar waves to determine vehicle velocities.

**SOLVED PROBLEM 15.13**

The siren of a fire engine has a frequency of 500 Hz. (*a*) The fire engine approaches a stationary car at 20 m/s. What frequency does a person in the car hear? (*b*) The fire engine stops and the car drives away from it at 20 m/s. What frequency does the person in the car hear now?

(*a*)  Here $f_S = 500$ Hz, $v = 343$ m/s, $v_S = +20$ m/s, and $v_L = 0$. Hence the perceived frequency of the siren is

$$f_L = f_S \left( \frac{v + v_L}{v - v_S} \right) = (500 \text{ Hz}) \left[ \frac{343 \text{ m/s}}{(343 - 20) \text{ m/s}} \right] = 531 \text{ Hz}$$

(*b*)  In this case $v_s = 0$ and $v_L = -20$ m/s. The perceived frequency of the siren is

$$f_L = f_S \left( \frac{v + v_L}{v - v_S} \right) = (500 \text{ Hz}) \left[ \frac{(343 - 20) \text{ m/s}}{343 \text{ m/s}} \right] = 471 \text{ Hz}$$

**SOLVED PROBLEM 15.14**

A person arriving late at a concert hurries toward her seat so fast that the note middle C (262 Hz) appears 1 Hz higher in frequency to her. How fast is she moving?

The first step is to solve for $v_L$ the formula for the frequency the listener hears. Since $v_S = 0$,

$$f = f_S \left( \frac{v + v_L}{v} \right)$$

$$\frac{f}{f_S} = \frac{v + v_L}{v} = 1 + \frac{v_L}{v}$$

$$\frac{v_L}{v} = \frac{f}{f_S} - 1$$

$$v_L = v \left( \frac{f}{f_S} - 1 \right)$$

Here $v = +343$ m/s, $f = 262$ Hz, and $f_L = 263$ Hz, so

$$v_L = (343 \text{ m/s})\left(\frac{263 \text{ Hz}}{262 \text{ Hz}} - 1\right) = 1.3 \text{ m/s}$$

**SOLVED PROBLEM 15.15**

A distant galaxy of stars in the constellation Hydra is moving away from the earth at $6.1 \times 10^7$ m/s. One of the characteristic wavelengths in the light the galaxy emits is $5.5 \times 10^{-7}$ m. What is the corresponding wavelength measured by astronomers on the earth?

Since $f = c/\lambda$ and $f_S = c/\lambda_S$, we have

$$f = f_S\left[\frac{1 + (v/c)}{1 - (v/c)}\right]^{1/2}$$

$$\frac{c}{\lambda} = \frac{c}{\lambda_S}\left[\frac{1 + (v/c)}{1 - (v/c)}\right]^{1/2}$$

$$\lambda = \lambda_S\left[\frac{1 - (v/c)}{1 + (v/c)}\right]^{1/2}$$

Here the galaxy is receding, so $v = -6.1 \times 10^7$ m/s and

$$\lambda = (5.5 \times 10^{-7} \text{ m})\left[\frac{1 - (-6.1 \times 10^7 \text{ m/s})(3 \times 10^8 \text{ m/s})}{1 + (-6.1 \times 10^7 \text{ m/s})(3 \times 10^8 \text{ m/s})}\right]^{1/2}$$

$$= (5.5 \times 10^{-7} \text{ m})\left(\frac{1 + 0.203}{1 - 0.203}\right)^{1/2}$$

$$= 8.3 \times 10^{-7} \text{ m}$$

## *Multiple-Choice Questions*

**15.1.** The lower the frequency of a wave,

    (*a*)   the higher its velocity        (*c*)   the smaller its amplitude
    (*b*)   the longer its wavelength    (*d*)   the shorter its period

**15.2.** An entirely longitudinal wave is

    (*a*)   a water wave    (*c*)   an electromagnetic wave
    (*b*)   a sound wave    (*d*)   a wave in a stretched string

**15.3.** Sound cannot travel through a

    (*a*)   vacuum    (*c*)   liquid
    (*b*)   gas        (*d*)   solid

**15.4.** A spacecraft is approaching the earth. Relative to the radio signals it sends out, the signals received on the earth have

    (*a*)   a lower frequency    (*c*)   a higher velocity
    (*b*)   a shorter wavelength    (*d*)   all of the above

**15.5.** A radio station broadcasts at a frequency of 660 kHz. The wavelength of these waves is

(a)  2.2 mm      (c)  4.55 km
(b)  455 m       (d)  $1.98 \times 10^{14}$ m

**15.6.** Waves whose crests are 30 m apart reach an anchored boat once every 3.0 s. The wave velocity is

(a)  0.1 m/s     (c)  10 m/s
(b)  5 m/s       (d)  900 m/s

**15.7.** An 80-dB sound relative to a 30-dB sound is more intense by a factor of

(a)  5       (c)  500
(b)  50      (d)  $10^5$

**15.8.** A sound intensity level of 55 dB is produced by 10 flutes. The number of flutes needed to produce a level of 65 dB under the same circumstances is

(a)  20      (c)  100
(b)  60      (d)  200

# Supplementary Problems

**15.1.** What quantity is carried by all types of waves from their source to the place where they are eventually absorbed?

**15.2.** Find the wavelength in air of a sound wave whose frequency is 440 Hz.

**15.3.** A tuning fork vibrating at 600 Hz is immersed in a tank of water, and the resulting sound waves in the water are found to have a wavelength of 8.2 ft. What is the velocity of sound in the water?

**15.4.** A wave of frequency $f_1$ and wavelength $\lambda_1$ goes from a medium in which its velocity is $v$ to another medium in which its velocity is $2v$. Find the frequency and wavelength of the wave in the second medium.

**15.5.** How many times more intense than the 60-dB sound of a person talking loudly is the 100-dB sound of a power lawn mower?

**15.6.** Sound waves whose intensities exceed about 1 W/m$^2$ cause damage to the ear. How many decibels is this equivalent to?

**15.7.** A certain jet aircraft produces sound of 140-dB intensity at a distance of 100 ft. How many miles away is the intensity 90 dB?

**15.8.** What is the power gain in decibels of an amplifier whose power input is 0.15 W and whose power output is 6 W?

**15.9.** What is the ratio between the output and input powers of a 30-dB amplifier?

**15.10.** A 60-dB amplifier has an output of 25 W. What is the input power?

**15.11.** A 1.2-kW radio transmitter is coupled to an antenna by a cable whose attenuation is 1.8 dB. How much power reaches the antenna?

**15.12.** A preamplifier whose gain is 20 dB is used with a 50-dB amplifier. What is the total gain of the system?

**15.13.** The frequency of a train's whistle is 800 Hz. The train is moving south at 120 km/h, and on a parallel road a car is moving north toward the train at 80 km/h. What frequency is heard by the people in the car?

**15.14.** The car and train of Prob. 15.13 pass each other and continue moving at the same velocities. What frequency is heard by the people in the car now?

**15.15.** A spacecraft moving away from the earth at 97 percent of the velocity of light transmits data at the rate of $1 \times 10^4$ pulses/s. At what rate is the data received on the earth? (Actual spacecraft do not travel this fast; the highest velocity of the *Apollo 11* spacecraft that went to the moon was only 0.0036 percent of the velocity of light.)

## *Answers to Multiple-Choice Questions*

| | | | |
|---|---|---|---|
| **15.1.** | (b) | **15.5.** | (b) |
| **15.2.** | (b) | **15.6.** | (c) |
| **15.3.** | (a) | **15.7.** | (d) |
| **15.4.** | (b) | **15.8.** | (c) |

## Answers to Supplementary Problems

| | | | |
|---|---|---|---|
| **15.1.** | Energy | **15.9.** | 1000 |
| **15.2.** | 78 cm | **15.10.** | 25 $\mu$W |
| **15.3.** | 4920 ft/s | **15.11.** | 793 W |
| **15.4.** | $f_2 = f_1$, $\lambda_2 = 2\lambda_1$ | **15.12.** | 70 dB |
| **15.5.** | $10^4 = 10{,}000$ times more intense | **15.13.** | 939 Hz |
| **15.6.** | 120 dB | **15.14.** | 678 Hz |
| **15.7.** | 6 mi | **15.15.** | 1.23 pulses/s |
| **15.8.** | 16 dB | | |

# Chapter 16

# Fluids at Rest

## DENSITY

The *density d* of a substance is its mass per unit volume. The SI unit of density is kilograms per cubic meter ($kg/m^3$); the density of aluminum, for instance, is 2700 $kg/m^3$. Another common unit of density is grams per cubic centimeter ($g/cm^3$). Since 1 kg = 1000 g and 1 $m^3$ = $(100\ cm)^3$ = $10^6\ cm^3$,

$$1\ g/cm^3 = 10^3\ kg/m^3$$

Hence the density of aluminum can also be given as 2.7 $g/cm^3$.

In British units density is properly expressed in slugs per cubic foot. The density of aluminum in these units is 5.3 $slugs/ft^3$. Because weight rather than mass is normally specified in this system, the quantity *weight density* is customarily used. The weight density of a substance is its weight per unit volume. Thus the weight density of aluminum is 170 $lb/ft^3$. There is no special symbol for weight density, and either $w/V$ or $dg$ can be used for it.

## SPECIFIC GRAVITY

The *specific gravity* (or *relative density*) of a substance is its density relative to that of pure water, which is

$$d(\text{water}) = 1000\ kg/m^3 = 1.00\ g/cm^3 = 1.94\ slugs/ft^3$$

The weight density of water is

$$dg(\text{water}) = 62\ lb/ft^3$$

Since the density of water is 1 $g/cm^3$, the specific gravity of a substance is the same as the numerical value of its density given in grams per cubic centimeter. Thus the specific gravity of aluminum is 2.7.

## SOLVED PROBLEM 16.1

The specific gravity of gold is 19. (*a*) What is the mass of 1 $cm^3$ of gold? (*b*) What is the weight of 1 $in.^3$ of gold?

(*a*) Since the density of water is 1 $g/cm^3$, the density of gold is 19 $g/cm^3$ and 1 $cm^3$ has a mass of 19 g.

(*b*) Since the weight density of water is 62 $lb/ft^3$, the weight density of gold is $dg$ = (19)(62) $lb/ft^3$ = 1200 $lb/ft^3$. Because 1 $ft^3$ = (12 in.)(12 in.)(12 in.) = 1728 $in.^3$, a cubic inch of gold weighs

$$w = (dg)V = \left(1200\ \frac{lb}{ft^3}\right)\left(\frac{1\ ft^3}{1728\ in^3}\right) = 0.7\ lb$$

## SOLVED PROBLEM 16.2

The density of mammals is roughly the same as that of water. Find the volume of a 250-kg lion.

$$V = \frac{m}{d} = \frac{250\ kg}{1000\ kg/m^3} = 0.25\ m^3$$

## SOLVED PROBLEM 16.3

An oak beam 10 cm by 20 cm by 4 m has a mass of 58 kg. Find the density and specific gravity of oak.

The volume of the beam is $V = (0.1 \text{ m})(0.2 \text{ m})(4 \text{ m}) = 0.08 \text{ m}^3$, and so its density is

$$d = \frac{m}{V} = \frac{58 \text{ kg}}{0.08 \text{ m}^3} = 725 \text{ kg/m}^3$$

Since the density of water is $1000 \text{ kg/m}^3$, the specific gravity (sp gr) of oak is

$$\text{sp gr} = \frac{d_\text{oak}}{d_\text{water}} = \frac{725}{1000} = 0.725$$

## SOLVED PROBLEM 16.4

How much does the air in a room 12 ft square and 10 ft high weigh? The weight density of air is $0.08 \text{ lb/ft}^3$ at sea level.

The volume of the room is $V = (12 \text{ ft})(12 \text{ ft})(10 \text{ ft}) = 1440 \text{ ft}^3$. Hence the weight of the air is

$$w = (dg)V = (0.08 \text{ lb/ft}^3)(1440 \text{ ft}^3) = 115 \text{ lb}$$

## PRESSURE

When a force acts perpendicular to a surface, the *pressure* exerted is the ratio between the magnitude of the force and the area of the surface:

$$p = \frac{F}{A}$$

$$\text{Pressure} = \frac{\text{force}}{\text{area}}$$

Pressures are properly expressed in pascals (1 Pa = 1 $\text{N/m}^2$) or in pounds per square foot, but other units are often used:

$$1 \text{ lb/in}^2 = 144 \text{ lb/ft}^2$$

1 atmosphere (atm) = average pressure exerted by earth's atmosphere at sea level

$$= 1.013 \times 10^5 \text{ Pa} = 14.7 \text{ lb/in.}^2$$

$$1 \text{ bar} = 10^5 \text{ Pa} \qquad \text{(slightly less than 1 atm)}$$

$$1 \text{ millibar (mb)} = 100 \text{ Pa} \qquad \text{(widely used in meteorology)}$$

$$1 \text{ torr} = 133 \text{ Pa} \qquad \text{(widely used in medicine for blood pressures)}$$

## GAUGE PRESSURE

Pressure gauges measure the difference between an unknown pressure and atmospheric pressure. What they measure is known as *gauge pressure*, and the true pressure is known as *absolute pressure*:

$$p = p_\text{gauge} + p_\text{atm}$$

$$\text{Absolute pressure} = \text{gauge pressure} + \text{atmospheric pressure}$$

A tire whose gauge pressure is 2 bar contains air at an absolute pressure of about 3 bar, since sea-level atmospheric pressure is about 1 bar.

## SOLVED PROBLEM 16.5

A 65-kg woman balances on the heel of her right shoe, which has a circular base 1 cm in radius. How much pressure does she exert on the ground?

The area of the heel is $A = \pi r^2 = 3.14$ cm$^2 = 3.14 \times 10^{-4}$ m$^2$. Hence the pressure is

$$p = \frac{F}{A} = \frac{mg}{A} = \frac{(65 \text{ kg})(9.8 \text{ m/s}^2)}{3.14 \times 10^{-4} \text{ m}^2} = 2.03 \times 10^6 \text{ Pa} = 20.3 \text{ bar}$$

Since atmospheric pressure is 1.013 bar, this pressure is 20 times atmospheric pressure.

## SOLVED PROBLEM 16.6

The weight of a car is equally supported by its four tires. The gauge pressure of the air in the tires is 2.0 bar and each tire has an area of 140 cm$^2$ in contact with the ground. What is the mass of the car?

The load on each tire consists of one-quarter of the car's weight plus the weight of the column of air directly above the area of the tire in contact with the ground, since this part of the tire has no air under it to provide an equal upward force. Therefore, only the gauge pressure of the air in the tires, which is the excess over atmospheric pressure, acts to support the car's weight. Since $p_{\text{gauge}} = 2.0$ bar $= 2.0 \times 10^5$ Pa, each tire supports a weight of

$$w = mg = p_{\text{gauge}} A = 2.0 \text{ bar} = (2.0 \times 10^5 \text{ Pa})(140 \text{ cm}^2)(10^{-4} \text{ m}^2/\text{cm}^2) = 2800 \text{ N}$$

The weight of the entire car is $4w$ and its mass is

$$M = \frac{4w}{g} = \frac{(4)(2800 \text{ N})}{9.8 \text{ m/s}^2} = 1143 \text{ kg}$$

## SOLVED PROBLEM 16.7

The flat roof of a house is 30 ft long and 25 ft wide and weighs 15,000 lb. Before a severe storm the doors and windows of the house are closed so tightly that the air pressure inside remains at a normal 14.7 lb/in.$^2$ even when the outside pressure falls to 14.3 lb/in.$^2$. Compare the upward force on the roof with its weight.

The area of the roof is $A = (30 \text{ ft})(25 \text{ ft}) = 750$ ft$^2$. The difference between the pressure on the inside and outside of the roof is $\Delta p = (14.7 - 14.3) \text{ ft/in.}^2 = 0.4 \text{ lb/in.}^2$. Because the pressure on the inside of the roof is greater, the net force on it is upward with the magnitude

$$F = A \, \Delta p = (750 \text{ ft}^2)\left(0.4 \, \frac{\text{lb}}{\text{in.}^2}\right)\left(\frac{144 \text{ in.}^2}{\text{ft}^2}\right) = 43,200 \text{ lb}$$

This is nearly three times the roof's weight. If the roof is not securely attached to the walls of the house and if the windows do not break first, the roof will be lifted off during the storm. Evidently a builidng should not be sealed when a large drop in pressure due to a storm is forecast.

## PRESSURE IN A FLUID

Pressure is a useful quantity where fluids (gases and liquids) are concerned because of the following properties of fluids:

1.  The forces that a fluid exerts on the walls of its container, and those that the walls exert on the fluid, always act perpendicular to the walls.
2.  The force exerted by the pressure in a fluid is the same in all directions at a given depth.
3.  An external pressure exerted on a fluid is transmitted uniformly throughout the fluid. This does not mean that pressures in a fluid are the same everywhere, because the weight of the fluid itself exerts pressures that increase with increasing depth. The pressure at a depth $h$ in a fluid of density $d$ due to the weight of fluid above is

$$p = dgh$$

Hence the total pressure at that depth is

$$p = p_{external} + dgh$$

When a body of fluid is in an open container, the atmosphere exerts an external pressure on it.

## SOLVED PROBLEM 16.8

The interior of a submarine located at a depth of 50 m in seawater is maintained at sea-level atmospheric pressure. Find the force acting on a window 20 cm square. The density of seawater is $1.03 \times 10^3$ kg/m$^3$.

The pressure outside the submarine is $p = p_{atm} + dgh$, and the pressure inside is $p_{atm}$. Hence the net pressure $p'$ acting on the window is

$$p' = dgh = (1.03 \times 10^3 \text{ kg/m}^3)(9.8 \text{ m/s}^2)(50 \text{ m}) = 5.05 \times 10^5 \text{ Pa}$$

Since the area of the window is $A = (0.2 \text{ m})(0.2 \text{ m}) = 0.04 \text{ m}^2$, the force acting on it is

$$F = p'A = (5.05 \times 10^5 \text{ Pa})(4 \times 10^{-2} \text{ m}^2) = 2.02 \times 10^4 \text{ N}$$

## SOLVED PROBLEM 16.9

What is the pressure at the bottom of a swimming pool 6 ft deep that is filled with freshwater? Express the answer in pounds per square inch.

$$p = p_{atm} + (dg)h$$

$$= 14.7 \frac{\text{lb}}{\text{in.}^2} + \left(62 \frac{\text{lb}}{\text{ft}^3}\right)(6 \text{ ft})\left(\frac{1 \text{ ft}^2}{144 \text{ in.}^2}\right) = 17.3 \text{ lb/in.}^2$$

Note the use of the conversion factor 1 ft$^2$/144 in.$^2$.

## ARCHIMEDES' PRINCIPLE

An object immersed in a fluid is acted on by an upward force that arises because pressures in a fluid increase with depth. Hence the upward force on the bottom of the object is more than the downward force on its top. The difference between the two, called the *buoyant force*, is equal to the weight of a body of the fluid whose volume is the same as that of the object. This is *Archimedes' principle*: The buoyant force on a submerged object is equal to the weight of fluid the object displaces.

If the buoyant force is less than the weight of the object itself, the object sinks; if the buoyant force equals the weight of the object, the object floats in equilibrium at any depth in the fluid; if the buoyant force is more than the weight of the object, the object floats with part of its volume above the surface.

**SOLVED PROBLEM 16.10**

A wooden block is on the bottom of a tank when water is poured in. The contact between the block and the tank is so good that no water gets between them. Is there a buoyant force on the block?

There is no buoyant force since there is no water under the block to exert an upward force on it.

**SOLVED PROBLEM 16.11**

How much force is needed to support a 100-kg iron anchor when it is immersed in seawater? The density of iron is $7.8 \times 10^3$ kg/m$^3$ and that of seawater is $1.03 \times 10^3$ kg/m$^3$.

The volume of the anchor is

$$V = \frac{m}{d} = \frac{100 \text{ kg}}{7.8 \times 10^3 \text{ kg/m}^3} = 0.0128 \text{ m}^3$$

The weight of seawater displaced by the anchor is

$$w = mg = dVg = (1.03 \times 10^3 \text{ kg/m}^3)(0.0128 \text{ m}^3)(9.8 \text{ m/s}^2) = 129 \text{ N}$$

Thus the buoyant force on the anchor is 129 N, and the net force needed to support it in seawater is

$$F_{\text{net}} = mg - F_{\text{buoyant}} = (100 \text{ kg})(9.8 \text{ m/s}^2) - 129 \text{ N} = 851 \text{ N}$$

**SOLVED PROBLEM 16.12**

A 70-kg person dives off a raft 2 m square moored in a freshwater lake. By how much does the raft rise?

The volume of water that must be displaced by the raft to support the diver is

$$V = \frac{m}{d} = \frac{70 \text{ kg}}{10^3 \text{ kg/m}^3} = 0.07 \text{ m}^3$$

The area of the raft is $A = (2 \text{ m})(2 \text{ m}) = 4 \text{ m}^2$. Since volume = (height)(area), the raft rises by

$$h = \frac{V}{A} = \frac{0.07 \text{ m}^3}{4 \text{ m}^2} = 0.018 \text{ m} = 1.8 \text{ cm}$$

**SOLVED PROBLEM 16.13**

The density of ice is 920 kg/m$^3$ and that of seawater is 1030 kg/m$^3$. What percentage of the volume of an iceberg is submerged?

When an iceberg of volume $V$ floats, its weight of $d_{ice}gV$ is balanced by the buoyant force on it, which is equal to the weight of water displaced. If $V_{sub}$ is the volume of the iceberg that is submerged, the weight of water displaced is $d_{water}gV_{sub}$. Hence

$$\text{Weight of iceberg} = \text{weight of displaced water}$$

$$d_{ice}gV = d_{water}gV_{sub}$$

$$\frac{V_{sub}}{V} = \frac{d_{ice}}{d_{water}} = \frac{920 \text{ kg/m}^3}{1030 \text{ kg/m}^3} = 0.89 = 89\%$$

Eighty-nine percent of the volume of an iceberg is under the water's surface.

## SOLVED PROBLEM 16.14

A 100-gal steel tank weighs 50 lb when empty. Will it float in seawater when it is filled with gasoline? The weight density of gasoline is 42 lb/ft$^3$, that of seawater is 64 lb/ft$^3$, and 1 gal = 0.134 ft$^3$.

The volume of the tank is $V = (100 \text{ gal})(0.134 \text{ ft}^3/\text{gal}) = 13.4 \text{ ft}^3$. The total weight of the tank when it is filled with gasoline is

$$w = 50 \text{ lb} + (dg)_{gasoline}V = 50 \text{ lb} + (42 \text{ lb/ft}^3)(13.4 \text{ ft}^3)$$
$$= 50 \text{ lb} + 563 \text{ lb} = 613 \text{ lb}$$

The maximum buoyant force on the tank is exerted when the tank is completely submerged. Thus

$$F_{max} = (dg)_{water}V = (64 \text{ lb/ft}^3)(13.4 \text{ ft}^3) = 858 \text{ lb}$$

Since the weight of the filled tank is less than 858 lb, it will float.

## HYDRAULIC PRESS

The *hydraulic press* is a basic machine which uses the fact that an external pressure exerted on a fluid is transmitted uniformly throughout the fluid. In a hydraulic press (Fig. 16-1), a piston whose cross-sectional area is $A_{in}$ is moved through a distance $L_{in}$ by an applied force, and fluid in the cylinder transmits the applied pressure to a piston of area $A_{out}$ which moves the distance $L_{out}$. The ideal mechanical advantage (IMA) of the system is

$$\text{IMA} = \frac{L_{in}}{L_{out}} = \frac{A_{out}}{A_{in}}$$

Since the area of a piston is proportional to the square of its diameter, a ratio of piston diameters of only 5, for instance, will yield an IMA of 25.

In Fig. 16-1, valve 1 is closed and valve 2 is open on the downstroke of the input piston. Valve 1 is open and valve 2 is closed on the upstroke, when additional fluid is drawn into the input cylinder to enable its piston to make another stroke.

## SOLVED PROBLEM 16.15

A hydraulic press has in input cylinder 2 cm in diameter and an output cylinder 12 cm in diameter. (*a*) Assuming 100 percent efficiency, find the force exerted by the output piston when a force of 80 N is applied to the input piston. (*b*) If the input piston is moved through 10 cm, how much is the output piston moved?

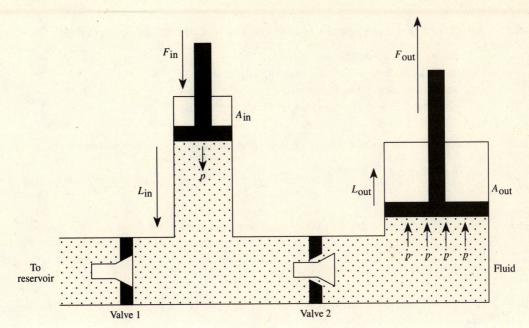

**Fig. 16-1**

(a)   At 100 percent efficiency the actual mechanical advantage equals the IMA, and

$$\frac{F_{out}}{F_{in}} = \frac{A_{out}}{A_{in}}$$

Since $A = \pi r^2 = \pi d^2/4$,

$$F_{out} = F_{in}\left(\frac{d_{out}^2}{d_{in}^2}\right) = (80 \text{ N})\frac{(12 \text{ cm})^2}{(2 \text{ cm})^2} = 2880 \text{ N}$$

(b)   $$L_{out} = L_{in}\left(\frac{A_{in}}{A_{out}}\right) = L_{in}\left(\frac{d_{in}^2}{d_{out}^2}\right) = (10 \text{ cm})\frac{(2 \text{ cm})^2}{(12 \text{ cm})^2} = 0.28 \text{ cm}$$

# Multiple-Choice Questions

**16.1.** The pressure of the atmosphere at sea level corresponds to which one or more of the following?

(a)  1.013 Pa       (c)  1013 mbar
(b)  98 N/m$^2$      (d)  14.7 lb/in.$^2$

**16.2.** The output piston of a hydraulic press cannot exceed the input piston's

(a)  displacement      (c)  work
(b)  velocity          (d)  force

**16.3.** A 2-N stone dropped into a lake sinks. The buoyant force on the stone is

(a)  0                 (c)  2 N
(b)  less than 2 N     (d)  more than 2 N

**16.4.** Seawater has a density 3 percent greater than that of freshwater. A boat will float

    (*a*)   higher in seawater than in freshwater

    (*b*)   at the same level in seawater and in freshwater

    (*c*)   lower in seawater than in freshwater

    (*d*)   any of the above, depending on the shape of the boat's hull

**16.5.** The density of copper is $8.9 \times 10^3$ kg/m$^3$. The volume of a 15-g copper bracelet is

    (*a*)   0.6 cm$^3$    (*c*)   17 cm$^3$

    (*b*)   1.7 cm$^3$    (*d*)   134 cm$^3$

**16.6.** Ethanol has a specific gravity of 0.79. One liter of ethanol weighs

    (*a*)   0.79 N    (*c*)   7.7 N

    (*b*)   1.3 N    (*d*)   12.4 N

**16.7.** A gallon of water and a gallon of antifreeze solution weigh, respectively, 8.4 and 9.2 lb. The antifreeze solution has a specific gravity of

    (*a*)   0.095    (*c*)   0.91

    (*b*)   0.80    (*d*)   1.1

**16.8.** The air in a vertical cylinder 20 cm in diameter that is open at the top supports a 20-kg piston. The absolute pressure on the air is

    (*a*)   0.0064 bar    (*c*)   1.019 bar

    (*b*)   0.062 bar    (*d*)   1.075 bar

**16.9.** A restaurant lobster tank is filled with seawater of density 1.03 g/cm$^3$ to a depth of 60 cm. If there are no lobsters in the tank, the gauge pressure on its bottom is

    (*a*)   6.1 Pa    (*c*)   6.1 kPa

    (*b*)   0.62 kPa    (*d*)   107.4 kPa

**16.10.** Six lobsters are put in the tank of Question 16.9 and enough water is removed to keep its depth at 60 cm. The gauge pressure on the tank bottom is now

    (*a*)   smaller    (*c*)   greater

    (*b*)   the same    (*d*)   any of the above, depending on the mass of the lobsters

**16.11.** A 20-g spoon is put into a dish filled with water to the brim, and 3.0 cm$^3$ of water overflows. The weight of the spoon in the water is

    (*a*)   0.03 N    (*c*)   0.20 N

    (*b*)   0.17 N    (*d*)   0.23 N

**16.12.** A wooden plank 2.5 m long, 40 cm wide, and 60 mm thick floats in water with 15 mm of its thickness above the surface. The plank's mass is

    (*a*)   15 kg    (*c*)   60 kg

    (*b*)   45 kg    (*d*)   441 kg

**16.13.** A hydraulic press has an input piston 10 mm in diameter and an output piston 50 mm in diameter. An input force of 80 N gives an output force of

(a)  3.2 N     (c)  400 N
(b)  16 N      (d)  2000 N

**16.14.** Pushing the input piston of the press in Question 16.13 through 40 mm causes the output piston to be pushed by

(a)  1.6 mm    (c)  40 mm
(b)  8 mm      (d)  200 mm

# Supplementary Problems

**16.1.** A sailboat has a lead keel to help keep it upright despite the pressure wind exerts on its sails. What difference, if any, is there between the stability of a sailboat in freshwater and in seawater?

**16.2.** Dam A and dam B are identical in size and shape, and the water levels at both are the same height above their bases. Dam A holds back a lake that contains 1 $km^3$ of water, and dam B holds back a lake that contains 2 $km^3$ of water. What is the ratio between the total force exerted on dam A and that exerted on dam B?

**16.3.** A 50-g gold bracelet is dropped into a full glass of water, and 2.6 $cm^3$ of water overflows. What is the density of gold? What is its specific gravity?

**16.4.** What is the mass of water in a swimming pool 7 m long, 3 m wide, and 2 m deep?

**16.5.** The density of iron is $7.8 \times 10^3$ $kg/m^3$. (a) What is the specific gravity of iron? (b) How many cubic meters does 1 metric ton (1000 kg) of iron occupy?

**16.6.** The weight density of ice is 58 $lb/ft^3$. What is its specific gravity?

**16.7.** A nail 2 mm in diameter is embedded in a tire in which the gauge pressure is 1.8 bar. How much force tends to push the nail out?

**16.8.** A 130-lb woman balances on the heel of one shoe, which is 1 in. in radius. How much pressure does she exert on the floor?

**16.9.** A phonograph needle whose point is 0.1 mm ($10^{-4}$ m) in radius exerts a downward force of 0.02 N. What is the pressure on the record groove? How many atmospheres is this?

**16.10.** What is the pressure at a depth of 100 m in the ocean? How many atmospheres is this? The density of seawater is $1.03 \times 10^3$ $kg/m^3$.

**16.11.** What pressure is experienced by a skin diver 20 ft below the surface of a freshwater lake?

**16.12.** (a) How much force is required to raise a 1000-kg block of concrete to the surface of a freshwater lake? (b) How much force is needed to lift it out of the water? The density of concrete is $2.3 \times 10^3$ $kg/m^3$.

**16.13.** An aluminum bar weighs 17 lb in air. How much force is required to support the bar when it is immersed in gasoline? The weight density of aluminum is 170 lb/ft$^3$ and that of gasoline is 42 lb/ft$^3$.

**16.14.** A raft 8 ft wide, 12 ft long, and 2 ft high is made from solid balsa wood ($dg = 8$ lb/ft$^3$). How much weight can it support in seawater ($dg = 64$ lb/ft$^3$)?

**16.15.** People have roughly the same density as freshwater. Find the buoyant force exerted by the atmosphere on a 50-kg woman at sea level where the density of air is 1.3 kg/m$^3$.

**16.16.** A balloon weighing 100 kg has a capacity of 1000 m$^3$. If it is filled with hydrogen, how great a payload in kilograms can it support? At sea level the density of hydrogen is 0.09 kg/m$^3$ and that of air is 1.3 kg/m$^3$.

**16.17.** A lever with a mechanical advantage of 10 is used to apply force to the input piston of a hydraulic jack whose input piston is 1 in. in diameter and whose output piston is 4 in. in diameter. (*a*) If the jack is 90 percent efficient, how much weight can it lift when a force of 50 lb is applied to the lever? (*b*) If each stroke of the lever moves the input piston 3 in., how many strokes are needed to raise the output piston 1 ft?

## *Answers to Multiple-Choice Questions*

| | | | |
|---|---|---|---|
| **16.1.** | (*c*), (*d*) | **16.8.** | (*d*) |
| **16.2.** | (*c*) | **16.9.** | (*c*) |
| **16.3.** | (*b*) | **16.10.** | (*b*) |
| **16.4.** | (*a*) | **16.11.** | (*b*) |
| **16.5.** | (*b*) | **16.12.** | (*b*) |
| **16.6.** | (*c*) | **16.13.** | (*d*) |
| **16.7.** | (*d*) | **16.14.** | (*a*) |

## **Answers to Supplementary Problems**

**16.1.** The boat is more stable in freshwater because the buoyancy of the lead keel is less there.

**16.2.** The forces are the same.

**16.3.** 19 g/cm$^3$; 19

**16.4.** 42,000 kg

**16.5.**   (*a*)   7.8      (*b*)   0.128 m$^3$

**16.6.**   0.93

**16.7.**   0.565 N

**16.8.**   41.4 lb/in.$^2$

**16.9.**   6.37 × 10$^5$ Pa; 6.3 atm

**16.10.**  1.11 × 10$^6$ Pa; 11.0 atm

**16.11.**  23.3 lb/in.$^2$

**16.12.**  (*a*)   5539 N      (*b*)   9800 N

**16.13.**  12.8 lb

**16.14.**  10,752 lb

**16.15.**  0.64 N

**16.16.**  1110 kg

**16.17.**  (*a*)   7200 lb      (*b*)   64 strokes

# Chapter 17

## Fluids in Motion

### FLUID FLOW

In the *streamline flow* of a fluid, the direction of motion of the individual particles is the same as that of the fluid as a whole. Each particle of the fluid that passes any point follows the same path as those particles which passed that point before. *Turbulent flow*, however, is characterized by the presence of irregular whirls and eddies; it occurs at high velocities and when the fluid's path changes direction sharply, for instance near an obstruction.

The rate at which a fluid whose velocity is $v$ flows through a pipe or channel of cross-sectional area $A$ is

$$R = vA$$

$$\text{Rate of flow} = (\text{velocity})(\text{cross-sectional area})$$

It is common for $R$ to be expressed in such units as gallons per minute and liters per second instead of the proper units of cubic feet per second and cubic meters per second (1 U.S. gal = 0.134 ft$^3$ and 1 L = $10^{-3}$ m$^3$ = $10^3$ cm$^3$).

When a fluid is incompressible, which is approximately true for most liquids, its rate of flow $R$ is constant even though the size of the pipe or channel varies. Thus if a liquid's velocity is $v_1$ when the cross-sectional area is $A_1$ and $v_2$ when it is $A_2$, then

$$v_1 A_1 = v_2 A_2$$

### SOLVED PROBLEM 17.1

A garden hose has an inside diameter of 12 mm, and water flows through it at 2.5 m/s. (*a*) What nozzle diameter is needed for the water to emerge at 10 m/s? (*b*) At what rate does water leave the nozzle?

(*a*)   The cross-sectional areas of hose and nozzle are in the same ratio as the squares of their diameters, since $A = \pi r^2 = \pi d^2/4$. From $v_1 A_1 = v_2 A_2$ we obtain

$$v_1 d_1^2 = v_2 d_2^2$$

$$d_2 = d_1 \sqrt{\frac{v_1}{v_2}} = (12 \text{ mm}) \sqrt{\frac{2.5 \text{ m/s}}{10 \text{ m/s}}} = 6 \text{ mm}$$

(*b*)   The rate of flow is

$$R = v_1 A_1 = v_1 \pi r_1^2 = (2.5 \text{ m/s})(\pi)(0.006 \text{ m})^2 = 2.83 \times 10^{-4} \text{ m}^3/\text{s}$$

Because 1 m$^3$ = $10^3$ liters (L), $R = 0.283$ L/s. The same result, of course, would be obtained from $R = v_2 A_2$.

### SOLVED PROBLEM 17.2

(*a*) A fluid leaves a pump at the rate of flow $R$ and velocity $v$. What is the power output of the pump? (*b*) A runner's heart pumps 25 L/min of blood at an average pressure of 140 torr. What is the power output of her heart?

(a)   A pump does work on a fluid passing through it at the rate $P = Fv$, where $F$ is the force applied to the fluid. Since $F = pA$, where $p$ is the fluid pressure,

$$P = Fv = pAv$$

But $R = vA$ is the rate of flow, and so

$$P = pR$$

$$\text{Power} = (\text{pressure})(\text{rate of flow})$$

(b)   The rate of flow is

$$R = \left(25 \; \frac{\text{L}}{\text{min}}\right)\left(10^{-3} \; \frac{\text{m}^3}{\text{L}}\right)\left(\frac{1}{60 \; \text{s/min}}\right) = 4.17 \times 10^{-4} \; \text{m}^3/\text{s}$$

and the pressure is

$$p = (140 \; \text{torr})(133 \; \text{Pa/torr}) = 1.86 \times 10^4 \; \text{Pa}$$

Hence the power output of the runner's heart is

$$P = pR = (1.86 \times 10^4 \; \text{Pa})(4.17 \times 10^{-4} \; \text{m}^3/\text{s}) = 7.8 \; \text{W}$$

## SOLVED PROBLEM 17.3

Water emerges horizontally at 10 m/s from the nozzle of a hose held 1.2 m above the ground. How far away does the water stream reach?

From Chapter 4 the time required for the water to strike the ground from a height $h$ is $t = \sqrt{2h/g}$. During this time the water will travel horizontally the distance

$$s = vt = v\sqrt{\frac{2h}{g}} = \left(10 \; \frac{\text{m}}{\text{s}}\right)\sqrt{\frac{(2)(1.2 \; \text{m})}{9.8 \; \text{m/s}^2}} = 4.9 \; \text{m}$$

## SOLVED PROBLEM 17.4

The nozzle of a hose has a radius of 25 mm. Water emerges from this nozzle at the rate of 750 L/min. (a) Find the force with which the nozzle must be held. (b) The water strikes a window of a house and moves off parallel to the window. Find the force exerted on the window.

(a)   The rate of flow is

$$R = \frac{750 \; \text{L/min}}{(60 \; \text{s/min})(10^3 \; \text{L/m}^3)} = 0.0125 \; \text{m}^3/\text{s}$$

Hence the mass of water that flows through the nozzle per second is

$$\frac{m}{t} = dR = (1000 \; \text{kg/m}^3)(0.0125 \; \text{m}^3/\text{s}) = 12.5 \; \text{kg/s}$$

Since $R = vA$ and $A = \pi r^2$, the velocity of the stream is

$$v = \frac{R}{A} = \frac{R}{\pi r^2} = \frac{0.0125 \; \text{m}^3/\text{s}}{\pi(0.025 \; \text{m})^2} = 6.34 \; \text{m/s}$$

Setting equal the impulse $Ft$ with the momentum change $mv$ of the water gives

$$Ft = mv$$

$$F = \frac{mv}{t} = \frac{m}{t}v = (12.5 \; \text{kg/s})(6.34 \; \text{m/s}) = 79 \; \text{N}$$

(b)    The change in momentum of the water stream per second is the same as in (a); hence the force on the window is also 79 N.

## SOLVED PROBLEM 17.5

The bilge pump on a boat is required to lift 0.4 m³/min of seawater through a height of 1.5 m. If the overall efficiency is 50 percent, what should the power rating of the pump motor be? The density of seawater is $1.03 \times 10^3$ kg/m³.

The work done in lifting $V = 0.4$ m³ of seawater through $h = 1.5$ m is

$$W = mgh = dVgh = (1.03 \times 10^3 \text{ kg/m}^3)(0.4 \text{ m}^3)(9.8 \text{ m/s}^2)(1.5 \text{ m}) = 6056 \text{ J}$$

Since $t = 1$ min $= 60$ s here, the required power at 50 percent efficiency is

$$P = \left(\frac{1}{\text{Eff}}\right)\left(\frac{W}{t}\right) = \frac{6056 \text{ J}}{(0.5)(60 \text{ s})} = 202 \text{ W}$$

## SOLVED PROBLEM 17.6

A 100-hp motor is used to power the pump of a fire engine. At 60 percent efficiency, how many tons of water per minute can be delivered to a height of 100 ft?

The available power is

$$P = (0.6)(100 \text{ hp})\left(550 \frac{\text{ft} \cdot \text{lb/s}}{\text{hp}}\right) = 3.3 \times 10^4 \text{ ft} \cdot \text{lb/s}$$

The power needed to raise a weight $w$ to a height $h$ in the time $t$ is

$$P = \frac{W}{t} = \frac{wh}{t}$$

and so, with $t = 1$ min $= 60$ s,

$$w = \frac{Pt}{h} = \frac{(3.3 \times 10^4 \text{ ft} \cdot \text{lb/s})(60 \text{ s})}{100 \text{ ft}} = 1.98 \times 10^4 \text{ lb}$$

Since 1 ton = 2000 lb,

$$w = \frac{1.98 \times 10^4 \text{ lb}}{2000 \text{ lb/ton}} = 9.9 \text{ tons}$$

## BERNOULLI'S EQUATION

*Bernoulli's equation* applies to the streamline flow of an incompressible fluid of density $d$ with negligible viscosity (internal friction). According to this equation, which is derived from the law of conservation of energy, the quantity $p + dhg + \frac{1}{2}dv^2$ has the same value at all points in the motion of such a fluid, where $p$ is the absolute pressure, $h$ is the height above an arbitrary reference level, and $v$ is the fluid velocity. Thus at the two locations 1 and 2

$$p_1 + dgh_1 + \tfrac{1}{2}dv_1^2 = p_2 + dgh_2 + \tfrac{1}{2}dv_2^2$$

The quantity $dgh$ is the potential energy of the fluid per unit volume, and $\frac{1}{2}dv^2$ is its kinetic energy per unit volume. Each term of this equation has the units of pressure.

Another form of Bernoulli's equation is obtained by dividing each term of the above equation by $dg$, which gives

$$\frac{p_1}{dg} + \frac{v_1^2}{2g} + h_1 = \frac{p_2}{dg} + \frac{v_2^2}{2g} + h_2$$

Each term of this equation has the dimensions of a length and is called a *head*, where

$$\frac{p}{dg} = \text{pressure head} \qquad \frac{v^2}{2g} = \text{velocity head} \qquad h = \text{elevation head}$$

## TORRICELLI'S THEOREM

In the case of a liquid flowing out of an open tank through a small orifice a distance $h$ below the surface, as in Fig. 17-1, the pressure on the liquid is the same at the surface and at the orifice, and the downward velocity of the fluid surface is negligible compared with the outward velocity of the fluid. Here Bernoulli's equation reduces to

$$v = \sqrt{2gh}$$

which is *Torricelli's theorem*. This is the same velocity an object would have if it were dropped from rest from a height $h$.

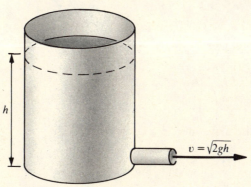

**Fig. 17-1**

## SOLVED PROBLEM 17.7

A barrel 80 cm high is filled with kerosene. When a tap at the bottom of the barrel is opened, with what velocity does the kerosene emerge?

$$v = \sqrt{2gh} = \sqrt{(2)(9.8 \text{ m/s}^2)(0.8 \text{ m})} = 3.96 \text{ m/s}$$

## SOLVED PROBLEM 17.8

At what velocity should water emerge from the nozzle of a fire hose if it is to reach a height of 80 ft when the hose is aimed vertically upward?

The velocity needed to reach a height $h$ is the same as the velocity $v = \sqrt{2gh}$ that would be acquired in free fall from that height. Hence

$$v = \sqrt{2gh} = \sqrt{(2)(32 \text{ ft/s}^2)(80 \text{ ft})} = 72 \text{ ft/s}$$

## SOLVED PROBLEM 17.9

A boat strikes an underwater rock that punctures a hole 20 cm$^2$ in area in its hull 1.5 m below the waterline. At what rate does water enter the hull?

From Torricelli's theorem, the velocity with which water enters the hull is $v = \sqrt{2gh}$. Since the rate of flow through an orifice of area $A$ is $R = vA$ when the fluid velocity is $v$, $R = A\sqrt{2gh}$. Here $A = 20 \text{ cm}^2 = 2 \times 10^{-3} \text{ m}^2$, so water enters the hull at the rate

$$R = A\sqrt{2gh} = (2 \times 10^{-3} \text{ m}^2)\sqrt{(2)(9.8 \text{ m/s}^2)(1.5 \text{ m})} = 1.08 \times 10^{-2} \text{ m}^3/\text{s} = 108 \text{ L/s}$$

This is over 3 tons/min—a serious leak.

## SOLVED PROBLEM 17.10

Water flows through the pipe shown in Fig. 17-2 at the rate of 80 L/s. If the pressure at point 1 is 180 kPa, find (*a*) the velocity at point 1, (*b*) the velocity at point 2, and (*c*) the pressure at point 2.

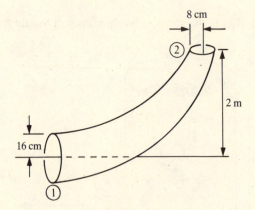

**Fig. 17-2**

(*a*)   Since $A = \pi r^2$ and $R = v_1 A_1$,

$$v_1 = \frac{R}{A_1} = \frac{R}{\pi r_1^2} = \frac{(80 \text{ L/s})(10^{-3} \text{ m}^3/\text{L})}{\pi (0.16 \text{ m})^2} = 0.99 \text{ m/s}$$

(*b*)   From $v_1 A_1 = v_2 A_2$ we have

$$v_2 = v_1 \frac{A_1}{A_2} = v_1 \frac{r_1^2}{r_2^2} = (0.99 \text{ m/s}) \frac{(0.16 \text{ m})^2}{(0.08 \text{ m})^2} = 3.96 \text{ m/s}$$

(*c*)   We now substitute the known quantities $p_1$, $v_1$, and $v_2$ with $h_1 = 0$ and $h_2 = 2$ m into Bernoulli's equation

$$p_1 + dgh_1 + \tfrac{1}{2}dv_1^2 = p_2 + dgh_2 + \tfrac{1}{2}dv_2^2$$

This yields

$$\begin{aligned} p_2 &= p_1 + \tfrac{1}{2}d(v_1^2 - v_2^2) - dgh_2 \\ &= 1.8 \times 10^5 \text{ Pa} + \tfrac{1}{2}(10^3 \text{ kg/m}^3)[(0.99 \text{ m/s})^2 - (3.96 \text{ m/s})^2] - (10^3 \text{ kg/m}^3)(9.8 \text{ m/s}^2)(2 \text{ m}) \\ &= 1.53 \times 10^5 \text{ Pa} \end{aligned}$$

## PRESSURE AND VELOCITY

When the flow is horizontal, so $h_1 = h_2$, Bernoulli's equation becomes

$$p_1 + \tfrac{1}{2}dv_1^2 = p_2 + \tfrac{1}{2}dv_2^2$$

The pressure is greatest where the fluid velocity is least, and the pressure is least where the fluid velocity is greatest. The lift developed by an airplane wing is an example of this effect: The air moving across the curved upper surface of the wing travels faster than that moving across the lower surface because it must cover a greater distance. Hence the pressure is less on the upper surface, and the net result is an upward force on the wing.

### SOLVED PROBLEM 17.11

An airplane whose mass is 40,000 kg and whose total wing area is 120 m$^2$ is in level flight. What is the difference in pressure between the upper and lower surfaces of its wings?

The lift developed by the wings in level flight is equal to the airplane's weight. Hence

$$F = w = mg = (4 \times 10^4 \text{ kg})(9.8 \text{ m/s}^2) = 3.92 \times 10^5 \text{ N}$$

The pressure difference is therefore

$$\Delta p = \frac{F}{A} = \frac{3.92 \times 10^5 \text{ N}}{120 \text{ m}^2} = 3.27 \times 10^3 \text{ Pa}$$

which is equal to 0.032 atm or 0.47 lb/in.$^2$.

### SOLVED PROBLEM 17.12

At what velocity does water emerge from an orifice in a tank (a) in which the gauge pressure is $3 \times 10^5$ Pa and (b) in which the gauge pressure is 40 lb/in.$^2$ ($d_{\text{water}} = 10^3$ kg/m$^3$ = 1.94 slugs/ft$^3$)?

(a)  Let point 1 be at the orifice and point 2 be inside the tank at the same level. In this situation $h_1 = h_2$ and $v_2 = 0$ (very nearly). Substituting in Bernoulli's equation yields

$$p_1 + \tfrac{1}{2}dv_1^2 = p_2 \qquad v_1 = \sqrt{\frac{2(p_2 - p_1)}{d}}$$

The quantity $p_2 - p_1$ is the gauge pressure of $3 \times 10^5$ Pa since $p_1$ is atmospheric pressure. Hence

$$v_1 = \sqrt{\frac{(2)(3 \times 10^5 \text{ Pa})}{10^3 \text{ kg/m}^3}} = 24.5 \text{ m/s}$$

(b)  The same procedure is followed here, but first it is necessary to convert 40 lb/in.$^2$ to its equivalent in pounds per square foot:

$$p_2 - p_1 = \left(40 \; \frac{\text{lb}}{\text{in.}^2}\right)\left(144 \; \frac{\text{in.}^2}{\text{ft}^2}\right) = 5760 \text{ lb/ft}^2$$

Hence

$$v_1 = \sqrt{\frac{(2)(5760) \text{ lb/ft}^2}{1.94 \text{ slugs/ft}^3}} = 77 \text{ ft/s}$$

### SOLVED PROBLEM 17.13

A horizontal pipe 1 in. in radius is joined to a pipe 4 in. in radius, as in Fig. 17-3. (a) If the velocity of seawater ($d = 2.00$ slugs/ft$^3$) in the small pipe is 20 ft/s and the pressure there is 30 lb/in.$^2$, find the velocity and pressure in the large pipe. (b) What is the rate of flow through the pipe in pounds per minute?

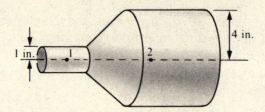

**Fig. 17-3**

(*a*)   The cross-sectional areas of the pipes are in the same ratio as the squares of their radii, since $A = \pi r^2$. From $v_1 A_1 = v_2 A_2$ we obtain

$$v_2 = v_1 \frac{A_1}{A_2} = v_1 \frac{r_1^2}{r_2^2} = \left(20\,\frac{\text{ft}}{\text{s}}\right)\frac{(1\ \text{in.})^2}{(4\ \text{in.})^2} = 1.25\ \text{ft/s}$$

Because both pipes are horizontal, $h_1 = h_2$, and Bernoulli's equation becomes

$$p_2 = p_1 + \tfrac{1}{2}d(v_1^2 - v_2^2) = \left(30\,\frac{\text{lb}}{\text{in.}^2}\right)\left(144\,\frac{\text{in.}^2}{\text{ft}^2}\right) + \left(\frac{1}{2}\right)\left(2\,\frac{\text{slugs}}{\text{ft}^3}\right)\left[\left(20\,\frac{\text{ft}}{\text{s}}\right)^2 - \left(1.25\,\frac{\text{ft}}{\text{s}}\right)^2\right]$$

$$= 4718\ \text{lb/ft}^2$$

which is
$$p_2 = \frac{4718\ \text{lb/ft}^2}{144\ \text{in.}^2/\text{ft}^2} = 33\ \text{lb/in.}^2$$

(*b*)                 $R = v_1 A_1 = (v_1)(\pi r_1^2) = (20\ \text{ft/s})(\pi)\left(\tfrac{1}{12}\ \text{ft}\right)^2 = 0.436\ \text{ft}^3/\text{s}$

Since $dg = 64\ \text{lb/ft}^3$ and $1\ \text{min} = 60\ \text{s}$, the rate of flow in the required units is

$$(0.436\ \text{ft}^3/\text{s})(64\ \text{lb/ft}^3)(60\ \text{s/min}) = 1674\ \text{lb/min}$$

## VISCOSITY

The *viscosity* of a fluid is an internal friction that prevents adjacent layers of the fluid from sliding freely past each other. The symbol of viscosity is $\eta$, the Greek letter *eta*, and its SI unit is the poise (P), where $1\ \text{P} = 1\ \text{N·s/m}^2$. The viscosities of liquids decrease with temperature; those of gases increase.

The rate of flow of a liquid through a pipe obeys *Poiseuille's law*

$$R = \frac{\pi r^4\,\Delta p}{8\eta L}$$

where $L$ is the length of the pipe, $r$ is its radius, $\Delta p$ is the pressure difference between the ends of the pipe, and $\eta$ is the viscosity of the liquid. Evidently the rate of flow depends most strongly on the radius of the pipe.

## SOLVED PROBLEM 17.14

Bernoulli's equation holds for incompressible, nonviscous fluids. How is this relationship changed when the viscosity of a fluid is not negligible?

The effect of viscosity, which is a kind of friction, is to dissipate mechanical energy into heat. Hence the quantity $p + dgh + \tfrac{1}{2}dv^2$ decreases along the direction of flow of a viscous fluid.

**SOLVED PROBLEM 17.15**

The grease nipple on a bearing has a hole 1 mm in diameter and 6 mm long. The grease being used has a viscosity of 80 P. How much pressure is needed to force 0.2 cm³ of grease into the nipple in 5 s?

Here $r = 0.5$ mm $= 5 \times 10^{-4}$ m, $L = 6$ mm $= 6 \times 10^{-3}$ m, and

$$R = \frac{(0.2 \text{ cm}^3)}{(10^6 \text{ cm}^3/\text{m}^3)(5 \text{ s})} = 4 \times 10^{-8} \text{ m}^3/\text{s}$$

Hence the required pressure is

$$p = \frac{8\eta LR}{\pi r^4} = \frac{(8)(80 \text{ P})(6 \times 10^{-3} \text{ m})(4 \times 10^{-8} \text{ m}^3/\text{s})}{\pi(5 \times 10^{-4})^4} = 7.8 \times 10^5 \text{ Pa}$$

This is about 7.7 atm.

## REYNOLDS NUMBER

The nature of the fluid flow in a particular situation (that is, whether laminar or turbulent) depends on the density $d$ and viscosity $\eta$ of the fluid, its average velocity $v$, and a characteristic dimension $D$ of the system according to the *Reynolds number* $N_R$ given by

$$N_R = \frac{dvD}{\eta}$$

The Reynolds number has no units associated with it. For fluid flow in a pipe, $D$ is the pipe diameter. In a pipe, $N_R < 2000$ corresponds to laminar flow, and $N_R > 3000$ corresponds to turbulent flow. If $N_R$ is between 2000 and 3000, the flow may be of either kind and may shift back and forth between them.

The Reynolds number is an important quantity because it provides a basis for experiments that use a small model system to replace a full-size one. If $N_R$ is the same for both, they are said to be *dynamically similar*, and the pattern of fluid flow will be the same for both. Wind tunnel tests of airplane models and towing tank tests of ship models give useful results only when dynamic similarity is obeyed.

**SOLVED PROBLEM 17.16**

Water at 20°C flows at 1.5 m/s through a tube whose inside diameter is 3 mm. The viscosity of water at this temperature is $1.0 \times 10^{-3}$ P. (*a*) Determine the nature of the flow in the tube by calculating its Reynolds number. (*b*) Find the maximum velocity for laminar flow in the tube.

(*a*)   The Reynolds number here is

$$N_R = \frac{dvD}{\eta} = \frac{(1.0 \times 10^3 \text{ kg/m}^3)(1.5 \text{ m/s})(3 \times 10^{-3} \text{ m})}{1.0 \times 10^{-3} \text{ P}} = 4500$$

Since 1P $= 1$ N·s/m² $= 1$ kg/(m·s), the units cancel, to leave a pure number. Because $N_R$ here is greater than 3000, the motion of the water in the tube is turbulent.

(*b*)   The maximum velocity for laminar flow corresponds to a Reynolds number of 2000, so

$$v_{max} = \frac{\eta N_R}{dD} = \frac{(1.0 \times 10^{-3} \text{ P})(2000)}{(1.0 \times 10^3 \text{ kg/m}^3)(3 \times 10^{-3} \text{ m})} = 0.67 \text{ m/s}$$

## *Multiple-Choice Questions*

**17.1.**   When a fast train goes through a station, someone standing at the edge of the platform is

  (*a*)   pulled toward the train
  (*b*)   pushed away from the train
  (*c*)   either pulled or pushed, depending on the ratio between the velocity of the train and the velocity of sound
  (*d*)   not affected by the train's passage

**17.2.**   Water leaks out of a hole in the bottom of a water tank. The rate at which water flows out depends on which one or more of the following?

  (*a*)   the density of water
  (*b*)   the acceleration of gravity
  (*c*)   the area of the hole
  (*d*)   the height of the water

**17.3.**   The Reynolds number for fluid flow in a pipe is independent of

  (*a*)   the viscosity of the fluid
  (*b*)   the velocity of the fluid
  (*c*)   the length of the pipe
  (*d*)   the diameter of the pipe

**17.4.**   Blood flows at 0.10 L/s through the circulatory system of a person whose capillaries have a total cross-sectional area of 0.25 $m^2$. The average velocity of blood in the capillaries is

  (*a*)   0.4 mm/s        (*c*)   25 mm/s
  (*b*)   4 mm/s          (*d*)   40 cm/s

**17.5.**   The average blood pressure of the person in Question 17.4 is 14 kPa. The power developed by the person's heart, which beats 1.2 times per second, in pumping blood against this pressure is

  (*a*)   1.2 W           (*c*)   1.7 W
  (*b*)   1.4 W           (*d*)   12 W

**17.6.**   A boiler in which the gauge pressure is 4.0 bar develops a leak. The velocity at which water comes out is

  (*a*)   89 mm/s         (*c*)   28 m/s
  (*b*)   20 m/s          (*d*)   0.80 km/s

**17.7.**   Water comes out of a valve at the bottom of a tank at 20 ft/s. The height of water in the tank is

  (*a*)   6.25 ft         (*c*)   20 ft
  (*b*)   12.5 ft         (*d*)   25 ft

**17.8.**   Water flows through a pipe whose cross-sectional area is 50 $cm^2$ into another pipe whose cross-sectional area is 10 $cm^2$. Both pipes are horizontal. If the velocity of the water in the large pipe is 1.2 m/s, the velocity in the small pipe is

  (*a*)   0.24 m/s        (*c*)   6.0 m/s
  (*b*)   1.2 m/s         (*d*)   30 m/s

**17.9.**  Water flows through the small pipe of Question 17.8 at

    (*a*)  0.6 L/s    (*c*)  3.0 L/s

    (*b*)  1.2 L/s    (*d*)  6.0 L/s

**17.10.**  If the gauge pressure in the large pipe of Question 17.8 is 200 kPa, the gauge pressure in the small pipe is

    (*a*)  181 kPa    (*c*)  183 kPa

    (*b*)  182 kPa    (*d*)  217 kPa

# Supplementary Problems

**17.1.**  An airplane's wings are designed to have a pressure difference of 0.4 lb/in.$^2$ between their upper and lower surfaces. If the area of the wings is 300 ft$^2$, what is the design weight of the airplane when it is fully loaded?

**17.2.**  A water tank is on the roof of an apartment house. A faucet is opened on the ground floor, 140 ft below the water level in the tank, and there is no frictional resistance in the plumbing system. With what velocity will water emerge from the faucet?

**17.3.**  What gauge pressure is required in a fountain if the jet of water is to be (*a*) 20 ft high and (*b*) 20 m high?

**17.4.**  Water leaves the nozzle of a hose at a velocity of 40 ft/s. (*a*) If the nozzle is held vertically, how high does the water reach? (*b*) If the nozzle is held horizontally 5 ft above the ground, how far does the water stream go before striking the ground?

**17.5.**  Water emerges from a faucet 5 mm in diameter at 2 m/s. (*a*) If the pipe leading to the faucet is 3 cm in diameter, find the water velocity in the pipe. (*b*) At what rate does water leave the faucet?

**17.6.**  Water flows at a velocity of 5 m/s through a hose whose inner diameter is 20 mm. (*a*) What should the nozzle diameter be for water to emerge at 15 m/s? (*b*) Find the rate of water through the hose in liters per minute. (*c*) Find the force needed to hold the nozzle.

**17.7.**  Water emerges from the 2-in.-diameter nozzle of a hose at the rate of 200 gal/min. Find the force with which the nozzle must be held.

**17.8.**  Water passes through a turbine at a rate of 3000 L/s. If the water enters the turbine at 7 m/s and leaves at 1.5 m/s, find the averge force exerted on the turbine blades.

**17.9.**  A pressure gauge in a water supply system reads 40 lb/in.$^2$ when no water flows and 34 lb/in.$^2$ when a tap is opened. What is the flow velocity?

**17.10.**  A tank of gasoline has a crack 1 mm wide and 5 cm long at its base. If the liquid level is 2 m above the base, find the rate at which gasoline leaks out in cubic meters per second and liters per second.

**17.11.** How many kilograms of milk can a $\frac{1}{3}$-hp pump in a dairy transfer per hour from one tank to another in which the liquid level is 10 m higher? Assume 70 percent efficiency.

**17.12.** A pump in a fuel station is required to deliver gasoline ($dg = 42$ lb/ft$^3$) at the rate of 30 gal/min. If the gasoline must be raised through 6 ft and the efficiency is 75 percent, find the minimum power rating of the pump motor.

**17.13.** Water leaves the safety valve of a boiler at a velocity of 100 ft/s. What is the gauge pressure in the boiler?

**17.14.** A horizontal pipe 15 cm in diameter has a constriction 7.5 cm in diameter. If the velocity of water in the constriction is 10 m/s, find the difference between the pressure there and that in the rest of the pipe. Which pressure is the greater?

**17.15.** Water enters the pipe shown in Fig. 17-4 from the left at a velocity of 10 ft/s and a pressure of 22 lb/in.$^2$. Find (a) the velocity with which water leaves the pipe at the right, (b) the pressure there, and (c) the rate of flow of water through the pipe in pounds per minute.

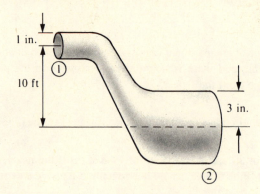

**Fig. 17-4**

**17.16.** A horizontal pipe whose cross-sectional area is 10 cm$^2$ is connected to another horizontal pipe whose cross-sectional area is 50 cm$^2$. Water flows into the small pipe at 6 m/s under a pressure of 200 kPa. Find (a) the speed of the water in the large pipe, (b) the pressure there, and (c) the rate of flow.

**17.17.** A typical capillary blood vessel is 1 mm long and has a radius of 2 $\mu$m. If the pressure difference between its ends is 20 torr, at what velocity does blood flow through the capillary? At body temperature, the viscosity of blood is about $2 \times 10^{-3}$ P.

**17.18.** Establish whether the flow of water in the hose of Prob. 17.6 is laminar or turbulent by calculating its Reynolds number.

## *Answers to Multiple-Choice Questions*

**17.1.**   (a)               **17.3.**    (c)

**17.2.**   (b), (c), (d)     **17.4.**    (a)

**17.5.** $(a)$          **17.8.**    $(c)$

**17.6.** $(c)$          **17.9.**    $(d)$

**17.7.** $(a)$          **17.10.**   $(c)$

# Answers to Supplementary Problems

**17.1.**   17,280 lb

**17.2.**   28 m/s

**17.3.**   $(a)$   1240 lb/ft$^2$ = 8.6 lb/in.$^2$     $(b)$   $196 \times 10^5$ N/m$^2$

**17.4.**   $(a)$   25 ft     $(b)$   22 ft

**17.5.**   $(a)$   0.0556 m/s     $(b)$   $3.93 \times 10^{-5}$ m$^3$/s

**17.6.**   $(a)$   11.5 mm     $(b)$   94 L/min     $(c)$   24 N

**17.7.**   17.8 lb

**17.8.**   15.6 kN

**17.9.**   5.28 ft/s

**17.10.**  $3.13 \times 10^{-4}$ m$^3$/s; 0.313 L/s

**17.11.**  6394 kg

**17.12.**  0.041 hp

**17.13.**  9700 lb/ft$^2$ = 67 lb/in.$^2$

**17.14.**  46.9 kPa; the pressure in the large part of the pipe is greater

**17.15.**  $(a)$   1.11 ft/s     $(b)$   27 lb/in.$^2$     $(c)$   812 lb/min

**17.16.**  $(a)$   1.2 m/s     $(b)$   217 Pa     $(c)$   6 L/s

**17.17.**  0.67 mm/s

**17.18.**  $N_R = 10^5$, so the flow is turbulent

# Chapter 18

## Heat

## INTERNAL ENERGY

Every body of matter, whether solid, liquid, or gas, consists of atoms or molecules in rapid motion. The kinetic energies of these particles constitute the *internal energy* of the body of matter. The *temperature* of the body is a measure of the average kinetic energy of its particles. *Heat* may be thought of as internal energy in transit. When heat is added to a body, its internal energy increases and its temperature rises; when heat is removed from a body, its internal energy decreases and its temperature falls.

## TEMPERATURE

Temperature is familiar as the property of a body of matter responsible for sensations of hot or cold to the touch. Temperature indicates the direction of internal energy flow: When two objects are in contact, internal energy goes from the one at higher temperature to the one at lower temperature, regardless of the total amounts of internal energy in each. Thus if hot coffee is poured into a cold cup, the coffee becomes cooler and the cup becomes warmer.

A *thermometer* is a device for measuring temperature. Matter usually expands when heated and contracts when cooled, the relative amount of change being different for different substances. This behaviour is the basis of most thermometers, which make use of the different rates of expansion of mercury and glass, or of two joined metal strips, to indicate temperature.

## TEMPERATURE SCALES

The *Celsius* (or *centigrade*) temperature scale assigns 0°C to the freezing point of water and 100°C to its boiling point. On the *Fahrenheit* scale these points are, respectively, 32 and 212°F. A Fahrenheit degree is therefore five-ninths as large as a Celsius degree. The following formulas give the procedure for converting a temperature expressed in one scale to the corresponding value in the other:

$$T_F = \tfrac{9}{5}T_C + 32°$$
$$T_C = \tfrac{5}{9}(T_F - 32°)$$

## SOLVED PROBLEM 18.1

What is the Celsius equivalent of 80°F?

$$T_C = \tfrac{5}{9}(T_F - 32°) = \tfrac{5}{9}(80° - 32°) = 26.7°C$$

## SOLVED PROBLEM 18.2

What is the Fahrenheit equivalent of 80°C?

$$T_F = \tfrac{9}{5}T_C + 32° = \left(\tfrac{9}{5}\right)(80°) + 32° = 176°F$$

**SOLVED PROBLEM 18.3**

Oxygen freezes at $-362°F$. What is the Celsius equivalent of this temperature?

$$T_C = \tfrac{5}{9}(T_F - 32°) = \tfrac{5}{9}(-362° - 32°) = -219°C$$

**SOLVED PROBLEM 18.4**

Nitrogen freezes at $-210°C$. What is the Fahrenheit equivalent of this temperature?

$$T_F = \tfrac{9}{5}(T_C + 32°) = \tfrac{9}{5}(-210°) + 32° = -346°F$$

## HEAT

Heat is a form of energy that, when it is added to a body of matter, increases the internal energy content of the body and thereby causes its temperature to rise. The customary symbol for heat is $Q$.

Because heat is a form of energy, the proper SI unit of heat is the joule. However, the *kilocalorie* is sometimes used with SI units: 1 kilocalorie (kcal) is the amount of heat needed to raise the temperature of 1 kg of water by 1°C. The calorie itself is the amount of heat needed to raise the temperature of 1 g of water by 1°C; hence 1 kcal = 1000 cal. (The calorie used by dieticians to measure the energy content of foods is the same as the kilocalorie.)

The British unit of heat is the *British thermal unit* (Btu): 1 Btu is the amount of heat needed to raise the temperature of 1 lb of water 1°F. To convert heat figures from one system to the other, we note that

$$1 \text{ J} = 2.39 \times 10^{-4} \text{ kcal} = 9.48 \times 10^{-4} \text{ Btu}$$

$$1 \text{ kcal} = 3.97 \text{ Btu} = 4185 \text{ J} = 3077 \text{ ft·lb}$$

$$1 \text{ Btu} = 0.252 \text{ kcal} = 778 \text{ ft·lb} = 1054 \text{ J}$$

Although weight rather than mass is specified in the British system when we are dealing with heat, in practice this makes no difference in the various calculations. Whenever $m$ appears in the equations of heat, it refers to mass in kilograms when metric units are used and to weight in pounds when British units are used.

## SPECIFIC HEAT CAPACITY

Different substances respond differently to the addition or removal of heat. For instance, 1 kg of water increases in temperature by 1°C when 1 kcal of heat is added, but 1 kg of aluminum increases in temperature by 4.5°C when this is done. The *specific heat capacity* of a substance is the amount of heat needed to change the temperature of a unit quantity of it by 1°. The symbol of specific heat capacity is $c$; its SI unit is J/(kg·°C) [although kcal/(kg·°C) is still sometimes used], and its British unit is Btu/(lb·°F).

Among common materials, water has the highest specific heat capacity, namely,

$$c_{\text{water}} = 4185 \text{ J/(kg·°C)} = 4.185 \text{ kJ/(kg·°C)} = 1.00 \text{ kcal/(kg·°C)} = 1.00 \text{ Btu/(lb·°F)}$$

Ice and steam have lower specific heat capacities than water:

$$c_{\text{ice}} = 2090 \text{ J/(kg·°C)} = 2.09 \text{ kJ/(kg·°C)} = 0.50 \text{ kcal/(kg·°C)} = 0.50 \text{ Btu/(lb·°F)}$$

$$c_{\text{steam}} = 2010 \text{ J/(kg·°C)} = 2.01 \text{ kJ/(kg·°C)} = 0.48 \text{ kcal/(kg·°C)} = 0.48 \text{ Btu/(lb·°F)}$$

Metals usually have low specific heat capacities; thus lead and iron have $c = 130$ and $460$ J/(kg·°C), respectively.

When an amount of heat $Q$ is transferred to or from a mass $m$ of a substance whose specific heat capacity is $c$, the resulting temperature change $\Delta T$ is related to $Q$, $m$, and $c$ by the formula

$$Q = mc\,\Delta T$$

Heat transferred = (mass)(specific heat capacity)(temperature change)

### SOLVED PROBLEM 18.5

How much heat must be added to 3 kg of water to raise its temperature from 20 to 80°C?

The temperature change is $\Delta T = 80°C - 20°C = 60°C$. Hence

$$Q = mc\,\Delta T = (3 \text{ kg})[4.185 \text{ kJ}/(\text{kg·°C})](60°C) = 753 \text{ kJ}$$

### SOLVED PROBLEM 18.6

From a 50-lb block of ice initially at 25°F, 200 Btu of heat is removed. What is its final temperature?

$$Q = mc\,\Delta T$$

$$\Delta T = \frac{Q}{mc} = \frac{200 \text{ Btu}}{(50 \text{ lb})[0.5 \text{ Btu}/(\text{lb·°F})]} = 8°F$$

The final temperature is therefore $25°F - 8°F = 17°F$.

### SOLVED PROBLEM 18.7

To a 1-kg sample of wood 40 kJ of heat is added, and its temperature is found to rise from 20 to 44°C. What is the specific heat capacity of the wood?

$$Q = mc\,\Delta T$$

$$c = \frac{Q}{m\,\Delta T} = \frac{40 \text{ kJ}}{(1 \text{ kg})(24°C)} = 1.67 \text{ kJ}/(\text{kg·°C})$$

### SOLVED PROBLEM 18.8

A total of 0.8 kg of water at 20°C is placed in a 1.2-kW electric kettle. How long a time is needed to raise the temperature of the water to 100°C?

The required heat is

$$Q = mc\,\Delta T = (0.8 \text{ kg})[4.185 \text{ kJ}/(\text{kg·°C})](100°C - 20°C) = 268 \text{ kJ}$$

Since $P = E/t$ and $P = 1.2$ kW $= 1.2$ kJ/s,

$$t = \frac{E}{P} = \frac{268 \text{ kJ}}{1.2 \text{ kJ/s}} = 223 \text{ s} = 3.7 \text{ min}$$

### SOLVED PROBLEM 18.9

In preparing tea, 600 g of water at 90°C is poured into a 200-g china pot [$c_{pot} = 0.84$ kJ/(kg·°C)] at 20°C. What is the final temperature of the water?

If $T$ is the final temperature, then the 600 g = 0.6 kg of water initially at 90°C undergoes a temperature change of $\Delta T_{water} = 90°C - T$. At the same time the 200 g = 0.2 kg pot initially at 20°C undergoes a temperature change of $\Delta T_{pot} = T - 20°C$. We proceed as follows:

$$\text{Heat gained by pot} = \text{heat lost by water}$$

$$m_{pot}c_{pot}\,\Delta T_{pot} = m_{water}c_{water}\,\Delta T_{water}$$

$$(0.2 \text{ kg})[0.84 \text{ kJ}/(\text{kg·°C})](T - 20 \text{ °C}) = (0.6 \text{ kg})[4.185 \text{ kJ}/(\text{kg·°C})](90°C - T)$$

$$(0.168T - 3.36) \text{ kJ} = (225.99 - 2.511T) \text{ kJ}$$

$$2.679T = 222.63$$

$$T = 83°C$$

## SOLVED PROBLEM 18.10

Three pounds of water at 100°F is added to 5 lb of water at 40°F. What is the final temperature of the mixture?

If $T$ is the final temperature, then the 5 lb of water initially at 40°F undergoes a temperature change of $\Delta T_1 = T - 40°F$ and the 3 lb of water initially at 100°F undergoes a temperature change of $\Delta T_2 = 100°F - T$. We proceed as in Prob. 18.9:

$$\text{Heat gained} = \text{heat lost}$$

$$m_1 c_1\,\Delta T_1 = m_2 c_2\,\Delta T_2$$

$$(5 \text{ lb})[1 \text{ Btu}/(\text{lb·°F})](T - 40°F) = (3 \text{ lb})[1 \text{ Btu}/(\text{lb·°F})](100°F - T)$$

$$(5T - 200) \text{ Btu} = (300 - 3T) \text{ Btu}$$

$$8T = 500$$

$$T = 62.5°F$$

## SOLVED PROBLEM 18.11

To raise the temperature of 5 kg of water from 20 to 30°C, a 2-kg iron bar is heated and then dropped into the water. What should the temperature of the bar be $[c_{iron} = 0.46 \text{ kJ}/(\text{kg·°C})]$?

Let the temperature of the iron bar be $T$. Then the change in the water's temperature is $\Delta T_w = 30°C - 20°C = 10°C$, and the change in the bar's temperature is $\Delta T_{iron} = T - 30°C$. We proceed in the usual way:

$$(5 \text{ kg})[(4.185 \text{ kJ}/(\text{kg·°C})](10°C) = (2 \text{ kg})[(0.46 \text{ kJ}/(\text{kg·°C})](T - 30°C)$$

$$209.3 \text{ kJ} = (0.92T - 27.6) \text{ kJ}$$

$$0.92T = 236.9$$

$$T = 258°C$$

## CHANGE OF STATE

When heat is continuously added to a solid, the solid grows hotter and hotter and finally begins to melt (Fig. 18-1). While it is melting, the material remains at the same temperature and the absorbed heat goes into changing its state from solid to liquid. After all the solid is melted, the temperature of the resulting liquid then increases as more heat is supplied until it begins to boil. Now the material again stays at a constant temperature until all of it has become a gas, after which the gas temperature rises.

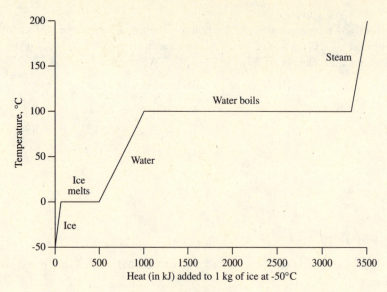

**Fig. 18-1**

The amount of heat that must be added to a unit quantity (1 kg or 1 lb) of a substance at its melting point to change it from a solid to a liquid is called its *heat of fusion $L_f$*. The same amount of heat must be removed from a unit quantity of the substance when it is a liquid at its melting point to change it to a solid.

The amount of heat that must be added to a unit quantity of a substance at its boiling point to change it from a liquid to a gas is called its *heat of vaporization $L_v$*. The same amount of heat must be removed from a unit quantity of the substance when it is a gas at its boiling point to change it to a liquid.

The heat of fusion of water is $L_f = 335$ kJ/kg $= 80$ kcal/kg $= 144$ Btu/lb, and its heat of vaporization is $L_v = 2260$ kJ/kg $= 540$ kcal/kg $= 972$ Btu/lb.

## PRESSURE AND BOILING POINT

The boiling point of a liquid depends on the pressure applied to it: The higher the pressure, the higher the boiling point. Thus water under a pressure of 2 atm boils at 121°C instead of at 100°C, as it does at sea-level atmospheric pressure. At high altitudes, where the atmospheric pressure is less than at sea level, water boils at a lower temperature than 100°C. At an elevation of 2000 m, for instance, atmospheric pressure is about three-quarters of its sea-level value and water boils at 93°C there.

## SOLVED PROBLEM 18.12

A person is dissatisfied with the rate at which eggs cook in a pan of boiling water. Would they cook faster if the person (*a*) turns up the gas flame or (*b*) uses a pressure cooker?

(*a*)   No. The maximum temperature that water can have while in the liquid state is its boiling point. Increasing the rate at which the heat is supplied to a pan of water increases the rate at which steam is produced, but does not raise the temperature of the water beyond 100°C (212°F).

(*b*)   Yes. In a pressure cooker, the pressure is greater than normal atmospheric pressure, which elevates the boiling point and so causes the eggs to cook faster.

**SOLVED PROBLEM 18.13**

An ice cube at 0°C is dropped on the ground and melts to water at 0°C. If all the kinetic energy of the ice went into melting it, from what height did it fall?

$$\text{(Mass of ice)(heat of fusion)} = \text{initial potential energy of ice}$$

$$mL_f = mgh$$

$$h = \frac{L_f}{g} = \frac{3.35 \times 10^5 \text{ J/kg}}{9.8 \text{ m/s}^2} = 3.4 \times 10^4 \text{ m}$$

**SOLVED PROBLEM 18.14**

Five kilograms of water at 40°C is poured on a large block of ice at 0°C. How much ice melts?

$$\text{Heat lost by water} = \text{heat gained by ice}$$

$$m_w c \, \Delta T = m_{\text{ice}} L_f$$

$$(5 \text{ kg})[4.185 \text{ kJ/(kg·°C)}](40°C) = (m_{\text{ice}})(335 \text{ kJ/kg})$$

$$m_{\text{ice}} = \frac{837}{335} \text{ kg} = 2.5 \text{ kg}$$

**SOLVED PROBLEM 18.15**

Five hundred kilocalories of heat is added to 2 kg of water at 80°C. How much steam is produced?

The heat needed to raise the temperature of the water from 80°C to the boiling point of 100°C is

$$Q_1 = m_1 c \, \Delta T = (2 \text{ kg})[1 \text{ kcal/(kg·°C)}](20°C) = 40 \text{ kcal}$$

Hence $Q_2 = 500 \text{ kcal} - 40 \text{ kcal} = 460 \text{ kcal}$ of heat is available to convert water at 100°C to steam at the same temperature. Since $Q_2 = m_{\text{steam}} L_v$, the steam produced is

$$m_{\text{steam}} = \frac{Q_2}{L_v} = \frac{460 \text{ kcal}}{540 \text{ kcal/kg}} = 0.85 \text{ kg}$$

**SOLVED PROBLEM 18.16**

How much heat must be added to 200 lb of lead at 70°F to cause it to melt? The specific heat capacity of lead is 0.03 Btu/(lb·°F), it melts at 626°F, and its heat of fusion is 10.6 Btu/lb.

Here $\Delta T = 626°F - 70°F = 556°F$. Hence

$$Q = mc \, \Delta T + mL_f$$
$$= (200 \text{ lb})[0.03 \text{ Btu/(lb·°F)}](556°F) + (200 \text{ lb})(10.6 \text{ Btu/lb})$$
$$= 3336 \text{ Btu} + 2120 \text{ Btu} = 5456 \text{ Btu}$$

**SOLVED PROBLEM 18.17**

Find the minimum amount of ice at $-10°C$ needed to bring the temperature of 500 g of water at 20°C down to 0°C.

Here $\Delta T_{\text{ice}} = 10°C$ and $\Delta T_{\text{water}} = 20°C$. Therefore

$$\text{Heat gained by ice} = \text{heat lost by water}$$

$$m_{\text{ice}}c_{\text{ice}}\,\Delta T_{\text{ice}} + m_{\text{ice}}L_{f\text{ice}} = m_{\text{water}}c_{\text{water}}\,\Delta T_{\text{water}}$$

$$m_{\text{ice}} = \frac{m_{\text{water}}c_{\text{water}}\,\Delta T_{\text{water}}}{c_{\text{ice}}\,\Delta T_{\text{ice}} + L_{f\text{ice}}} = \frac{(0.5\text{ kg})[4.185\text{ kJ}/(\text{kg}\cdot°C)](20°C)}{[2.09\text{ kJ}/(\text{kg}\cdot°C)](10°C) + 335\text{ kJ/kg}}$$

$$= 0.12\text{ kg}$$

## SOLVED PROBLEM 18.18

A 30-kg ice cube at 0°C is dropped into 200 g of water at 30°C. What is the final temperature?

If $T$ is the final temperature, then $\Delta T_{\text{ice}} = T - 0°C$ and $\Delta T_{\text{water}} = 30°C - T$. Therefore

$$\text{Heat gained by ice} = \text{heat lost by water}$$

$$m_{\text{ice}}L_f + m_{\text{ice}}c_{\text{water}}\,\Delta T_{\text{ice}} = m_{\text{water}}c_{\text{water}}\,\Delta T_{\text{water}}$$

$$(0.03\text{ kg})\left(335\ \frac{\text{kJ}}{\text{kg}}\right) + (0.03\text{ kg})\left(4.185\ \frac{\text{kJ}}{\text{kg}\cdot°C}\right)(T - 0°C) = (0.2\text{ kg})\left(4.185\ \frac{\text{kJ}}{\text{kg}\cdot°C}\right)(30°C - T)$$

$$(10.05 + 0.126T)\text{ kJ} = (25.11 - 0.837T)\text{ kJ}$$

$$0.963T = 15.06$$

$$T = 15.6°C$$

## SOLVED PROBLEM 18.19

How much steam at 150°C is needed to melt 1 kg of ice at 0°C?

Here $\Delta T_1 = 150°C - 100°C = 50°C$ and $\Delta T_2 = 100°C - 0°C = 100°C$. If $m_s$ is the mass of steam,

$$\text{Heat gained by ice} = \text{heat lost by steam}$$

$$m_{\text{ice}}L_f = m_s c_s\,\Delta T_1 + m_s L_v + m_s c_{\text{water}}\,\Delta T_2$$

$$(1\text{ kg})(335\text{ kJ/kg}) = (m_s)[2.01\text{ kJ}/(\text{kg}\cdot°C)](50°C) + (m_s)[2260\text{ kJ}/(\text{kg}\cdot°C)]$$

$$+ (m_s)[4.185\text{ kJ}/(\text{kg}\cdot°C)](100°C)$$

$$335\text{ kJ} = (100.5 + 2260 + 418.5)m_s\text{ kJ} = 2779m_s\text{ kJ}$$

$$m_s = \frac{335}{2779}\text{ kg} = 0.12\text{ kg}$$

# *Multiple-Choice Questions*

**18.1.** Two thermometers, one calibrated in °F and the other in °C, are used to measure the same temperature. The reading in °C

(a)  is proportional to that in °F
(b)  is less than that in °F
(c)  is greater than that in °F
(d)  may be less or greater than that in °F

**18.2.**   Nitrogen boils at $-196°C$. The Fahrenheit equivalent of this temperature is

  (*a*)   $-228°F$       (*c*)   $-321°F$
  (*b*)   $-295°F$       (*d*)   $-385°F$

**18.3.**   Lead melts at $626°F$. The Celsius equivalent of this temperature is

  (*a*)   $316°C$       (*c*)   $366°C$
  (*b*)   $330°C$       (*d*)   $1069°C$

**18.4.**   Steam at $100°C$ is more dangerous than the same mass of water at $100°C$ because the steam

  (*a*)   moves faster
  (*b*)   is less dense
  (*c*)   contains more internal energy
  (*d*)   has a higher specific heat capacity

**18.5.**   When a liquid evaporates

  (*a*)   it gives off heat       (*c*)   its temperature drops
  (*b*)   it absorbs heat         (*d*)   its temperature rises

**18.6.**   In an hour a 1000-W electric heater produces

  (*a*)   3.4 Btu       (*c*)   3416 Btu
  (*b*)   1054 Btu      (*d*)   3600 Btu

**18.7.**   When 20 kJ of heat is removed from 1.2 kg of ice originally at $-15°C$, its new temperature is

  (*a*)   $-18°C$       (*c*)   $-26°C$
  (*b*)   $-23°C$       (*d*)   $-35°C$

**18.8.**   A hot liquid at $80°C$ is added to 600 g of the same liquid originally at $10°C$. When the mixture reaches $30°C$, the total mass of liquid is

  (*a*)   825 g
  (*b*)   840 g
  (*c*)   857 g
  (*d*)   Impossible to calculate without knowing the specific heat capacity of the liquid.

**18.9.**   If 400 g of water at $10°C$ is poured into a 600-g pitcher $[c = 0.80$ kJ/(kg·°C)] at $20°C$, the final temperature of the water is

  (*a*)   $11°C$       (*c*)   $14°C$
  (*b*)   $12°C$       (*d*)   $17°C$

**18.10.**   A 1.0-kg iron bar $[c = 0.11$ kcal/(kg·°C)] at $100°C$ is placed in 3.0 kg of water at $15°C$. The temperature of the water increases by

  (*a*)   $0.7°C$       (*c*)   $5°C$
  (*b*)   $3°C$         (*d*)   $18°C$

**18.11.**   When 10 lb of water at $50°F$ is poured over 1.0 lb of ice at $0°F$, the resulting mixture is at

  (*a*)   $19°F$       (*c*)   $32°F$
  (*b*)   $31°F$       (*d*)   $34°F$

**18.12.** Which one or more of the following combinations will result in water at 50°C?

(a)  1 kg each of ice at 0°C and steam at 100°C
(b)  1 kg each of ice at 0°C and water at 100°C
(c)  1 kg each of water at 0°C and steam at 100°C
(d)  1 kg each of water at 0°C and water at 100°C

**18.13.** If 3 MJ of heat is removed from 1 kg of steam at 200°C, the result is

(a)  ice                    (c)  water
(b)  water and ice          (d)  water and steam

# Supplementary Problems

**18.1.** A glass of water is stirred and then allowed to stand until the water stops moving. What has happened to the kinetic energy of the moving water?

**18.2.** Why is an ice cube at 0°C more effective in cooling a drink than the same mass of water at 0°C?

**18.3.** Ethyl alcohol melts at $-114$°C and boils at 78°C. What are the Fahrenheit equivalents of these temperatures?

**18.4.** Bromine melts at 19°F and boils at 140°F. What are the Celsius equivalents of these temperatures?

**18.5.** How much heat must be removed from 4 lb of water to 200°F to reduce its temperature to 50°F?

**18.6.** How many joules of heat must be removed from 2 kg of water at 90°C to reduce its temperature to 20°C?

**18.7.** How many kilocalories of heat must be added to a 20-kg block of ice to raise its temperature from $-20$ to $-5$°C?

**18.8.** How much heat is lost by a 50-g silver spoon when it is cooled from 20 to 0°C?

**18.9.** Three thousand kilojoules of heat is added to a 100-kg statue of Isaac Newton initially at 18°C. What is its final temperature [$c_{statue} = 0.58$ kJ/(kg·°C)]?

**18.10.** Two kilojoules of heat is added to a 0.88-kg glass jar at 20°C, and its temperature is found to rise to 28°C. What is the specific heat capacity of the glass?

**18.11.** Ten kilograms of water at 5°C is added to 100 kg of water at 80°C. What is the final temperautre of the mixture?

**18.12.** A 600-g copper dish contains 1500 g of water at 20°C. A 100-g iron bar at 120°C is dropped into the water. What is the final temperature of the water [$c_{copper} = 0.39$ kJ/(kg·°C); $c_{iron} = 0.46$ kJ/(kg·°C)]?

**18.13.** Four pounds of soup at 140°F is poured into a 4-lb china serving dish at 70°F. What is the final temperature of the soup [$c_{soup} = 0.9$ Btu/(lb·°F); $c_{dish} = 0.2$ Btu/(lb·°F)]?

**18.14.** How much water at 20°C must be added to 5 kg of punch at 70°C that is in a 1.5-kg silver punch bowl to lower its temperature to 60°C [$c_{punch}$ = 2.93 kJ/(kg·°C); $c_{silver}$ = 0.234 kJ/(kg·°C)]?

**18.15.** How much heat must be removed from 200 g of water at 30°C to convert it to ice at 0°C?

**18.16.** Three pounds of water at 100°F is poured on a large block of ice at 32°F. How much ice melts?

**18.17.** Two thousand kilojoules of heat is added to 10 kg of zinc at 20°C. How much zinc melts? The specific heat capacity of zinc is 0.39 kJ/(kg·°C), it melts at 420°C, and its heat of fusion is 100 kJ/kg.

**18.18.** How much ice at 0°C must be added to 200 g of water at 30°C to lower its temperature to 20°C?

**18.19.** Ten kilograms of ice at 0°C is mixed with 2 kg of steam at 100°C. Find the temperature of the resulting water.

**18.20.** Three thousand Btu of heat is added to 3 lb of water at 100°F. How much steam is produced? What is the temperature of the steam?

**18.21.** River water at 10°C is used to condense spent steam at 120°C to water at 50°C in an electric generating plant. If the cooling water leaves the condenser at 30°C, how many kilograms of river water are needed per kilogram of spent steam?

**18.22.** (*a*) How many kilocalories per hour are given off by a 100-W light bulb? (*b*) How many Btu per hour?

**18.23.** A typical gumdrop contains 35 kcal of energy. If this energy were used to raise an 80-kg person above the ground, how high would that person go?

**18.24.** A lead bullet traveling at 200 m/s strikes a tree and comes to a stop. If half the heat produced is retained by the bullet, by how much does its temperature increase [$c_{lead}$ = 0.13 kJ/(kg·°C)]?

**18.25.** Some 1.5 kg of water at 10°C in a 300-g aluminum kettle is placed on a 2-kW electric hot plate. What is the temperature of the water after 3 min [$c_{aluminum}$ = 0.92 kJ/(kg·°C)]?

## *Answers to Multiple-Choice Questions*

**18.1.**  (*d*)      **18.6.**  (*c*)      **18.11.**  (*d*)

**18.2.**  (*c*)      **18.7.**  (*b*)      **18.12.**  (*d*)

**18.3.**  (*b*)      **18.8.**  (*b*)      **18.13.**  (*b*)

**18.4.**  (*c*)      **18.9.**  (*b*)

**18.5.**  (*b*)      **18.10.**  (*b*)

# Answers to Supplementary Problems

**18.1.** The kinetic energy of the moving water becomes dissipated into internal energy, and the water therefore has a higher temperature than before it was stirred.

**18.2.** The ice absorbs heat from the drink in order to melt to water at 0°C.

**18.3.** −173°F; 172°F

**18.4.** −7°C; 60°C

**18.5.** 600 Btu

**18.6.** 586 kJ                    **18.16.** 1.4 lb

**18.7.** 150 kcal                  **18.17.** 4.4 kg

**18.8.** 0.23 kJ                   **18.18.** 20 g

**18.9.** 70°C                      **18.19.** 40°C

**18.10.** 0.83 kJ/(kg·°C)          **18.20.** 3 lb; 158°F

**18.11.** 73°C                     **18.21.** 30 kg

**18.12.** 21°C                     **18.22.** (a)  86 kcal     (b)  342 Btu

**18.13.** 127°F                    **18.23.** 187 m

**18.14.** 0.62 kg                  **18.24.** 80°C

**18.15.** 92 kJ                    **18.25.** 65°C

# Chapter 19

# Expansion of Solids, Liquids, and Gases

## LINEAR EXPANSION

A change in temperature $\Delta T$ causes most solids to change in length (or other linear dimension) by an amount $\Delta L$ that is proportional to both the original length $L_0$ and $\Delta T$:

$$\Delta L = aL_0 \, \Delta T$$
Change in length = $(a)$(original length)(temperature change)

The quantity $a$ is a constant that depends on the nature of the material; it is called the *coefficient of linear expansion*.

## VOLUME EXPANSION

The change in volume $\Delta V$ of a solid or liquid whose original volume is $V_0$ when its temperature is changed by $\Delta T$ is

$$\Delta V = bV_0 \, \Delta T$$
Change in volume = $(b)$(original volume)(temperature change)

The quantity $b$ is the *coefficient of volume expansion*. Generally $b = 3a$ for a given material.

## SOLVED PROBLEM 19.1

A surveyor's steel tape measure is calibrated at 20°C. A reading of 50 m is found when the tape measure is used to determine the width of a building lot when the temperature is −10°C. How great an error does the temperature difference introduce? The coefficient of linear expansion of steel is $1.2 \times 10^{-5}/$°C.

Since $\Delta T = 20$°C $- (-10$°C$) = 30$°C, the 50-m length of steel tape shrinks by

$$\Delta L = aL_0 \, \Delta T = (1.2 \times 10^{-5}/°C)(50 \text{ m})(30°C) = 1.8 \times 10^{-2} \text{ m} = 18 \text{ mm}$$

## SOLVED PROBLEM 19.2

A wooden wagon wheel has an outside diameter of 3750 mm. The iron tire for this wheel is deliberately made smaller so that it can be shrunk in place to be a tight fit. If the tire's inside diameter is 3737 mm at 20°C, find the temperature to which it must be heated to fit over the wheel. The coefficient of linear expansion of steel is $1.2 \times 10^{-5}/$°C.

Here $L_0 = 3737$ mm and $\Delta L = 13$ mm. Hence

$$\Delta T = \frac{\Delta L}{aL_0} = \frac{13 \text{ mm}}{(1.2 \times 10^{-5}/°C)(3737 \text{ mm})} = 290°C$$

The required temperature is therefore 20°C + 290°C = 310°C.

## SOLVED PROBLEM 19.3

Find the force associated with the expansion of a steel beam whose cross-sectional area is 200 cm² when its temperature increases from 5 to 30°C. Young's modulus for steel is $Y = 2.0 \times 10^{11}$ N/m², and $a = 1.2 \times 10^{-5}$/°C.

The change in the beam's length $\Delta L$ due to thermal expansion is given by

$$\Delta L = aL_0 \, \Delta T$$

To obtain the same change in length by mechanical means would involve applying a tension force $F$ such that

$$F = \frac{YA \, \Delta L}{L_0}$$

according to Chapter 13. Thus the amount of force associated with the expansion is

$$F = \frac{YA \, \Delta L}{L_0} = \frac{YA(aL_0 \, \Delta T)}{L_0} = YaA \, \Delta T$$

The force depends on the cross section of the beam but not on its length. For the beam under consideration here

$$F = YaA \, \Delta T = (2.0 \times 10^{11} \text{ N/m}^2)(1.2 \times 10^{-5}/°\text{C})(0.02 \text{ m}^2)(25°\text{C}) = 1.2 \times 10^6 \text{ N} = 1.2 \text{ MN}$$

which is equivalent to nearly 135 tons.

## SOLVED PROBLEM 19.4

How much water overflows when a Pyrex vessel filled to the brim with 1 L (1000 cm³) of water at 20°C is heated to 90°C? The coefficients of volume expansion of Pyrex and water are $9 \times 10^{-6}$/°C and $2.1 \times 10^{-4}$/°C, respectively.

A cavity in an object expands or contracts by the same amount as a solid body with the composition of the object and the same volume as the cavity. Hence the volume of water that overflows is

$$V_w - V_p = b_w V_0 \, \Delta T - b_P V_0 \, \Delta T = (b_w - b_P)v_0 \, \Delta T$$
$$= (210 \times 10^{-6}/°\text{C} - 9 \times 10^{-6}/°\text{C})(1000 \text{ cm}^3)(70°\text{C})$$
$$= 14.1 \text{ cm}^3$$

## SOLVED PROBLEM 19.5

Lead has a coefficient of linear expansion of $3 \times 10^{-5}$/°C, and its density at 20°C is 11.0 g/cm³. Find the density of lead at 200°C.

Since $b = 3a$, the coefficient of volume expansion of lead is

$$b = (3)(3 \times 10^{-5}/°\text{C}) = 9 \times 10^{-5}/°\text{C}$$

At 20°C the volume of a mass $m$ of lead is $V_0 = m/d_0$, and at 200°C it is

$$V = \frac{m}{d} = V_0 + \Delta V = V_0 + bV_0 \, \Delta T$$

Substituting $V_0 = m/d_0$ yields

$$\frac{m}{d} = \frac{m}{d_0} + \frac{bm \, \Delta T}{d_0} = \frac{m}{d_0} \, (1 + b \, \Delta T)$$
$$d = \frac{d_0}{1 + b \, \Delta T}$$

Here $\Delta T = 200°C - 20°C = 180°C$, and so

$$d = \frac{11 \text{ g/cm}^3}{1 + (9 \times 10^{-5}/°C)(180°C)} = 10.8 \text{ g/cm}^3$$

## BOYLE'S LAW

At constant temperature, the volume of a sample of gas is inversely proportional to the absolute pressure applied to the gas. The greater the pressure, the smaller the volume. This relationship is known as *Boyle's law*. If $p_1$ is the gas pressure when its volume is $V_1$ and $p_2$ is its pressure when its volume is $V_2$, then Boyle's law states that

$$p_1 V_1 = p_2 V_2 \qquad T = \text{constant}$$

## SOLVED PROBLEM 19.6

A 1-L sample of nitrogen at 0°C and 1-atm pressure is compressed to 0.5 L. If the temperature is unchanged, what happens to the pressure of the sample?

Since $p_2 = p_1 V_1/V_2$ and here $V_1/V_2 = 2$, the pressure doubles to 2 atm.

## SOLVED PROBLEM 19.7

A scuba diver's 12-L tank is filled with air at an absolute pressure of 150 bar. If the diver uses 30 L of air per minute at the same 2.5-bar absolute pressure as the water pressure at her depth of 15 m below the surface, how long can she stay at that depth?

From Boyle's law the volume of air available at a pressure of 2.5 bar is

$$V_2 = \frac{p_1 V_1}{p_2} = \frac{(150 \text{ bar})(12 \text{ L})}{2.5 \text{ bar}} = 720 \text{ L}$$

However, 12 L of air remains in the tank, so she can use only 708 L. Hence

$$t = \frac{708 \text{ L}}{30 \text{ L/min}} = 23.6 \text{ min}$$

## SOLVED PROBLEM 19.8

A steel cylinder contains 3 m$^3$ of air at a gauge pressure of 15 bar. What volume would this amount of air occupy at sea-level atmospheric pressure of about 1 bar?

The absolute pressure of the air in the tank is

$$p_1 = \text{gauge pressure } + \text{ atmospheric pressure}$$
$$= 15 \text{ bar} + 1 \text{ bar} = 16 \text{ bar}$$

When $p_2 = 1$ bar,

$$V_2 = \frac{p_1 V_1}{p_2} = \frac{(16 \text{ bar})(3 \text{ m}^3)}{1 \text{ bar}} = 48 \text{ m}^2$$

## SOLVED PROBLEM 19.9

What additional volume of air at 1 bar pressure must be pumped into the tank of Prob. 19.8 in order to raise the gauge pressure to 25 bar?

When the absolute pressure of the air in the tank is

$$p_1 = 25 \text{ bar} + 1 \text{ bar} = 26 \text{ bar}$$

the equivalent volume of air at $p_2 = 1$ bar is

$$V_2 = \frac{p_1 V_1}{p_2} = \frac{(26 \text{ bar})(3 \text{ m}^3)}{1 \text{ bar}} = 78 \text{ m}^3$$

The additional volume of air is $78 \text{ m}^3 - 48 \text{ m}^3 = 30 \text{ m}^3$.

## ABSOLUTE TEMPERATURE SCALES

When the temperature of a sample of gas is changed while the pressure on it is held constant, its volume changes by 1/273 of its volume at 0°C for each temperature change of 1°C. If it were possible to cool a gas sample to −273°C, its volume would diminish to zero. Since all gases condense into liquids at temperatures above −273°C, this experiment cannot be carried out; nevertheless, −273°C is a significant temperature.

On the *absolute temperature scale*, the zero point is set at −273°C. Temperatures in this scale are expressed in *kelvins* (K); these units are equal to Celsius degrees. Thus

$$T_K = T_C + 273$$

The freezing point of water on the absolute scale is 273 K, and its boiling point is 373 K.

The *Rankine scale* is an absolute temperature scale based on the Fahrenheit scale. Absolute zero in the Rankine scale is −460°F, and

$$T_R = T_F + 460°$$

The freezing point of water in the Rankine scale is 460°R, and its boiling point is 672°R.

## CHARLES'S LAW

Because of the way the absolute temperature scales are defined, the relationship between the temperature and volume of a gas sample at constant pressure can be expressed as

$$\frac{V_1}{T_1} = \frac{V_2}{T_2} \qquad p = \text{constant}$$

In this formula, which is called *Charles's law*, $V_1$ is the volume of the sample at the absolute temperature $T_1$ and $V_2$ is its absolute temperature $T_2$; the formula only holds when the temperatures are expressed in an absolute scale.

## IDEAL GAS LAW

Boyle's law and Charles's law can be combined to form the *ideal gas law*:

$$\frac{p_1 V_1}{T_1} = \frac{p_2 V_2}{T_2}$$

This law is obeyed fairly well by all gases through a wide range of pressures and temperatures. An *ideal gas* is one for which $pV/T = $ constant under all circumstances. Although no such gas actually exists, the fact that a real gas behaves approximately as an ideal one provides a specific target for theories of the gaseous state.

The ideal gas law is further discussed in Chapter 20.

**SOLVED PROBLEM 19.10**

Nitrogen boils at $-196°C$. What is this temperature on the absolute scale?

$$T_K = T_C + 273 = -196 + 273 = 77 \text{ K}$$

**SOLVED PROBLEM 19.11**

The surface temperature of the sun is about 6000 K. What is the Celsius equivalent of this temperature?

$$T_C = T_K - 273 = 5727°C$$

**SOLVED PROBLEM 19.12**

Ethyl alcohol freezes at $-173°F$ and boils at $172°F$. What are these temperatures on the Rankine scale?

$$T_R = T_F + 460° = -173° + 460° = 287°R$$
$$T_R = 172° + 460° = 632°R$$

**SOLVED PROBLEM 19.13**

To what temperature must a gas sample initially at $0°C$ and atmospheric pressure be heated if its volume is to double while its pressure remains the same?

Since $T_1 = 0°C = 273$ K and $V_2 = 2V_1$, from Charles's law

$$T_2 = \frac{T_1 V_2}{V_1} = \frac{(273 \text{ K})(2V_1)}{V_1} = 546 \text{ K} = 273°C$$

**SOLVED PROBLEM 19.14**

The tire of a car contains air at an absolute pressure of 35 lb/in.$^2$ when its temperature is $50°F$. If the tire's volume does not change, what is the pressure when the temperature is $120°F$?

The first step is to convert the temperature to their Rankine equivalents:

$$T_1 = 50° + 460° = 510°R \qquad T_2 = 120° + 460° = 580°R$$

Since $p_1 = 35$ lb/in.$^2$ and $V_1 = V_2$, from the ideal gas law $p_1 V_1 / T_1 = p_2 V_2 / T_2$, we have

$$p_2 = \frac{T_2 p_1}{T_1} = \frac{(580°R)(35 \text{ lb/in.}^2)}{510°R} = 40 \text{ lb/in.}^2$$

**SOLVED PROBLEM 19.15**

A tank whose capacity is 0.1 m$^3$ contains helium at an absolute pressure of 10 bar and a temperature of $20°C$. A rubber weather balloon is inflated with this helium. (a) The gas cools as it expands, and when the pressure of the helium in the balloon is 1 bar, its temperature is $-40°C$. Find the volume of the balloon. (b) Eventually the helium in the balloon absorbs heat from the air around it and returns to $20°C$. Find the volume of the balloon at this time.

(a)  Here $T_1 = 20°C = 293$ K, $T_2 = -40°C = 233$ K, $V_1 = 0.1$ m$^3$, $p_1 = 10$ bar, $p_2 = 1$ bar. From the ideal gas law,

$$V_2 = \frac{T_2 p_1 V_1}{T_1 p_2} = \frac{(233 \text{ K})(10 \text{ bar})(0.1 \text{ m}^3)}{(293 \text{ K})(1 \text{ bar})} = 0.8 \text{ m}^3$$

The volume of the balloon is therefore 0.7 m$^3$ at this time, since the tank retains 0.1 m$^3$ of the helium after the expansion.

(b)  Here $p_1 = 10$ bar, $p_3 = 1$ bar, $V_1 = 0.1$ m$^3$, and, since $T_1 = T_3$, Boyle's law can be used. We have

$$V_3 = \frac{p_1 V_1}{p_2} = \frac{(10 \text{ bar})(0.1 \text{ m}^3)}{1 \text{ bar}} = 1 \text{ m}^3$$

The volume of the balloon is therefore 0.9 m$^3$, assuming it is still attached to the tank.

## SOLVED PROBLEM 19.16

A gas sample occupies 5 ft$^3$ at 60°F and atmospheric pressure. (a) Find its volume at 200°F and a gauge pressure of 50 lb/in.$^2$. (b) Find its gauge pressure when it has been compressed to 1 ft$^3$ and the temperature has been reduced to 0°F.

(a)  Here $T_1 = 60° + 460° = 520°$R, $V_1 = 5$ ft$^3$, and $p_1 = 15$ lb/in.$^2$. When

$$T_2 = 200° + 460° = 660°\text{R} \qquad \text{and} \qquad p_2 = 50 \text{ lb/in.}^2 + 15 \text{ lb/in.}^2 = 65 \text{ lb/in.}^2$$

the volume $V_2$ is

$$V_2 = \frac{T_2 p_1 V_1}{T_1 p_2} = \frac{(660°\text{R})(15 \text{ lb/in.}^2)(5 \text{ ft}^3)}{(520°\text{R})(65 \text{ lb/in.}^2)} = 1.46 \text{ ft}^3$$

(b)  Now $T_2 = 0° + 460° = 460°$R and $V_2 = 1$ ft$^3$. Hence the new absolute pressure is

$$p_2 = \frac{T_2 p_1 V_1}{T_1 V_2} = \frac{(460°\text{R})(15 \text{ lb/in.}^2)(5 \text{ ft}^3)}{(520°\text{R})(1 \text{ ft}^3)} = 66 \text{ lb/in.}^2$$

The new gauge pressure is $66 - 15 = 51$ lb/in.$^2$.

## SOLVED PROBLEM 19.17

The density of air at 0°C and 1 bar of pressure is 1.293 kg/m$^3$. Find its density at 100°C and 2-bar pressure.

At a pressure of $p_1 = 1$ bar and an absolute temperature of $T_1 = 0 + 273 = 273$ K, the mass of a volume of air $V_1$ of density $d_1$ is $m = d_1 V_1$. At $p_2 = 2$ bar and $T_2 = 100 + 273 = 373$ K, the volume $V_2$ is

$$V_2 = \frac{m}{d_2} = \frac{T_2 p_1 V_1}{T_1 p_2}$$

Since $m = d_1 V_1$,

$$\frac{d_1 V_1}{d_2} = \frac{T_2 p_1 V_1}{T_1 p_2}$$

and so
$$d_2 = \frac{d_1 T_1 p_2}{T_2 p_1} = \frac{(1.293 \text{ kg/m}^3)(273 \text{ K})(2 \text{ bar})}{(373 \text{ K})(1 \text{ bar})} = 1.893 \text{ kg/m}^3$$

## Multiple-Choice Questions

**19.1.** At what temperature will an iron bar ($a = 1.2 \times 10^{-5}/°C$) be longer by 0.10 percent than it is at 20°C?

 (*a*)  63°C       (*c*)  103°C
 (*b*)  83°C       (*d*)  120°C

**19.2.** An aluminum pot whose volume is 1000 cm$^3$ at 20°C has a volume of 1006 cm$^3$ at 100°C. The coefficient of linear expansion of aluminum is

 (*a*)  $2.5 \times 10^{-5}/°C$       (*c*)  $7.5 \times 10^{-5}/°C$
 (*b*)  $6.0 \times 10^{-5}/°C$       (*d*)  $2.25 \times 10^{-4}/°C$

**19.3.** The Celsius equivalent of 200 K is

 (*a*)  −73°C       (*c*)  232°C
 (*b*)  73°C       (*d*)  473°C

**19.4.** An object's temperature is raised by 100°C. The resulting increase in its absolute temperature is

 (*a*)  32K       (*c*)  180 K
 (*b*)  100 K       (*d*)  373 K

**19.5.** Nitrogen boils at −320°F. On the Rankine scale this temperature is

 (*a*)  −47°R       (*c*)  140°R
 (*b*)  108°R       (*d*)  172°R

**19.6.** The pressure and absolute temperature of a gas sample whose volume is fixed are related by which one or more of the following formulas?

 (*a*)  $\dfrac{p_1}{T_2} = \dfrac{p_2}{T_1}$       (*c*)  $\dfrac{p_1}{p_2} = \dfrac{T_2}{T_1}$

 (*b*)  $\dfrac{p_1}{T_1} = \dfrac{p_2}{T_2}$       (*d*)  $\dfrac{p_1}{T_1} = \dfrac{T_2}{p_2}$

**19.7.** At constant temperature the absolute pressure on 10 ft$^3$ of air is increased from 20 to 80 lb/in.$^2$. The volume of the air is now

 (*a*)  2.5 ft$^3$       (*c*)  40 ft$^3$
 (*b*)  5 ft$^3$       (*d*)  80 ft$^3$

**19.8.** A sample of oxygen whose volume at 0°C and 200 kPa of pressure is 4.00 L is compressed to 1.00 L and its temperature is raised to 273°C. The pressure of the gas is now

 (*a*)  100 kPa       (*c*)  800 kPa
 (*b*)  400 kPa       (*d*)  1600 kPa

## Supplementary Problems

**19.1.** A steel bridge is 500 m long at 0°C. By how much does it expand when the temperature becomes 35°C? The coefficient of linear expansion of steel is $1.2 \times 10^{-5}/°C$.

**19.2.** A brass rod 4 ft long expands by $\frac{1}{16}$ in. when heated from 70 to 200°F. Find the coefficient of linear expansion of brass.

**19.3.** How much mercury ($b = 1.8 \times 10^{-4}$/°C) overflows when a glass vessel ($b = 2 \times 10^{-5}$/°C) filled to the brim with 50 cm$^3$ of mercury at 15°C is heated to 75°C?

**19.4.** The density of mercury is 830 lb/ft$^3$ at 70°F. Find its density at 0°F. The coefficient of volume expansion of mercury is $1.0 \times 10^{-4}$/°F.

**19.5.** A load of 2000 kg is placed on a vertical steel beam 5 m high whose cross-sectional area is 30 cm$^2$. The temperature is 20°C. (*a*) By how much is the beam compressed? (*b*) At what temperature will the beam return to its original height? For steel, $Y = 2 \times 10^{11}$ Pa and $a = 1.2 \times 10^{-5}$/°F; 1 m$^2 = 10^4$ cm$^2$.

**19.6.** Aluminum has a coefficient of linear expansion of $2.4 \times 10^{-5}$/°C and a Young's modulus of $7 \times 10^{10}$ Pa. For the steel respective values are $1.2 \times 10^{-5}$/°C and $20 \times 10^{10}$ Pa. If identical bars of the two metals undergo the same temperature change, which bar has the greater force associated with its change in length?

**19.7.** A compressor pumps 50 L of air at a pressure of 1 bar into an 8-L tank. What is the absolute pressure (in atmospheres) of the air in the tank?

**19.8.** How much air at a pressure of 1 bar can be stored in a 2-m$^3$ tank which can safely withstand a pressure of 5 bar?

**19.9.** What is the Celsius equivalent of 500 K?

**19.10.** What is the Kelvin equivalent of 500°C?

**19.11.** What is the Fahrenheit equivalent of 500°R?

**19.12.** What is the Rankine equivalent of 500°F?

**19.13.** A tire contains 1 ft$^3$ of air at a gauge pressure of 28 lb/in.$^2$. How much additional air at atmospheric pressure must be pumped into the tire to raise the pressure to 36 lb/in.$^2$ at the same temperature?

**19.14.** A tire contains air at a gauge pressure of 2 bar at 15°C. If the tire's volume does not change, what will the gauge pressure be when its temperature is 38°C?

**19.15.** The weight density of carbon dioxide is 0.1234 lb/ft$^3$ at 32°F and 1-atm pressure. Find its weight density at 80°F and 10-atm pressure.

**19.16.** A gas sample occupies 4 m$^3$ at an absolute pressure of 2 bar and a temperature of 320 K. Find its volume (*a*) at the same pressure and a temperature of 400 K and (*b*) at the same temperature and a pressure of 0.4 bar.

**19.17.** A gas sample occupies a volume of 1 m$^3$ at a temperature of 27°C and a pressure of 1 bar. Find its volume (*a*) at 127°C and 0.5 bar; (*b*) at 127°C and 2 bar; (*c*) at −73°C and 0.5 bar; and (*d*) at −73°C and 2 bar.

## Answers to Multiple-Choice Questions

**19.1.** (c)     **19.5.** (c)

**19.2.** (a)     **19.6.** (b)

**19.3.** (a)     **19.7.** (a)

**19.4.** (b)     **19.8.** (d)

## Answers to Supplementary Problems

**19.1.** 21 cm                          **19.10.** 773 K

**19.2.** $1.0 \times 10^{-5}/°F$        **19.11.** 40°F

**19.3.** 0.48 cm$^3$                    **19.12.** 960°R

**19.4.** 836 lb/ft$^3$                  **19.13.** 0.53 ft$^3$

**19.5.** (a) 0.163 mm    (b) 22.7°C     **19.14.** 2.24 bar

**19.6.** The steel bar                  **19.15.** 1.124 lb/ft$^3$

**19.7.** 6.25 bar                       **19.16.** (a) 5 m$^3$     (b) 20 m$^3$

**19.8.** 9.87 m$^3$                     **19.17.** (a) 2.67 m$^3$  (c) 1.33 m$^3$

**19.9.** 227°C                                  (b) 0.67 m$^3$  (d) 0.33 m$^3$

# Chapter 20

# Kinetic Theory of Matter

## KINETIC THEORY OF GASES

The *kinetic theory of gases* holds that a gas is composed of very small particles, called *molecules*, which are in constant random motion. The molecules are far apart relative to their dimensions and do not interact with one another except in collisions.

The pressure a gas exerts is due to the impacts of its molecules; there are so many molecules in even a small gas sample that the individual blows appear as a continuous force. Boyle's law is readily understood in terms of the kinetic theory of gases. Expanding a gas sample means that its molecules must travel farther between successive impacts on the container walls and that the impacts are spread over a larger area. Hence an increase in volume means a decrease in pressure, and vice versa.

## MOLECULAR ENERGY

According to the kinetic theory of gases, the average kinetic energy of the molecules of a gas is proportional to the absolute temperature of the gas. This relationship is usually expressed in the form

$$KE_{av} = \tfrac{3}{2}kT$$

where $k$ = Boltzmann's constant = $1.38 \times 10^{-23}$ J/K. Actual molecular energies vary considerably on either side of $KE_{av}$.

At absolute zero, 0 K, gas molecules would be at rest, which is why this is such a significant temperature. At any temperature, all gases have the same average molecular energy. Therefore, in a gas whose molecules are heavy, the molecules move more slowly on the average than do those in a gas at the same temperature whose molecules are light.

Charles's law follows directly from the above interpretation of temperature. Compressing a gas causes its temperature to rise because molecules rebound from the inward-moving walls of the container with increased energy, just as a tennis ball rebounds with greater energy when it is struck by a moving racket. Similarly, expanding a gas causes its temperature to fall because molecules rebound from the outward-moving walls with decreased energy.

## SOLIDS AND LIQUIDS

The molecules of a solid are close enough together to exert forces on one another that hold the entire assembly to a definite size and shape. As in the case of a gas, the molecules are in constant motion, but they vibrate about fixed locations instead of moving randomly. The molecules of a liquid continually move around past one another more or less freely, which enables the liquid to flow, but their spacing does not change, and so the volume of a given liquid sample does not vary.

When a solid melts, the original ordered arrangement of its molecules changes to the random arrangement of molecules in a liquid. To accomplish the change, the molecules must be pulled apart against the forces holding them in place, which requires energy. The heat of fusion of a solid represents this energy. When a liquid boils, the heat of vaporization represents the energy needed to pull its molecules entirely free of one another so that a gas is formed.

**SOLVED PROBLEM 20.1**

Gas molecules have velocities comparable with those of rifle bullets, yet we all know that a gas with a strong odor, such as ammonia, takes several seconds to diffuse through a room. Why?

   Gas molecules collide frequently with one another, which means that a particular molecule follows a long, very complicated path in going from one place to another.

**SOLVED PROBLEM 20.2**

Explain the evaporation of a liquid at a temperature below its boiling point on the basis of the kinetic theory of matter.

   At any moment in a liquid, some molecules are moving faster and others are moving slower than the average. The fastest ones are able to escape from the liquid surface despite the attractive forces exerted by the other molecules; this constitutes evaporation. The warmer the liquid, the greater the number of very fast molecules, and the more rapidly evaporation takes place. Since the molecules that remain behind are the slower ones, the liquid has a lower temperature than before (unless heat has been added to it from an outside source during the process).

**SOLVED PROBLEM 20.3**

What is the average kinetic energy of the molecules of any gas at 100°C?

   The absolute temperature corresponding to 100°C is

$$T_K = T_C + 273 = 373 \text{ K}$$

The average kinetic energy at this temperature is

$$\text{KE}_{av} = \tfrac{3}{2}kT = (\tfrac{3}{2})(1.38 \times 10^{-23} \text{ J/K})(373 \text{ K}) = 7.72 \times 10^{-21} \text{ J}$$

**SOLVED PROBLEM 20.4**

What is the average velocity of the molecules in a sample of oxygen at 100°C? The mass of an oxygen molecule is $5.3 \times 10^{-26}$ kg.

   Since $\text{KE}_{av} = \tfrac{1}{2}mv_{av}^2 = \tfrac{3}{2}kT$,

$$v_{av} = \sqrt{\frac{3kT}{m}}$$

Here $T = 100°C = 373$ K, and so

$$v_{av} = \sqrt{\frac{3kT}{m}} = \sqrt{\frac{(3)(1.38 \times 10^{-23} \text{ J/K})(373 \text{ K})}{5.3 \times 10^{-26} \text{ kg}}} = \sqrt{29.1 \times 10^4} \text{ m/s}$$
$$= 5.4 \times 10^2 \text{ m/s} = 540 \text{ m/s}$$

**SOLVED PROBLEM 20.5**

A certain tank holds 1 g of hydrogen at 0°C, and another identical tank holds 1 g of oxygen at 0°C. The mass of an oxygen molecule is 16 times greater than that of a hydrogen molecule. (*a*) Which tank contains more molecules? How many more? (*b*) Which gas exerts the greater pressure? How much greater? (*c*) In which gas do the molecules have greater average energies? How much greater? (*d*) In which gas do the molecules have greater average velocities? How much greater?

(a) There are 16 times more hydrogen molecules.

(b) The hydrogen pressure is 16 times greater because there are 16 times more molecules to exert force on the container walls.

(c) The average molecular energies are the same in both gases because their temperatures are the same.

(d) The average velocity of the hydrogen molecules is $\sqrt{16} = 4$ times more than that of the oxygen molecules because the hydrogen molecules are 16 times lighter and $v_{av} = \sqrt{3kT/m}$.

## RELATIVE HUMIDITY

The *humidity* of air refers to the amount of water vapor it contains. Air is *saturated* when it contains the maximum amount of water vapor possible; the higher the temperature, the greater the water vapor density at saturation, as shown in Fig. 20-1. The *relative humidity* of a volume of air describes its degree of saturation. A relative humidity of 0 means perfectly dry air; 50 percent means that the air contains half the maximum water vapor possible; 100 percent means that the air is saturated.

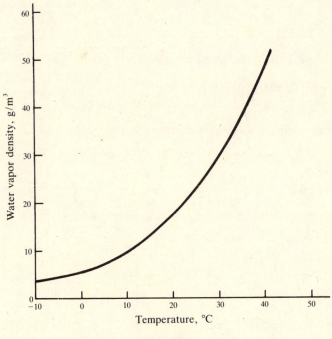

**Fig. 20-1**

## SOLVED PROBLEM 20.6

Find the water vapor density of air at 25°C whose relative humidity is 50 percent.

From Fig. 20-1 we see that the saturated water vapor density at 25°C is about 23 g/m$^3$. Since

$$\text{Relative humidity} = \frac{\text{actual vapor density}}{\text{saturation vapor density}}$$

we have here

$$\text{Actual vapor density} = (\text{relative humidity})(\text{saturation vapor density})$$
$$= (0.50)(23 \text{ g/m}^3) = 11.5 \text{ g/m}^3$$

**SOLVED PROBLEM 20.7**

The *dew point* is the temperature at which air with a certain water vapor density would be saturated and moisture would start to condense. Dew point is useful in predicting fog: When the air temperature is near the dew point and is decreasing, fog is likely to occur. What is the dew point of air at 25°C whose relative humidity is 50 percent?

The water vapor density of the air is 11.5 g/m$^3$ as found in Prob. 20.6. From Fig. 20-1 air with this vapor density is saturated at about 13°C, so this temperature is the dew point here.

## ATOMS AND MOLECULES

*Elements* are the fundamental substances of which all matter is composed. There are over 100 known elements, of which a number are not found in nature but have been prepared in the laboratory. Elements cannot be transformed into one another by ordinary chemical or physical means, but two or more elements can combine to form a *compound*, which is a substance whose properties are different from those of its constituent elements.

The ultimate particles of an element are called *atoms*, and those of a compound that exists in the gaseous state are called *molecules*. The molecules of a compound consist of the atoms of the elements that compose it joined in a specific arrangement. Each molecule of water, for instance, contains two hydrogen atoms and one oxygen atom, as its symbol $H_2O$ indicates. Many compounds in the solid and liquid states do not consist of individual molecules, as discussed later. Elemental gases may consist of atoms (helium, He; argon, Ar) or of molecules (hydrogen, $H_2$; oxygen, $O_2$).

The masses of atoms and molecules are expressed in *atomic mass units* (u), where

$$1 \text{ atomic mass unit} = 1 \text{ u} = 1.660 \times 10^{-27} \text{ kg}$$

The mass $m$ of a molecule is the sum of the masses of the atoms of which it is composed; thus $m(H_2O) = 2m(H) + m(O)$.

## THE MOLE

The samples of matter used in both industry and the laboratory involve so many atoms or molecules that counting them is out of the question. Instead the mass of a particular sample is used as a measure of its quantity, and it is necessary to relate this mass to the number of atoms or molecules in the sample.

When the masses of two samples of different substances stand in the same proportion as their molecular masses, they contain the same number of molecules. To use this fact, the mole is defined as follows: A *mole* of any substance is that amount of it whose mass is equal to its molecular mass expressed in grams instead of atomic mass units.

Thus a mole of water has a mass of 18 g since a water molecule has a mass of 18 u. A mole of an elemental substance that consists of individual atoms rather than molecules is that amount of it whose mass is equal to its atomic mass expressed in grams. In SI units the amount of a substance corresponding to a mole is taken as a basic unit and written as 1 mol.

The number of molecules in a mole of any substance is *Avogadro's number N*, whose value is

$$N = 6.023 \times 10^{23} \text{ molecules/mol}$$

The number of molecules in a sample of a substance is the number of moles it contains multiplied by $N$.

## SOLVED PROBLEM 20.8

Find the mass of (*a*) the water molecule $H_2O$ and (*b*) the ethyl alcohol molecule $C_2H_6O$. The atomic masses of H, C, and O are, respectively, 1.008, 12.01, and 16.00 u.

(*a*)
$$2(H) = (2)(1.008) \text{ u} = \underline{\phantom{0}2.02 \text{ u}}$$
$$1(O) = (1)(16.00) \text{ u} = \underline{16.00 \text{ u}}$$
$$18.02 \text{ u}$$

$$m(H_2O) = (18.02 \text{ u})(1.66 \times 10^{-27} \text{ kg/u}) = 2.99 \times 10^{-26} \text{ kg}$$

(*b*)
$$2(C) = (2)(12.01) \text{ u} = 24.02 \text{ u}$$
$$6(H) = (6)(1.008) \text{ u} = \underline{\phantom{0}6.05 \text{ u}}$$
$$1(O) = (1)(16.00) \text{ u} = \underline{16.00 \text{ u}}$$
$$46.07 \text{ u}$$

$$m(C_2H_6O) = (46.07 \text{ u})(1.66 \times 10^{-27} \text{ kg/u}) = 7.65 \times 10^{-26} \text{ kg}$$

## SOLVED PROBLEM 20.9

How many $H_2O$ molecules are present in 1 kg of water?

$$\text{Molecules of } H_2O = \frac{\text{mass of } H_2O}{\text{mass of } H_2O \text{ molecule}}$$
$$= \frac{1 \text{ kg}}{2.99 \times 10^{-26} \text{ kg}} = 3.34 \times 10^{25} \text{ molecules}$$

## SOLVED PROBLEM 20.10

Find the mass of 75 mol of uranium (U).

The atomic mass of uranium is 238.03 u, which means that a mole of U has a mass of 238.03 g. Hence

$$\text{Mass of U} = (\text{moles of U})(\text{atomic mass of U})$$
$$= (75 \text{ mol})(238.02 \text{ g/mol}) = 1.785 \times 10^4 \text{ g} = 17.85 \text{ kg}$$

## SOLVED PROBLEM 20.11

How many moles of Cu are present in 100 g of copper?

The atomic mass of Cu is 63.54 u = 63.54 g/mol. Hence

$$\text{Moles of Cu} = \frac{\text{mass of Cu}}{\text{atomic mass of Cu}} = \frac{100 \text{ g}}{63.54 \text{ g/mol}} = 1.574 \text{ mol}$$

## SOLVED PROBLEM 20.12

(*a*) Find the mass of 9.4 mol of ethylene ($C_2H_4$). (*b*) How many carbon atoms are present?

(a)  The molecular mass of ethylene is

$$2(C) = (2)(12.01) \text{ u} = 24.02 \text{ u}$$
$$4(H) = (4)(1.008) \text{ u} = \underline{\phantom{0}4.03 \text{ u}}$$
$$28.05 \text{ u} = 28.05 \text{ g/mol}$$

The required mass is

Mass of $C_2H_4$ = (moles of $C_2H_4$)(molecular mass of $C_2H_4$) = (9.4 mol)(28.05 g/mol) = 264 g

(b)  Two moles of carbon are present in each mole of $C_2H_4$. Hence there is (2) (9.4 mol) = 18.8 mol of carbon present in 9.4 mol of $C_2H_4$. The number of carbon atoms is

Atoms of C = (moles of C)(Avogadro's number)
$$= (18.8 \text{ mol})(6.023 \times 10^{23} \text{ atoms/mol}) = 1.13 \times 10^{25} \text{ atoms}$$

## MOLAR VOLUME

Equal volumes of all gases, under the same conditions of temperature and pressure, contain the same number of molecules and therefore the same number of moles. This observation is most useful stated in reverse: Under given conditions of temperature and pressure, the volume of a gas is proportional to the number of moles present.

For convenience, a temperature of 0°C (273 K) and a pressure of 1 atm ($1.013 \times 10^5$ N/m$^2$ = 14.7 lb/in.$^2$) are taken as the standard temperature and pressure (STP); Charles's and Boyle's laws permit measurements made at other temperatures and pressures to be reduced to their equivalents at STP. Experimentally it is found that 1 mol of any gas at STP occupies a volume of 22.4 L. Thus the *molar volume* of a gas is 22.4 L at STP. This observation makes it possible to deal with gas volumes in chemical reactions. If a certain reaction is known to produce 2.5 mol of a gas, for instance, we know that at STP the volume of the gas will be

$$V = (\text{moles of gas})(\text{molar volume}) = (2.5 \text{ mol})(22.4 \text{ L/mol}) = 56 \text{ L}$$

## UNIVERSAL GAS CONSTANT

According to the ideal gas law (Chapter 19), the pressure, volume, and temperature of a gas sample obey the relationship $pV/T$ = constant. We can find the value of the constant in terms of the number of moles $n$ of gas in the sample by making use of the fact that the molar volume at STP is 22.4 L. At STP we have $T = 0°C = 273$ K, $p = 1$ atm, and $V = (n)(22.4 \text{ L/mol})$ so that

$$\frac{pV}{T} = \frac{(n)(1 \text{ atm})(22.4 \text{ L/mol})}{273 \text{ K}} = nR$$

where $R$, the *universal gas constant*, has the value

$$R = 0.0821 \text{ atm·L/(mol·K)}$$

In SI units, in which $p$ is in newtons per square meter and $V$ is in cubic meters,

$$R = 8.31 \text{ J/(mol·K)}$$

The complete ideal gas law is usually written in the form

$$pV = nRT$$

**SOLVED PROBLEM 20.13**

(a) What volume does 1 g of ammonia ($NH_3$) occupy at STP? (b) What volume does it occupy at 100°C and a pressure of 1.2 atm?

(a)  The molecular mass of $NH_3$ is

$$1(N) = (1)(14.01)\ u = 14.01\ u$$
$$3(H) = (3)(1.008)\ u = \underline{\phantom{0}3.02\ u}$$
$$17.03\ u = 17.03\ g/mol$$

so the number of moles in 1 g of $NH_3$ is

$$\text{Moles of } NH_3 = \frac{\text{mass of } NH_3}{\text{molecular mass of } NH_3} = \frac{1\ g}{17.03\ g/mol} = 0.0587\ mol$$

The volume at STP is therefore

$$\text{Volume of } NH_3 = (\text{moles of } NH_3)(\text{molar volume}) = (0.0587\ mol)(22.4\ L/mol) = 1.32\ L$$

(b)  From the ideal gas law,

$$\frac{p_1 V_1}{T_1} = \frac{p_2 V_2}{T_2} \qquad \text{or} \qquad V_2 = \frac{p_1 V_1 T_2}{p_2 T_1}$$

Here $p_1 = 1$ atm, $V_1 = 1.32$ L, $T_1 = 0°C = 273$ K and $p_2 = 1.2$ atm, $V_2 = ?$, $T_2 = 100°C = 373$ K. Hence

$$V_2 = \frac{(1\ atm)(1.32\ L)(373\ K)}{(1.2\ atm)(273\ K)} = 1.50\ L$$

**SOLVED PROBLEM 20.14**

What is the mass of 40 L of uranium hexafluoride ($UF_6$) at 500°C and 4 atm of pressure?

The most direct way to solve this problem is to use the ideal gas law to find the number of moles of $UF_6$ in the sample. Since $pV = nRT$ and $T = 500°C = 773$ K, we have

$$n = \frac{pV}{RT} = \frac{(4\ atm)(40\ L)}{[0.0821\ atm \cdot L/(mol \cdot K)](773\ K)} = 2.52\ mol$$

The molecular mass of $UF_6$ is

$$1(U) = (1)(238.03)\ u = 238.03\ u$$
$$6(F) = (6)(19.00)\ u = \underline{114.00\ u}$$
$$352.03\ u = 352.03\ g/mol$$

so the mass of $UF_6$ is

$$\text{Mass of } UF_6 = (\text{moles of } UF_6)(\text{molecular mass of } UF_6) = (2.52\ mol)(352.03\ g/mol) = 887\ g$$

**SOLVED PROBLEM 20.15**

Find the density in grams per liter of ethylene ($C_2H_4$) at STP.

At STP 1 mol of any gas occupies 22.4 L. The molecular mass of $C_2H_4$ is

$$2(C) = (2)(12.01)\ u = 24.02\ u$$
$$4(H) = (4)(1.008)\ u = \underline{\phantom{0}4.03\ u}$$
$$28.05\ u = 28.05\ g/mol$$

One mole of $C_2H_4$ therefore has a density at STP of

$$d = \frac{m}{V} = \frac{28.05 \text{ g}}{22.4 \text{ L}} = 1.25 \text{ g/L}$$

## SOLVED PROBLEM 20.16

What is the density of oxygen at 20°C and 5 atm of pressure?

It is simplest here to use the ideal gas law to find the mass of 1 L of $O_2$ under the specified conditions. The number of moles in 1 L of $O_2$ at $T = 20°C = 293$ K and $p = 5$ atm is, from $pV = nRT$,

$$n = \frac{pV}{RT} = \frac{(5 \text{ atm})(1 \text{ L})}{[0.0821 \text{ atm} \cdot \text{L/(mol} \cdot \text{K)}](293 \text{ K})} = 0.208 \text{ mol}$$

Since the molecular mass of $O_2$ is (2)(16.00) u = 32.00 u = 32.00 g/mol, the mass here is

Mass of $O_2$ = (moles of $O_2$)(molecular mass of $O_2$) = (0.208 mol)(32.00 g/mol) = 6.66 g

Hence the density of the gas is

$$g = \frac{m}{V} = \frac{6.66 \text{ g}}{1 \text{ L}} = 6.66 \text{ g/L}$$

There are $10^3$ grams per kilogram and $10^3$ liters per cubic meter, which means that the density in SI units is 6.66 kg/m$^3$.

## SOLVED PROBLEM 20.17

A sample of an unknown gas has a mass of 28.1 g and occupies 4.8 L at STP. What is its molecular mass?

Since 1 mol of any gas occupies 22.4 L at STP, this sample must consist of

$$\frac{4.8 \text{ L}}{22.4 \text{ L/mol}} = 0.214 \text{ mol}$$

Hence

$$\text{Molecular mass} = \frac{\text{mass of sample}}{\text{moles of sample}} = \frac{28.1 \text{ g}}{0.214 \text{ mol}} = 131 \text{ g/mol} = 131 \text{ u}$$

# *Multiple-Choice Questions*

**20.1.** Molecular motion in a gas stops

(*a*)  at absolute zero
(*b*)  when the gas becomes a liquid
(*c*)  when the gas becomes a solid
(*d*)  when the pressure on it exceeds a certain value

**20.2.** At a given temperature

(*a*)  the molecules in a gas all have the same average velocity
(*b*)  the molecules in a gas all have the same average energy
(*c*)  light gas molecules have lower average energies than heavy gas molecules
(*d*)  heavy gas molecules have lower average energies than light gas molecules

**20.3.** The temperature of a gas sample in a container of fixed volume is raised. The gas exerts a higher pressure on the walls of its container because its molecules

 (*a*) lose more PE when they strike the walls.
 (*b*) lose more KE when they strike the walls
 (*c*) are in contact with the walls for a shorter time
 (*d*) have higher average velocities and strike the walls more often

**20.4.** The volume of a gas sample is increased while its temperature is held constant. The gas exerts a lower pressure on the walls of its container partly because its molecules strike the walls

 (*a*) less often    (*c*) with less energy
 (*b*) with lower velocities (*d*) with less force

**20.5.** When evaporation occurs, the liquid that remains is cooler because

 (*a*) the pressure on the liquid decreases
 (*b*) the volume of the liquid decreases
 (*c*) the slowest molecules remain behind
 (*d*) the fastest molecules remain behind

**20.6.** When a volume of air is heated,

 (*a*) it can hold less water vapor
 (*b*) it can hold more water vapor
 (*c*) the amount of water vapor it can hold does not change
 (*d*) its relative humidity increases

**20.7.** Cooling saturated air causes

 (*a*) its relative humidity to decrease
 (*b*) its relative humidity to increase
 (*c*) its ability to take up water vapor to increase
 (*d*) some of its water content to condense out

**20.8.** Which one or more of the following quantities are the same for both a mole of oxygen molecules and a mole of nitrogen molecules at the same temperature and pressure?

 (*a*) the number of molecules present
 (*b*) the average velocities of the molecules
 (*c*) the volume of the gas
 (*d*) the density of the gas

**20.9.** The mass of a nitrogen atom is 14 u. The number of moles of molecular nitrogen ($N_2$) in 56 g of nitrogen is

 (*a*) 2  (*c*) 28
 (*b*) 4  (*d*) 64

**20.10.** A gas sample at 200 K is heated until its temperature is 400 K. If the original average velocity of the gas molecules was *v*, their new average velocity is

 (*a*) *v*   (*c*) 2*v*
 (*b*) $\sqrt{2}v$  (*d*) 4*v*

**20.11.** The molecules of a gas at 10°C would have twice as much average KE at

    (*a*)  20°C    (*c*)  566°C
    (*b*)  293°C   (*d*)  859°C

**20.12.** An oxygen molecule has 16 times the mass of a hydrogen molecule. A sample of hydrogen gas whose molecules have the same average KE as the molecules in a sample of oxygen at 400 K is at a temperature of

    (*a*)  25 K    (*c*)  1600 K
    (*b*)  400 K   (*d*)  6400 K

# Supplementary Problems

**20.1.** The volume of a gas sample is enlarged. Why does the pressure the gas exerts decrease?

**20.2.** The temperature of a gas sample is raised. Why does the pressure the gas exerts increase?

**20.3.** (*a*) Find the relative humidity of air at 20°C whose water vapor density is 8 g/m$^3$. (*b*) What is its dew point?

**20.4.** Air at 25°C has a dew point of 15°C. What is its relative humidity?

**20.5.** Find the mass of the propane molecule $C_3H_8$ and that of the glucose molecule $C_6H_{12}O_6$.

**20.6.** The atomic mass of copper is 63.54 u. Find the number of atoms present in 100 g of copper.

**20.7.** A gas sample at 0°C is heated until the average energy of its molecules doubles. What is its new temperature?

**20.8.** A gas sample at 0°C is heated until the average velocity of its molecules doubles. What is its new temperature?

**20.9.** At room temperature oxygen molecules have an averge velocity of about 1000 mi/h. (*a*) What is the average velocity of hydrogen molecules, whose mass is one-sixteenth that of oxygen molecules, at this temperature? (*b*) What is the average velocity of sulfur dioxide molecules, whose mass is twice that of oxygen molecules, at this temperature?

**20.10.** Mercury is a gas at 500°C. (*a*) What is the average energy of mercury atoms at this temperature? (*b*) What is the average velocity of mercury atoms at this temperature? The mass of a mercury atom is $3.3 \times 10^{-25}$ kg.

**20.11.** How many moles are present in 120 kg of boric acid ($H_3BO_3$)? The atomic mass of boron is 10.82 u.

**20.12.** (*a*) Find the mass of 80 mol of sulfuric acid ($H_2SO_4$). (*b*) How many oxygen atoms are present?

**20.13.** (*a*) What volume does 8.2 mol of fluorine ($F_2$) occupy at STP? (*b*) What volume does it occupy at 40°C and a pressure of 2.5 atm?

**20.14.** (*a*) What volume does 5 g of methane ($CH_4$) occupy at STP? (*b*) What volume does it occupy at $0°C$ and a pressure of 0.5 atm? (*c*) What volume does it occupy at $80°C$ and a pressure of 2 atm?

**20.15.** (*a*) What volume does 20 g of $CO_2$ occupy at STP? (*b*) What volume does it occupy at $-20°C$ and a pressure of 4 atm?

**20.16.** What is the mass of 4 L of ammonia ($NH_3$) at STP?

**20.17.** (*a*) Find the mass of 12 L of chlorine ($Cl_2$) at $40°C$ and 0.8 atm of pressure. (*b*) What is its density under those conditions?

**20.18.** A balloon is filled with 50 $m^3$ ($5 \times 10^4$ L) of hydrogen at STP. What is the mass of the hydrogen?

**20.19.** Find the density of sulfur dioxide ($SO_2$) at STP.

**20.20.** Find the density of nitrogen ($N_2$) at $120°C$ and a pressure of 66,600 Pa.

**20.21.** One liter of an unknown gas has a mass of 2.9 g at STP. Find its molecular mass.

## *Answers to Multiple-Choice Questions*

**20.1.**  (*a*)        **20.7.**  (*d*)

**20.2.**  (*b*)        **20.8.**  (*a*), (*c*)

**20.3.**  (*d*)        **20.9.**  (*a*)

**20.4.**  (*a*)        **20.10.**  (*b*)

**20.5.**  (*c*)        **20.11.**  (*b*)

**20.6.**  (*b*)        **20.12.**  (*b*)

## Answers to Supplementary Problems

**20.1.**     The pressure decreases partly because the gas molecules now must travel farther between impacts on the container walls and partly because these impacts are now distributed over a larger area.

**20.2.**     The pressure increases partly because the gas molecules move faster than before and therefore strike the walls more often and partly because each impact yields a greater force than before.

**20.3.**     (*a*)   46 percent        (*b*)   $7°C$

**20.4.**   56 percent

**20.5.**   $7.32 \times 10^{-26}$ kg; $2.99 \times 10^{-25}$ kg

**20.6.**   $9.48 \times 10^{23}$ atoms

**20.7.**   273°C

**20.8.**   819°C

**20.9.**   (a)  4000 mi/h      (b)  707 mi/h

**20.10.**   (a)  $1.6 \times 10^{-20}$ J      (b)  311 m/s

**20.11.**   1940 mol

**20.12.**   (a)  7.85 kg      (b)  $1.93 \times 10^{26}$ atoms

**20.13.**   (a)  184 L      (b)  84.4 L

**20.14.**   (a)  6.98 L      (b)  13.96 L      (c)  4.51 L

**20.15.**   (a)  10.2 L      (b)  2.36 L

**20.16.**   3.04 g

**20.17.**   (a)  26.5 g      (b)  2.21 g/L = 2.21 kg/m$^3$

**20.18.**   4.5 kg

**20.19.**   2.86 g/L = 2.86 kg/m$^3$

**20.20.**   0.571 kg/m$^3$

**20.21.**   65 u

# Chapter 21

# Thermodynamics

## FIRST LAW OF THERMODYNAMICS

To convert internal energy to mechanical energy is much more difficult than the reverse, and perfect efficiency is impossible. A *heat engine* is a device or system that can perform this conversion; the human body and the earth's atmosphere are heat engines, as are gasoline and diesel motors, aircraft jet engines, and steam turbines. All heat engines operate by absorbing heat from a reservoir of some kind at a high temperature, performing work, and then giving off heat to a reservoir of some kind at a lower temperature (Fig. 21-1).

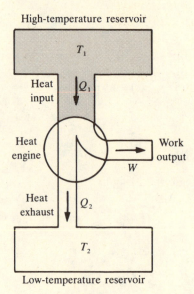

High-temperature reservoir

$T_1$

Heat input $Q_1$

Heat engine

Work output $W$

Heat exhaust $Q_2$

$T_2$

Low-temperature reservoir

**Fig. 21-1**

Two general principles apply to all heat engines. The *first law of thermodynamics* is an expression of the principle of conservation of energy. According to this law, in any process that a system of some kind (such as a heat engine) undergoes, we have

$$Q = \Delta U + W$$

Net heat input = change in internal energy + net work output

Here $Q$ is the net heat added to the system during the process; if the system gives off heat, $Q$ is negative. When the internal energy of the system $U$ increases, $\Delta U$ is positive; when $U$ decreases, $\Delta U$ is negative. The net work done *by* the system during the process is $W$; if work is done *on* the system, $W$ is negative.

If the system is a heat engine that operates in a cycle, energy may be stored and released from storage, but the engine does not undergo a net change in its internal energy during each cycle. In this case

Net heat input = net work output

The net heat input is the amount of heat $Q_1$ the engine takes in from the high-temperature reservoir minus the amount of heat $Q_2$ the engine gives off to the low-temperature reservoir, as in Fig. 21-1, so that

$$Q_1 - Q_2 = W$$

## WORK DONE BY AND ON A GAS

The work output of most heat engines is produced by an expanding gas. If the volume of the gas changes from $V_1$ to $V_2$ at the constant pressure $p$, the work done is

$$W = p(V_2 - V_1) \qquad p = \text{constant}$$

If the gas is compressed rather than expanded, $V_2$ is less than $V_1$ and $W$ is negative. This means that work is done *on* the gas during a compression. In the *p-V* (pressure-volume) diagram of Fig. 21-2 the expansion of a gas at constant pressure appears as a horizontal line from $V_1$ to $V_2$. The area under the line is equal to $p(V_2 - V_1)$ and so equals the work $W$ done in the expansion. If the gas pressure varies during the expansion, the expansion appears as a curved line on a *p-V* diagram, as in Fig. 21-3. We can imagine the region under the curve as divided into thin strips, each corresponding to a small expansion at a different constant pressure so that the total area under the curve equals the work done in this situation also.

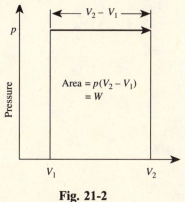

Fig. 21-2

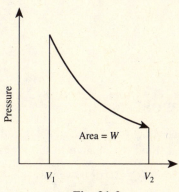

Fig. 21-3

Three important kinds of expansion and comparison that can occur in a gas are as follows:

1.  An *isobaric* process is one that takes place at constant pressure.

2.  An *isothermal* process is one that takes place at constant temperature. The expansions and compressions of a gas in a container that is surrounded by a constant-temperature heat reservoir are approximately isothermal.

3.  An *adiabatic* process is one that takes place in a system so isolated from its surroundings that heat neither enters nor leaves the system during the process. Most rapid thermodynamic processes are approximately adiabatic because heat transfer takes time and a rapid process may be completed before much heat has passed through the walls of the system.

## SOLVED PROBLEM 21.1

Show that the work done by a gas expanding at the constant pressure $p$ from $V_1$ to $V_2$ is given by $W = p(V_2 - V_1)$.

Figure 21-4 shows a gas-filled cylinder with a movable piston. The cross-sectional area of the cylinder is $A$, and the gas pressure is $p$. Since $p = F/A$, the force the gas exerts on the piston is

$$F = pA$$

The work done by the gas in moving the piston through the distance $\Delta s$ is

$$W = (\text{force})(\text{distance}) = F\,\Delta s = pA\,\Delta s$$

But $A\,\Delta s$ is the change $\Delta V = V_2 - V_1$ in the volume of the gas, so

$$W = p(V_2 - V_1)$$

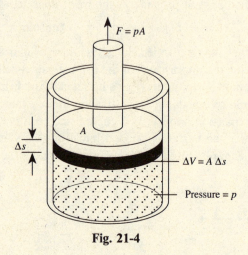

**Fig. 21-4**

## SOLVED PROBLEM 21.2

A gas expands by 1.2 L at a constant pressure of 2.5 bar. During the expansion 500 J of heat is added. Find the change in the internal energy of the gas.

From $Q = \Delta U + W$ we have

$$\Delta U = Q - W$$

Here $W = p(V_2 - V_1)$ since the expansion is isobaric. Since $p = (2.5\ \text{bar})\,(10^5\ \text{Pa/bar}) = 2.5 \times 10^5$ Pa and $V_2 - V_1 = (1.2\ \text{L})/(10^3\ \text{L/m}^3) = 1.2 \times 10^{-3}\ \text{m}^3$, we have

$$\Delta U = Q - p(V_2 - V_1) = 500\ \text{J} - (2.5 \times 10^5\ \text{Pa})(1.2 \times 10^{-3}\ \text{m}^3) = 500\ \text{J} - 300\ \text{J} = 200\ \text{J}$$

## SOLVED PROBLEM 21.3

A sample of gas expands from $V_1$ to $V_2$. Is the work done by the gas greatest when the expansion is (a) isobaric, (b) isothermal, or (c) adiabatic? How does the temperature vary during each expansion?

(a)   At the constant pressure $p$ the work done is $p(V_2 - V_1)$ and is the greatest of the three expansions. The temperature must increase during the expansion in order to maintain the pressure constant despite the increase in volume.

(b)   Since $pV/T = $ constant, during an expansion at constant temperature the pressure must drop as $V$ increases, and the work done is accordingly less than in (a).

(c)   In an adiabatic expansion the temperature must drop since all the work done is at the expense of the
internal energy of the gas. The final pressure is therefore lower than in (a) or (b), and the least
amount of work is done. See Fig. 21-5.

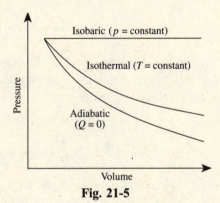

**Fig. 21-5**

## SOLVED PROBLEM 21.4

At 100°C and atmospheric pressure the heat of vaporization of steam is $L_v = 2260$ kJ/kg, the
density of water is $10^3$ kg/m$^3$, and the density of steam is 0.6 kg/m$^3$. What proportion of $L_v$
represents work done to expand water into steam against the pressure of the atmosphere?

The volumes of 1 kg of water and 1 kg of steam at 100°C and atmospheric pressure are,
respectively,

$$V_{water} = \frac{m}{d_{water}} = \frac{1 \text{ kg}}{10^3 \text{ kg/m}^3} = 0.001 \text{ m}^3$$

$$V_{steam} = \frac{m}{d_{steam}} = \frac{1 \text{ kg}}{0.6 \text{ kg/m}^3} = 1.667 \text{ m}^3$$

Atmospheric pressure is $p = 1.013 \times 10^5$ Pa. Hence the work done in the expansion is

$$W = p(V_{steam} - V_{water}) = (1.013 \times 10^5 \text{ Pa})(1.667 - 0.001) \text{ m}^3 = 1.69 \times 10^5 \text{ J} = 169 \text{ kJ}$$

This is

$$\frac{W}{L_v} = \frac{169 \text{ kJ}}{2260 \text{ kJ}} = 0.075 = 7.5\%$$

of the heat of vaporization of water. The remainder of the heat of vaporization goes into pulling the water
molecules apart to create a gas from a liquid and so becomes internal energy of the steam.

## SOLVED PROBLEM 21.5

Show that the power output of each cylinder of a reciprocating engine of any kind (steam,
gasoline, diesel) is given by the formula $P_c = pLAn$, where

$p =$ average pressure on piston during each power stroke
$L =$ length of piston travel
$A =$ cross-sectional area of piston
$n =$ number of power strokes per second

In general, $P = Fs/t$. Here $F$ is the force exerted on the piston during each power stroke by the pressure $p$, so since $p = F/A$, $F = pA$. The distance traveled by the piston per power stroke is $L$, and the distance it covers per second is therefore $s/t = Ln$. Hence

$$P_c = \frac{Fs}{t} = pLAn$$

## SOLVED PROBLEM 21.6

The four-cylinder, four-stroke diesel engine of a car develops 60 kW at 2600 rev/min. The pistons of this engine are 100 mm in diameter, and they travel 130 mm. Find the average pressure on the pistons during each power stroke.

The area of each piston is

$$A = \frac{\pi d^2}{4} = \frac{\pi (0.1 \text{ m})^2}{4} = 0.00785 \text{ m}^2$$

In a four-stroke engine a power stroke occurs in each cylinder once every 2 rev, so there are 1300 power strokes per minute or $1300/60 = 21.7$ strokes per second. Because the engine has four cylinders, its total power output is $P = 4P_c = 4pLAn$ and

$$p = \frac{P}{4LAn} = \frac{60 \times 10^3 \text{ W}}{(4)(0.13 \text{ m})(0.00785 \text{ m}^2)(21.7/\text{s})} = 6.8 \times 10^5 \text{ Pa} = 6.8 \text{ bar}$$

## SECOND LAW OF THERMODYNAMICS

Internal energy resides in the kinetic energies of randomly moving atoms and molecules, whereas the output of a heat engine appears in the ordered motions of a piston or a wheel. Since all physical systems in the universe tend to go in the opposite direction, from order to disorder, no heat engine can completely convert heat to mechanical energy or, in general, to work. This fundamental principle leads to the *second law of thermodynamics*: It is impossible to construct a continuously operating engine that takes heat from a source and performs an exactly equivalent amount of work.

Because some of the heat input to a heat engine must be wasted and because heat flows from a hot reservoir to a cold one, every heat engine must have a low-temperature reservoir for exhaust heat to go to as well as a high-temperature reservoir from which the input heat is to come, as in Fig. 21-1.

## CARNOT ENGINE

An ideal heat engine is one in which every process that occurs is reversible without any loss of energy. Such an engine is not subject to such practical mechanisms of energy loss as friction and heat conduction to the outside world. An example of an ideal heat engine is the imaginary *Carnot engine* which consists of a cylinder filled with an ideal gas that has a movable piston at one end. Figure 21-6 shows the four stages in the operation of a Carnot engine. The engine does work during the two expansions, and work is done on the engine during the two compressions; the net work done per cycle is the area enclosed by the curve.

A Carnot engine is the most efficient engine that can operate between the temperatures $T_1$ and $T_2$ at which heat is absorbed and exhausted. If heat $Q_1$ is absorbed at the absolute temperature $T_1$ and heat $Q_2$ is given off at the absolute temperature $T_2$, then $Q_1/Q_2 = T_1/T_2$ in such an engine. Its efficiency is therefore

$$\text{Efficiency (ideal)} = \frac{\text{work output}}{\text{heat input}} = \frac{W}{Q_1} = \frac{Q_1 - Q_2}{Q_1} = 1 - \frac{Q_2}{Q_1} = 1 - \frac{T_2}{T_1}$$

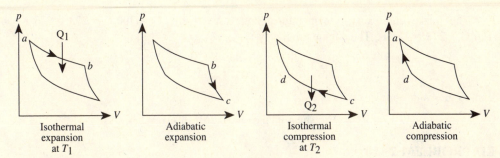

**Fig. 21-6**

The smaller the ratio between $T_2$ and $T_1$, the more efficient the engine. Because no reservoir can exist at a temperature of 0 K or 0°R, which is absolute zero, no heat engine can be 100 percent efficient.

### SOLVED PROBLEM 21.7

The Carnot engine uses only isothermal and adiabatic processes in its operating cycle. Why?

An engine has maximum efficiency when all the processes that occur in its operation are reversible without the performance of work, since any other processes must necessarily involve the waste of energy. Heat flow from a hot reservoir to a cooler one is not reversible in this sense because the natural direction of heat flow is from hot to cold (in fact, this is an alternate statement of the second law of thermodynamics). However, in an isothermal process, the heat flow occurs at a constant temperature, so the process can be reversed without any work being lost. An adiabatic process is also reversible in the same sense because no heat enters or leaves a system during such a process. Hence an engine that uses only isothermal and adiabatic processes is the most efficient possible.

### SOLVED PROBLEM 21.8

A 1-MW ($10^6$-W) generating plant has an overall efficiency of 40 percent. How much fuel oil whose heat of combustion is 45 MJ/kg does the plant burn each day?

In 1 day the plant produces

$$W = Pt = (10^6 \text{ W})(3600 \text{ s/h})(24 \text{ h/day}) = 8.64 \times 10^{10} \text{ J}$$

of electric energy. Since its efficiency is 0.4 and Eff = work output/heat input, we have

$$\text{Heat input} = \frac{\text{work output}}{\text{Eff}} = \frac{8.64 \times 10^{10} \text{ J}}{0.4} = 2.16 \times 10^{11} \text{ J}$$

The mass of fuel required to supply this amount of heat is

$$m = \frac{2.16 \times 10^{11} \text{ J}}{45 \times 10^6 \text{ J/kg}} = 4.8 \times 10^3 \text{ kg}$$

### SOLVED PROBLEM 21.9

A Carnot engine absorbs 1 MJ of heat from a reservoir at 300°C and exhausts heat to a reservoir at 150°C. Find the work it does.

The intake and exhaust temperatures are, respectively, $T_1 = 300°C + 273 = 573$ K and $T_2 = 150°C + 273 = 423$ K. The engine efficiency is

$$\text{Eff} = 1 - \frac{T_2}{T_1} = 1 - \frac{423 \text{ K}}{573 \text{ K}} = 1 - 0.74 = 0.26$$

and so the work done is

$$W = (\text{Eff})(\text{heat intake}) = (0.26)(10^6 \text{ J}) = 2.6 \times 10^5 \text{ J}$$

## SOLVED PROBLEM 21.10

Three designs are proposed for an engine that is to operate between 500 and 300 K. Design A is claimed to produce 750 J of work per kilojoule of heat input, B is claimed to produce 500 J, and C is claimed to produce 250 J. Which design would you choose?

The efficiency of an ideal engine operating between $T_1 = 500$ K and $T_2 = 300$ K is

$$\text{Eff} = 1 - \frac{T_2}{T_1} = 1 - \frac{300 \text{ K}}{500 \text{ K}} = 0.40 = 40\%$$

The claimed efficiencies of the proposed engines are

$$\text{Eff (A)} = \frac{\text{work output}}{\text{heat input}} = \frac{750 \text{ J}}{1000 \text{ J}} = 0.75 = 75\%$$

$$\text{Eff (B)} = \frac{500 \text{ J}}{1000 \text{ J}} = 0.50 = 50\%$$

$$\text{Eff (C)} = \frac{250 \text{ J}}{1000 \text{ J}} = 0.25 = 25\%$$

Both A and B claim efficiencies greater than that of an ideal engine and hence could not possibly work as stated. Design C is therefore the only possible choice.

## SOLVED PROBLEM 21.11

A steam engine is being planned that is to use steam at 400°F and whose efficiency is to be 20 percent. Find the maximum temperature at which the spent steam can emerge.

The intake temperature is $T_1 = 400°F + 460° = 860°R$. We proceed as follows:

$$\text{Eff} = 1 - \frac{T_2}{T_1} \qquad \frac{T_2}{T_1} = 1 - \text{Eff}$$

$$T_2 = T_1(1 - \text{Eff}) = (860°R)(1 - 0.20) = 688°R$$

The maximum exhaust temperature is therefore $T_2 = 688°R - 460° = 228°F$.

## REFRIGERATION

A *refrigerator* is a heat engine that operates backward to extract heat from a low-temperature reservoir and transfer it to a high-temperature reservoir (Fig. 21-7). Because the natural tendency of heat is to flow from a hot region to a cold one, energy must be provided to a refrigerator to reverse the flow, and this energy adds to the heat exhausted by the refrigerator.

If a refrigerator absorbs heat $Q_1$ at the absolute temperature $T_1$ and ejects heat $Q_2$ at the absolute temperature $T_2$, its *coefficient of performance* (CP) is given by

$$\text{CP} = \frac{\text{heat absorbed}}{\text{work done}} = \frac{Q_1}{W} = \frac{Q_1}{Q_2 - Q_1}$$

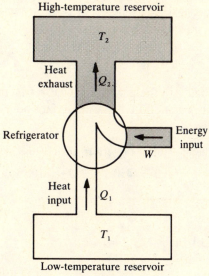

**Fig. 21-7**

In an ideal refrigerator, which is a Carnot engine run backward, $Q_1/Q_2 = T_1/T_2$ and

$$CP(ideal) = \frac{T_1}{T_2 - T_1}$$

### SOLVED PROBLEM 21.12

In an effort to cool a kitchen during the summer, the refrigerator door is left open and the kitchen's door and windows are closed. What will happen?

Since no refrigerator can be completely efficient, more heat is exhausted by the refrigerator into the kitchen than is extracted from the kitchen. The net effect, then, is to increase the kitchen's temperature.

### SOLVED PROBLEM 21.13

A 1-kW refrigerator whose coefficient of performance is 2.0 takes heat from a freezer compartment at $-20°C$ and exhausts it at $40°C$. How does its CP compare with that of an ideal refrigerator?

Here $T_1 = -20°C + 273 = 253$ K and $T_2 = 40°C + 273 = 313$ K. The CP of an ideal refrigerator operating between these temperatures is

$$CP = \frac{T_1}{T_2 - T_1} = \frac{253 \text{ K}}{313 \text{ K} - 253 \text{ K}} = 4.2$$

The actual refrigerator is $2.0/4.2 = 0.48 = 48$ percent as efficient as the ideal refrigerator.

### SOLVED PROBLEM 21.14

At what rate does the refrigerator of Prob. 21.13 remove heat from the freezer compartment?

Since $P = W/t$ and $Q_1 = (CP)(W)$,

$$\frac{Q_1}{t} = (CP)\left(\frac{W}{t}\right) = (CP)(P) = (2.0)(1 \text{ kW}) = 2 \text{ kW}$$

For every joule of energy supplied to the refrigerator, it removes 2 J of heat from the freezer compartment.

**SOLVED PROBLEM 21.15**

A refrigerator which is half as efficient as an ideal refrigerator extracts heat from a storage chamber at 0°F and exhausts it at 100°F. How many foot-pounds of work per Btu extracted does this refrigerator require?

We begin by finding the coefficient of performance of the ideal refrigerator. Since

$$T_1 = 0°F = 460° = 460°R \qquad T_2 = 100°F + 460° = 560°R$$

$$\frac{Q_1}{W} = \frac{T_2}{T_2 - T_1} = \frac{560°R}{560°R - 460°R} = 5.6$$

Here $Q_1 = 1$ Btu and, since 1 Btu = 778 ft·lb,

$$W = \frac{Q_1}{5.6} = \frac{778 \text{ ft·lb}}{5.6} = 139 \text{ ft·lb}$$

This refrigerator is half as efficient as an ideal refrigerator, so the required work per Btu is twice as great, or 278 ft·lb.

**SOLVED PROBLEM 21.16**

The British unit of refrigeration capacity is the *ton*, which is that rate of heat extraction that can freeze 1 ton of water at 32°F to ice at 32°F per day. Since the heat of fusion of water is 144 Btu/lb,

$$1 \text{ refrigeration ton} = 12,000 \text{ Btu/h}$$

The refrigerator of Prob. 21.15 has a capacity of 2 tons. How much power is required to operate its compressor?

Since the refrigerator requires 278 ft·lb of work per Btu of heat extracted,

$$P = \frac{W}{t} = (2 \text{ tons})\left(12,000 \frac{\text{Btu/h}}{\text{ton}}\right)\left(278 \frac{\text{ft·lb}}{\text{Btu}}\right)\left(\frac{1}{3600 \text{ s/h}}\right) = 1853 \text{ ft·lb/s}$$

In terms of horsepower,

$$P = \frac{1853 \text{ ft·lb/s}}{550 \text{ (ft·lb/s)/hp}} = 3.37 \text{ hp}$$

# *Multiple-Choice Questions*

**21.1.**    A heat engine takes in heat at one temperature and turns

  (*a*)   all of it into work
  (*b*)   some of it into work and rejects the rest at a lower temperature
  (*c*)   some of it into work and rejects the rest at the same temperature
  (*d*)   some of it into work and rejects the rest at a higher temperature

**21.2.** A process that can be reversed without energy input from an outside source is one that takes place at constant

(a) pressure      (c) velocity
(b) density       (d) temperature

**21.3.** In an adiabatic process in a system,

(a) its temperature stays the same
(b) its pressure stays the same
(c) no heat enters or leaves it
(d) no work is done by or on it

**21.4.** Without work being done on it, a gas cannot be

(a) compressed isobarically      (c) compressed adiabatically
(b) compressed isothermally      (d) expanded adiabatically

**21.5.** To be completely efficient (which is impossible), the exhaust temperature of a frictionless heat engine would have to be

(a) 0 K
(b) 273 K
(c) less than its intake temperature
(d) the same as its intake temperature

**21.6.** The Carnot cycle does not include an

(a) isobaric expansion       (c) adiabatic expansion
(b) isothermal expansion     (d) adiabatic compression

**21.7.** The efficiency of a Carnot engine operating between the absolute temperatures $T_1$ and $T_2$ is

(a) equal to $T_2/T_1$
(b) 100%
(c) the maximum possible between these temperatures
(d) the same as that of an actual engine operating between these temperatures

**21.8.** The heat a refrigerator absorbs from its contents is

(a) less than it gives off
(b) the same amount it gives off
(c) more than it gives off
(d) any of the above, depending on its design

**21.9.** Four kilojoules of heat is given off by a gas when it is compressed from 0.08 to 0.05 m$^3$ under a pressure of 200 kPa. The internal energy of the gas

(a) decreases
(b) is unchanged
(c) increases
(d) any of the above, depending on the initial and final temperatures

**21.10.** A Carnot engine that absorbs heat at 300°C and exhausts heat at 100°C has an efficiency of

(a) 33%    (c) 65%
(b) 35%    (d) 67%

**21.11.** If a Carnot engine absorbs 10 kJ of heat per cycle when it operates between 500 and 400 K, the work it does per cycle is

(a)  2 kJ       (c)  8 kJ
(b)  2.5 kJ     (d)  10 kJ

**21.12.** To have an efficiency of 40 percent, a heat engine that exhausts heat at 350 K must absorb heat at no less than

(a)  210 K      (c)  875 K
(b)  583 K      (d)  1038 K

# Supplementary Problems

**21.1.** Why is it impossible for a ship to use the internal energy of seawater to operate its engine?

**21.2.** One kilojoule of heat is added to a gas as it expands from 8 to 10 L at a constant pressure of 2 bar. What is the change in the internal energy of the gas?

**21.3.** A 2400-hp diesel locomotive burns 160 gal/h of fuel. If the heat of combustion of the diesel oil used is $1.2 \times 10^5$ Btu/gal, find the efficiency of the engine.

**21.4.** Steam enters a turbine engine at 550°C and emerges at 90°C. The engine has an actual overall efficiency of 35 percent. What percentage of its ideal efficiency is this?

**21.5.** The six-cylinder, four-cycle gasoline engine of a car has pistons 3.4 in. in diameter whose stroke (length of travel) is 4 in. If the mean effective pressure on the pistons during the power stroke is 70 lb/in.$^2$, find the number of horsepower developed by the engine when it operates at 2000 rev/min.

**21.6.** A Carnot engine absorbs 200 kJ of heat at 500 K and exhausts 150 kJ. What is its exhaust temperature?

**21.7.** A Carnot engine absorbs 1 MJ of heat at 327°C and exhausts heat to a reservoir at 127°C. How much work does it do?

**21.8.** A Carnot engine whose efficiency is 35 percent absorbs heat at 500°C. What must its intake temperature be if its efficiency is to be 50 percent with the same exhaust temperature?

**21.9.** A Carnot engine absorbs 500 Btu of heat at 500°F and performs $1 \times 10^5$ ft·lb of work. What is its exhaust temperature?

**21.10.** Three designs for a refrigerator to operate between −20 and 30°C are proposed. Design A is claimed to need 100 J of work per kilojoule of heat extracted, design B to need 200 J, and design C to need 300 J. Which design would you choose and why?

**21.11.** An ideal refrigerator extracts heat from a freezer at −20°C and exhausts it at 50°C. How many joules of heat are extracted per joule of work input?

**21.12.** How much work must an ideal refrigerator perform to make 1 kg of ice at $-10°C$ from 1 kg of water at $20°C$ when $20°C$ is also its exhaust temperature?

**21.13.** The *energy efficiency ratio* (EER) of a refrigerator or air conditioner is the ratio between the heat in Btu extracted per hour and the power in watts the machine uses. Find the relationship between EER and coefficient of performance.

**21.14.** What horsepower rating is needed for the motor of a 5-ton air conditioner whose coefficient of performance is 2.5?

## *Answers to Multiple-Choice Questions*

**21.1.**   (*b*)          **21.7.**   (*c*)

**21.2.**   (*d*)          **21.8.**   (*a*)

**21.3.**   (*c*)          **21.9.**   (*c*)

**21.4.**   (*a*), (*b*), (*c*)   **21.10.**   (*b*)

**21.5.**   (*a*)          **21.11.**   (*a*)

**21.6.**   (*a*)          **21.12.**   (*b*)

## Answers to Supplementary Problems

**21.1.**    There would be no suitable low-temperature reservoir to absorb the waste heat from the engine.

**21.2.**    The internal energy increases by 600 J.

**21.3.**    32 percent

**21.4.**    63 percent

**21.5.**    38.2 hp

**21.6.**    375 K

**21.7.**    0.33 MJ

**21.8.**    732°C

**21.9.**    253°F

**21.10.** A is more efficient than a Carnot refrigerator, and B has the same efficiency, so both are impossible as practical refrigerators. C is less efficient than a Carnot refrigerator and so might work as claimed.

**21.11.** 4.6 J

**21.12.** 50 kJ

**21.13.** EER = 3.42(CP)

**21.14.** 9.43 hp

# Chapter 22

# Heat Transfer

## CONDUCTION

The three mechanisms by which heat can be transferred from one place to another are conduction, convection, and radiation.

In *conduction*, heat is carried by means of collisions between rapidly moving molecules at the hot end of a body of matter and the slower molecules at the cold end. Some of the kinetic energy of the fast molecules passes to the slow molecules, and the result of successive collisions is a flow of heat through the body of matter. Solids, liquids, and gases all conduct heat. Conduction is poorest in gases because their molecules are relatively far apart and so interact less frequently than in solids and liquids. Metals are the best conductors of heat because some of their electrons are able to move about relatively freely and can travel past many atoms between collisions.

The rate at which heat is conducted through a slab of a particular material is proportional to the area $A$ of the slab and to the temperature difference $\Delta T$ between its sides and inversely proportional to the slab's thickness $d$ (Fig. 22-1). The amount of heat $Q$ that flows through the slab in the time $t$ is given by

$$\text{Rate of heat conduction} = \frac{Q}{t} = \frac{kA\,\Delta T}{d}$$

where $k$, the *thermal conductivity* of the material, is a measure of its ability to conduct heat. In SI, the correct unit of $k$ is the W/(m·°C), but the kcal/(m·s·°C) is also often used. In the British system, the usual unit of $k$ is the Btu/(ft²·h·°F/in.) since $A$ is customarily expressed in square feet and $d$ in inches.

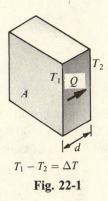

$$T_1 - T_2 = \Delta T$$

**Fig. 22-1**

## SOLVED PROBLEM 22.1

The thermal conductivities of brick and pine wood are, respectively, 0.6 and 0.13 W/(m·°C). What thickness of brick has the same insulating ability as 5 cm of pine?

When the ratio $k/d$ is the same for the two materials, their insulating abilities will also be the same. Hence

$$d_{\text{brick}} = \frac{k_{\text{brick}}}{k_{\text{pine}}}\,d_{\text{pine}} = \left(\frac{0.6}{0.13}\right)(5\text{ cm}) = 23\text{ cm}$$

## SOLVED PROBLEM 22.2

The thermal conductivity of ice is $5.2 \times 10^{-4}$ kcal/(m·s·°C). At what rate is heat lost by the water in a 6-m by 10-m outdoor swimming pool covered by a layer of ice 1 cm thick if the water is at a temperature of 0°C and the surrounding air is at a temperature of $-10°C$?

Here $A = (6\text{ m})(10\text{ m}) = 60\text{ m}^2$, $d = 0.01$ m, and $\Delta T = 10°C$. Hence

$$\frac{Q}{t} = \frac{kA\,\Delta T}{d} = \left(5.2 \times 10^{-4}\ \frac{\text{kcal}}{\text{m·s·°C}}\right)\left[\frac{(60\text{ m}^2)(10°C)}{0.01\text{ m}}\right] = 31.2\text{ kcal/s}$$

## SOLVED PROBLEM 22.3

The handle of a freezer door 5 in. thick is attached by two brass bolts $\frac{1}{4}$ in. in diameter that pass through the entire door and are secured on the inside by nuts. The interior of the freezer is maintained at 0°F, and the room temperature is 65°F; the thermal conductivity of brass is 730 Btu/(ft²·h·°F/in.). Find the heat lost per hour through the bolts.

The total cross-sectional area of the two bolts is

$$A = (2)(\pi r^2) = (2)(\pi)\left(\frac{0.125\text{ in.}}{12\text{ in./ft}}\right)^2 = 6.82 \times 10^{-4}\text{ ft}^2$$

and their length is $d = 5$ in. Since $\Delta T = 65°F$,

$$\frac{Q}{t} = \frac{kA\,\Delta T}{d} = \left(730\ \frac{\text{Btu}}{\text{ft}^2\text{·h·°F/in.}}\right)\left[\frac{(6.82 \times 10^{-4}\text{ ft}^2)(65°F)}{5\text{ in.}}\right] = 6.47\text{ Btu/h}$$

## THERMAL RESISTANCE

The *thermal resistance* $R$ of a layer of a material of thickness $d$ and of thermal conductivity $k$ is given by

$$R = \frac{d}{k}$$

$$\text{Thermal resistance} = \frac{\text{thickness}}{\text{thermal conductivity}}$$

The greater the value of $R$, the greater the resistance to the flow of heat. In British units, thermal resistance is given in ft²·°F/(Btu/h), but it is common for $R$-values to be stated without units. Thus an $R$-value of 1.5 ft²·°F/(Btu/h) would normally be given as just "$R$–1.5." In terms of $R$, the rate of heat conduction is

$$\frac{Q}{t} = \frac{A\,\Delta T}{R}$$

An advantage of using thermal resistance is that the $R$-value for a sandwich consisting of two or more layers is just the sum of the $R$-values of the separate layers:

$$R = R_1 + R_2 + R_3 + \cdots$$

## SOLVED PROBLEM 22.4

An icebox 3 ft by 3 ft by 2 ft is built of $R$–0.3 plywood on the inside and outside, with $R$–10 plastic foam insulation between the plywood layers. If the box is filled with ice at 32°F and the outside temperature is 85°F, how much ice melts per hour?

The thermal resistance of the box walls is

$$R = R_1 + R_2 + R_3 = 0.3 + 10 + 0.3 = 10.6$$

The area of the box walls is

$$A = (2)(3 \text{ ft})(3 \text{ ft}) + (4)(3 \text{ ft})(2 \text{ ft}) = 42 \text{ ft}^2$$

and the temperature difference is $\Delta T = 85°F - 32°F = 53°F$. Hence the heat flow is

$$\frac{Q}{t} = \frac{A \, \Delta T}{R} = \frac{(42 \text{ ft}^2)(53°F)}{10.6 \text{ ft}^2 \cdot °F/(\text{Btu/h})} = 210 \text{ Btu/h}$$

The heat of fusion of water is 144 Btu/lb, so the rate at which ice in the box melts is

$$\frac{m}{t} = \frac{Q/t}{L_f} = \frac{210 \text{ Btu/h}}{144 \text{ Btu/lb}} = 1.5 \text{ lb/h}$$

## CONVECTION

In *convection*, heat is transferred from one place to another by the actual motion of a hot fluid. Convection is usually the chief mechanism of heat transfer in fluids.

In natural convection, the buoyancy of a heated fluid leads to its motion. When a pan of water is heated on a stove, for instance, the hot water at the bottom expands slightly so that its density decreases. The buoyancy of this water causes it to rise to the surface while colder, denser water sinks to the bottom. In forced convection, a pump or blower is responsible for the motion of the heated fluid. An example is the cooling system of a car, in which water is circulated between the hot engine block and the radiator. In the radiator the convected heat is conducted through thin-walled metal tubes to the atmosphere.

The rate $Q/t$ at which a hot object transfers heat to a surrounding fluid by convection is approximately proportional to the area $A$ of the object in contact with the fluid and to the temperature difference $\Delta T$ between them:

$$\frac{Q}{t} = hA \, \Delta T$$

The convection coefficient $h$ depends on the shape and orientation of the object and on the properties and state of motion of the fluid. For instance, $h$ is greater for the upper surface of a horizontal plate than for its lower surface, and increases with the speed of the fluid in forced convection.

## SOLVED PROBLEM 22.5

A spotlight has a 50-W bulb in a metal hood whose temperature is 60°C when the room temperature is 20°C. If the 50-W bulb is replaced by a 75-W bulb, what is the new temperature of the hood?

Because $h$ and $A$ are the same in both cases,

$$\frac{(Q/t)_1}{(Q/t)_2} = \frac{\Delta T_1}{\Delta T_2}$$

When $(Q/t)_1 = 50$ W, the temperature difference between the hood and the surrounding air is $\Delta T_1 = 60°C - 20°C = 40°C$. When $(Q/t)_2 = 75$ W,

$$\Delta T_2 = \Delta T_1 \frac{(Q/t)_2}{(Q/t)_1} = (40°C)\left(\frac{75 \text{ W}}{50 \text{ W}}\right) = 60°C$$

Hence the new hood temperature is $20°C + 60°C = 80°C$.

**SOLVED PROBLEM 22.6**

If a cup of coffee cools from 85 to 75°C in 2 min in a room at 20°C, how long will it take to cool from 45 to 35°C?

The average temperature difference between the coffee and the air in the first period of cooling is $T_1 = 80°C - 20°C = 60°C$, and in the second period it is $T_2 = 40°C - 20°C = 20°C$. Since $T_2 = \frac{1}{3}T_1$, according to the solution of Prob. 22.5 $(Q/t)_2 = \frac{1}{3}(Q/t)_1$, so the rate of cooling will be one-third as great. The coffee will therefore take (3)(2 min) = 6 min to cool from 45 to 35°C.

## RADIATION

In *radiation*, energy is carried by the *electromagnetic waves* emitted by every object. Electromagnetic waves, of which light, radio waves, and X-rays are examples, travel at the velocity of light ($3 \times 10^8$ m/s = 186,000 mi/s) and require no material medium for their passage. The better an object absorbs radiation, the better it emits radiation. A perfect absorber of radiation is called a *blackbody*, and it is accordingly the best radiator.

The rate at which an object whose surface area is $A$ and whose absolute temperature is $T$ emits radiation is given by the *Stefan-Boltzmann law*:

$$R = \frac{P}{A} = e\sigma T^4$$

The constant $\sigma$ (Greek letter *sigma*) has the value $5.67 \times 10^{-8}$ W/(m$^2$·K$^4$). The *emissivity e* has a value between 0 (for a perfect reflector, hence a nonradiator) and 1 (for a blackbody), depending on the nature of the radiating surface.

With increasing temperature, the predominant wavelength of the radiation emitted by a body decreases. Thus a hot body that glows red is cooler than one that glows bluish white since red light has a longer wavelength than blue light (see Chapter 30). A body at room temperature emits radiation that is chiefly in the infrared part of the spectrum, to which the eye is not sensitive.

**SOLVED PROBLEM 22.7**

If all objects radiate electromagnetic energy, why don't the objects around us in everyday life grow colder and colder?

Every object also absorbs electromagnetic energy from its surroundings, and if both object and surroundings are at the same temperature, energy is emitted and absorbed at the same rate. When an object is at a higher temperature than its surroundings and heat is not supplied to it, the object radiates more energy than it absorbs and cools down to the temperature of its surroundings.

**SOLVED PROBLEM 22.8**

A copper ball 2 cm in radius is heated in a furnace to 400°C. If its emissivity is 0.3, at what rate does it radiate energy?

The surface area of the ball is

$$A = 4\pi r^2 = (4\pi)(0.02 \text{ m})^2 = 0.005 \text{ m}^2$$

and its absolute temperature is $T = 400°C + 273 = 673$ K. Hence

$$P = e\sigma AT^4 = (0.3)[5.67 \times 10^{-8} \text{ W/(m}^2\cdot\text{K}^4)](0.005 \text{ m}^2)(673 \text{ K})^4 = 17.4 \text{ W}$$

**SOLVED PROBLEM 22.9**

The sun radiates energy at the rate of $6.5 \times 10^7$ W/m$^2$ from its surface. Assuming that the sun radiates as a blackbody (which is approximately true), find its surface temperature.

The emissivity of a blackbody is $e = 1$. From $R = \sigma T^4$ we have

$$T = \sqrt[4]{\frac{R}{\sigma}} = \sqrt[4]{\frac{6.5 \times 10^7 \text{ W/m}^2}{5.67 \times 10^{-8} \text{ W/(m}^2 \cdot \text{K}^4)}} = 5800 \text{ K}$$

**SOLVED PROBLEM 22.10**

In the operation of a *thermograph*, the radiation from each small area of a person's skin is measured and shown by different shades of gray or by different colors in a *thermogram*. Because the skin over a tumor is warmer than elsewhere, thermograms are used in screening for various cancers. What is the percentage difference between the radiation rates from skin at 34 and 35°C?

The emissivity of skin is the same at both temperatures, so $R_1/T_1^4 = R_2/T_2^4$. Here $T_1 = 34°C + 273 = 307$ K and $T_2 = 35°C + 273 = 308$ K. Hence

$$\frac{R_2 - R_1}{R_1} = \frac{T_2^4 - T_1^4}{T_1^4} = \frac{(308 \text{ K})^4 - (307 \text{ K})^4}{(307 \text{ K})^4} = 0.013 = 1.3\%$$

which is large enough to detect.

# *Multiple-Choice Questions*

**22.1.**  Heat transfer by conduction occurs

    (*a*)  only in liquids    (*c*)  only in liquids and solids
    (*b*)  only in solids    (*d*)  in solids, liquids, and gases

**22.2.**  Heat transfer by convection occurs

    (*a*)  only in liquids    (*c*)  only in liquids and gases
    (*b*)  only in gases    (*d*)  in solids, liquids, and gases

**22.3.**  Radiation occurs

    (*a*)  only from liquids
    (*b*)  only from solids
    (*c*)  only from liquids and solids
    (*d*)  from solids, liquids, and gases

**22.4.**  Heat flows through a slab of some material at a rate that depends on which one or more of the following?

    (*a*)  the thickness of the slab
    (*b*)  the area of the slab
    (*c*)  the specific heat capacity of the material
    (*d*)  the temperature difference between the faces of the slab

**22.5.**  The temperature of an object that emits electromagnetic radiation must be

  (a)  higher than 0°C
  (b)  higher than 0 K
  (c)  higher than that of its surroundings
  (d)  high enough for it to glow

**22.6.**  A concrete wall 6.0 m long, 3.5 m high, and 25 cm thick has a conductivity of 0.80 W/(m·°C). When the wall has an outside temperature of 5°C and an inside temperature of 20°C, heat flows through it at

  (a)  10 W        (c)  1.6 kW
  (b)  1.0 kW      (d)  1.7 kW

**22.7.**  Heat flows through a wooden board 30 mm thick at 0.0086 W/cm$^2$ when one of its sides is 20°C warmer than the other. The thermal conductivity of the wood is

  (a)  0.013 W/(m·°C)     (c)  0.52 W/(m·°C)
  (b)  0.13 W/(m·°C)      (d)  0.78 W/(m·°C)

**22.8.**  Glass wool has a thermal conductivity of 0.30 Btu/(ft$^2$·h·°F/in.). An $R$–3.0 layer of glass wool has a thickness of

  (a)  1 in.       (c)  3 in.
  (b)  0.9 in.     (d)  10 in.

**22.9.**  The bare skin of a certain person is at an average temperature of 33°C in a room whose temperature is 20°C. If $e = 1$, the net rate at which the person loses energy by radiation is

  (a)  0.16 $\mu$W/cm$^2$     (c)  50 mW/cm$^2$
  (b)  7.9 mW/cm$^2$          (d)  79 W/cm$^2$

**22.10.**  An iron bar at 200°C radiates energy at 50 W. At 250°C the same bar will radiate at

  (a)  55 W        (c)  75 W
  (b)  63 W        (d)  122 W

# Supplementary Problems

**22.1.**  Outdoors in the winter, why does a piece of metal feel colder than a piece of wood?

**22.2.**  Why is it desirable to paint hot-water pipes with aluminum paint?

**22.3.**  An icebox whose walls consist of a 100-mm thickness of pine wood is to be replaced by a more modern one using glass wool insulation between pine inner and outer walls 12 mm thick. What thickness of glass wool will give the same degree of insulation as before? The thermal conductivities of pine and glass wool are, respectively, 0.13 and 0.04 W/(m·°C).

**22.4.**  A yacht has an aluminum hull 5 mm thick whose underwater area is 100 m$^2$. How much heat is conducted from the yacht's interior to the water per hour if the interior is at a temperature of 20°C and the water is at 12°C? The thermal conductivity of aluminum is 205 W/(m·°C).

**22.5.** How many Btu/day are conducted through a glass window 5 ft by 8 ft by 0.3 in. whose inner and outer faces are at the respective temperatures of 65 and 40°F? The thermal conductivity of glass is 5.5 Btu/(ft$^2$·h·°F/in.).

**22.6.** A sheet of $R$–0.06 plasterboard is $\frac{3}{8}$ in. thick. What is the thermal conductivity of plasterboard?

**22.7.** A double-glazed window consists of two panes of glass separated by an airspace. To see why this is such a good idea, compute the thermal resistances of two 0.2-in. glass panes (a) when they are in contact and (b) when they are 0.3 in. apart. The thermal conductivities of glass and air are, respectively, 5.5 and 0.17 Btu/(ft$^2$·h·°F/in.).

**22.8.** A 3-kg beef roast requires 13 min to be heated from 40 to 45°C in an oven maintained at 200°C. How long will the same roast take to be heated from 60 to 65°C in the same oven?

**22.9.** A small hole leading into a cavity behaves as a blackbody because any radiation that falls on it is trapped inside by multiple reflections until it is absorbed. How many watts are radiated from a hole 1 cm in diameter in the wall of a furnace whose interior temperature is 650°C?

**22.10.** A blackbody is at a temperature of 500°C. What should its temperature be in order that it radiates twice as much energy per second?

**22.11.** Sunspots appear dark even though they are actually very hot (typically 5000 K) because the rest of the solar surface is even hotter (about 6000 K). Find the ratio between the radiation rates of a sunspot and the normal solar surface.

## *Answers to Multiple-Choice Questions*

| | | | |
|---|---|---|---|
| **22.1.** | (d) | **22.6.** | (b) |
| **22.2.** | (c) | **22.7.** | (b) |
| **22.3.** | (d) | **22.8.** | (b) |
| **22.4.** | (a), (b), (d) | **22.9.** | (b) |
| **22.5.** | (b) | **22.10.** | (c) |

## **Answers to Supplementary Problems**

**22.1.** Metals are much better conductors of heat than wood and therefore conduct heat away from the hand more rapidly.

**22.2.** Such paint gives a finish that reflects most of the light that falls on it. Since a poor absorber of radiation is also a poor emitter of radiation, a pipe painted in this way radiates relatively little heat.

**22.3.** 23 mm

**22.4.** 118 GJ

**22.5.** $4.4 \times 10^5$ Btu

**22.6.** 6.25 Btu/(ft$^2$·h·°F/in.)

**22.7.** (*a*)  *R*–0.07    (*b*)  *R*–4.18

**22.8.** 15 min

**22.9.** 3.23 W

**22.10.** 646°C

**22.11.** $R_{\text{spot}}/R_{\text{sun}} = 0.48$

# Chapter 23

# Electricity

## ELECTRIC CHARGE

*Electric charge*, like mass, is one of the basic properties of certain of the elementary particles of which all matter is composed. There are two kinds of charge, *positive charge* and *negative charge*. The positive charge in ordinary matter is carried by *protons*, the negative charge by *electrons*. Charges of the same sign repel each other, charges of opposite sign attract each other.

The unit of charge is the *coulomb* (C). The charge of the proton is $+1.6 \times 10^{-19}$ C, and the charge of the electron is $-1.6 \times 10^{-19}$ C. All charges in nature occur in multiples of $\pm e = \pm 1.6 \times 10^{-19}$ C.

According to the principle of *conservation of charge*, the net electric charge in an isolated system always remains constant. (Net charge means the total positive charge minus the total negative charge.) When matter is created from energy, equal amounts of positive and negative charge always come into being, and when matter is converted to energy, equal amounts of positive and negative charge disappear.

## ATOMS AND IONS

An atom of any element consists of a small, positively charged *nucleus* with a number of electrons some distance away. The nucleus is composed of protons (charge $+e$, mass $= 1.673 \times 10^{-27}$ kg) and neutrons (uncharged, mass $= 1.675 \times 10^{-27}$ kg). The number of protons in the nucleus is normally equal to the number of electrons around it, so the atom as a whole is electrically neutral. The forces between atoms that hold them together as solids and liquids are electric in origin. The mass of the electron is $9.1 \times 10^{-31}$ kg.

Under certain circumstances an atom may lose one or more electrons and become a *positive ion*, or it may gain one or more electrons and become a *negative ion*. Many solids consist of positive and negative ions rather than of atoms or molecules. An example is ordinary table salt, which is made up of positive sodium ions ($Na^+$) and negative chlorine ions ($Cl^-$). Solutions of such solids in water also contain ions. Sparks, flames, and X-rays have the ability to ionize gases. Ions of opposite sign in a gas come together soon after being formed, and the excess electrons on the negative ions pass to the positive ones to form neutral molecules. A gas can be maintained in an ionized state by passing an electric current through it (as in a neon sign) or by bombarding it with X-rays or ultraviolet light (as in the upper atmosphere of the earth, where the radiation comes from the sun).

## COULOMB'S LAW

The force one charge exerts on another is given by *Coulomb's law*:

$$\text{Electric force} = F = k\,\frac{q_1 q_2}{r^2}$$

where $q_1$ and $q_2$ are the magnitudes of the charges, $r$ is the distance between them, and $k$ is a constant whose value in free space is

$$k = 9.0 \times 10^9 \text{ N·m}^2/\text{C}^2$$

The value of $k$ in air is slightly greater. The constant $k$ is sometimes replaced by

$$k = \frac{1}{4\pi\varepsilon_0}$$

where $\varepsilon_0$, the *permittivity of free space*, has the value

$$\varepsilon_0 = 8.85 \times 10^{-12} \text{ C}^2/(\text{N}\cdot\text{m}^2)$$

($\varepsilon$ is the Greek letter *epsilon*.)

### SOLVED PROBLEM 23.1

An electric charge $Q$ is placed on a metal object. How does the charge distribute itself on the object?

Metals are good conductors of electricity, so the mutual repulsions of the individual charges that make up $Q$ cause them to spread out over the object's surface in order to be as far apart as possible.

### SOLVED PROBLEM 23.2

(*a*) When two objects attract each other electrically, must both be charged? (*b*) When two objects repel each other electrically, must both be charged?

(*a*)   No. A charged object can produce a separation of charge in a nearly uncharged object because atomic electrons can shift around to some extent even without leaving their parent atoms. Figure 23-1 shows how a comb that has been given a negative charge by being run through someone's hair affects a small bit of paper. Electric forces vary inversely with distance. Hence the attraction between the comb and the adjacent positive charges in the paper is greater than the repulsion between the comb and the more distance negative charges. As a result, the paper moves toward the comb. Only a small amount of charge separation can occur in this way, and so only very light objects can be picked up; the charge separation in Fig. 23-1 is greatly exaggerated.
(*b*)   Yes.

### SOLVED PROBLEM 23.3

An iron atom has 26 protons in its nucleus. (*a*) How many electrons does this atom contain? (*b*) How many electrons does the $Fe^{3+}$ ion contain?

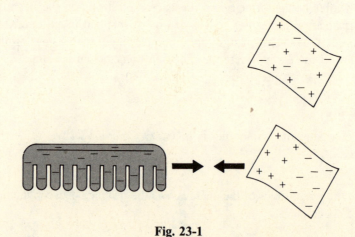

**Fig. 23-1**

(a)   Since a normal atom is electrically neutral, the number of negatively charged electrons it contains equals the number of positively charged protons in its nucleus. The iron atom therefore contains 26 electrons.

(b)   The symbol $Fe^{3+}$ represents an iron atom that has a net charge of $+3e$, which means that it has lost three of its usual complement of electrons. The $Fe^{3+}$ ion therefore contains 23 electrons.

## SOLVED PROBLEM 23.4

A plastic ball has a charge of $+10^{-12}$ C. (a) Does it contain an excess or a deficiency of electrons compared with its normal state of electrical neutrality? (b) How many such electrons are involved?

(a)   Since the pith ball is positively charged, it has fewer electrons than are needed to balance the positive charge of its nuclear protons.

(b)   The charge on an electron is $e = 1.6 \times 10^{-19}$ C. Hence

$$\text{Number of electrons} = \frac{q}{e} = \frac{10^{-12}\ \text{C}}{1.6 \times 10^{-19}\ \text{C/electron}} = 6.25 \times 10^6 \text{ electrons}$$

## SOLVED PROBLEM 23.5

What is the magnitude and direction of the force on a charge of $+4 \times 10^{-9}$ C that is 5 cm from a charge of $+5 \times 10^{-8}$ C?

Since 5 cm $= 5 \times 10^{-2}$ m, we have from Coulomb's law

$$F = k\,\frac{q_1 q_2}{r^2} = \frac{(9 \times 10^9\ \text{N·m}^2/\text{C}^2)(4 \times 10^{-9}\ \text{C})(5 \times 10^{-8}\ \text{C})}{(5 \times 10^{-2}\ \text{m})^2} = 7.2 \times 10^{-4}\ \text{N}$$

The force is directed away from the $+5 \times 10^{-8}$ C charge since both charges are positive.

## SOLVED PROBLEM 23.6

Two charges, one of $+5 \times 10^{-7}$ and the other of $-2 \times 10^{-7}$ C, attract each other with a force of 100 N. How far apart are they?

From Coulomb's law we have

$$r = \sqrt{\frac{k q_1 q_2}{F}} = \sqrt{\frac{(9 \times 10^9\ \text{N·m}^2/\text{C}^2)(5 \times 10^{-7}\ \text{C})(2 \times 10^{-7}\ \text{C}}{10^2\ \text{N}}}$$

$$= \sqrt{90 \times 10^{-7}}\ \text{m} = \sqrt{9 \times 10^{-6}}\ \text{m} = 3 \times 10^{-3}\ \text{m} = 3\ \text{mm}$$

## SOLVED PROBLEM 23.7

Two charges repel each other with a force of $10^{-5}$ N when they are 20 cm apart. (a) What is the force on each when they are 5 cm apart? (b) When they are 100 cm apart?

(a)   Since $F$ is proportional to $1/r^2$, the force increases when the charges are brought closer together to $(20/5)^2 = 16$ times what it was before, namely, to $1.6 \times 10^{-4}$ N.

(b)   The force decreases when the charges are moved apart to $(20/100)^2 = 0.04$ times what it was before, namely, to $4 \times 10^{-7}$ N.

## SOLVED PROBLEM 23.8

Under what circumstances, if any, is the gravitational attraction between two protons equal to their electric repulsion?

Since the proton mass is $1.67 \times 10^{-27}$ kg, the gravitational force between two protons that are a distance $r$ apart is

$$F_{grav} = \frac{Gm_1m_2}{r^2} = \frac{(6.67 \times 10^{-11} \text{ N·m}^2/\text{kg}^2)(1.67 \times 10^{-27} \text{ kg})^2}{r^2} = \frac{1.86 \times 10^{-64}}{r^2} \text{N·m}^2$$

The electric force between the protons is

$$F_{elec} = \frac{kq_1q_2}{r^2} = \frac{(9 \times 10^9 \text{ N·m}^2/\text{C}^2)(1.6 \times 10^{-19} \text{ C})^2}{r^2} = \frac{2.3 \times 10^{-28}}{r^2} \text{N·m}^2$$

At every separation $r$, the electric force between the protons is greater than the gravitational force between them by a factor of more than $10^{36}$; the forces are never equal.

## SOLVED PROBLEM 23.9

A test charge of $+1 \times 10^{-6}$ C is placed halfway between a charge of $+5 \times 10^{-6}$ and a charge of $+3 \times 10^{-6}$ C that are 20 cm apart (Fig. 23-2). Find the magnitude and direction of the force on the test charge.

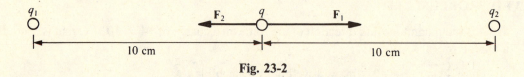

**Fig. 23-2**

The force exerted on the test charge $q$ by the charge $q_1$ is

$$F_1 = \frac{kqq_1}{r_1^2} = \frac{(9 \times 10^9 \text{ N·m}^2/\text{C}^2)(1 \times 10^{-6} \text{ C})(5 \times 10^{-6} \text{ C})}{(0.1 \text{ m})^2} = +4.5 \text{ N}$$

This force is taken to be positive because it acts to the right. The force exerted by the charge $q_2$ on $q$ is

$$F_2 = \frac{kqq_2}{r_2^2} = \frac{(9 \times 10^9 \text{ N·m}^2/\text{C}^2)(1 \times 10^{-6} \text{ C})(3 \times 10^{-6} \text{ C})}{(0.1 \text{ m})^2} = -2.7 \text{ N}$$

This force is taken to be negative because it acts to the left. The net force on the test charge $q$ is

$$F = F_1 + F_2 = +4.5 \text{ N} - 2.7 \text{ N} = +1.8 \text{ N}$$

and it acts to the right, that is, toward the $+3 \times 10^{-6}$ C charge.

## ELECTRIC FIELD

An *electric field* is a region of space in which a charge would be acted upon by an electric force. An electric field may be produced by one or more charges, and it may be uniform or it may vary in magnitude and/or direction from place to place.

If a charge $q$ at a certain point is acted on by the force **F**, the electric field **E** at that point is defined as the ratio between **F** and $q$:

$$\mathbf{E} = \frac{\mathbf{F}}{q}$$

$$\text{Electric field} = \frac{\text{force}}{\text{charge}}$$

Electric field is a vector quantity whose direction is that of the force on a positive charge. The unit of **E** is the newton per coulomb (N/C) or, more commonly, the equivalent unit volt per meter (V/m).

The advantage of knowing the electric field at some point is that we can at once establish the force on *any* charge $q$ placed there, which is

$$\mathbf{F} = q\mathbf{E}$$

$$\text{Force} = (\text{charge})(\text{electric field})$$

## ELECTRIC FIELD LINES

*Field lines* are a means of describing a force field, such as an electric field, by using imaginary lines to indicate the direction and magnitude of the field. The direction of an electric field line at any point is the direction in which a positive charge would move if placed there, and field lines are drawn close together where the field is strong and far apart where the field is weak (Fig. 23-3).

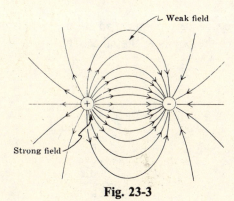

**Fig. 23-3**

## SOLVED PROBLEM 23.10

(*a*) What is the electric field a distance $r$ from a charge $q$? (*b*) What is the electric field that acts on the electron in a hydrogen atom, which is $5.3 \times 10^{-11}$ m from the proton that is the atom's nucleus?

(*a*)    The force $F$ that the charge $q$ exerts on a test charge $q_0$ when it is $r$ away is $F = kqq_0/r^2$. From the definition of electric field, then,

$$E = \frac{F}{q_0} = k\,\frac{q}{r^2}$$

(*b*)    Here $q = e = 1.6 \times 10^{-19}$ C, and so

$$E = k\,\frac{q}{r^2} = \frac{(9 \times 10^9 \text{ N·m}^2/\text{C}^2)(1.6 \times 10^{-19} \text{ C})}{(5.3 \times 10^{-11} \text{ m})^2} = 5.1 \times 10^{11} \text{ V/m}$$

## SOLVED PROBLEM 23.11

The electric field in a certain neon sign is 5000 V/m. (*a*) What force does this field exert on a neon ion of mass $3.3 \times 10^{-26}$ kg and charge $+e$? (*b*) What is the acceleration of the ion?

(a)   The force on the neon ion is

$$F = qE = eE = (1.6 \times 10^{-19} \text{ C})(5 \times 10^3 \text{ V/m}) = 8 \times 10^{-16} \text{ N}$$

(b)   According to the second law of motion $F = ma$, and so here

$$a = \frac{F}{m} = \frac{8 \times 10^{-16} \text{ N}}{3.3 \times 10^{-26} \text{ kg}} = 2.4 \times 10^{10} \text{ m/s}^2$$

## SOLVED PROBLEM 23.12

How strong an electric field is required to exert a force on a proton equal to its weight at sea level?

The electric force on the proton is $F = eE$, and its weight is $mg$. Hence $eE = mg$, and

$$E = \frac{mg}{e} = \frac{(1.67 \times 10^{-27} \text{ kg})(9.8 \text{ m/s}^2)}{1.6 \times 10^{-19} \text{ C}} = 1.02 \times 10^{-7} \text{ V/m}$$

## POTENTIAL DIFFERENCE

The *potential difference V* between two points in an electric field is the amount of work needed to take a charge of 1 C from one of the points to the other. Thus

$$V = \frac{W}{q}$$

$$\text{Potential difference} = \frac{\text{work}}{\text{charge}}$$

The unit of potential difference is the *volt* (V):

$$1 \text{ volt} = 1 \frac{\text{joule}}{\text{coulomb}}$$

The potential difference between two points in a uniform electric field $\mathbf{E}$ is equal to the product of $E$ and the distance $s$ between the points in a direction parallel to $\mathbf{E}$:

$$V = Es$$

Since an electric field is usually produced by applying a potential difference between two metal plates $s$ apart, this equation is most useful in the form

$$E = \frac{V}{s}$$

$$\text{Electric field} = \frac{\text{potential difference}}{\text{distance}}$$

A battery uses chemical reactions to produce a potential difference between its terminals; a generator uses electromagnetic induction (Chapter 28) for this purpose.

## SOLVED PROBLEM 23.13

The potential difference between a certain thundercloud and the ground is $7 \times 10^6$ V. Find the energy dissipated when a charge of 50 C is transferred from the cloud to the ground in a lightning stroke.

$$W = qV = (50 \text{ C})(7 \times 10^6 \text{ V}) = 3.5 \times 10^8 \text{ J}$$

## SOLVED PROBLEM 23.14

A potential difference of 20 V is applied across two parallel metal plates, and an electric field of 500 V/m is produced. How far apart are the plates?

Since $E = V/s$, here

$$s = \frac{V}{E} = \frac{20 \text{ V}}{500 \text{ V/m}} = 0.04 \text{ m} = 4 \text{ cm}$$

## SOLVED PROBLEM 23.15

(a) What potential difference must be applied across two metal plates 15 cm apart if the electric field between them is 600 V/m? (b) What is the force on a charge of $10^{-10}$ C in this field? (c) How much kinetic energy will the charge have when it has moved through 5 cm in the field, starting from rest?

(a) $$V = Es = (600 \text{ V/m})(0.15 \text{ m}) = 90 \text{ V}$$

(b) $$F = qE = (10^{-10} \text{ C})(600 \text{ V/m}) = 6 \times 10^{-8} \text{ N}$$

(c) Since the KE of the charge is equal to the work done on it by the electric field when it travels 0.05 m,

$$\text{KE} = W = Fs = (6 \times 10^{-8} \text{ N})(0.05 \text{ m}) = 3 \times 10^{-9} \text{ J}$$

## SOLVED PROBLEM 23.16

What potential difference must be applied to produce an electric field that can accelerate an electron to a velocity of $10^7$ m/s?

The kinetic energy of such an electron is

$$\text{KE} = \tfrac{1}{2}mv^2 = (\tfrac{1}{2})(9.1 \times 10^{-31} \text{ kg})(10^7 \text{ m/s})^2 = 4.6 \times 10^{-17} \text{ J}$$

This KE is equal to the work $W$ that must be done on the electron by the electric field, and so, since $W = qV$ in general, we have here

$$V = \frac{W}{q} = \frac{\text{KE}}{e} = \frac{4.6 \times 10^{-17} \text{ J}}{1.6 \times 10^{-19} \text{ C}} = 2.9 \times 10^2 \text{ V} = 290 \text{ V}$$

## SOLVED PROBLEM 23.17

A 12-V storage battery is being charged at the rate of 15 C/s. (a) How much power is being used to charge the battery? (b) How much energy is stored in the battery if it is charged at this rate for 1 h?

(a) The work done to transfer the charge $q$ from one set of the battery's electrodes to the other set against the potential difference $V$ is $W = Vq$. Since power is the rate at which work is being done, here

$$P = \frac{W}{t} = \frac{Vq}{t} = (12 \text{ V})(15 \text{ C/s}) = 180 \text{ W}$$

(b) The work done in $t = 1 \text{ h} = 3600 \text{ s}$ is

$$W = Pt = (180 \text{ W})(3600 \text{ s}) = 6.48 \times 10^5 \text{ J}$$

If the charging process is perfectly efficient, this amount of energy will be stored in the battery as a result.

# *Multiple-Choice Questions*

**23.1.** Which one or more of the following statements is true?

(*a*)   All protons have the same charge.
(*b*)   Electrons and protons have equal masses.
(*c*)   Protons and neutrons have equal masses.
(*d*)   Atomic nuclei contain only protons and neutrons.

**23.2.** Electric charge occurs only in separate parcels

(*a*)   of $\pm 1.6 \times 10^{-19}$ C
(*b*)   of $\pm 1$ C
(*c*)   whose value is proportional to the mass of the particle carrying the charge
(*d*)   that are different for positive and negative particles

**23.3.** The electric force between two protons

(*a*)   is weaker than the gravitational force between them
(*b*)   is equal to the gravitational force between them
(*c*)   is stronger than the gravitational force between them
(*d*)   any of the above, depending on the distance between the protons

**23.4.** The same potential difference is used to accelerate a proton and an electron. Afterward

(*a*)   the proton has the higher velocity
(*b*)   the electron has the higher velocity
(*c*)   the proton has more KE
(*d*)   the electron has more KE

**23.5.** If $10^5$ electrons are added to a neutral object, its charge will be

(*a*)   $-1.6 \times 10^{-24}$ C        (*c*)   $+1.6 \times 10^{-24}$ C
(*b*)   $-1.6 \times 10^{-14}$ C        (*d*)   $+1.6 \times 10^{-14}$ C

**23.6.** When they are 80 mm apart, two charges attract each other with a force of $4.0 \times 10^{-6}$ N. When the same charges are 20 mm apart, the force between them is

(*a*)   $1.6 \times 10^{-6}$ N        (*c*)   $1.6 \times 10^{-5}$ N
(*b*)   $8.0 \times 10^{-6}$ N        (*d*)   $6.4 \times 10^{-5}$ N

**23.7.** Two charges of $+2.0 \times 10^{-9}$ C that are 3.0 mm apart repel each other with a force of

(*a*)   $4 \times 10^{-9}$ N        (*c*)   $6 \times 10^3$ N
(*b*)   $4 \times 10^{-3}$ N        (*d*)   $3.6 \times 10^7$ N

**23.8.** The force on a proton in an electric field of 1.0 kV/m is

(*a*)   $2.6 \times 10^{-35}$ N        (*c*)   $1.6 \times 10^{-16}$ N
(*b*)   $1.6 \times 10^{-22}$ N        (*d*)   $6.3 \times 10^{23}$ N

**23.9.** Ten millimeters from a certain charge its electric field is 10 kV/m. The magnitude of the field 20 mm from the charge is

    (a)   2.5 kV/m    (c)   7.1 kV/m
    (b)   5.0 kV/m    (d)   10 kV/m

**23.10.** An electric force of $1.0 \times 10^{-5}$ N acts on a charge of $5.0 \times 10^{-10}$ C that is between two parallel metal plates 4.0 mm apart. The potential difference between the plates is

    (a)   $1.25 \times 10^{-12}$ V    (c)   80 V
    (b)   32 V    (d)   $1.6 \times 10^{11}$ V

**23.11.** The potential difference between a certain thundercloud and the ground is 4 MW. During a lightning stroke, 80 C of charge is transferred between the cloud and the ground. The energy dissipated during the stroke is

    (a)   $5 \times 10^{-6}$ J    (c)   $3.2 \times 10^{7}$ J
    (b)   $2 \times 10^{5}$ J    (d)   $3.2 \times 10^{8}$ J

**23.12.** A storage battery is being charged at a rate of 75 W. If the potential difference across its terminals is 13.6 V, charge is being transferred between its plates at

    (a)   0.18 C/s    (c)   5.5 C/s
    (b)   2.8 C/s    (d)   1020 C/s

# Supplementary Problems

**23.1.** An oxygen atom has 8 protons in its nucleus. (a) How many electrons does this atom contain? (b) How many electrons does the $O^{2-}$ ion contain?

**23.2.** What information is provided by a sketch of the field lines of an electric field?

**23.3.** Why is it impossible for the field lines of an electric field to cross one another?

**23.4.** A rod with a charge of $+q$ at one end and $-q$ at the other is placed in a uniform electric field whose direction is parallel to the rod. How does the rod behave?

**23.5.** The rod of Prob. 23.4 is placed in a uniform electric field whose direction is perpendicular to the rod. How does the rod behave?

**23.6.** A billion ($10^9$) electrons are added to a neutral pith ball. What is its charge?

**23.7.** What is the force between two $+1$-C charges located 1 m apart?

**23.8.** What are the magnitude and direction of the force on a charge of $+2 \times 10^{-7}$ C that is 0.3 m from a charge of $-5 \times 10^{-7}$ C?

**23.9.** Two electrons repel each other with a force of $10^{-8}$ N. How far apart are they?

**23.10.** Two charges attract each other with a force of $10^{-6}$ N when they are 1 cm apart. How far apart should they be for the force between them to be (*a*) $10^{-4}$ N and (*b*) $10^{-8}$ N?

**23.11.** A test charge of $+2 \times 10^{-7}$ C is located 5 cm to the right of a charge of $+1 \times 10^{-6}$ C and 10 cm to the left of a charge of $-1 \times 10^{-6}$ C. The three charges lie on a straight line. Find the force on the test charge.

**23.12.** A charge of $+1 \times 10^{-7}$ C and a charge of $+3 \times 10^{-7}$ C are 40 cm apart. (*a*) Where should a charge of $+q$ be placed on the line between these charges so that no net force acts on it? (*b*) Where should a charge of $-q$ be so placed?

**23.13.** How much force is exerted on a charge of $10^{-6}$ C by an electric field of 50 V/m?

**23.14.** An electron is present in an electric field of $10^4$ V/m. (*a*) Find the force on the electron. (*b*) Find the electron's acceleration.

**23.15.** Two charges of $+10^{-6}$ C are located 1 cm apart. (*a*) What is the force on a charge of $+10^{-8}$ C halfway between them? (*b*) What is the force on a charge of $-10^{-8}$ C at the same place? (*c*) What must be true of the strength of the electric field halfway between the two $+10^{-6}$-C charges?

**23.16.** A potential difference of 100 V is applied by a battery across a pair of metal plates 5 cm apart. (*a*) What is the electric field between the plates? (*b*) How much force does a charge of $+10^{-8}$ C experience in this field? (*c*) How much kinetic energy does this charge acquire when it goes from the positive plate to the negative plate?

**23.17.** A charge of $-2 \times 10^{-9}$ C in an electric field between two parallel metal plates 4 cm apart is acted on by a force of $10^{-4}$ N. (*a*) What is the strength of the field? (*b*) What is the potential difference between the plates?

**23.18.** A proton is accelerated by a potential difference of 15,000 V. What is its kinetic energy?

**23.19.** A particle of charge $10^{-12}$ C starts to move from rest in an electric field of 500 V/m. (*a*) What is the force on the particle? (*b*) How much kinetic energy will it have when it has moved 1 cm in the field?

**23.20.** When a certain 12-V storage battery is charged, a total of $10^5$ C is transferred from one set of its electrodes to the other set. Find the energy stored in the battery.

## *Answers to Multiple-Choice Questions*

| | | | |
|---|---|---|---|
| **23.1.** | (*a*), (*d*) | **23.5.** | (*b*) |
| **23.2.** | (*a*) | **23.6.** | (*d*) |
| **23.3.** | (*c*) | **23.7.** | (*b*) |
| **23.4.** | (*b*) | **23.8.** | (*c*) |

**23.9.**   (*a*)          **23.11.**   (*d*)

**23.10.**  (*c*)          **23.12.**   (*c*)


# Answers to Supplementary Problems

**23.1.**   (*a*)  8 electrons     (*b*)  10 electrons

**23.2.**   Such a sketch shows how the magnitude and direction of the field vary in space. At a given point, the direction of the field is given by the direction of the nearest lines of force, and the relative magnitude of the field is indicated by how close together the lines of force are in the vicinity of the point.

**23.3.**   By definition, a line of force represents the path a positively charged particle would follow in an electric field, and such a particle can travel in only one direction at any point.

**23.4.**   The force exerted by the field on the $-q$ charge is equal and opposite to the force exerted on the $+q$ charge, so the rod does not move since the two forces have the same line of action.

**23.5.**   The equal and opposite forces exerted on the charges now cause the rod to rotate until it is parallel to the electric field.

**23.6.**   $-1.6 \times 10^{-10}$ C

**23.7.**   $9 \times 10^{9}$ N; repulsive

**23.8.**   $10^{-2}$ N directed toward the other charge

**23.9.**   $1.5 \times 10^{-10}$ m

**23.10.**  (*a*)  0.1 m     (*b*)  10 cm

**23.11.**  0.9 N directed to the right

**23.12.**  (*a*)  14.6 cm from the charge of $+1 \times 10^{-7}$ C     (*b*)  The same location

**23.13.**  $5 \times 10^{-5}$ N

**23.14.**  (*a*)  $1.6 \times 10^{-15}$ N     (*b*)  $1.8 \times 10^{15}$ m/s$^2$

**23.15.**  (*a*)  0     (*b*)  0     (*c*)  $E = 0$

**23.16.**  (*a*)  2000 V     (*b*)  $2 \times 10^{-5}$ N   (*c*)  $10^{-6}$ J

**23.17.**  (*a*)  $5 \times 10^{4}$ V/m     (*b*)  2000 V

**23.18.**  $2.4 \times 10^{-15}$ J

**23.19.**  (*a*)  $5 \times 10^{-10}$ N     (*b*)  $5 \times 10^{-12}$ J

**23.20.**  $1.2 \times 10^{6}$ J

# Chapter 24

## Electric Current

### ELECTRIC CURRENT

A flow of charge from one place to another constitutes an *electric current*. An *electric circuit* is a closed path in which an electric current carries energy from a source (such as a battery or generator), to a load (such as a motor or a lamp). In such a circuit (see Fig. 24-1), electric current is assumed to go from the positive terminal of the battery (or generator) through the circuit and back to the negative terminal of the battery. The direction of a current is conventionally considered to be that in which positive charge would have to move to produce the same effects as the actual current. Thus a current is always supposed to go from the positive terminal of a battery or generator to its negative terminal.

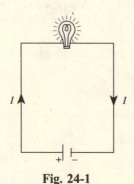

**Fig. 24-1**

A *conductor* is a substance through which charge can flow easily, and an *insulator* is one through which charge can flow only with great difficulty. Metals, many liquids, and *plasmas* (gases whose molecules are charged) are conductors; nonmetallic solids, certain liquids, and gases whose molecules are electrically neutral are insulators. A number of substances, called *semiconductors*, are intermediate in their ability to conduct charge.

Electric currents in metal wires always consist of flows of electrons; such currents are assumed to occur in the direction opposite to that in which the electrons move. Since a positive charge going one way is for most purposes equivalent to a negative charge going the other way, this assumption makes no practical difference. Both positive and negative charges move when a current is present in a liquid or gaseous conductor.

If an amount of charge $q$ passes a given point in a conductor in the time interval $t$, the current in the conductor is

$$I = \frac{q}{t}$$

$$\text{Electric current} = \frac{\text{charge}}{\text{time interval}}$$

The unit of electric current is the *ampere* (A), where

$$1 \text{ ampere} = 1 \frac{\text{coulomb}}{\text{second}}$$

278

## OHM'S LAW

For a current to exist in a conductor, there must be a potential difference between its ends, just as a difference in height between source and outlet is necessary for a river current to exist. In the case of a metallic conductor, the current is proportional to the applied potential difference: Doubling $V$ causes $I$ to double, tripling $V$ causes $I$ to triple, and so forth. This relationship is known as *Ohm*'s law and is expressed in the form

$$I = \frac{V}{R}$$

$$\text{Electric current} = \frac{\text{potential difference}}{\text{resistance}}$$

The quantity $R$ is a constant for a given conductor and is called its *resistance*. The unit of resistance is the *ohm* ($\Omega$), where

$$1 \text{ ohm} = 1 \ \frac{\text{volt}}{\text{ampere}}$$

The greater the resistance of a conductor, the less the current when a certain potential difference is applied.

Ohm's law is not a physical principle but is an experimental relationship that most metals obey over a wide range of values of $V$ and $I$.

## SOLVED PROBLEM 24.1

Since electric current is a flow of charge, why are two wires rather than a single one used to carry current?

If a single wire were used, charge of one sign or the other (depending on the situation) would be permanently transferred from the source of current to the appliance at the far end of the wire. In a short time so much charge would have been transferred that the source would be unable to shift further charge against the repulsive force of the charge piled up at the appliance. Thus a single wire cannot carry a current continuously. The use of two wires, however, enables charge to be circulated from source to appliance and back, so that a continuous one-way flow of energy can take place.

## SOLVED PROBLEM 24.2

Which solids are good electric conductors and which are good insulators? How well do these substances conduct heat?

All metals are good electric conductors. All nonmetallic solids are good insulators, for instance, glass, wood, plastics, rubber. In general, solids that are good conductors of electricity are also good conductors of heat, and solids that are good electric insulators are poor conductors of heat. Metals are good conductors of heat and electricity because both are transferred through a metal by the freely moving electrons that are a characteristic feature of its structure.

## SOLVED PROBLEM 24.3

A wire carries a current of 1 A. How many electrons pass any point in the wire each second?

The electron charge is of magnitude $e = 1.6 \times 10^{-19}$ C, and so a current of 1 A = 1 C/s corresponds to a flow of

$$\frac{1 \text{ C/s}}{1.6 \times 10^{-19} \text{ C/electron}} = 6.3 \times 10^{18} \text{ electrons/s}$$

## SOLVED PROBLEM 24.4

A 120-V toaster has a resistance of 12 $\Omega$. What must be the minimum rating of the fuse in the electric circuit to which the toaster is connected?

The current in the toaster is

$$I = \frac{V}{R} = \frac{120 \text{ V}}{12 \text{ } \Omega} = 10 \text{ A}$$

so this must be the rating of the fuse.

## SOLVED PROBLEM 24.5

A 120-V electric heater draws a current of 25 A. What is its resistance?

$$R = \frac{V}{I} = \frac{120 \text{ V}}{25 \text{ A}} = 4.8 \text{ } \Omega$$

## RESISTIVITY

The resistance of a conductor that obeys Ohm's law is given by

$$R = \rho \, \frac{L}{A}$$

where $L$ is the length of the conductor, $A$ is its cross-sectional area, and $\rho$ (Greek letter *rho*), is the *resistivity* of the material of the conductor. In SI, the unit of resistivity is the ohm-meter.

The resistivities of most materials vary with temperature. If $R$ is the resistance of a conductor at a particular temperature, then the change in its resistance $\Delta R$ when the temperature changes by $\Delta T$ is approximately proportional to both $R$ and $\Delta T$ so that

$$\Delta R = \alpha R \, \Delta T$$

The quantity $\alpha$ is the *temperature coefficient of resistance* of the material.

## SOLVED PROBLEM 24.6

What is the resistance of a copper wire 0.5 mm in diameter and 20 m long? The resistivity of copper is $1.7 \times 10^{-8}$ $\Omega \cdot$m.

The wire's cross-sectional area is $\pi r^2$, where $r = 0.25$ mm $= 2.5 \times 10^{-4}$. Hence

$$R = \rho \, \frac{L}{A} = \frac{(1.7 \times 10^{-8} \text{ } \Omega \cdot \text{m})(20 \text{ m})}{(\pi)(2.5 \times 10^{-4} \text{ m})^2} = 1.73 \text{ } \Omega$$

## SOLVED PROBLEM 24.7

A platinum wire 80 cm long is to have a resistance of 0.1 $\Omega$. What should its diameter be? The resistivity of platinum is $1.1 \times 10^{-7}$ $\Omega \cdot$m.

Since $R = \rho L/A = \rho L/\pi r^2$,

$$r = \sqrt{\frac{\rho L}{\pi R}} = \sqrt{\frac{(1.1 \times 10^{-7} \text{ } \Omega \cdot \text{m})(0.8 \text{ m})}{(\pi)(0.1 \text{ } \Omega)}} = 5.3 \times 10^{-4} \text{ m} = 0.53 \text{ mm}$$

The wire's diameter should therefore be $2r = 1.06$ mm.

## SOLVED PROBLEM 24.8

A copper wire has a resistance of 10.0 Ω at 20°C. (*a*) What will its resistance be at 80°C? (*b*) At 0°C? The temperature coefficient of resistance of copper is 0.004/°C.

(*a*)   Here $R = 10.0$ Ω and $\Delta T = 60°C$. Hence the wire's change in resistance is

$$\Delta R = \alpha R \, \Delta T = (0.004/°C)(10.0 \, \Omega)(60°C) = 2.4 \, \Omega$$

and the resistance at 80°C will be $R + \Delta R = 12.4$ Ω.

(*b*)   Here $\Delta T = 20°C$, and so

$$\Delta R = \alpha R \, \Delta T = (0.004/°C)(10.0 \, \Omega)(-20°C) = -0.8 \, \Omega$$

The resistance at 0°C will be $R + \Delta R = 9.2$ Ω.

## SOLVED PROBLEM 24.9

A *resistance thermometer* makes use of the variation of the resistance of a conductor with temperature. If the resistance of such a thermometer with a platinum element is 5 Ω at 20°C and 9 Ω when it is inserted in a furnace, find the temperature of the furnace. The value of $\alpha$ for platinum is 0.0036/°C.

Here $R = 5$ Ω and $\Delta R = 9 \, \Omega - 5 \, \Omega = 4$ Ω. Since $\Delta R = \alpha R \, \Delta T$,

$$\Delta T = \frac{\Delta R}{\alpha R} = \frac{4 \, \Omega}{(0.0036/°C)(5 \, \Omega)} = 222°C$$

The temperature of the furnace is $T + \Delta T = 20°C + 222°C = 242°C$.

## CIRCULAR MIL

In engineering practice the unit of area of a round conductor is often the *circular mil*, or *cmil*. The *mil* is a unit of length equal to 0.001 in., which is $\frac{1}{1000}$ in. A circular mil is a unit of area equal to the area of a circle whose diameter is 1 mil, as in Fig. 24-2. The area $A_{cm}$ in circular mils of a circle whose diameter in mils is $D_m$ is equal to $D_m^2$:

$$A_{cm} = D_m^2$$

The advantage of using the circular mil as a unit of area is that it avoids multiplication and division by $\pi$. When the length of a conductor is given in feet and its area in circular mils, the unit of resistivity is the ohm-cmil per foot.

## SOLVED PROBLEM 24.10

The specific resistance of the copper used in electric wires is 10.4 Ω·cmil/ft. Find the resistance of 1500 ft of copper wire whose diameter is 0.080 in.

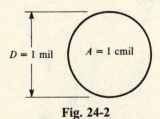

$D = 1$ mil    $A = 1$ cmil

**Fig. 24-2**

Since 0.080 in. = 80 mils,

$$A_{cm} = D_m^2 = (80)^2 \text{ cmil} = 6400 \text{ cmil}$$

and

$$R = \frac{\rho l}{A} = \frac{(10.4 \ \Omega \cdot \text{cmil/ft})(1500 \text{ ft})}{6400 \text{ cmil}} = 2.44 \ \Omega$$

## SOLVED PROBLEM 24.11

The specific resistance of Nichrome is $600 \ \Omega \cdot$cmil/ft. How long should a Nichrome wire 20 mils in diameter be for it to have a resistance of 5 $\Omega$?

The cross-sectional area of the wire is

$$A_{cm} = D_m^2 = (20)^2 \text{ cmil} = 400 \text{ cmil}$$

Next we solve for $l$ and substitute $\rho = 600 \ \Omega \cdot$cmil/ft, $R = 5 \ \Omega$, and $A = 400$ cmil to find the value of the length $l$:

$$l = \frac{RA}{\rho} = \frac{(5 \ \Omega)(400 \text{ cmil})}{600 \ \Omega \cdot \text{cmil/ft}} = 3.33 \text{ ft}$$

## ELECTRIC POWER

The rate at which work is done to maintain an electric current is given by the product of the current $I$ and the potential difference $V$:

$$P = IV$$
$$\text{Power} = (\text{current})(\text{potential difference})$$

When $I$ is in amperes and $V$ is in volts, $P$ will be in watts.

If the conductor or device through which a current passes obeys Ohm's law, the power consumed may be expressed in the alternative forms

$$P = IV = I^2 R = \frac{V^2}{R}$$

Table 24.1 is a summary of the various formulas for potential difference $V$, current $I$, resistance $R$, and power $P$ that follow from Ohm's law $I = V/R$ and from the power formula $P = VI$.

**Table 24.1**

| Unknown Quantity | Known Quantities | | | | | |
|---|---|---|---|---|---|---|
|  | $V, I$ | $I, R$ | $V, R$ | $P, I$ | $P, V$ | $P, R$ |
| $V =$ |  | $IR$ |  | $P/I$ |  | $\sqrt{PR}$ |
| $I =$ |  |  | $V/R$ |  | $P/V$ | $\sqrt{P/R}$ |
| $R =$ | $V/I$ |  |  | $P/I^2$ | $V^2/P$ |  |
| $P =$ | $VI$ | $I^2 R$ | $V^2/R$ |  |  |  |

**SOLVED PROBLEM 24.12**

The current through a 50-$\Omega$ resistance is 2 A. How much power is dissipated as heat?

$$P = I^2 R = (2 \text{ A})^2 (50 \, \Omega) = 200 \text{ W}$$

**SOLVED PROBLEM 24.13**

A 2-kW water heater is to be connected to a 240-V power line whose circuit breaker is rated at 10 A. Will the breaker open when the heater is switched on?

The heater draws a current of

$$I = \frac{P}{V} = \frac{2000 \text{ W}}{240 \text{ V}} = 8\tfrac{1}{3} \text{ A}$$

Since this current is less than 10 A, the breaker will not open.

**SOLVED PROBLEM 24.14**

A generator driven by a diesel engine that develops 12 hp delivers 30 A at 240 V. What is the efficiency of the generator?

Since 1 hp = 746 W and $P = IV$,

$$\text{Input power} = (12 \text{ hp})(746 \text{ W/hp}) = 8952 \text{ W}$$
$$\text{Output power} = (30 \text{ A})(240 \text{ V}) = 7200 \text{ W}$$

The efficiency of the generator is therefore

$$\text{Eff} = \frac{\text{output}}{\text{input}} = \frac{7200 \text{ W}}{8952 \text{ W}} = 0.80 = 80\%$$

**SOLVED PROBLEM 24.15**

A 12-V storage battery is charged by a current of 20 A for 1 h. (*a*) How much power is required to charge the battery at this rate? (*b*) How much energy has been provided during the process?

(*a*)                 $P = IV = (20 \text{ A})(12 \text{ V}) = 240 \text{ W}$

(*b*)             $W = Pt = (240 \text{ W})(3600 \text{ s}) = 8.64 \times 10^5 \text{ J}$

**SOLVED PROBLEM 24.16**

The *kilowatthour* (kWh) is an energy unit equal to the energy delivered by a source whose power is 1 kW in 1 h of operation. How much energy in kilowatthours does a 240-V clothes dryer that draws 15 A use in 45 min of operation?

The power of the dryer is

$$P = IV = (15 \text{ A})(240 \text{ V}) = 3600 \text{ W} = 3.6 \text{ kW}$$

and the time interval is

$$t = \frac{45 \text{ min}}{60 \text{ min/h}} = 0.75 \text{ h}$$

Hence                                 $W = Pt = (3.6 \text{ kW})(0.75 \text{ h}) = 2.7 \text{ kWh}$

## SOLVED PROBLEM 24.17

The 12-V battery of a certain car has a capacity of 80 Ah, which means that it can furnish a current of 80 A for 1 h, a current of 40 A for 2 h, and so forth. (a) How much energy is stored in the battery? (b) If the car's lights require 60 W of power, how long can the battery keep them lighted when the engine (and hence its generator) is not running?

(a)   The 80-Ah capacity of the battery is a way to express the amount of charge it can transfer from one of its terminals to the other. Here the amount of charge is

$$q = (80 \text{ Ah})(3600 \text{ s/h}) = 2.88 \times 10^5 \text{A} \cdot \text{s} = 2.88 \times 10^5 \text{ C}$$

and so the energy the battery can provide is

$$W = qV = (2.88 \times 10^5 \text{ C})(12 \text{ V}) = 3.46 \times 10^6 \text{ J}$$

(b)   Since $P = W/t$,

$$t = \frac{W}{P} = \frac{3.46 \times 10^6 \text{ J}}{60 \text{ W}} = 5.8 \times 10^4 \text{ s} = 16 \text{ h}$$

# *Multiple-Choice Questions*

**24.1.**   The same volume of copper can have any of the following combinations of length $L$ and cross-sectional area $A$. Which combination has the most resistance?

(a)   $L$ and $A$
(b)   $L/2$ and $2A$
(c)   $2L$ and $A/2$
(d)   All have the same resistance because the volume of copper is the same.

**24.2.**   The area in circular mils of a circle whose diameter is $d$ mils is

(a)   $d$          (c)   $\pi d$
(b)   $d^2$        (d)   $\pi d^2$

**24.3.**   When a copper wire is heated, its resistance

(a)   decreases
(b)   remains the same
(c)   increases
(d)   any of the above, depending on the temperatures

**24.4.**   Which one or more of the following combinations of units is equal to the watt?

(a)  J/s          (c)  $A^2\Omega$
(b)  $V^2/A$      (d)  $V^2\Omega$

**24.5.**   The diameter of a wire whose cross-sectional area is 2500 cmil is

(a)  0.016 in.    (c)  0.050 in.
(b)  0.025 in.    (d)  0.056 in.

**24.6.**   The resistivity of copper is $1.7 \times 10^{-8}$ $\Omega\cdot$m. The diameter of a 50-m length of copper wire whose resistance is 1.2 $\Omega$ is

(a)  0.47 mm      (c)  0.90 mm
(b)  0.84 mm      (d)  0.95 mm

**24.7.**   The resistance of a 120-V light bulb that draws a current of 1.25 A is

(a)  9.8 $\Omega$     (c)  96 $\Omega$
(b)  77 $\Omega$      (d)  150 $\Omega$

**24.8.**   A 120-V appliance that draws a current of 20 A has a power rating of

(a)  6 W          (c)  2.4 kW
(b)  0.72 kW      (d)  48 kW

**24.9.**   A light bulb rated at 240 V, 150 W is connected to a 120-V power line. The current that flows in it is

(a)  0.313 A      (c)  1.25 A
(b)  0.625 A      (d)  1.88 A

**24.10.**   A fully charged 24-V, 100-Ah storage battery has an energy content of

(a)  2.4 kJ       (c)  144 kJ
(b)  57.6 kJ      (d)  8.64 MJ

# Supplementary Problems

**24.1.**   Bends in a pipe slow down the flow of water through it. Do bends in a wire increase its electric resistance?

**24.2.**   How many electrons pass through the filament of a 75-W, 120-V light bulb per second?

**24.3.**   Find the current in a 200-$\Omega$ resistor when the potential difference across it is 40 V.

**24.4.**   An electric water heater draws 10 A of current from a 240-V power line. What is its resistance?

**24.5.**   The resistance of a 10-ft length of 20-mil iron wire is found to be 1.8 $\Omega$. What is the resistivity of iron?

**24.6.** Find the resistance of 8 m of aluminum wire 0.1 mm in diameter. The resistivity of aluminum is $2.6 \times 10^{-8}$ $\Omega \cdot$m.

**24.7.** How long should a copper wire 0.4 mm in diameter be for it to have a resistance of 10 $\Omega$? The resistivity of copper is $1.7 \times 10^{-8}$ $\Omega \cdot$m.

**24.8.** The diameter of No. 14 wire is 0.06408 in. (*a*) Find its cross-sectional area in circular mils. (*b*) The area of No. 10 wire is 10,380 cmil. How does the resistance of a given length of No. 14 wire compare with that of the same length of No. 10 wire?

**24.9.** The specific resistance of iron is 72 $\Omega \cdot$cmil/ft. Find the resistance of 250 ft of iron wire whose diameter is $\frac{1}{16}$ in.

**24.10.** The temperature coefficient of resistance of carbon is $-0.0005/°$C. If the resistance of a carbon resistor is 1000 $\Omega$ at 0°C, find its resistance at 120°C.

**24.11.** An iron wire has a resistance of 0.20 $\Omega$ at 20°C and a resistance of 0.30 $\Omega$ at 110°C. Find the temperature coefficient of resistance of the iron used in the wire.

**24.12.** What is the resistance of a 750-W, 120-V electric iron?

**24.13.** How much power is developed by an electric motor that draws a current of 4 A when it is operated at 240 V? How many horsepower is this?

**24.14.** What is the current in a 100-W light bulb when it is operated at 120 V?

**24.15.** A 75-W hall light is left on continuously. At $0.09/kWh, find the cost per week.

**24.16.** A 32-V storage battery has a capacity of $10^6$ J. How long can it supply a current of 5 A?

**24.17.** The 12-V battery of a car is required to be able to operate the 1.5-kW starting motor for a total of at least 10 min. (*a*) What should the minimum capacity of the battery be (in ampere-hours)? (*b*) How much energy is stored in such a battery?

**24.18.** A light bulb whose power is 100 W when it is operated at 240 V is instead connected to a 120-V source. (*a*) What is the current in the bulb? (*b*) How much power does it dissipate?

**24.19.** Currents of 5 A pass through two resistors, one of which has a potential difference of 100 V across it and the other of which has a potential difference of 300 V across it. (*a*) Compare the rates at which charge passes through each resistor. (*b*) Compare the rates at which energy is dissipated by each resistor.

## Answers to Multiple-Choice Questions

**24.1.** (*c*)        **24.6.** (*d*)

**24.2.** (*b*)        **24.7.** (*c*)

**24.3.** (*c*)        **24.8.** (*c*)

**24.4.** (*a*), (*c*)   **24.9.** (*a*)

**24.5.** (*c*)        **24.10.** (*d*)

## Answers to Supplementary Problems

**24.1.**    Bends in a wire have no effect on its electric resistance because the electrons whose motion constitutes an electric current are extremely small with very little mass and therefore can change direction readily.

**24.2.**    $3.9 \times 10^{18}$ electrons

**24.3.**    0.2 A

**24.4.**    24 Ω

**24.5.**    72 Ω·cmil/ft

**24.6.**    26.5 Ω

**24.7.**    74 m

**24.8.**    (*a*)  4106 cmil    (*b*)  2.53 times as much

**24.9.**    4.6 Ω

**24.10.**   940 Ω

**24.11.**   0.0056/°C

**24.12.**   19.2 Ω

**24.13.**   960 W; 1.3 hp

**24.14.**   0.83 A

**24.15.**   $1.13

**24.16.**   6250 s = 1 h 44 min 10 s

**24.17.**   (*a*)   21 Ah      (*b*)   $9.0 \times 10^5$ J

**24.18.**   (*a*)   0.21 A      (*b*)   25 W

**24.19.**   (*a*) Since the currents are the same, charge flows at the same rate through each resistor.
(*b*) Since $P = IV$, energy is dissipated by the second resistor three times as fast as by the first resistor.

# Chapter 25

## Direct-Current Circuits

### RESISTORS IN SERIES

The equivalent resistance of a set of resistors depends on the way in which they are connected as well as on their values. If the resistors are joined in *series*, that is, consecutively (Fig. 25-1), the equivalent resistance $R$ of the combination is the sum of the individual resistances:

$$R = R_1 + R_2 + R_3 + \cdots \qquad \text{series resistors}$$

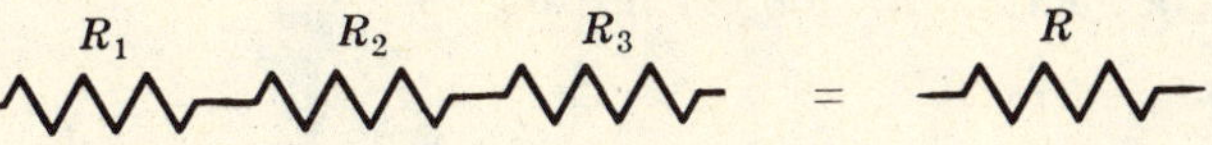

**Fig. 25-1**

### SOLVED PROBLEM 25.1

Show that the equivalent resistance of three resistors in series is given by $R = R_1 + R_2 + R_3$.

To find the equivalent resistance, we start from the fact that the potential difference $V$ across the set is the sum of the potential differences across the individual resistors:

$$V = V_1 + V_2 + V_3$$

Because the current in each resistor is $I$, the potential differences across them are

$$V_1 = IR_1 \qquad V_2 = IR_2 \qquad V_3 = IR_3$$

The potential difference across the equivalent resistance $R$ is

$$V = IR$$

Substituting for the $V$'s in $V = V_1 + V_2 + V_3$ gives

$$IR = IR_1 + IR_2 + IR_3$$

Now we divide both sides of this equation by $I$ and find that

$$R = R_1 + R_2 + R_3$$

### SOLVED PROBLEM 25.2

It is desired to limit the current in a 50-$\Omega$ resistor to 10 A when it is connected to a 600-V power source. (*a*) How should an auxiliary resistor be connected in the circuit and what should its resistance be? (*b*) What is the voltage drop across each resistor?

(*a*) For a current of 10 A, the total resistance in the circuit should be

$$R = \frac{V}{I} = \frac{600 \text{ V}}{10 \text{ A}} = 60 \ \Omega$$

Hence a 10-$\Omega$ resistor should be connected in series with the 50-$\Omega$ resistor to give a total of 60 $\Omega$.

(*b*) $\qquad V_1 = IR_1 = (10 \text{ A})(50 \ \Omega) = 500 \text{ V} \qquad V_2 = IR_2 = (10 \text{ A})(10 \ \Omega) = 100 \text{ V}$

## SOLVED PROBLEM 25.3

(a) What is the equivalent resistance of three 5-$\Omega$ resistors connected in series? (b) If a potential difference of 60 V is applied across the combination, what is the current in each resistor?

(a)                         $$R = R_1 + R_2 + R_3 = 5 \ \Omega + 5 \ \Omega + 5 \ \Omega = 15 \ \Omega$$

(b)   The current in the entire circuit is

$$I = \frac{V}{R} = \frac{60 \text{ V}}{15 \ \Omega} = 4 \text{ A}$$

Since the resistors are in series, this current flows through each of them.

## SOLVED PROBLEM 25.4

A 2000- and a 5000-$\Omega$ resistor are in series as part of a larger circuit. A voltmeter shows the potential difference across the 2000-$\Omega$ resistor to be 2 V. Find (a) the current in each resistor and (b) the potential difference across the 5000-$\Omega$ resistor.

(a)   The current in the 2000-$\Omega$ resistor is

$$I = \frac{V_1}{R_1} = \frac{2 \text{ V}}{200 \ \Omega} = 0.001 \text{ A}$$

This current flows through the other resistor as well.

(b)   The potential difference across $R_2$ is

$$V_2 = IR_2 = (0.001 \text{ A})(5000 \ \Omega) = 5 \text{ V}$$

## RESISTORS IN PARALLEL

In a *parallel* set of resistors, the corresponding terminals of the resistors are connected (Fig. 25-2). The reciprocal $1/R$ of the equivalent resistance of the combination is the sum of the reciprocals of the individual resistances:

$$\frac{1}{R} = \frac{1}{R_1} + \frac{1}{R_2} + \frac{1}{R_3} + \cdots \qquad \text{parallel resistors}$$

A calculator with a reciprocal $(1/X)$ key makes it easy to find the equivalent resistance of a set of resistors in parallel. The key sequence would be

$$[R_1][1/X][+][R_2][1/X][+][R_3][1/X][+] \cdots [=][1/X]$$

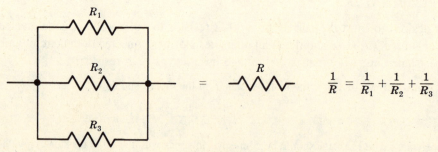

**Fig. 25-2**

This method is much faster than working out the calculation one term at a time. What is being done here is to replace the formula $1/R = 1/R_1 + 1/R_2 + 1/R_3 + \cdots$ by its reciprocal

$$R = \frac{1}{1/R_1 + 1/R_2 + 1/R_3 + \cdots}$$

If only two resistors are connected in parallel,

$$\frac{1}{R} = \frac{1}{R_1} + \frac{1}{R_2} = \frac{R_1 + R_2}{R_1 R_2} \qquad \text{and so} \qquad R = \frac{R_1 R_2}{R_1 + R_2}$$

## SOLVED PROBLEM 25.5

Show that the equivalent resistance of three resistors in parallel is given by $1/R = 1/R_1 + 1/R_2 + 1/R_3$.

To find the equivalent resistance, we start from the fact that the total current $I$ is equal to the sum of the currents through the separate resistors:

$$I = I_1 + I_2 + I_3$$

Because the potential difference $V$ is the same across all the resistors, their respective currents are

$$I_1 = \frac{V}{R_1} \qquad I_2 = \frac{V}{R_2} \qquad I_3 = \frac{V}{R_3}$$

The smaller the resistance, the greater the current through a resistor in a parallel set. The total current is given in terms of the equivalent resistance $R$ by

$$I = \frac{V}{R}$$

Substituting for the $I$'s in $I = I_1 + I_2 + I_3$ gives

$$\frac{V}{R} = \frac{V}{R_1} + \frac{V}{R_2} + \frac{V}{R_3}$$

Now we divide both sides of this equation by $V$:

$$\frac{1}{R} = \frac{1}{R_1} + \frac{1}{R_2} + \frac{1}{R_3}$$

## SOLVED PROBLEM 25.6

(a) What is the equivalent resistance of three 5-$\Omega$ resistors connected in parallel? (b) If a potential difference of 60 V is applied across the combination, what is the current in each resistor?

(a)
$$\frac{1}{R} = \frac{1}{R_1} + \frac{1}{R_2} + \frac{1}{R_3} = \frac{1}{5\,\Omega} + \frac{1}{5\,\Omega} + \frac{1}{5\,\Omega} = \frac{3}{5\,\Omega}$$
$$R = \tfrac{5}{3}\,\Omega = 1.67\,\Omega$$

(b)   Since each resistor has a potential difference of 60 V across it, the current in each one is

$$I = \frac{V}{R} = \frac{60\text{ V}}{5\,\Omega} = 12\text{ A}$$

### SOLVED PROBLEM 25.7

Two 240-$\Omega$ light bulbs are to be connected to a 120-V power source. To determine whether they will be brighter when connected (a) in series or (b) in parallel, calculate the power they dissipate in each arrangement.

(a)  The equivalent resistance of the two bulbs in series is

$$R = R_1 + R_2 = 240\ \Omega + 240\ \Omega = 480\ \Omega$$

The current in the circuit is therefore

$$I = \frac{V}{R} = \frac{120\ \text{V}}{480\ \Omega} = 0.25\ \text{A}$$

Since the bulbs are in series, this current passes through each of them. The power each bulb dissipates is

$$P = I^2 R = (0.25\ \text{A})^2 (240\ \Omega) = 15\ \text{W}$$

(b)  When the bulbs are in parallel, the potential difference across each is 120 V. Hence the power each bulb dissipates is

$$P = \frac{V^2}{R} = \frac{(120\ \text{V})^2}{240\ \Omega} = 60\ \text{W}$$

The bulbs will be brighter when they are connected in parallel.

### SOLVED PROBLEM 25.8

A circuit has a resistance of 50 $\Omega$. How can it be reduced to 20 $\Omega$?

To obtain an equivalent resistance of $R = 20\ \Omega$, a resistor $R_2$ must be connected in parallel with the circuit of $R_1 = 50\ \Omega$. To find $R_2$ we proceed as follows:

$$\frac{1}{R} = \frac{1}{R_1} + \frac{1}{R_2} \qquad \frac{1}{R_2} = \frac{1}{R} - \frac{1}{R_1} = \frac{R_1 - R}{R_1 R}$$

$$R_2 = \frac{R_1 R}{R_1 - R} = \frac{(50\ \Omega)(20\ \Omega)}{50\ \Omega - 20\ \Omega} = 33.3\ \Omega$$

### SOLVED PROBLEM 25.9

Find the equivalent resistance of the circuit shown in Fig. 25-3(a).

Figure 25-3(b) shows how the original circuit is decomposed into its series and parallel parts, each of which is treated in turn. The equivalent resistance of $R_1$ and $R_2$ is

$$R' = \frac{R_1 R_2}{R_1 + R_2} = \frac{(10\ \Omega)(10\ \Omega)}{10\ \Omega + 10\ \Omega} = 5\ \Omega$$

This equivalent resistance is in series with $R_3$, and so

$$R'' = R' + R_3 = 5\ \Omega + 3\ \Omega = 8\ \Omega$$

Finally $R''$ is in parallel with $R_4$; hence the equivalent resistance of the entire circuit is

$$R = \frac{R'' R_4}{R'' + R_4} = \frac{(8\ \Omega)(12\ \Omega)}{8\ \Omega + 12\ \Omega} = 4.8\ \Omega$$

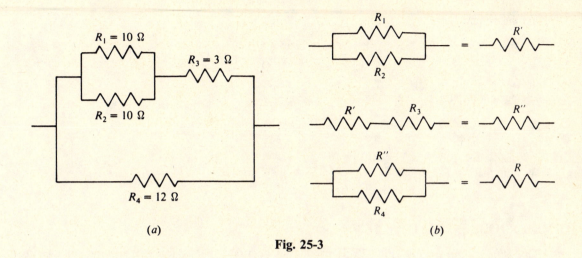

$(a)$                                                                          $(b)$

**Fig. 25-3**

### SOLVED PROBLEM 25.10

A potential difference of 20 V is applied to the circuit of Fig. 25-4. Find the current through each resistor and the current through the entire circuit.

Because resistor $R_4$ has the full 20-V potential difference across it,

$$I_4 = \frac{V}{R_4} = \frac{20\text{ V}}{12\text{ }\Omega} = 1.67\text{ A}$$

From Fig. 25-4 we see that the current $I_3$ also flows through the entire upper branch of the circuit, whose equivalent resistance is $R'' = 8$ $\Omega$. Hence

$$I_3 = \frac{V}{R''} = \frac{20\text{ V}}{8\text{ }\Omega} = 2.5\text{ A}$$

The potential difference $V'$ across $R_1$ and $R_2$ is

$$V' = V - I_3 R_3 = 20\text{ V} - (2.5\text{ A})(3\text{ }\Omega) = 12.5\text{ A}$$

Hence the current $I_1$ is

$$I_1 = \frac{V'}{R_1} = \frac{12.5\text{ V}}{10\text{ }\Omega} = 1.25\text{ V}$$

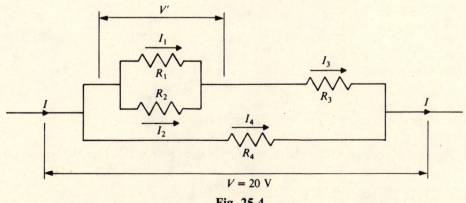

**Fig. 25-4**

and the current $I_2$ is

$$I_2 = \frac{V'}{R_2} = \frac{12.5 \text{ V}}{10 \text{ }\Omega} = 1.25 \text{ A}$$

The current through the entire circuit is

$$I = \frac{V}{R} = \frac{20 \text{ V}}{4.8 \text{ }\Omega} = 4.17 \text{ A}$$

We note that $I = I_3 + I_4$ and that $I_3 = I_1 + I_2$, as they should.

## SOLVED PROBLEM 25.11

The *Wheatstone bridge* (Fig. 25-5) proves an accurate means for determining an unknown resistance $R_x$ with the help of the fixed resistors $R_1$ and $R_2$ and the calibrated variable resistor $R_3$. Resistance $R_3$ is varied until the galvanometer $G$ shows no deflection, a situation described by saying that the bridge is *balanced*. Find the value of $R_x$ in terms of $R_1$, $R_2$, and $R_3$.

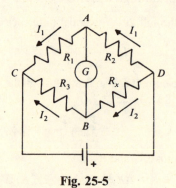

**Fig. 25-5**

No current passes through the galvanometer when the bridge is balanced, so the same current $I_1$ passes through $R_1$ and $R_2$ and the same current $I_2$ passes through $R_3$ and $R_x$. Also, points $A$ and $B$ must have no potential difference between them; hence $V_{AC} = V_{BC}$ and $V_{AD} = V_{BD}$. Since

$$V_{AC} = I_1 R_1 \qquad V_{AD} = I_1 R_2$$
$$V_{BC} = I_2 R_3 \qquad V_{BD} = I_2 R_x$$

we have the equations

$$V_{AC} = V_{BC} \qquad V_{AD} = V_{BD}$$
$$I_1 R_1 = I_2 R_3 \qquad I_1 R_2 = I_2 R_x$$

Dividing the last equation in the first column by the last equation in the second column gives

$$\frac{I_1 R_1}{I_1 R_2} = \frac{I_2 R_3}{I_2 R_x}$$

The currents $I_1$ and $I_2$ cancel, and so

$$\frac{R_1}{R_2} = \frac{R_3}{R_x} \qquad R_x = \frac{R_2 R_3}{R_1}$$

## EMF AND INTERNAL RESISTANCE

The work done per coulomb on the charge passing through a battery, generator, or other source of electric energy is called the *electromotive force*, or *emf*, of the source. The emf is equal to the potential difference across the terminals of the source when no current flows. When a current $I$ flows, this potential difference is less than the emf because of the *internal resistance* of the source. If the internal resistance is $r$, then a potential drop of $Ir$ occurs within the source. The terminal voltage $V$ across a source of emf $V_e$ whose internal resistance is $r$ when it provides a current of $I$ is therefore

$$V = V_e - Ir$$

Terminal voltage = emf − potential drop due to internal resistance

When a battery or generator of emf $V_e$ is connected to an external resistance $R$, the total resistance in the circuit is $R + r$, and the current that flows is

$$I = \frac{V_e}{R + r}$$

$$\text{Current} = \frac{\text{emf}}{\text{external resistance} + \text{internal resistance}}$$

## BATTERIES

The emf of a set of batteries connected in series is the sum of the emf's of the individual cells. The internal resistance of the set is the sum of the individual internal resistances. Thus

$$V_{e,\text{series}} = V_{e,1} + V_{e,2} + V_{e,3} + \cdots$$

$$r_{\text{series}} = r_1 + r_2 + r_3 + \cdots$$

Batteries that are intended to be connected in parallel should always have the same emf; otherwise currents would circulate among them, wasting energy. If $V_e$ is the emf of each battery, the emf of a set of batteries connected in parallel is also $V_e$. The internal resistance of the set is

$$\frac{1}{r_{\text{parallel}}} = \frac{1}{r_1} + \frac{1}{r_2} + \frac{1}{r_3} + \cdots$$

## SOLVED PROBLEM 25.12

A dry cell of emf 1.5 V and internal resistance 0.05 Ω is connected to a flashlight bulb whose resistance is 0.4 Ω. Find the current in the circuit.

$$I = \frac{V_e}{R + r} = \frac{1.5 \text{ V}}{0.4 \text{ } \Omega + 0.05 \text{ } \Omega} = 3.33 \text{ A}$$

## SOLVED PROBLEM 25.13

A battery whose emf is 45 V is connected to a 20-Ω resistance, and a current of 2.1 A flows. (*a*) Find the internal resistance of the battery. (*b*) Find the terminal voltage of the battery.

(a)  From $I = V_e/(R + r)$ we obtain

$$r = \frac{V_e}{I} - R = \frac{45 \text{ V}}{2.1 \text{ A}} - 20 \ \Omega = 21.4 \ \Omega - 20 \ \Omega = 1.4 \ \Omega$$

(b)                     $V = V_e - Ir = 45 \text{ V} - (2.1 \text{ A})(1.4) = 42 \text{ V}$

## SOLVED PROBLEM 25.14

A generator has an emf of 120 V and an internal resistance of 0.2 $\Omega$. (a) How much current does the generator supply when the terminal voltage is 115 V? (b) How much power does it supply? (c) How much power is dissipated in the generator itself?

(a)  From $V = V_e - Ir$ we obtain

$$I = \frac{V_e - V}{r} = \frac{120 \text{ V} - 115 \text{ V}}{0.2 \ \Omega} = 25 \text{ A}$$

(b)                     $P = IV = (25 \text{ A})(115 \text{ V}) = 2875 \text{ W}$

(c)                     $P = I^2 r = (25 \text{ A})^2 (0.2 \ \Omega) = 125 \text{ W}$

## SOLVED PROBLEM 25.15

A source of what potential difference is required to charge a battery of $V_e = 6$ V and $r = 0.1 \ \Omega$ at a rate of 10 A?

The required potential difference must equal the emf of the battery *plus* the $Ir$ drop in its internal resistance. Hence

$$V_{\text{applied}} = V_e + Ir = 6 \text{ V} + (10 \text{ A})(0.1 \ \Omega) = 7 \text{ V}$$

## SOLVED PROBLEM 25.16

What are the advantages of connecting a set of batteries in parallel?

(1) The reduced internal resistance of the set enables more current to be drawn by a load. (2) The ampere-hour capacity of the set is the sum of the capacities of the individual batteries and so is greater than that of any of them.

## SOLVED PROBLEM 25.17

Three batteries, each of emf 12 V and internal resistance 0.1 $\Omega$, are connected in series with a load of 2 $\Omega$. What is the current in the load?

The emf of the set of batteries is

$$V_e = 12 \text{ V} + 12 \text{ V} + 12 \text{ V} = 36 \text{ V}$$

and its internal resistance is

$$r = 0.1 \ \Omega + 0.1 \ \Omega + 0.1 \ \Omega = 0.3 \ \Omega$$

Hence the current in the load is

$$I = \frac{V_e}{R + r} = \frac{36 \text{ V}}{(2 + 0.3) \ \Omega} = 15.7 \text{ A}$$

**SOLVED PROBLEM 25.18**

The batteries of Prob. 25.17 are connected in parallel with the same load. What is the current in the load now?

The emf of the set of batteries is 12 V, and its internal resistance is found as follows:

$$\frac{1}{r} = \frac{1}{0.1\ \Omega} + \frac{1}{0.1\ \Omega} + \frac{1}{0.1\ \Omega} = \frac{3}{0.1\ \Omega}$$

$$r = \frac{0.1\ \Omega}{3} = 0.033\ \Omega$$

## IMPEDANCE MATCHING

When a source of electric energy is connected to a load, the power transfer is a maximum when both source and load have the same resistance. This is an example of *impedance matching*: When energy is being transferred from one system to another (here from an emf source to a load), the efficiency is greatest when both systems have the same *impedance*, which is a general term for resistance to the flow of energy. Another example of impedance matching occurs when a moving body strikes a stationary one. When both bodies have the same mass (which is a measure of the resistance of a body to a change in its state of motion or of rest), the energy transfer to the stationary body is a maximum.

Figure 25-6 shows an emf source of internal resistance $r$ connected to a load resistance $R$. If the current that flows is $I$,

$$I = \frac{V_{emf}}{R + r}$$

The power in the load is

$$P = I^2 R = \frac{V_{emf}^2 R}{(R + r)^2}$$

The maximum value of $P$ corresponds to $R = r$, when

$$P_{max} = \frac{V_{emf}^2}{4R}$$

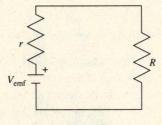

**Fig. 25-6**

**SOLVED PROBLEM 25.19**

Verify that $P_{max}$ occurs when $R = r$ by calculating the power delivered by a battery of emf 10 V and internal resistance 0.5 $\Omega$ when it is connected to (*a*) 0.25-$\Omega$, (*b*) 0.5-$\Omega$ and (*c*) 1-$\Omega$ resistors.

(a)
$$I_1 = \frac{V_e}{R_1 + r} = \frac{10 \text{ V}}{0.25 \ \Omega + 0.5 \ \Omega} = 13.3 \text{ A}$$

$$P_1 = I_1^2 R_1 = (13.3 \text{ A})^2 (0.25 \ \Omega) = 44 \text{ W}$$

(b)
$$I_2 = \frac{V_e}{R_2 + r} = \frac{10 \text{ V}}{0.5 \ \Omega + 0.5 \ \Omega} = 10 \text{ A}$$

$$P_2 = I_2^2 R_2 = (10 \text{ A})^2 (0.5 \ \Omega) = 50 \text{ W}$$

(c)
$$I_3 = \frac{V_e}{R_3 + r} = \frac{10 \text{ V}}{1 \ \Omega + 0.5 \ \Omega} = 6.7 \text{ A}$$

$$P_3 = I_3^2 R_3 = (6.7 \text{ A})^2 (1 \ \Omega) = 44 \text{ W}$$

## KIRCHHOFF'S RULES

The current that flows in each branch of a complex circuit can be found by applying *Kirchhoff's rules* to the circuit. The first rule applies to *junctions* of three or more wires (Fig. 25-7) and is a consequence of conservation of charge. The second rule applies to *loops*, which are closed conducting paths in the circuit, and is a consequence of conservation of energy. The rules are:

1. The sum of the currents that flow into a junction is equal to the sum of the currents that flow out of the junction.

2. The sum of the emf's around a loop is equal to the sum of the *IR* potential drops around the loop.

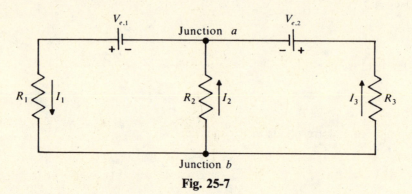

**Fig. 25-7**

The procedure for applying Kirchhoff's rules is as follows:

1. Choose a direction for the current in each branch of the circuit, as in Fig. 25-7. (A *branch* is a section of a circuit between two junctions.) If the choice is correct, the current will turn out to be positive. If not, the current will turn out to be negative, which means that the actual current is in the opposite direction. The current is the same in all the resistors and emf sources in a given branch. Of course, the currents will usually be different in the different branches.

2. Apply the first rule to the currents at the various junctions. This gives as many equations as the number of junctions. However, one of these equations is always a combination of the others and so gives no new information. (If there are only two junction equations, they will be the same.) Thus the number of usable junction equations is equal to one less than the number of junctions.

3. Apply the second rule to the emf's and *IR* drops in the loops. In going around a loop (which can be done either clockwise or counterclockwise) an emf is considered positive if the − terminal of

its source is met first. If the + terminal is met first, the emf is considered negative. An $IR$ drop is considered positive if the current in the resistor $R$ is in the same direction as the path being followed. If the current direction is opposite to the path, the $IR$ drop is considered negative.

In the case of the circuit shown in Fig. 25-7, Kirchhoff's first rule, applied to either junction $a$ or junction $b$, yields

$$I_1 = I_2 + I_3$$

The second rule applied to loop 1, shown in Fig. 25-8($a$), and proceeding counterclockwise, yields

$$V_{e,1} = I_1 R_1 + I_2 R_2$$

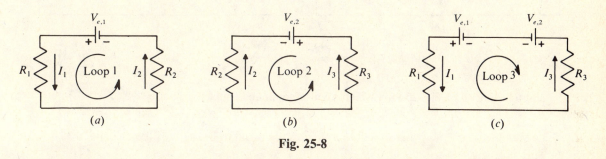

(a)                                          (b)                                          (c)

**Fig. 25-8**

The rule applied to loop 2, shown in Fig. 25-8($b$), and again proceeding counterclockwise, yields

$$-V_{e,2} = -I_2 R_2 + I_3 R_3$$

There is also a third loop, namely, the outside one shown in Fig. 25-8($c$), which must similarly obey Kirchhoff's second rule. For the sake of variety we now proceed clockwise and obtain

$$-V_{e,1} + V_{e,2} = -I_3 R_3 - I_1 R_1$$

Note that this last equation is just the negative sum of the two preceding equations. Thus, we may use the junction equation and *any two* of the loop equations to solve for the unknown currents $I_1$, $I_2$, and $I_3$. In general, the number of useful loop equations equals the number of separate areas enclosed by branches of the circuit. A helpful way to think of these areas is as adjacent plots of land. Loops (1) and (2) of Fig. 25-8 fit this description, but the outer loop (3) does not. Hence there are two separate areas here and we can use only two of the three loop equations, as we found above.

## SOLVED PROBLEM 25.20

Two batteries in parallel, one of emf 6 V and internal resistance 0.5 $\Omega$ and the other of emf 8 V and internal resistance 0.6 $\Omega$, are connected to an external 10-$\Omega$ resistor, as in Fig. 25-9($a$). Find the current in the external resistor.

Directions for the currents $I$, $I_1$, and $I_2$ are assumed as in Fig. 25-9($a$). At junction $a$ Kirchhoff's first law gives

$$I_2 = I + I_1$$

For the loop of Fig. 25-9($b$), which we go around counterclockwise,

$$V_{e,1} = IR - I_1 r_1$$

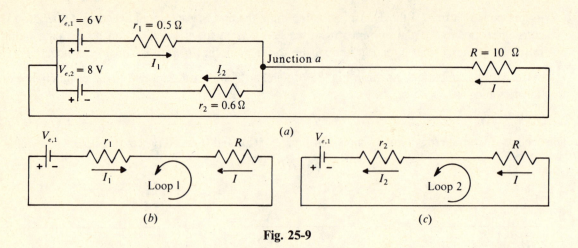

**Fig. 25-9**

and for the loop of Fig. 25-9(c), which we also go around counterclockwise,

$$V_{e,2} = IR + I_2 r_2$$

Since $I_2 = I + I_1$,

$$V_{e,2} = IR + Ir_2 + I_1 r_2$$

We now solve this equation and the first loop equation for $I_1$, set the two expressions for $I_1$ equal, and then solve for $I$:

$$I_1 = \frac{IR - V_{e,1}}{r_1} \quad \text{and} \quad I_1 = \frac{V_{e,2} - IR - Ir_2}{r_2}$$

$$\frac{IR - V_{e,1}}{r_1} = \frac{V_{e,2} - IR - Ir_2}{r_2}$$

$$I(Rr_2 + Rr_1 + r_2 r_1) = V_{e,2} r_1 + V_{e,1} r_2$$

$$I = \frac{V_{e,2} r_1 + V_{e,1} r_2}{Rr_2 + Rr_1 + r_2 r_1} = \frac{(8 \text{ V})(0.5 \ \Omega) + (6 \text{ V})(0.6 \ \Omega)}{(10 \ \Omega)(0.6 \ \Omega) + (10 \ \Omega)(0.5 \ \Omega) + (0.6 \ \Omega)(0.5 \ \Omega)} = 0.673 \text{ A}$$

## SOLVED PROBLEM 25.21

Find the currents in the three resistors of the circuit shown in Fig. 25-10(a). The internal resistances of the emf sources are included in $R_1$ and $R_3$.

We assume the current directions shown in the figure. Applying Kirchhoff's first rule to junction $a$ yields

$$I_3 = I_1 + I_2$$

Next we analyze the two inner loops with the help of Kirchhoff's second rule. Proceeding counterclockwise in loop 1 yields

$$V_{e,1} = I_1 R_1 - I_2 R_2$$

and proceeding counterclockwise in loop 2 yields

$$V_{e,2} = I_2 R_2 + I_3 R_3$$

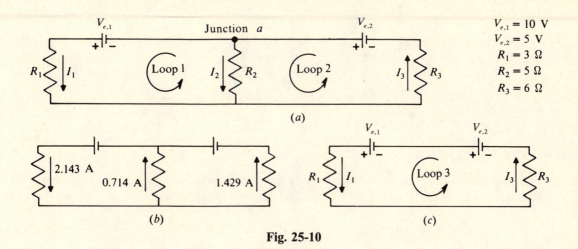

**Fig. 25-10**

We now have three equations that relate the unknown quantities $I_1$, $I_2$, and $I_3$. One way to proceed (there are others, equally suitable) is to substitute $I_3 = I_1 + I_2$ in the second loop equation, which gives

$$V_{e,2} = I_2R_2 + I_1R_3 + I_2R_3$$

$$I_1 = \frac{V_{e,2} - I_2R_2 - I_2R_3}{R_3}$$

From the first loop equation,

$$I_1 = \frac{V_{e,1} + I_2R_2}{R_1}$$

Since these two expressions must be equal,

$$\frac{V_{e,1} + I_2R_2}{R_1} = \frac{V_{e,2} - I_2R_2 - I_2R_3}{R_3}$$

At this point we substitute the values of the various emf's and resistances and solve for $I_2$. This substitution can also be done earlier or later in a calculation of this kind, whatever seems most convenient. The calculation proceeds as follows:

$$\frac{10\text{ V} + (5\ \Omega)(I_2)}{3\ \Omega} = \frac{5\text{ V} - (5\ \Omega)(I_2) - (6\ \Omega)(I_2)}{6\ \Omega}$$

$$\frac{10\text{ V}}{3\ \Omega} + \left(\frac{5\ \Omega}{3\ \Omega}\right)I_2 = \frac{5\text{ V}}{6\ \Omega} - \left(\frac{5\ \Omega}{6\ \Omega}\right)I_2 - \left(\frac{6\ \Omega}{6\ \Omega}\right)I_2$$

$$\left(\frac{5}{3} + \frac{5}{6} + \frac{6}{6}\right)I_2 = \left(\frac{5}{6} - \frac{10}{3}\right)\text{A}$$

$$I_2 = -0.714\text{ A}$$

The minus sign means that the current $I_2$ is in the opposite direction to the one shown in the figure. From the first loop equation,

$$I_1 = \frac{V_{e,1} + I_2R_2}{R_1} = \frac{10\text{ V} - (0.714\text{ A})(5\ \Omega)}{3\ \Omega} = 2.143\text{ A}$$

Finally we find $I_3$ from the junction equation:

$$I_3 = I_1 + I_2 = 2.143\text{ A} - 0.714\text{ A} = 1.429\text{ A}$$

The actual currents are shown in Fig. 25-10(b).

As a check on the calculation we can apply Kirchhoff's second rule to the outside loop of the circuit, shown in Fig. 25-10($c$). Proceeding counterclockwise, we get

$$V_{e,2} + V_{e,1} = I_1 R_1 + I_3 R_3$$
$$5 \text{ V} + 10 \text{ V} = (2.143 \text{ A})(3 \ \Omega) + (1.429 \text{ A})(6 \ \Omega)$$
$$15 \text{ V} = 15 \text{ V}$$

## SOLVED PROBLEM 25.22

Find the currents in the three resistors of the circuit shown in Fig. 25-11($a$). The internal resistances of the emf sources are included in the resistances shown.

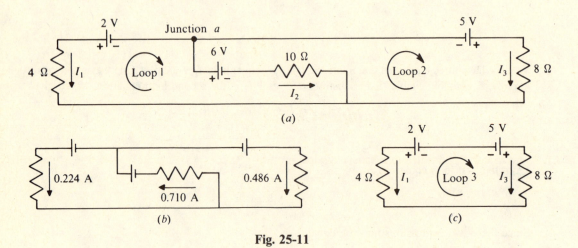

**Fig. 25-11**

Let us now work directly from the numerical values of the emf's and resistances in the figure. Applying Kirchhoff's first rule to junction $a$, with the current directions shown, yields

$$I_1 + I_2 + I_3 = 0$$

Clearly one or two current directions are incorrect, but this makes no difference since the result is a negative current in those cases. Next we apply Kirchhoff's second rule to loop 1 and proceed clockwise:

$$-2 \text{ V} - 6 \text{ V} = (10 \ \Omega)(I_2) - (4 \ \Omega)(I_1)$$

The emf's are considered negative because we encountered their + terminals first. Solving for $I_2$ gives

$$I_2 = \frac{-8 \text{ V}}{10 \ \Omega} + \left(\frac{4 \ \Omega}{10 \ \Omega}\right)(I_1) = -0.8 \text{ A} + 0.4 I_1$$

Now we proceed clockwise in loop 2 and obtain

$$6 \text{ V} + 5 \text{ V} = (8 \ \Omega)(I_3) - (10 \ \Omega)(I_2)$$

Substituting $I_3 = -I_1 - I_2$ and solving for $I_2$ yields

$$11 \text{ V} = -(8 \ \Omega)(I_1) - (8 \ \Omega)(I_2) - (10 \ \Omega)(I_2)$$
$$I_2 = \frac{-11 \text{ V}}{18 \ \Omega} - \left(\frac{8 \ \Omega}{18 \ \Omega}\right)(I_1) = -0.611 \text{ A} - 0.444 I_1$$

Setting equal the two expressions for $I_2$ and solving for $I_1$, we find

$$-0.8 \text{ A} + 0.4I_1 = -0.611 \text{ A} - 0.444I_1$$
$$0.844I_1 = 0.189 \text{ A}$$
$$I_1 = 0.224 \text{ A}$$

From the first loop equation

$$I_2 = -0.8 \text{ A} + 0.4I_1 = -0.710 \text{ A}$$

and so
$$I_3 = -I_1 - I_2 = -0.224 \text{ A} + 0.710 \text{ A} = 0.486 \text{ A}$$

The actual currents are shown in Fig. 25-11(b).

Again we check the results by using the outside loop of the circuit as shown in Fig. 25-11(c). Proceeding clockwise yields

$$-2 \text{ V} + 5 \text{ V} = (8 \text{ } \Omega)(0.486 \text{ A}) - (4 \text{ } \Omega)(0.224 \text{ A})$$
$$3 \text{ V} = 3 \text{ V}$$

## *Multiple-Choice Questions*

**25.1.**   A resistor $R$ connected to a battery dissipates energy at the rate $P$. If another resistor is connected in parallel with $R$, the power dissipated by $R$ is

(a)   less than $P$        (c)   more than $P$
(b)   $P$                  (d)   any of the above, depending on the values of the resistances

**25.2.**   The second resistor of Question 25.1 is connected in series with $R$. The power dissipated by $R$ is now

(a)   less than $P$        (c)   more than $P$
(b)   $P$                  (d)   any of the above, depending on the values of the resistances

**25.3.**   A network is being analyzed using Kirchhoff's rules. If the wrong direction is assumed for one of the currents $I$, the calculated current will be

(a)   0            (c)   $I$
(b)   $-I$         (d)   incorrect

**25.4.**   A 60-V potential difference is applied across a 5- and a 10-$\Omega$ resistor in series. The current in the 5-$\Omega$ resistor is

(a)   4 A          (c)   12 A
(b)   6 A          (d)   18 A

**25.5.**   The potential difference across the 5-$\Omega$ resistor in Question 25.4 is

(a)   20 V         (c)   40 V
(b)   30 V         (d)   60 V

**25.6.**   A 5- resistor and a 10-$\Omega$ resistor are connected in parallel. Their equivalent resistance is

(a)   0.3 $\Omega$        (c)   7.5 $\Omega$
(b)   3.3 $\Omega$        (d)   15 $\Omega$

**25.7.** A 60-V potential difference is applied across the resistors of Question 25.6. The current in the 5-$\Omega$ resistor is

(a)  4 A      (c)  12 A
(b)  6 A      (d)  18 A

**25.8.** The equivalent resistance of two identical resistors in parallel is 10 $\Omega$. The equivalent resistance of the same resistors in series would be

(a)  10 $\Omega$     (c)  40 $\Omega$
(b)  20 $\Omega$     (d)  100 $\Omega$

**25.9.** A network of three 5-$\Omega$ resistors cannot have an equivalent resistance of

(a)  1.67 $\Omega$     (c)  7.5 $\Omega$
(b)  2.5 $\Omega$      (d)  15 $\Omega$

**25.10.** A battery of emf 12 V whose internal resistance is negligible is connected across a 10-$\Omega$ resistor in parallel with a resistor of resistance $R$. If the battery current is 2 A, the value of $R$ is

(a)  3.75 $\Omega$     (c)  15 $\Omega$
(b)  6 $\Omega$        (d)  20 $\Omega$

**25.11.** A current of 9 A flows when a 120-V battery is connected across a 12-$\Omega$ resistor. The battery has an internal resistance of

(a)  1.3 $\Omega$      (c)  13.3 $\Omega$
(b)  12 $\Omega$       (d)  25.3 $\Omega$

**25.12.** A battery of emf 12 V and internal resistance 1 $\Omega$ is connected across a 3-$\Omega$ resistor. The potential difference across the resistor is

(a)  2.25 V     (c)  9 V
(b)  3 V        (d)  12 V

# Supplementary Problems

**25.1.** It is desired to have a current of 20 A in a 5-$\Omega$ resistor when it is connected to an 80-V battery. Is there any way in which an auxiliary resistor can be connected in the circuit to increase the current in the 5-$\Omega$ resistor to this value? If so, what should its resistance be?

**25.2.** (a) Find the equivalent resistance of four 60-$\Omega$ resistors connected in series. (b) If a potential difference of 12 V is applied across the combination, what is the current in each resistor?

**25.3.** (a) Find the equivalent resistance of four 60-$\Omega$ resistors connected in parallel. (b) If a potential difference of 12 V is applied across the combination, what is the current in each resistor?

**25.4.** You have three 2-$\Omega$ resistors. List the various resistances you can provide with them.

**25.5.** A 100- and a 200-$\Omega$ resistor are connected in series with a 40-V power source. (a) What is the current in each resistor? (b) How much power does each one dissipate?

**25.6.** A 100- and a 200-Ω resistor are connected in parallel with a 40-V power source. (*a*) What is the current in each resistor? (*b*) How much power does each one dissipate?

**25.7.** What resistance should be connected in parallel with a 1000-Ω resistor to produce an equivalent resistance of 200 Ω?

**25.8.** A 5-Ω resistor is connected in parallel with a 15-Ω resistor. When a potential difference is applied to the combination, which resistor will carry the greater current? What will the ratio of the currents be?

**25.9.** A 25-, a 40-, and a 60-Ω resistor are in series in a circuit such that the voltage across the 25-Ω resistor is 18 V. Find (*a*) the voltage across the other resistors and (*b*) the current in each.

**25.10.** (*a*) Find the equivalent resistance of the circuit shown in Fig. 25-12. (*b*) If a potential difference of 20 V is applied to the circuit, find the current in each resistor.

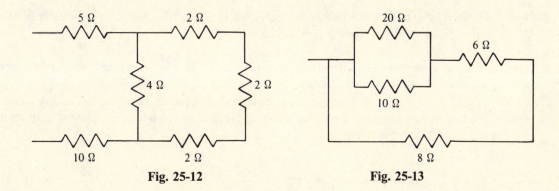

Fig. 25-12                              Fig. 25-13

**25.11.** (*a*) Find the equivalent resistance of the circuit shown in Fig. 25-13. (*b*) If a potential difference of 100 V is applied to the circuit, find the current in each resistor.

**25.12.** (*a*) Find the equivalent resistance of the circuit shown in Fig. 25-14. (*b*) A 6-V battery whose internal resistance is 1 Ω is connected to the circuit. Find the current in each resistor.

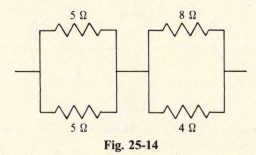

Fig. 25-14

**25.13.** Two batteries in parallel, each of emf 10 V and internal resistance 0.5 Ω, are connected to an external 20-Ω resistor. Find the current in the external resistor.

**25.14.** A dry cell has an emf of 1.5 V and an internal resistance of 0.08 Ω. (*a*) Find the current when the cell's terminals are connected together. (*b*) Find the current when the cell is connected to a 5-Ω resistance.

**25.15.** A certain "12-V" storage battery actually has an emf of 13.2 V and an internal resistance of 0.01 Ω. What is the terminal voltage of the battery when it delivers 80 A to the starter motor of a car engine?

**25.16.** A current of 2.2 A flows when a battery of emf 24 V is connected to a 10-Ω load. Find (*a*) the internal resistance of the battery and (*b*) its terminal voltage.

**25.17.** Two 12-V batteries, one with an internal resistance of 0.05 Ω and the other with an internal resistance of 0.15 Ω, are connected in parallel with a load of 0.5 Ω. Find the current in the load.

**25.18.** The batteries of Prob. 25.17 are connected in series with the same load. Find the current in the load now.

**25.19.** A generator whose emf is 240 V has a terminal voltage of 220 V when it delivers a current of 50 A. (*a*) Find the internal resistance of the generator. (*b*) Find the power supplied by the generator. (*c*) Find the power dissipated within the generator.

**25.20.** A storage battery of emf 34 V and internal resistance 0.1 Ω is to be charged at a rate of 20 A from a 110-V source. What series resistance is needed in the circuit?

**25.21.** Find the currents in the resistors of the circuit shown in Fig. 25-15. The internal resistances of the emf sources are included in the external resistances.

**25.22.** Find the current in the resistors of the circuit shown in Fig. 25-16. The internal resistances of the emf sources must be taken into account.

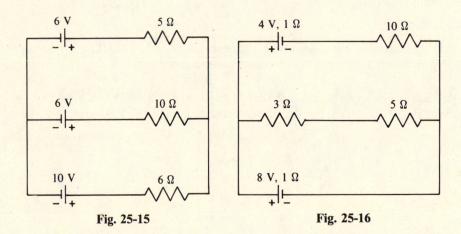

Fig. 25-15　　　　　　　　　　　　Fig. 25-16

**25.23.** Find the currents in the resistors of the circuit shown in Fig. 25-17. The internal resistances of the emf sources are included in the external resistances.

**25.24.** Find the currents in the resistors of the circuit shown in Fig. 25-18. The internal resistances of the emf sources are included in the external resistances.

**25.25.** (*a*) Find the current in the 5-Ω resistor in the circuit shown in Fig. 25-19. (*b*) Find the potential difference between points *A* and *B*. The internal resistances of the emf sources must be taken into account.

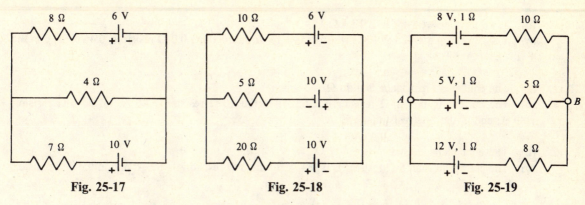

Fig. 25-17                        Fig. 25-18                        Fig. 25-19

## Answers to Multiple-Choice Questions

**25.1.** (b)      **25.7.** (c)

**25.2.** (a)      **25.8.** (c)

**25.3.** (b)      **25.9.** (b)

**25.4.** (a)      **25.10.** (c)

**25.5.** (a)      **25.11.** (a)

**25.6.** (b)      **25.12.** (c)

## Answers to Supplementary Problems

**25.1.**   There is no way in which an auxiliary resistor can be connected to increase the current.

**25.2.**   (a)  240 Ω      (b)  0.05 A

**25.3.**   (a)  15 Ω   (b)  0.2 A

**25.4.**   0.67 Ω; 1 Ω; 2 Ω; 3 Ω; 4 Ω; 6 Ω

**25.5.**   (a)  0.133 A; 0.133 A      (b)  1.78 W; 3.55 W

**25.6.**   (a)  0.4 A; 0.2 A      (b)  16 W; 8 W

**25.7.**   250 Ω

**25.8.**   The 5-Ω resistor will carry a current three times greater than that carried by the 15-Ω resistor.

**25.9.**   (a)  28.8 V across the 40-Ω resistor; 43.2 V across the 60-Ω resistor.      (b)  0.72 A in each resistor.

**25.10.**  (*a*)   The equivalent resistance is 17.4 $\Omega$.
(*b*)   The current in the 5- and 10-$\Omega$ resistors is 1.15 A; in the three 0.2-$\Omega$ resistors, 0.46 A; and in the 4-$\Omega$ resistor, 0.69 A.

**25.11.**  (*a*)   The equivalent resistance is 4.90 $\Omega$.
(*b*)   The current in the 20-$\Omega$ resistor is 2.63 A; in the 10-$\Omega$ resistor, 5.26 A; in the 6-$\Omega$ resistor, 7.89 A; and in the 8-$\Omega$ resistor, 12.50 A.

**25.12.**  (*a*)   The equivalent resistance is 5.167 $\Omega$.
(*b*)   The current in each 5-$\Omega$ resistor is 0.581 A; in the 8-$\Omega$ resistor, 0.387 A; and in the 4-$\Omega$ resistor, 0.774 A.

**25.13.**  0.494 A

**25.14.**  (*a*)   19 A     (*b*)   0.3 A

**25.15.**  12.4 V

**25.16.**  (*a*)   0.9 $\Omega$     (*b*)   22 V

**25.17.**  22.3 A

**25.18.**  34.3 A

**25.19.**  (*a*)   0.4 $\Omega$     (*b*)   11 kW     (*c*)   1 kW

**25.20.**  3.7 $\Omega$

**25.21.**  The current in the 5-$\Omega$ resistor is 0.286 A to the left, the current in the 10-$\Omega$ resistor is 0.143 A to the left, and the current in the 6-$\Omega$ resistor is 0.429 A to the right.

**25.22.**  The current in the 10-$\Omega$ resistor is 0.935 A to the left, and the currents in the 3-$\Omega$ and 5-$\Omega$ resistors are both 0.785 A to the left.

**25.23.**  The current in the 8-$\Omega$ resistor is 0.22 A to the left, the current in the 4-$\Omega$ resistor is 1.05 A to the right, and the current in the 7-$\Omega$ resistor is 0.83 A to the left.

**25.24.**  The current in the 10-$\Omega$ resistor is 0.857 A to the left, the current in the 5-$\Omega$ resistor is 1.486 A to the right, and the current in the 20-$\Omega$ resistor is 0.629 A to the left.

**25.25.**  (*a*)   0.475 A to the right     (*b*)   7.85 V

# Chapter 26

# Capacitance

## CAPACITANCE

A *capacitor* is a system that stores energy in the form of an electric field. In its simplest form, a capacitor consists of a pair of parallel metal plates separated by air or other insulating material.

The potential difference $V$ between the plates of a capacitor is directly proportional to the charge $Q$ on either of them, so the ratio $Q/V$ is always the same for a particular capacitor. This ratio is called the *capacitance C* of the capacitor:

$$C = \frac{Q}{V}$$

$$\text{Capacitance} = \frac{\text{charge on either plate}}{\text{potential difference between plates}}$$

The unit of capacitance is the *farad* (F), where 1 farad = 1 coulomb/volt. Since the farad is too large for practical purposes, the *microfarad* and *picofarad* are commonly used, where

$$1 \text{ microfarad} = 1 \ \mu\text{F} = 10^{-6} \text{ F}$$
$$1 \text{ picofarad} = 1 \ p\text{F} = 10^{-12} \text{ F}$$

A charge of $10^{-6}$ C on each plate of 1-$\mu$F capacitor will produce a potential difference of $V = Q/C = 1$ V between the plates.

## PARALLEL-PLATE CAPACITOR

A capacitor that consists of parallel plates each of area $A$ separated by the distance $d$ has a capacitance of

$$C = K\varepsilon_0 \ \frac{A}{d}$$

The constant $\varepsilon_0$ is the permittivity of free space mentioned in Chapter 23; its value is

$$\varepsilon_0 = 8.85 \times 10^{-12} \text{ C}^2/(\text{N·m}^2) = 8.85 \times 10^{-12} \text{ F/m}$$

The quantity $K$ is the *dielectric constant* of the material between the capacitor plates; the greater $K$ is, the more effective the material is in diminishing an electric field. For free space, $K = 1$; for air, $K = 1.0006$; a typical value for glass is $K = 6$; and for water, $K = 80$.

## SOLVED PROBLEM 26.1

A 200-pF capacitor is connected to a 100-V battery. Find the charge on the capacitor's plates.

$$Q = CV = (200 \times 10^{-12} \text{ F})(100 \text{ V}) = 2 \times 10^{-8} \text{ C}$$

**SOLVED PROBLEM 26.2**

A capacitor has a charge of $5 \times 10^{-4}$ C when the potential difference across its plates is 300 V. Find its capacitance.

$$C = \frac{Q}{V} = \frac{5 \times 10^{-4} \text{ C}}{300 \text{ V}} = 1.67 \times 10^{-6} \text{ F} = 1.67 \ \mu\text{F}$$

**SOLVED PROBLEM 26.3**

A parallel-plate capacitor has plates 5 cm square and 0.1 mm apart. Find its capacitance (*a*) in air and (*b*) with mica of $K = 6$ between the plates.

(*a*)   In air $K$ is very nearly 1, and so

$$C = K\varepsilon_0 \frac{A}{d} = (1)\left(8.85 \times 10^{-12} \ \frac{\text{F}}{\text{m}}\right) \frac{(0.05 \text{ m})^2}{(10^{-4} \text{ m})} = 2.21 \times 10^{-10} \text{ F} = 221 \text{ pF}$$

(*b*)   With mica between the plates the capacitance will be $K = 6$ times greater, or

$$C = (16)(221 \text{ pF}) = 1326 \text{ pF}$$

**SOLVED PROBLEM 26.4**

A parallel-plate capacitor has a capacitance of 2 $\mu$F in air and 4.6 $\mu$F when it is immersed in benzene. What is the dielectric constant of benzene?

Since $C$ is proportional to $K$, in general

$$\frac{C_1}{K_1} = \frac{C_2}{K_2}$$

for the same capacitor. Here, with $K_1 = K_{\text{air}} = 1$, the dielectric constant $K_2$ of benzene is

$$K_2 = K_1 \frac{C_2}{C_1} = (1)\left(\frac{4.6 \ \mu\text{F}}{2 \ \mu\text{F}}\right) = 2.3$$

**SOLVED PROBLEM 26.5**

A 10-$\mu$F capacitor with air between its plates is connected to a 50-V source and then disconnected. (*a*) What are the charge on the capacitor and the potential difference across it? (*b*) The space between the plates of the charged capacitor is filled with Teflon ($K = 2.1$). What are the charge on the capacitor and the potential difference across it now?

(*a*)   The capacitor's charge is

$$Q = CV = (10 \times 10^{-6} \text{ F})(50 \text{ V}) = 5 \times 10^{-4} \text{ C}$$

The potential difference across it remains 50 *V* after it is disconnected.

(*b*)   The presence of another dielectric does not change the charge on the capacitor. Since its capacitance is now

$$C_2 = \frac{K_2}{K_1} C_1$$

and $V = Q/C$, the new potential difference is

$$V_2 = \frac{Q}{C_2} = \frac{K_1}{K_2} \frac{Q}{C_1} = \frac{K_1}{K_2} V_1 = \left(\frac{1}{2.1}\right)(50 \text{ V}) = 23.8 \text{ V}$$

## CAPACITORS IN COMBINATION

The *equivalent capacitance* of a set of connected capacitors is the capacitance of the single capacitor that can replace the set without changing the properties of any circuit it is part of. The equivalent capacitance of a set of capacitors joined in series (Fig. 26-1) is

$$\frac{1}{C} = \frac{1}{C_1} + \frac{1}{C_2} + \frac{1}{C_3} + \cdots \qquad \text{capacitors in series}$$

When using a calculator with a reciprocal $(1/X)$ key, it is easiest to proceed in the way described in Chapter 25 in the case of the similar formula for the equivalent resistance of a set of resistors in parallel. The key sequence here would be

$$[C_1][1/X][+][C_2][1/X][+][C_3][1/X][+] \cdots [=][1/X]$$

If there are only two capacitors in series,

$$\frac{1}{C} = \frac{1}{C_1} + \frac{1}{C_2} = \frac{C_1 + C_2}{C_1 C_2} \qquad \text{and so} \qquad C = \frac{C_1 C_2}{C_1 + C_2}$$

In a parallel set of capacitors (Fig. 26-2), the equivalent capacitance is the sum of the individual capacitances:

$$C = C_1 + C_2 + C_3 + \cdots \qquad \text{capacitors in parallel}$$

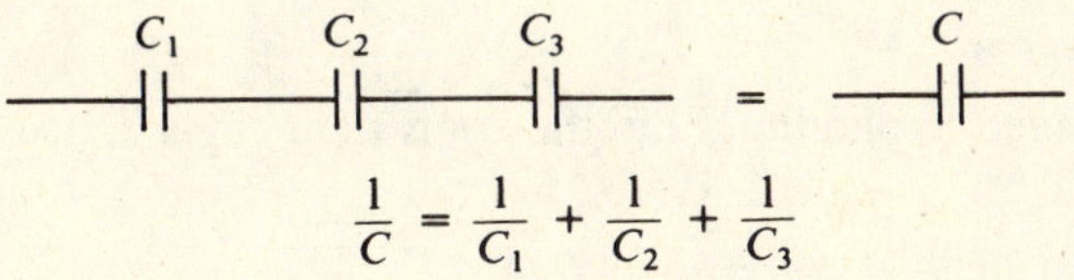

**Fig. 26-1**

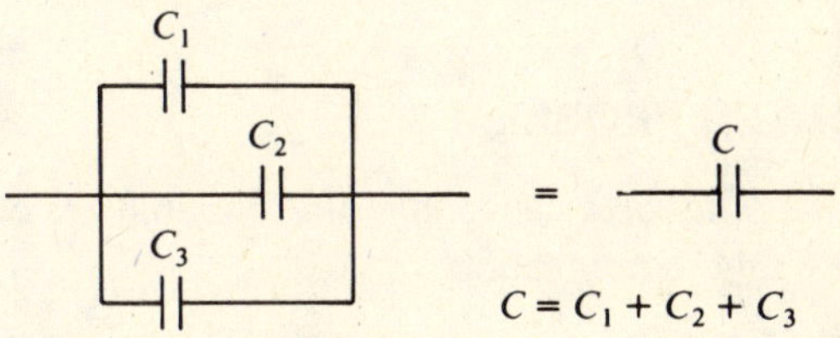

**Fig. 26-2**

## SOLVED PROBLEM 26.6

Show that the equivalent capacitance $C$ of three capacitors connected in series is given by $1/C = 1/C_1 + 1/C_2 + 1/C_3$.

Each capacitor in a series connection has charges of the same magnitude $Q$ on its plates, so the voltages across them are, respectively,

$$V_1 = \frac{Q}{C_1} \qquad V_2 = \frac{Q}{C_2} \qquad V_3 = \frac{Q}{C_3}$$

If $C$ is the equivalent capacitance of the set, then

$$V = V_1 + V_2 + V_3 \qquad \frac{Q}{C} = \frac{Q}{C_1} + \frac{Q}{C_2} + \frac{Q}{C_3}$$

Dividing through by the charge $Q$ gives

$$\frac{1}{C} = \frac{1}{C_1} + \frac{1}{C_2} + \frac{1}{C_3}$$

## SOLVED PROBLEM 26.7

Show that the equivalent capacitance $C$ of three capacitors connected in parallel is given by $C = C_1 + C_2 + C_3$.

Now the same voltage $V$ is across all the capacitors, and their respective charges are

$$Q_1 = C_1 V \qquad Q_2 = C_2 V \qquad Q_3 = C_3 V$$

The total charge $Q_1 + Q_2 + Q_3$ on either the $+$ or $-$ plates of the capacitors is equal to the charge $Q$ on the corresponding plate of the equivalent capacitor $C$, and so

$$Q = Q_1 + Q_2 + Q_3 \qquad CV = C_1 V = C_2 V + C_3 V$$

Dividing through by $V$ gives

$$C = C_1 + C_2 + C_3$$

## SOLVED PROBLEM 26.8

Three capacitors whose capacitances are 1, 2, and 3 $\mu$F are connected in series. Find the equivalent capacitance of the combination.

$$\frac{1}{C} = \frac{1}{C_1} + \frac{1}{C_2} + \frac{1}{C_3} = \frac{1}{1 \ \mu F} + \frac{1}{2 \ \mu F} + \frac{1}{3 \ \mu F} = \frac{11}{6 \ \mu F}$$

Hence

$$C = \frac{6}{11} \ \mu F = 0.545 \ \mu F$$

## SOLVED PROBLEM 26.9

The three capacitors of Prob. 26.8 are connected in parallel. Find the equivalent capacitance of the combination.

$$C = C_1 + C_2 + C_3 = 1 \ \mu F + 2 \ \mu F + 3 \ \mu F = 6 \ \mu F$$

## SOLVED PROBLEM 26.10

A 2- and 3-$\mu$F capacitor are connected in series. (a) What is their equivalent capacitance? (b) A potential difference of 500 V is applied to the combination. Find the charge on each capacitor and the potential difference across it.

(a)
$$C = \frac{C_1 C_2}{C_1 + C_2} = \frac{(2 \ \mu F)(3 \ \mu F)}{2 \ \mu F + 3 \ \mu F} = 1.2 \ \mu F$$

(b) The charge on the combination is

$$Q = CV = (1.2 \times 10^{-6} \ F)(500 \ V) = 6 \times 10^{-4} \ C$$

The same charge is present on each capacitor (Fig. 26-3). Hence the potential difference across the 2-$\mu$F capacitor is

$$V_1 = \frac{Q}{C_1} = \frac{6 \times 10^{-4} \ C}{2 \times 10^{-6} \ F} = 300 \ V$$

and that across the 3-$\mu$F capacitor is

$$V_2 = \frac{Q}{C_2} = \frac{6 \times 10^{-4} \ C}{3 \times 10^{-6} \ F} = 200 \ V$$

As a check we note that $V_1 + V_2 = 500$ V.

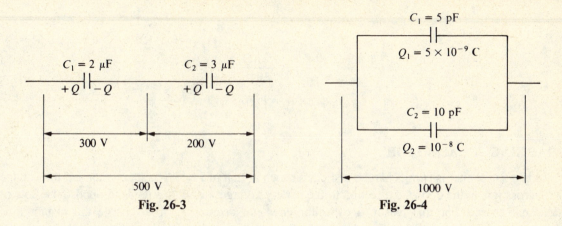

**Fig. 26-3**                                    **Fig. 26-4**

## SOLVED PROBLEM 26.11

A 5- and 10-pF capacitor are connected in parallel. (*a*) What is their equivalent capacitance? (*b*) A potential difference of 1000 V is applied to the combination. Find the charge on each capacitor and the potential difference across it.

(*a*)                                 $$C = C_1 + C_2 = 5 \text{ pF} + 10 \text{ pF} = 15 \text{ pF}$$

(*b*)    The same potential difference $V = 1000$ V is across each capacitor (Fig. 26-4). The charge on the 5-pF capacitor is

$$Q_1 = C_1 V = (5 \times 10^{-12} \text{ F})(10^3 \text{ V}) = 5 \times 10^{-9} \text{ C}$$

and that on the 10-pF capacitor is

$$Q_2 = C_2 V = (10 \times 10^{-12} \text{ F})(10^3 \text{ V}) = 10^{-8} \text{ C}$$

## ENERGY OF A CHARGED CAPACITOR

To produce the electric field in a charged capacitor, work must be done to separate the positive and negative charges. This work is stored as electric potential energy in the capacitor. The potential energy $W$ of a capacitor of capacitance $C$ whose charge is $Q$ and whose potential difference is $V$ is given by

$$W = \tfrac{1}{2}QV = \tfrac{1}{2}CV^2 = \frac{1}{2}\left(\frac{Q^2}{C}\right)$$

## SOLVED PROBLEM 26.12

How much energy is stored in a 50-pF capacitor when it is charged to a potential difference of 200 V?

$$W = \tfrac{1}{2}CV^2 = (\tfrac{1}{2})(50 \times 10^{-12} \text{ F})(200 \text{ V})^2 = 10^{-6} \text{ J}$$

## SOLVED PROBLEM 26.13

A 100-$\mu$F capacitor is to have an energy content of 50 J to operate a flashlamp. (*a*) What voltage is required to charge the capacitor? (*b*) How much charge passes through the flashlamp?

(a)   Since $W = \frac{1}{2}CV^2$,

$$V = \sqrt{\frac{2W}{C}} = \sqrt{\frac{(2)(50 \text{ J})}{10^{-4} \text{ F}}} = 1000 \text{ V}$$

(b)                                 $Q = CV = (10^{-4} \text{ F})(10^3 \text{ V}) = 0.1 \text{ C}$

## CHARGING A CAPACITOR

When a capacitor is being charged in a circuit such as that of Fig. 26-5, at any moment the voltage $Q/C$ across it is in the opposite direction to the battery voltage $V$ and thus tends to oppose the flow of additional charge. For this reason a capacitor does not acquire its final charge the instant it is connected to a battery or other source of emf. The net potential difference when the charge on the capacitor is $Q$ is $V - (Q/C)$, and the current is then

$$I = \frac{\Delta Q}{\Delta t} = \frac{V - (Q/C)}{R}$$

As $Q$ increases, its rate of increase $I = \Delta Q/\Delta t$ decreases. Figure 26-6 shows how $Q$, measured in percent of final change, varies with time when a capacitor is being charged; the switch of Fig. 26-5 is closed at $t = 0$.

The product $RC$ of the resistance $R$ in the circuit and the capacitance $C$ governs the rate at which the capacitor reaches its ultimate charge of $Q_0 = CV$. The product $RC$ is called the *time constant T* of the circuit. After a time equal to $T$, the charge on the capacitor is 63 percent of its final value.

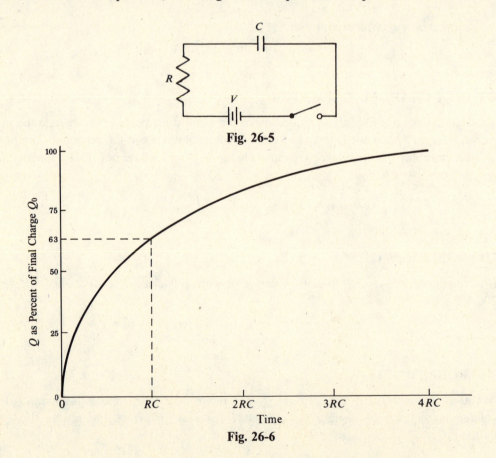

**Fig. 26-5**

**Fig. 26-6**

The formula that governs the growth of charge in the circuit of Fig. 26-5 is

$$Q = Q_0(1 - e^{-t/T})$$

where $Q_0$ is the final charge $CV$ and $T$ is the time constant $RC$. Figure 26-6 is a graph of this formula. The quantity $e$ has the value

$$e = 2.718 \cdots$$

and is often found in equations in engineering and science. A quantity that consists of $e$ raised to a power is called an *exponential*. To find the value of $e^x$ or $e^{-x}$, an electronic calculator or a suitable table can be used. Exponentials are sometimes written $\exp x$ or $\exp(-x)$ instead of $e^x$ or $e^{-x}$. The meaning is exactly the same.

It is easy to see why $Q$ reaches 63 percent of $Q_0$ in time $T$. When $t = T$, $t/T = 1$ and

$$Q = Q_0(1 - e^{-1}) = Q_0\left(1 - \frac{1}{e}\right) = Q_0(1 - 0.37) = 0.63Q_0$$

## DISCHARGING A CAPACITOR

When a charged capacitor is discharged through a resistance, as in Fig. 26-7, the decrease in charge is governed by the formula

$$Q = Q_0 e^{-t/T}$$

where again $T = RC$ is the time constant. The charge will fall to 37 percent of its original value after time $T$ (Fig. 26-8). The smaller the time constant $T$, the more rapidly a capacitor can be charged or discharged.

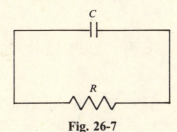

**Fig. 26-7**

## SOLVED PROBLEM 26.14

A 20-$\mu$F capacitor is connected to a 45-V battery through a circuit whose resistance is 2000 $\Omega$. (a) What is the final charge on the capacitor? (b) How long does it take for the charge to reach 63 percent of its final value?

(a)
$$Q = CV = (20 \times 10^{-6}\ \text{F})(45\ \text{V}) = 9 \times 10^{-4}\ \text{C}$$
(b)
$$t = RC = (2000\ \Omega)(20 \times 10^{-6}\ \text{F}) = 0.04\ \text{s}$$

## SOLVED PROBLEM 26.15

Find the charge on the capacitor of Prob. 26.14 at 0.01 and 0.1 s after the connection to the battery is made.

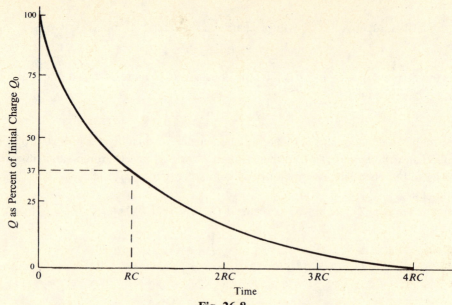

**Fig. 26-8**

When $t = 0.01$ s, $t/T = (0.01 \text{ s})/(0.04 \text{ s}) = 0.25$. Using a calculator or table of exponentials gives

$$e^{-t/T} = e^{-0.25} = 0.78$$

Hence

$$Q = Q_0(1 - e^{-t/T}) = (9 \times 10^{-4} \text{ C})(1 - 0.78) = (9 \times 10^{-4} \text{ C})(0.22) = 2.0 \times 10^{-4} \text{ C}$$

Similarly, when $t = 0.1$ s, $t/T = (0.1 \text{ s})/(0.04 \text{ s}) = 2.5$ and

$$e^{-t/T} = e^{-2.5} = 0.082$$

Hence

$$Q = Q_0(1 - e^{-t/T}) = (9 \times 10^{-4} \text{ C})(1 - 0.082) = (9 \times 10^{-4} \text{ C})(0.918) = 8.3 \times 10^{-4} \text{ C}$$

## SOLVED PROBLEM 26.16

A 5-$\mu$F capacitor is charged by being connected to a 3-V battery. The battery is then disconnected. (*a*) If the resistance of the dielectric material between the capacitor plates is $10^9$ $\Omega$, find the time required for the charge on the capacitor to drop to 37 percent of its original value. (*b*) What is the charge remaining on the capacitor 1 h after it has been disconnected? What is the charge 10 h afterward?

(*a*)
$$T = RC = (10^9 \ \Omega)(5 \times 10^{-6} \text{ F}) = 5 \times 10^3 \text{ s}$$

which is
$$\frac{5 \times 10^3 \text{ s}}{(60 \text{ s/min})(60 \text{ min/h})} = \frac{5000 \text{ s}}{3600 \text{ s/h}} = 1.4 \text{ h}$$

(*b*)   The initial charge on the capacitor is

$$Q_0 = CV = (5 \times 10^{-6} \text{ F})(3 \text{ V}) = 1.5 \times 10^{-5} \text{ C}$$

After $t = 1$ h, $t/T = (1 \text{ h})/(1.4 \text{ h}) = 0.71$, and

$$Q = Q_0 e^{-t/T} = (1.5 \times 10^{-5} \text{ C})(e^{-0.71}) = (1.5 \times 10^{-5} \text{ C})(0.49) = 7.4 \times 10^{-6} \text{ C}$$

After $t = 10$ h, $t/T = (10 \text{ h})/(1.4 \text{ h}) = 7.1$, and

$$Q = Q_0 e^{-t/T} = (1.5 \times 10^{-5} \text{ C})(e^{-7.1}) = (1.5 \times 10^{-5} \text{ C})(8.3 \times 10^{-4}) = 1.2 \times 10^{-8} \text{ C}$$

# *Multiple-Choice Questions*

**26.1.** An electric field $E$ is present between the plates of a charged capacitor. If a slab of insulating material is inserted between the plates, the new electric field there is

    (*a*)  smaller than $E$    (*c*)  larger than $E$

    (*b*)  $E$                (*d*)  any of the above, depending on the material

**26.2.** The distance between the plates of a parallel-plate capacitor of capacitance $C$ is doubled and their area is halved. The capacitor now has a capacitance of

    (*a*)  $C/4$    (*c*)  $2C$

    (*b*)  $C$     (*d*)  $4C$

**26.3.** A parallel-plate capacitor with air between its plates has an energy of $W$ when it is charged until a potential difference of $V$ appears across its plates. An otherwise identical capacitor has a material of dielectric constant $K = 2$ between its plates and is also charged to the potential difference $V$. The energy of the second capacitor is

    (*a*)  $W/2$    (*c*)  $2W$

    (*b*)  $W$     (*d*)  $4W$

**26.4.** A capacitor connected to a 24-V battery has a charge of 0.004 C. Its capacitance is

    (*a*)  $1.67\ \mu F$    (*c*)  $167\ \mu F$

    (*b*)  $60\ \mu F$     (*d*)  0.048 F

**26.5.** A charged 50-$\mu F$ capacitor has an energy of 1 J. The voltage across it is

    (*a*)  141 V    (*c*)  20 kV

    (*b*)  200 V    (*d*)  40 kV

**26.6.** The capacitance of a parallel-plate capacitor is 20 $\mu F$ in air and 42 $\mu F$ when Teflon is between its plates. Teflon has a dielectric constant of

    (*a*)  0.48    (*c*)  2.1

    (*b*)  1.4     (*d*)  4.2

**26.7.** The equivalent resistance of two 20-$\mu F$ capacitors in series is

    (*a*)  $0.1\ \mu F$    (*c*)  $30\ \mu F$

    (*b*)  $10\ \mu F$    (*d*)  $40\ \mu F$

**26.8.** The equivalent resistance of two 20-$\mu F$ capacitors in parallel is

    (*a*)  $0.1\ \mu F$    (*c*)  $30\ \mu F$

    (*b*)  $10\ \mu F$    (*d*)  $40\ \mu F$

**26.9.** A charged 50-$\mu F$ capacitor is discharged through a 1.0-k$\Omega$ resistor. The charge on the capacitor will drop to 37 percent of its original value in

    (*a*)  50 ns    (*c*)  25 ms

    (*b*)  20 ms    (*d*)  50 ms

# Supplementary Problems

**26.1.** Verify that $RC$ has the dimensions of time.

**26.2.** A 10-$\mu$F capacitor has a potential difference of 250 V across it. What is the charge on the capacitor?

**26.3.** A capacitor has a charge of 0.002 C when it is connected across a 100-V battery. Find its capacitance.

**26.4.** What is the potential difference across a 500-pF capacitor whose charge is 0.3 $\mu$C?

**26.5.** The plates of a parallel-plate capacitor have areas of 40 cm$^2$ and are separated by 0.2 mm of waxed paper ($K = 2.2$). Find the capacitance.

**26.6.** The waxed paper is removed from between the plates of the capacitor of Prob. 26.5. Find the new capacitance.

**26.7.** The plates of a parallel-plate capacitor of capacitance $C$ are moved closer together until they are half their original separation. What is the new capacitance?

**26.8.** The capacitance of a parallel-plate capacitor is increased from 8 to 50 $\mu$F when a sheet of glass is inserted between its plates. What is the dielectric constant $K$ of the glass?

**26.9.** Three capacitors whose capacitances are 5, 10, and 20 $\mu$F are connected in series. Find the equivalent capacitance of the combination.

**26.10.** The three capacitors of Prob. 26.9 are connected in parallel. Find the equivalent capacitance of the combination.

**26.11.** List the capacitance that can be obtained by combining three 10-$\mu$F capacitors in all possible ways.

**26.12.** A 20- and a 25-pF capacitor are connected in parallel, and a potential difference of 100 V is applied to the combination. Find the charge on each capacitor and the potential difference across it.

**26.13.** A 50- and a 75-pF capacitor are connected in series, and a potential difference of 250 V is applied to the combination. Find the charge on each capacitor and the potential difference across it.

**26.14.** A 5-$\mu$F capacitor has a potential difference of 1000 V. What is its potential energy?

**26.15.** (a) What potential difference must be applied across a 10-$\mu$F capacitor if it is to have an energy content of 1 J? (b) What is the charge on the capacitor under these circumstances?

**26.16.** The dielectric between the plates of a certain 80-$\mu$F capacitor has a resistance of $10^9$ $\Omega$. If the capacitor is charged and then disconnected, how long will it take for the charge on the capacitor to fall to 37 percent of its original value?

**26.17.** A 5-$\mu$F capacitor is connected to a 100-V battery through a circuit whose resistance is 800 $\Omega$. (a) What is the time constant of this arrangement? (b) What is the initial current that flows when the battery is connected? (c) What is the final charge on the capacitor?

**26.18.** Find the charge on the capacitor of Prob. 26.17 at 0.001, 0.005, and 0.01 s after the connection to the battery is made.

**26.19.** The resistance of the dielectric between the plates of the capacitor of Prob. 26.17 is 10 Ω. If the capacitor is disconnected from the battery, find the charge remaining on it 30 s, 1 min, and 10 min later.

# Answers to Multiple-Choice Questions

**26.1.** $(a)$  **26.6.** $(c)$

**26.2.** $(a)$  **26.7.** $(b)$

**26.3.** $(c)$  **26.8.** $(d)$

**26.4.** $(c)$  **26.9.** $(d)$

**26.5.** $(b)$

# Answers to Supplementary Problems

**26.1.** From their definitions, $R = V/I = Vt/Q$ and $C = Q/V$. Hence $RC = (Vt/Q)(Q/V) = t$.

**26.2.** 0.0025 C  **26.11.** 3.33 $\mu$F; 6.67 $\mu$F; 15 $\mu$F; 30 $\mu$F

**26.3.** 20 $\mu$F  **26.12.** $Q_1 = 2 \times 10^{-9}$ C; $Q_2 = 2.5 \times 10^{-9}$ C; $V_1 = V_2 = 100$ V

**26.4.** 600 V  **26.13.** $Q_1 = Q_2 = 7.5 \times 10^{-9}$ C; $V_1 = 150$ V; $V_2 = 100$ V

**26.5.** 3.9 pF  **26.14.** 2.5 J

**26.6.** 5.9 pF  **26.15.** $(a)$ 447 V  $(b)$ $4.47 \times 10^{-3}$ C

**26.7.** 2$C$  **26.16.** $8 \times 10^4$ s = 22.2 h

**26.8.** 6.25  **26.17.** $(a)$ 0.004 s  $(b)$ 0.125 A  $(c)$ $5 \times 10^{-4}$ C

**26.9.** 2.86 $\mu$F  **26.18** $1.106 \times 10^{-4}$ C; $3.567 \times 10^{-4}$ C; $4.590 \times 10^{-4}$ C

**26.10.** 35 $\mu$F  **26.19.** $2.744 \times 10^{-4}$ C; $1.506 \times 10^{-4}$ C; $3.072 \times 10^{-9}$ C

# Chapter 27

# Magnetism

## NATURE OF MAGNETISM

Two electric charges at rest exert forces on each other according to Coulomb's law. When the charges are in motion, the forces are different, and it is customary to attribute the differences to *magnetic forces* that occur between moving charges in addition to the electric forces between them. In this interpretation, the total force on a charge $Q$ at a certain time and place can be divided into two parts: an electric force that depends only on the value of $Q$ and a magnetic force that depends on the velocity $v$ of the charge as well as on $Q$.

In reality, there is only a single interaction between charges, the *electromagnetic interaction*. The theory of relativity provides the link between electric and magnetic forces: Just as the mass of an object moving with respect to an observer is greater than when it is at rest, so the electric force between two charges appears altered to an observer when the charges are moving with respect to the observer. Magnetism is not distinct from electricity in the way that, for example, gravitation is. Despite the unity of the electromagnetic interaction, it is convenient for many purposes to treat electric and magnetic effects separately.

## MAGNETIC FIELD

A *magnetic field* **B** is present wherever a magnetic force acts on a moving charge. The direction of **B** at a certain place is that along which a charge can move without experiencing a magnetic force; along any other direction the charge would be acted on by such a force. The magnitude of **B** is equal numerically to the force on a charge of 1 C moving at 1 m/s perpendicular to **B**.

The unit of magnetic field is the *tesla* (T), where

$$1 \text{ tesla} = 1 \ \frac{\text{newton}}{\text{ampere-meter}} = 1 \ \frac{\text{weber}}{(\text{meter})^2}$$

The *gauss* (G), equal to $10^{-4}$ T, is another unit of magnetic field sometimes used.

## MAGNETIC FIELD OF A STRAIGHT CURRENT

The magnetic field a distance $s$ from a long, straight current $I$ has the magnitude

$$B = \left(\frac{\mu}{2\pi}\right)\left(\frac{1}{s}\right) \qquad \text{straight current}$$

where $\mu$ is the *permeability* of the medium in which the magnetic field exists. The permeability of free space $\mu_0$ has the value

$$\mu_0 = 4\pi \times 10^{-7} \text{ T·m/A} = 1.257 \times 10^{-6} \text{ T·m/A}$$

The value of $\mu$ in air is very nearly the same as $\mu_0$ and is usually considered as equal to $\mu_0$.

The field lines of the magnetic field around a straight current are in the form of concentric circles around the current. To find the direction of **B**, place the thumb of the right hand in the direction of the current; the curled fingers of that hand then point in the direction of **B** (Fig. 27-1).

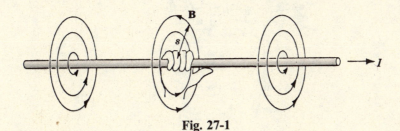

**Fig. 27-1**

## MAGNETIC FIELD OF A CURRENT LOOP

The magnetic field at the center of a current loop of radius $r$ has the magnitude

$$B = \left(\frac{\mu}{2}\right)\left(\frac{I}{r}\right) \qquad \text{current loop}$$

The field lines of **B** are perpendicular to the plane of the loop, as shown in Fig. 27-2(*a*). To find the direction of **B**, grasp the loop so the curled fingers of the right hand point in the direction of the current; the thumb of that hand then points in the direction of **B** [Fig. 27-2(*b*)].

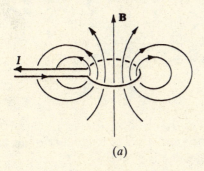

(*a*)

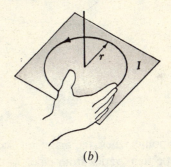

(*b*)

**Fig. 27-2**

A *solenoid* is a coil consisting of many loops of wire. If the turns are close together and the solenoid is long compared with its diameter, the magnetic field inside it is uniform and parallel to the axis with magnitude

$$B = \mu \frac{N}{L} I \qquad \text{solenoid}$$

In this formula $N$ is the number of turns, $L$ is the length of the solenoid, and $I$ is the current. The direction of **B** is as shown in Fig. 27-3.

## EARTH'S MAGNETIC FIELD

The earth has a magnetic field that arises from electric currents in its liquid iron core. The field is like that which would be produced by a current loop centered a few hundred miles from the earth's center whose plane is tilted by 11° from the plane of the equator (Fig. 27-4). The *geomagnetic poles*

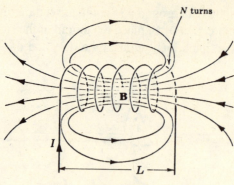

Fig. 27-3

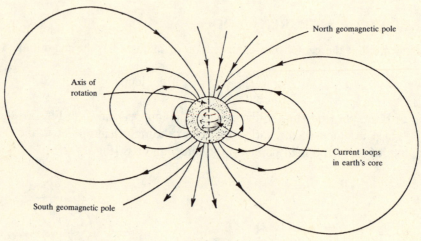

Fig. 27-4

are the points where the magnetic axis passes through the earth's surface. The magnitude of the earth's magnetic field varies from place to place; a typical sea-level value is $3 \times 10^{-5}$ T.

## SOLVED PROBLEM 27.1

In what ways are electric and magnetic fields similar? In what ways are they different?

*Similarities*: Both fields originate in electric charges, and both fields can exert forces on electric charges.

*Differences*: All electric charges give rise to electric fields, but only a charge in motion relative to an observer gives rise to a magnetic field. Electric fields exert forces on all charges, but magnetic fields exert forces only on moving charges.

## SOLVED PROBLEM 27.2

A cable 5 m above the ground carries a current of 100 A from east to west. Find the direction and magnitude of the magnetic field on the ground directly beneath the cable. (Neglect the earth's magnetic field.)

From the right-hand rule, the direction of the field is south. The magnitude of the field is

$$B = \left(\frac{\mu_0}{2\pi}\right)\left(\frac{I}{s}\right) = \frac{(4\pi \times 10^{-7} \text{ T·m/A})(100 \text{ A})}{(2\pi)(5 \text{ m})} = 4 \times 10^{-6} \text{ T}$$

## SOLVED PROBLEM 27.3

Two parallel wires 10 cm apart carry currents of 8 A in the same direction. What is the magnetic field halfway between them?

The magnetic field halfway between the wires is zero because the fields of the currents are opposite in direction and have the same magnitude there.

## SOLVED PROBLEM 27.4

Two parallel wires 10 cm apart carry currents of 8 A in opposite directions. What is the magnetic field halfway between them?

Here the magnetic fields of the two currents are in the same direction and hence add. Since $s = 5$ cm $= 5 \times 10^{-2}$ m halfway between the wires, the field of each current there is

$$B = \left(\frac{\mu_0}{2\pi}\right)\left(\frac{I}{s}\right) = \frac{(4\pi \times 10^{-7} \text{ T·m/A})(8 \text{ A})}{(2\pi)(5 \times 10^{-2} \text{ m})} = 3.2 \times 10^{-5} \text{ T}$$

and the total field is $2B = 6.4 \times 10^{-5}$ T.

## SOLVED PROBLEM 27.5

A 100-turn flat circular coil has a radius of 5 cm. Find the magnetic field at the center of the coil when the current is 4 A.

Each turn of the coil acts as a separate loop in contributing to the total magnetic field. If there are $N$ turns, the result is a field $N$ times stronger than each turn produces by itself. Hence

$$B = \left(\frac{\mu_0}{2\pi}\right)\left(\frac{NI}{r}\right) = \frac{(4\pi \times 10^{-7} \text{ T·m/A})(100)(4 \text{ A})}{(2\pi)(5 \times 10^{-2} \text{ m})} = 1.6 \times 10^{-3} \text{ T}$$

## SOLVED PROBLEM 27.6

A solenoid 0.2 m long has 1000 turns of wire and is oriented with its axis parallel to the earth's magnetic field at a place where the latter is $2.5 \times 10^{-5}$ T. What should the current in the solenoid be in order that its field exactly cancel the earth's field inside the solenoid?

The magnetic field inside an air-core solenoid is

$$B = \mu_0 \frac{N}{L} I$$

Here $N = 10^3$, $L = 0.2$ m, and $B = 2.5 \times 10^{-5}$ T, so

$$I = \frac{BL}{\mu_0 N} = \frac{(2.5 \times 10^{-5} \text{ T})(0.2 \text{ m})}{(4\pi)(10^{-7} \text{ T·m/A})(10^3)} = 0.004 \text{ A}$$

## MAGNETIC FORCE ON A MOVING CHARGE

The magnetic force on a moving charge $Q$ in a magnetic field varies with the relative directions of **v** and **B**. When **v** is parallel to **B**, $F = 0$; when **v** is perpendicular to **B**, $F$ has its maximum value of

$$F = QvB \qquad \mathbf{v} \perp \mathbf{B}$$

The direction of **F** in the case of a positive charge is given by the right-hand rule, shown in Fig. 27-5; **F** is in the opposite direction when the charge is negative.

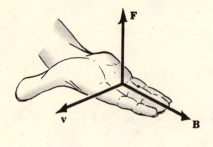
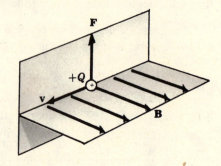

Fig. 27-5

## MAGNETIC FORCE ON A CURRENT

Since a current consists of moving charges, a current-carrying wire will experience no force when parallel to a magnetic field **B** and maximum force when perpendicular to **B**. In the latter case, $F$ has the value

$$F = ILB \qquad \mathbf{I} \perp \mathbf{B}$$

where $I$ is the current and $L$ is the length of wire in the magnetic field. The direction of the force is as shown in Fig. 27-6.

Owing to the different forces exerted on each of its sides, a current loop in a magnetic field always tends to rotate so that its plane is perpendicular to **B**. This effect underlies the operation of all electric motors.

## FORCE BETWEEN TWO CURRENTS

Two parallel electric currents exert magnetic forces on each other (Fig. 27-7). If the currents are in the same direction, the forces are attractive; if the currents are in opposite directions, the forces are

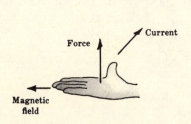
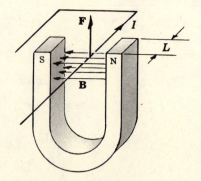

Fig. 27-6

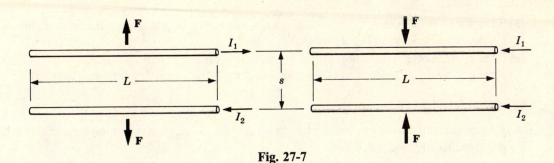

**Fig. 27-7**

repulsive. The force per unit length $F/L$ on each current depends on currents $I_1$ and $I_2$ and their separation $s$:

$$\frac{F}{L} = \left(\frac{\mu_0}{2\pi}\right)\left(\frac{I_1 I_2}{s}\right) \qquad \text{parallel currents}$$

### SOLVED PROBLEM 27.7

A positive charge is moving virtually upward when it enters a magnetic field directed to the north. In what direction is the force on the charge?

To apply the right-hand rule here, the fingers of the right hand are pointed north and the thumb of that hand is pointed upward. The palm of the hand faces west, which is therefore the direction of the force on the charge.

### SOLVED PROBLEM 27.8

A stream of protons is moving parallel to a stream of electrons. Do the streams tend to come together or to move apart?

The electric force between the streams is attractive, but the magnetic force is repulsive. Which of the forces is stronger depends on how fast the particles are moving.

### SOLVED PROBLEM 27.9

A proton is moving in a uniform magnetic field. Describe the path of the proton if its initial direction is (a) parallel to the field, (b) perpendicular to the field, and (c) at an intermediate angle to the field.

(a)   There is no magnetic force on the proton, so it continues to move in a straight line [Fig. 27-8(a)].

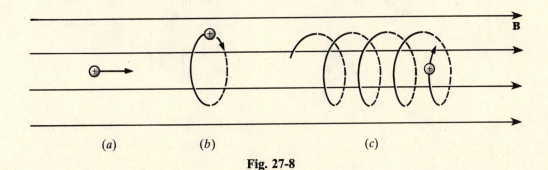

**Fig. 27-8**

(b) The force on the proton is perpendicular to its velocity **v** and also perpendicular to **B**; hence it moves in a circle, as in Fig. 27-8(b).

(c) The proton moves in the helical path of Fig. 27-8(c) because the component of **v** parallel to **B** is not changed while the component of **v** perpendicular to **B** leads to an inward force, as in (b).

## SOLVED PROBLEM 27.10

A charged particle moving perpendicular to a uniform magnetic field follows a circular path. Find the radius of the circle.

The magnetic force $QvB$ in the particle provides the centripetal force $mv^2/r$ that keeps it moving in a circle of radius $r$. Hence

$$F_{\text{magnetic}} = F_{\text{centripetal}}$$

$$QvB = \frac{mv^2}{r}$$

$$r = \frac{mv}{QB}$$

The radius is directly proportional to the particle's momentum $mv$ and inversely proportional to its charge $Q$ and the magnetic field $B$.

## SOLVED PROBLEM 27.11

Show that a current-carrying wire loop experiences a torque in a magnetic field provided the plane of the loop is not perpendicular to the field.

Figure 27-9(a) shows a current-carrying loop whose plane is parallel to a magnetic field **B**. Sides $A$ and $C$ of the loop are parallel to **B**, and so no magnetic force acts on them. Sides $B$ and $D$ are perpendicular to **B**, however, and each experiences the force shown. Since $\mathbf{F}_B$ is opposite in direction to $\mathbf{F}_D$ along different lines of action, they produce a torque on the loop. Such a torque will occur even if the plane of the loop is not parallel to **B**, although it will then be smaller since the moment arm will be shorter, provided the plane is not perpendicular to **B**.

In Fig. 27-9(b), the plane of the loop is perpendicular to **B**. Now $\mathbf{F}_A$ and $\mathbf{F}_C$ are equal and opposite along the same line of action, as are $\mathbf{F}_B$ and $\mathbf{F}_D$, so there is no net torque on the loop.

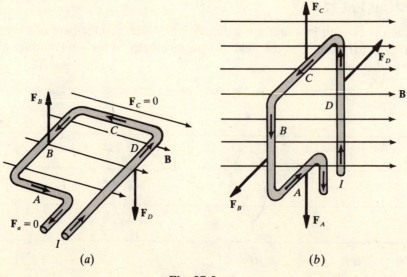

(a)                                                                          (b)

**Fig. 27-9**

## SOLVED PROBLEM 27.12

In a certain electric motor wires that carry a current of 5 A are perpendicular to a magnetic field of 0.8 T. What is the force on each centimeter of these wires?

$$F = ILB = (5 \text{ A})(0.01 \text{ m})(0.8 \text{ T}) = 0.04 \text{ N}$$

## SOLVED PROBLEM 27.13

The wires that supply current to a 120-V, 2-kW electric heater are 2 mm apart. What is the force per meter between the wires?

Since $P = IV$, the current in the wires is

$$I = I_1 = I_2 = \frac{P}{V} = \frac{2000 \text{ W}}{120 \text{ V}} = 16.7 \text{ A}$$

Since $s = 2 \text{ mm} = 2 \times 10^{-3}$ m, the force between the wires is

$$\frac{F}{L} = \left(\frac{\mu_0}{2\pi}\right)\left(\frac{I_1 I_2}{s}\right) = \frac{(4\pi \times 10^{-7} \text{ T·m/A})(16.7 \text{ A})^2}{(2\pi)(2 \times 10^{-3} \text{ m})} = 0.028 \text{ N/m}$$

The currents are in opposite directions, so the force is repulsive.

## FERROMAGNETISM

The magnetic field produced by a current is altered by the presence of a substance of any kind. Usually the change, which may be an increase or a decrease in **B**, is very small, but in certain cases there is an increase in **B** by hundreds or thousands of times. Substances that have the latter effect are called *ferromagnetic*; iron and iron alloys are familiar examples. An *electromagnet* is a solenoid with a ferromagnetic core to increase its magnetic field.

Ferromagnetism is a consequence of the magnetic properties of the electrons that all atoms contain. An electron behaves in some respects as though it is a spinning charged sphere, and it is therefore magnetically equivalent to a tiny current loop. In most substances the magnetic fields of the atomic electrons cancel, but in ferromagnetic substances the cancellation is not complete and each atom has a certain magnetic field of its own. The atomic magnetic fields align themselves in groups called *domains* with an external magnetic field to produce a much stronger total **B**. When the external field is removed, the atomic magnetic fields may remain aligned to produce a *permanent magnet*. The field of a bar magnet has the same form as that of a solenoid because both fields are due to parallel current loops (Fig. 27-10).

## MAGNETIC INTENSITY

A substance which decreases the magnetic field of a current is called *diamagnetic*; it has a permeability $\mu$ that is less than $\mu_0$. Copper and water are examples. A substance which increases the

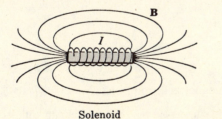

Solenoid          Bar magnet

**Fig. 27-10**

magnetic field of a current by a small amount is called *paramagnetic*; it has a permeability $\mu$ that is greater than $\mu_0$. Aluminum is an example. Ferromagnetic substances have permeabilities hundreds or thousands of times greater than $\mu_0$. Diamagnetic substances are repelled by magnets; paramagnetic and ferromagnetic ones are attracted by magnets.

Because different substances have different magnetic properties, it is useful to define a quantity called *magnetic intensity* **H** which is independent of the medium in which a magnetic field is located. The magnetic intensity in a place where the magnetic field is **B** and the permeability is $\mu$ is given by

$$\mathbf{H} = \frac{\mathbf{B}}{\mu}$$

$$\text{Magnetic intensity} = \frac{\text{magnetic field}}{\text{permeability of medium}}$$

The unit of **H** is the ampere per meter. Magnetic intensity is sometimes called *magnetizing force* or *magnetizing field*.

The permeability of a ferromagnetic material at a given value of $H$ varies both with $H$ and with the previous degree of magnetization of the material. The latter effect is known as *hysteresis*.

## SOLVED PROBLEM 27.14

The ends of a bar magnet are traditionally called its *poles*, with the end that tends to point north called the *north pole* and the end that tends to point south called the *south pole*. Like poles of nearby magnets repel each other, and unlike poles attract. Explain this behavior in terms of the interaction of current loops.

Bar magnets with poles facing each other are equivalent to parallel current loops whose currents are in opposite directions [Fig. 27-11(*a*)]. Such loops repel. Bar magnets with opposite poles facing each other are equivalent to parallel current loops whose current loops are in the same direction [Fig. 27-11(*b*)]. Such loops attract.

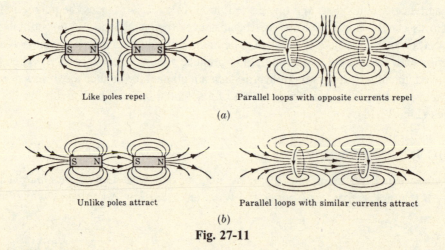

Like poles repel              Parallel loops with opposite currents repel

(*a*)

Unlike poles attract          Parallel loops with similar currents attract

(*b*)

**Fig. 27-11**

## SOLVED PROBLEM 27.15

How does a permanent magnet attract an unmagnetized iron object?

The presence of the magnet induces the atomic magnets in the object to line up with its field (Fig. 27-12), and the attraction of opposite poles then produces a net force on the object.

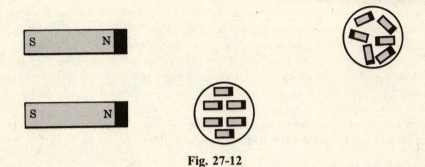

**Fig. 27-12**

## SOLVED PROBLEM 27.16

An unmagnetized iron rod is placed inside a solenoid. The current in the solenoid is then increased from zero to a maximum in one direction, decreased back to zero, increased to a maximum in the other direction, brought back to zero, and so on. Plot $B$ versus $H$ for the iron rod, and discuss the shape of the resulting curve.

The required curve is shown in Fig. 27-13. At $a$, both $H$ and $B$ are 0. As $H$ increases, $B$ increases slowly at first, then rapidly, and finally levels off at a maximum value at $b$. The rod is now *saturated*, and a further increase in $H$ will not change $B$. Saturation occurs when all the magnetic domains in the rod are aligned with $H$. The curve from $a$ to $b$ is called the *magnetization curve* of the material.

When $H$ is brought back to 0 from $H_b$, $B$ lags behind so that $B = B_c$ when $H = 0$. This is an example of hysteresis. The value of $B_c$ is called the *retentivity* of the material. To demagnetize the rod completely, $H$ must be reversed in direction and increased to $H_d$, the *coercive force*. The greater the retentivity, the stronger the residual magnetization of the rod; the greater the coercive force, the better able the rod will be to keep its magnetization despite the presence of strong magnetic fields. Thus a good material for a permanent magnet should have both a high retentivity and a high coercive force.

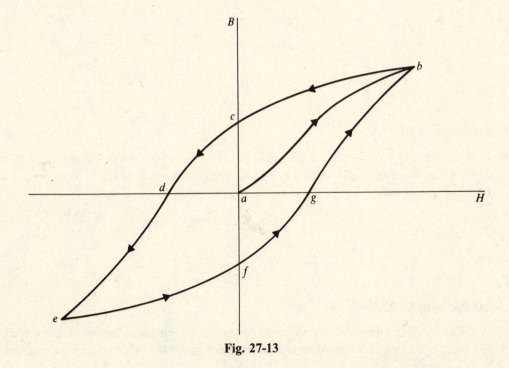

**Fig. 27-13**

Increasing $-H$ beyond $d$ produces an increasing $-B$ until the rod saturates again at $e$, where $B_e = -B_b$. When $H$ is returned to 0, again $B$ lags behind and this time has the value $B_f$ at $H = 0$, where $B_f = B_c$. Increasing $H$ then returns $B$ to $B_b$ to complete the *hysteresis loop bcdefgb*.

The area enclosed by a hysteresis loop is proportional to the energy dissipated as heat during each magnetization cycle. A broad hysteresis loop with high values of retentivity and coercive force is characteristic of a suitable material for a permanent magnet, since a great deal of work must be done to change its magnetization. However, a material with a narrow hysteresis loop is better for such applications as transformer cores which must undergo frequent reversals in magnetization; the smaller the area of the loop, the greater the efficiency of the transformer.

**SOLVED PROBLEM 27.17**

How can a magnetized piece of iron be demagnetized?

One method is to heat the iron, since all ferromagnetic materials lose their ability to retain magnetization beyond a certain temperature, which is about 760°C in the case of iron. Another method is to bring the iron through a succession of hysteresis loops of smaller size, as in Fig. 27-14. To do this, the iron can be placed in a solenoid connected to a source of alternating current, and then the current is gradually decreased to zero.

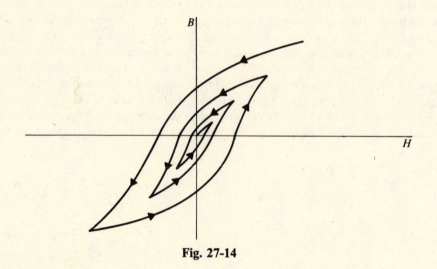

**Fig. 27-14**

**SOLVED PROBLEM 27.18**

A sample of cast iron exhibits a magnetic field of $B = 0.50$ T when the magnetic intensity is $H = 10$ A/m. (*a*) Find the permeability of cast iron at this value of $H$. (*b*) What would the field be in air at this value of $H$?

(*a*)
$$\mu = \frac{B}{H} = \frac{0.50 \text{ T}}{10 \text{ A/m}} = 0.05 \text{ T·m/A}$$

(*b*)
$$B = \mu_0 H = \left(4\pi \times 10^{-7} \frac{\text{T·m}}{\text{A}}\right)\left(10 \frac{\text{A}}{\text{m}}\right) = 1.257 \times 10^{-5} \text{ T}$$

**SOLVED PROBLEM 27.19**

A solenoid 20 cm long is wound with 300 turns of wire and carries a current of 1.5 A. (*a*) What is the magnetic intensity $H$ inside the solenoid? (*b*) What should the permeability of the core at this value of $H$ be so that the magnetic field inside is 0.6 T? How many times greater than $\mu_0$ is this?

(a)
$$H = \frac{B}{\mu} = \frac{N}{L} I = \left(\frac{300}{0.2 \text{ m}}\right)(1.5 \text{ A}) = 2250 \text{ A/m}$$

(b)
$$\frac{B}{H} = \frac{0.6 \text{ T}}{2250 \text{ A/m}} = 2.67 \times 10^{-4} \text{ T·m/A}$$

$$\frac{\mu}{\mu_0} = \frac{2.67 \times 10^{-4} \text{ T·m/A}}{4\pi \times 10^{-7} \text{ T·m/A}} = 212$$

## Multiple-Choice Questions

**27.1.** There is no interaction between a magnetic field and a

    (a) stationary electric charge    (c) stationary magnet
    (b) moving electric charge    (d) moving magnet

**27.2.** When an observer moves past a stationary electric charge, she detects

    (a) only an electric field
    (b) only a magnetic field
    (c) both an electric and a magnetic field
    (d) any of the above, depending on her velocity

**27.3.** Neglecting the earth's magnetic field, the direction of the magnetic field below a power cable in which a current is flowing north is

    (a) north    (c) south
    (b) east    (d) west

**27.4.** The magnetic field lines around a long, straight current are

    (a) straight lines parallel to the current
    (b) straight lines that radiate from the current like spokes of a wheel
    (c) concentric circles around the current
    (d) concentric helixes around the current

**27.5.** The magnetic field inside a current-carrying solenoid

    (a) is 0    (c) decreases away from the axis
    (b) is uniform    (d) increases away from the axis

**27.6.** The following statements concern the magnetic field of a solenoid that has an iron core. Which one or more of them are not always true?

    (a) $B = 0$ when $I = 0$
    (b) $B$ increases when $I$ increases
    (c) $B$ decreases when $I$ decreases
    (d) The direction of $B$ changes when the direction of $I$ changes

**27.7.** When a moving charged particle enters a uniform magnetic field in a direction parallel to the field lines, the particle's

    (a) direction is changed
    (b) velocity magnitude is changed
    (c) energy is changed
    (d) motion is unaffected

**27.8.** When a moving charged particle enters a uniform magnetic field in a direction perpendicular to the field lines, the particle's

    (a) direction is changed     (c) energy is changed
    (b) velocity magnitude is changed     (d) motion is unaffected

**27.9.** When a current-carrying wire loop is a magnetic field, it tends to turn so as to bring the plane of the loop

    (a) parallel to the field
    (b) perpendicular to the field
    (c) at a 45° angle to the field
    (d) any of the above, depending on the strength and direction of the current

**27.10.** If we travel around the earth, we would find that the earth's magnetic field

    (a) is the same in direction and magnitude everywhere
    (b) varies in direction but not in magnitude
    (c) varies in magnitude but not in direction
    (d) varies in both magnitude and direction

**27.11.** A long, straight wire carries a current of 1.2 A. The magnetic field 8.0 mm from the wire is

    (a) $3.0 \times 10^{-8}$ T     (c) $3.6 \times 10^{-5}$ T
    (b) $3.0 \times 10^{-5}$ T     (d) $1.9 \times 10^{-4}$ T

**27.12.** A 200-turn solenoid is 40 mm long. If the magnetic field inside it is to be 0.010 T, the current it carries should be

    (a) 1.6 A     (c) $1.6 \times 10^3$ A
    (b) 251 A     (d) $4.0 \times 10^7$ A

**27.13.** A magnetic force of 0.08 N acts on each centimeter of a wire that carries a 20-A current in an electric motor. If the wire is perpendicular to the magnetic field in the motor, the magnitude of the field is

    (a) 0.004 T     (c) 0.4 T
    (b) 0.016 T     (d) 40 T

**27.14.** Two parallel wires 800 cm long are 5 cm apart and carry currents of 20 A each in the same direction. Each wire exerts a force on the other of

    (a) $1.6 \times 10^{-3}$ N, attractive     (c) $1.6 \times 10^{-3}$ N, repulsive
    (b) $1.3 \times 10^{-2}$ N, attractive     (d) $1.3 \times 10^{-2}$ N, repulsive

# Supplementary Problems

**27.1.** An observer is able to measure electric, magnetic, and gravitational fields. Which of these does she detect when (*a*) a proton moves past her, and (*b*) she moves past a proton?

**27.2.** A beam of electrons that are moving slowly at first are accelerated to higher and higher velocities. What happens to the diameter of the beam as this happens?

**27.3.** An electric current is flowing south along a power line. What is the direction of the magnetic field above it? Below it?

**27.4.** A charged particle moves through a magnetic field perpendicular to **B**. Is the particle's energy affected? Is its momentum?

**27.5.** In a sketch of magnetic lines of force, what is the significance of lines of force that are closer in a particular region than they are elsewhere?

**27.6.** A negative charge is moving west when it enters a magnetic field directed vertically downward. In what direction is the force on the charge?

**27.7.** A wire carrying a current is placed in a magnetic field **B**. (*a*) Under what circumstances, if any, will the force on the wire be zero? (*b*) Under what circumstances will the force on the wire be a maximum?

**27.8.** Under what circumstances, if any, does a current-carrying wire loop not tend to rotate in a magnetic field?

**27.9.** The magnetic field 5 cm from a certain straight wire is $10^{-4}$ T. Find the current in the wire.

**27.10.** How far away from a compass should a wire carrying a 1-A current be located if its magnetic field at the compass is not to exceed 1 percent of the earth's magnetic field, which is typically $3 \times 10^{-5}$ T?

**27.11.** Two parallel wires 20 cm apart carry currents of 5 A in the same direction. Find the magnetic field between the wires 5 cm from one of them and 15 cm from the other.

**27.12.** What should the current be in a wire loop 1 cm in radius if the magnetic field in the center of the loop is to be 0.01 T?

**27.13.** What is the magnetic field inside a solenoid wound with 20 turns per centimeter when the current in it is 5 A?

**27.14.** An electron in a television picture tube travels at $3 \times 10^7$ m/s. Does the earth's gravitational field or its magnetic field exert the greater force on the electron? Assume **v** is perpendicular to **B**.

**27.15.** (*a*) A wire 1 m long is perpendicular to a magnetic field of 0.01 T. What is the force on the wire when it carries a current of 10 A? (*b*) What is the force on the wire when it is parallel to the magnetic field?

**27.16.** A horizontal wire 10 cm long whose mass is 1 g to be supported magnetically against the force of gravity. The current in the wire is 10 A and goes from north to south. (*a*) What should be the direction of the magnetic field? (*b*) What should its magnitude be?

**27.17.** The starting motor of a certain car is connected to the battery by a pair of cables that are 8 mm apart for a distance of 50 cm. Find the force between the cables when the current is 100 A.

**27.18.** An alternating current that varies with time according to the formula $I = I_{max} \sin \omega t$ is sent through a solenoid with an iron core. (*a*) Does the magnetic intensity in the core also vary sinusoidally with time? (*b*) Does the magnetic field?

**27.19.** The magnetic intensity $H$ inside an air-core solenoid 25 cm long that is wound with 300 turns of wire is 600 A/m. (*a*) What is the current in the solenoid? (*b*) What is the magnetic field $B$ inside the solenoid?

**27.20.** An air-core solenoid wound with 20 turns/cm carries a current of 0.1 A. (*a*) Find $H$ and $B$ inside the solenoid. (*b*) An iron core whose permeability is $6 \times 10^{-3}$ T·m/A is inserted in the solenoid. Find $H$ and $B$ now.

**27.21.** A sample of carbon steel has a permeability of 0.01 T·m/A when the magnetic intensity is 75 A/m. (*a*) Find the magnetic field in the sample at this value of $H$. (*b*) Find the field at this value of $H$.


# *Answers to Multiple-Choice Questions*

| | | | |
|---|---|---|---|
| **27.1.** | (*a*) | **27.8.** | (*a*) |
| **27.2.** | (*c*) | **27.9.** | (*b*) |
| **27.3.** | (*d*) | **27.10.** | (*d*) |
| **27.4.** | (*c*) | **27.11.** | (*b*) |
| **27.5.** | (*b*) | **27.12.** | (*a*) |
| **27.6.** | (*a*), (*b*), (*c*), (*d*) | **27.13.** | (*c*) |
| **27.7.** | (*d*) | **27.14.** | (*b*) |


# **Answers to Supplementary Problems**

**27.1.** All three fields are detected in both cases.

**27.2.** At first the mutual electric repulsion of the electrons causes the beam diameter to increase, but as they go faster the magnetic attraction becomes more significant and the beam diameter decreases.

**27.3.** West; east

**27.4.** The particle's energy is not changed since the magnetic force on it is perpendicular to its direction of motion, and so no work is done on it by the field. The particle's direction changes, however, and hence its momentum, which is a vector quantity, also changes.

**27.5.** The closer the lines of force are drawn in a particular region, the stronger the field there.

**27.6.** To the north

**27.7.** (*a*)  When the wire is parallel to **B**      (*b*)  When it is perpendicular to **B**

**27.8.** Such a loop does not tend to rotate when its plane is perpendicular to the direction of the magnetic field.

**27.9.** 25 A

**27.10.** 67 cm

**27.11.** Since the fields are in opposite directions, $B = 2 \times 10^{-5}$ T $- 0.67 \times 10^{-5}$ T $= 1.33 \times 10^{-5}$ T.

**27.12.** 159 A

**27.13.** $1.26 \times 10^{-2}$ T

**27.14.** The magnetic force $evB$ is more than $10^{13}$ times greater than the gravitational force $mg$.

**27.15.** (*a*)  0.1 N      (*b*)  0

**27.16.** (*a*)  The direction of the field should be toward the west so that the force on the wire is upward.

(*b*)  $9.8 \times 10^{-3}$ T

**27.17.** 0.125 N; the force is repulsive

**27.18.** (*a*)  Yes      (*b*)  No

**27.19.** (*a*)  0.5 A      (*b*)  $7.54 \times 10^{-4}$ T

**27.20.** (*a*)  200 A/m; $2.51 \times 10^{-4}$ T      (*b*)  200 A/m; 1.2 T

**27.21.** (*a*)  0.75 T      (*b*)  $9.4 \times 10^{-5}$ T

# Chapter 28

## Electromagnetic Induction

### ELECTROMAGNETIC INDUCTION

A current is produced in a conductor whenever it cuts across magnetic field lines, a phenomenon known as *electromagnetic induction*. If the motion is parallel to the field lines of force, there is no effect. Electromagnetic induction originates in the force a magnetic field exerts on a moving charge: When a wire moves across a magnetic field, the electrons it contains experience sideways forces which push them along the wire to cause a current. It is not even necessary for there to be relative motion of a wire and a source of magnetic field, since a magnetic field whose strength is changing has moving field lines associated with it and a current will be induced in a conductor that is in the path of these moving field lines.

When a straight conductor of length $l$ is moving across a magnetic field $\mathbf{B}$ with the velocity $\mathbf{v}$, the emf induced in the conductor is given by

$$\text{Induced emf} = V_e = Blv$$

when $\mathbf{B}$, $\mathbf{v}$, and the conductor are all perpendicular to one another.

### FARADAY'S LAW

Figure 28-1 shows a coil (called a *solenoid*) of $N$ turns that encloses an area $A$. The axis of the coil is parallel to a magnetic field $\mathbf{B}$. According to *Faraday's law of electromagnetic induction*, the emf induced in the coil when the product $BA$ changes by $\Delta(BA)$ in the time $\Delta t$ is given by

$$\text{Induced emf} = V_e = -N\,\frac{\Delta(BA)}{\Delta t}$$

The quantity $BA$ is called the *magnetic flux* enclosed by the coil and is denoted by the symbol $\Phi$ (Greek capital letter *phi*):

$$\Phi = BA$$

Magnetic flux = (magnetic field)(cross-sectional area)

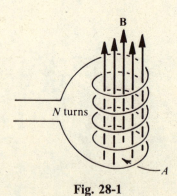

**Fig. 28-1**

336

The unit of magnetic flux is the *weber* (Wb), where 1 Wb = 1 T·m². Thus Faraday's law can be written

$$V_e = -N \frac{\Delta \Phi}{\Delta t}$$

## LENZ'S LAW

The minus sign in Faraday's law is a consequence of *Lenz's law*: An induced current is always in such a direction that its own magnetic field acts to oppose the effect that brought it about. For example, if **B** is decreasing in magnitude in the situation of Fig. 28-1, the induced current in the coil will be counterclockwise in order that its own magnetic field will add to **B** and so reduce the rate at which **B** is decreasing. Similarly, if **B** is increasing, the induced current in the coil will be clockwise so that its own magnetic field will subtract from **B** and thus reduce the rate at which **B** is increasing.

## SOLVED PROBLEM 28.1

The vertical component of the earth's magnetic field in a certain region is $3 \times 10^{-5}$ T. What is the potential difference between the rear wheels of a car, which are 1.5 m apart, when the car's velocity is 20 m/s?

The rear axle of the car may be considered as a rod 1.5 m long moving perpendicular to the magnetic field's vertical component. The potential difference between the wheels is therefore

$$V_e = Blv = (3 \times 10^{-5} \text{ T})(1.5 \text{ m})(20 \text{ m/s}) = 9 \times 10^{-4} \text{ V} = 0.9 \text{ mV}$$

## SOLVED PROBLEM 28.2

A square wire loop 8 cm on a side is perpendicular to a magnetic field of $5 \times 10^{-3}$ T. (*a*) What is the magnetic flux through the loop? (*b*) If the field drops to 0 in 0.1 s, what average emf is induced in the loop during this time?

(*a*)   The area of the loop is

$$A = \frac{(8 \text{ cm})(8 \text{ cm})}{(100 \text{ cm/m})^2} = 0.0064 \text{ m}^2 = 6.4 \times 10^{-3} \text{ m}^2$$

The flux through the loop is

$$\Phi = BA = (5 \times 10^{-3} \text{ T})(6.4 \times 10^{-3} \text{ m}^2) = 3.2 \times 10^{-5} \text{ Wb}$$

(*b*)   Since $N = 1$ for a single turn, we have from Faraday's law (disregarding the minus sign)

$$V_e = N \frac{\Delta \Phi}{\Delta t} = (1)\left(\frac{3.2 \times 10^{-5} \text{ Wb}}{0.1 \text{ s}}\right) = 3.2 \times 10^{-4} \text{ V}$$

## SOLVED PROBLEM 28.3

A 100-turn coil whose resistance is 6 Ω encloses an area of 80 cm². How rapidly should a magnetic field parallel to its axis change to induce a current of 1 mA in the coil?

The required emf here is

$$V_e = IR = (10^{-3} \text{ A})(6 \text{ } \Omega) = 6 \times 10^{-3} \text{ V}$$

and the coil's area is

$$A = \frac{80 \text{ cm}^2}{(100 \text{ cm/m})^2} = 8 \times 10^{-3} \text{ m}^2$$

Since $A$ is constant here,

$$V_e = N\ \frac{\Delta(BA)}{\Delta t} = NA\ \frac{\Delta B}{\Delta t}$$

and $\qquad \dfrac{\Delta B}{\Delta t} = \dfrac{V_e}{NA} = \dfrac{6 \times 10^{-3}\ \text{V}}{(100)(8 \times 10^{-3}\ \text{m}^2)} = 0.0075\ \text{T/s}$

## THE TRANSFORMER

A *transformer* consists of two coils of wire, usually wound on an iron core. Figure 28-2 shows an idealized transformer. When an alternating current is passed through one of the windings, the changing magnetic field it gives rise to induces an alternating current in the other winding. The potential difference per turn is the same in both primary and secondary windings, so the ratio of turns in the winding determines the ratio of voltages across them:

$$\frac{V_1}{V_2} = \frac{N_1}{N_2}$$

$$\frac{\text{Primary voltage}}{\text{Secondary voltage}} = \frac{\text{primary turns}}{\text{secondary turns}}$$

Since the power $I_1 V_1$ going into a transformer must equal the power $I_2 V_2$ going out, where $I_1$ and $I_2$ are the primary and secondary currents, respectively, the ratio of currents is inversely proportional to the ratio of turns:

$$\frac{I_1}{I_2} = \frac{N_2}{N_1}$$

$$\frac{\text{Primary voltage}}{\text{Secondary voltage}} = \frac{\text{secondary turns}}{\text{primary turns}}$$

## SOLVED PROBLEM 28.4

Alternating current is in wide use chiefly because its voltage can be so easily changed by a transformer. Since $P = IV$, we see that for given power $P$ the higher the voltage, the lower the current, and vice versa. In transmitting electric energy through long distances, a small current is

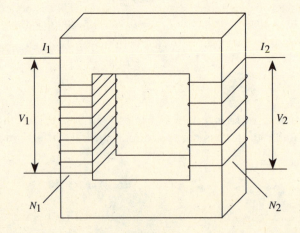

**Fig. 28-2**

desirable in order to minimize energy loss to heat, which is equal to $I^2R$ where $R$ is the resistance of the transmission line. However, both the generation and the final use of electric energy are best accomplished at moderate potential differences. Hence electricity is typically generated at 10,000 V or so, stepped up by transformers at the power station to 500,000 V or even more for transmission, and near the point of consumption other transformers reduce the potential difference to 240 or 120 V. To verify the advantage of high-voltage transmission, find the rate of energy loss to heat when a 5-$\Omega$ cable is used to transmit 1 kW of electricity at 100 V and at 100,000 V.

Since $P = IV$, the currents in the cable are, respectively,

$$I_A = \frac{P}{V_A} = \frac{1000\ \text{W}}{100\ \text{V}} = 10\ \text{A} \qquad I_B = \frac{P}{V_B} = \frac{1000\ \text{W}}{100,000\ \text{V}} = 0.01\ \text{A}$$

The rates of heat production per kilowatt are, respectively,

$$I_A^2 R = (10\ \text{A})^2 (5\ \Omega) = 500\ \text{W}$$

$$I_B^2 R = (0.01\ \text{A})^2 (5\ \Omega) = 0.0005\ \text{W} = 5 \times 10^{-4}\ \text{W}$$

Transmission at 100 V therefore means $10^6$—1 million—times more energy lost as heat than does transmission at 100,000 V.

## SOLVED PROBLEM 28.5

A transformer has 100 turns in its primary winding and 500 turns in its secondary winding. If the primary voltage and current are, respectively, 120 V and 3 A, what are the secondary voltage and current?

$$V_2 = \frac{N_2}{N_1}\ V_1 = \left(\frac{500\ \text{turns}}{100\ \text{turns}}\right)(120\ \text{V}) = 600\ \text{V}$$

$$I_2 = \frac{N_1}{N_2}\ I_1 = \left(\frac{100\ \text{turns}}{500\ \text{turns}}\right)(3\text{A}) = 0.6\ \text{A}$$

## SOLVED PROBLEM 28.6

A transformer rated at a maximum power of 10 kW is used to connect a 5000-V transmission line to a 240 V circuit. (a) What is the ratio of turns in the windings of the transformer? (b) What is the maximum current in the 240-V circuit?

(a)
$$\frac{N_1}{N_2} = \frac{V_1}{V_2} = \frac{5000\ \text{V}}{240\ \text{V}} = 20.8$$

(b)   Since $P = IV$, here

$$I_2 = \frac{P}{V_2} = \frac{10,000\ \text{W}}{240\ \text{V}} = 41.7\ \text{A}$$

## SOLVED PROBLEM 28.7

A transformer connected to a 120-V alternating-current (ac) power line has 200 turns in its primary winding and 50 turns in its secondary winding. The secondary is connected to a 100-$\Omega$ light bulb. How much current is drawn from the 120-V power line?

The voltage across the secondary is

$$V_2 = \frac{N_2}{N_1} \, V_1 = \left(\frac{50 \text{ turns}}{200 \text{ turns}}\right)(120 \text{ V}) = 30 \text{ V}$$

and so the current in the secondary circuit is

$$I_2 = \frac{V_2}{R} = \frac{30 \text{ V}}{100 \text{ } \Omega} = 0.3 \text{ A}$$

Hence the current in the primary circuit is

$$I_1 = \frac{N_2}{N_1} \, I_2 = \left(\frac{50 \text{ turns}}{200 \text{ turns}}\right)(0.3 \text{ A}) = 0.075 \text{ A}$$

## SELF-INDUCTANCE

When the current in a circuit changes, the magnetic field enclosed by the circuit also changes, and the resulting change in flux leads to a *self-induced emf* of

$$\text{Self-induced emf} = V_e = -L \, \frac{\Delta I}{\Delta t}$$

Here $\Delta I / \Delta t$ is the rate of change of the current, and $L$ is a property of the circuit called its *self-inductance*, or, more commonly, its *inductance*. The minus sign indicates that the direction of $V_e$ is such as to oppose the change in current $\Delta I$ that caused it.

The unit of inductance is the *henry* (H). A circuit or circuit element that has an inductance of 1 H will have a self-induced emf of 1 V when the current through it changes at the rate of 1 A/s. Because the henry is a rather large unit, the *millihenry* and *microhenry* are often used, where

$$1 \text{ millihenry} = 1 \text{ mH} = 10^{-3} \text{ H}$$

$$1 \text{ microhenry} = 1 \text{ } \mu\text{H} = 10^{-6} \text{ H}$$

A circuit element with inductance is called an *inductor*. A solenoid is an example of an inductor. The inductance of a solenoid is

$$L = \frac{\mu N^2 A}{l}$$

where $\mu$ is the permeability of the core material, $N$ is the number of turns, $A$ is the cross-sectional area, and $l$ is the length of the solenoid.

## INDUCTORS IN COMBINATION

When two or more inductors are sufficiently far apart for them not to interact electromagnetically, their equivalent inductances when they are connected in series and in parallel are as follows:

$$L = L_1 + L_2 + L_3 + \cdots \qquad \text{inductors in series}$$

$$\frac{1}{L} = \frac{1}{L_1} + \frac{1}{L_2} + \frac{1}{L_3} + \cdots \qquad \text{inductors in parallel}$$

Connecting coils in parallel reduces the total inductance to less than that of any of the individual coils. This is rarely done because coils are relatively large and expensive compared with other electronic components; a coil of the required smaller inductance would normally be used in the first place.

Because the magnetic field of a current-carrying coil extends beyond the inductor itself, the total inductance of two or more connected coils will be changed if they are close to one another. Depending on how the coils are arranged, the total inductance may be larger or smaller than if the coils were farther apart. This effect is called *mutual inductance* and is not considered in the above formula.

## ENERGY OF A CURRENT-CARRYING INDUCTOR

Because a self-induced emf opposes any change in an inductor, work has to be done against this emf to establish a current in the inductor. This work is stored as magnetic potential energy. If $L$ is the inductance of an inductor, its potential energy when it carries the current $I$ is

$$W = \tfrac{1}{2}LI^2$$

This energy powers the self-induced emf that opposes any decrease in the current through the inductor.

## SOLVED PROBLEM 28.8

Find the inductance in air of a 500-turn solenoid 10 cm long whose cross-sectional area is 20 cm$^2$.

$$L = \frac{\mu_0 N^2 A}{l} = \frac{(1.26 \times 10^{-6} \text{ H/m})(500)^2(2 \times 10^{-3} \text{ m}^2)}{10^{-1} \text{ m}} = 6.3 \times 10^{-3} \text{ H} = 6.3 \text{ mH}$$

## SOLVED PROBLEM 28.9

A solenoid 20 cm long and 2 cm in diameter has an inductance of 0.178 mH. How many turns of wire does it have?

The radius of the solenoid is $r = 1$ cm $= 0.01$ m, and so its cross-sectional area is

$$A = \pi r^2 = (\pi)(0.01 \text{ m})^2 = 3.14 \times 10^{-4} \text{ m}^2$$

Since $L = \mu_0 N^2 A/l$ in air,

$$N = \sqrt{\frac{Ll}{\mu_0 A}} = \sqrt{\frac{(0.178 \times 10^{-3} \text{ H})(0.2 \text{ m})}{(4\pi \times 10^{-7} \text{ T·m/A})(3.14 \times 10^{-4} \text{ m}^2)}} = 300 \text{ turns}$$

## SOLVED PROBLEM 28.10

An inductor consists of an iron ring 5 cm in diameter and 1 cm$^2$ in cross-sectional area that is wound with 1000 turns of wire. If the permeability of the iron is constant at 400 times that of free space at the magnetic intensities at which the inductor will be used, find its inductance.

An inductor of this kind is essentially a solenoid bent into a circle. The length of the equivalent solenoid is therefore

$$l = \pi d = (\pi)(0.05 \text{ m}) = 0.157 \text{ m}$$

and its cross-sectional area is $A = 1 \text{ cm}^2/(100 \text{ cm/m})^2 = 10^{-4} \text{ m}^2$. The permeability of the core is $\mu = 400 \, \mu_0$. The inductance of this inductor is therefore

$$L = \frac{\mu N^2 A}{l} = \frac{(400)(4\pi \times 10^{-7} \text{ T·m/A})(10^3)^2(10^{-4} \text{ m}^2)}{0.157 \text{ m}} = 0.32 \text{ H}$$

**SOLVED PROBLEM 28.11**

Find the equivalent inductances of a 5- and an 8-mH inductor when they are connected (*a*) in series and (*b*) in parallel.

(*a*) $$L = L_1 + L_2 + L_3 = 5 \text{ mH} + 8 \text{ mH} = 13 \text{ mH}$$

(*b*) $$\frac{1}{L} = \frac{1}{L_1} + \frac{1}{L_2} = \frac{1}{5 \text{ mH}} + \frac{1}{8 \text{ mH}} \qquad L = 3.08 \text{ mH}$$

**SOLVED PROBLEM 28.12**

The current in a circuit falls from 5 to 1 A in 0.2 s. If an average emf of 2 V is induced in the circuit while this is happening, find the inductance of the circuit.

Since $V_e = -L \, \Delta I/\Delta t$, we have (disregarding the minus sign)

$$L = \frac{V_e \, \Delta t}{\Delta I} = \frac{(2 \text{ V})(0.1 \text{ s})}{5 \text{ A} - 1 \text{ A}} = 0.05 \text{ H}$$

**SOLVED PROBLEM 28.13**

(*a*) How much magnetic potential energy is stored in a 20-mH coil when it carries a current of 0.2 A? (*b*) What should the current in the coil be in order that it contains 1 J of energy?

(*a*) $$W = \tfrac{1}{2}LI^2 = (\tfrac{1}{2})(20 \times 10^{-3} \text{ H})(0.2 \text{ A})^2 = 4 \times 10^{-4} \text{ J}$$

(*b*) $$I = \sqrt{\frac{2W}{L}} = \sqrt{\frac{(2)(1 \text{ J})}{20 \times 10^{-3} \text{ H}}} = 10 \text{ A}$$

**TIME CONSTANT**

Because the self-induced emf in a circuit such as that of Fig. 28-3 is always such as to oppose any changes in the current in the circuit, the current does not rise instantly to its final value of $I = V/R$ when the switch is closed. When the switch in Fig. 28-3 is closed, the current starts to build up; as a result, a self-induced emf $- L(\Delta I/\Delta t)$ occurs that opposes the battery voltage $V$. The net voltage in the circuit is therefore $V - L(\Delta I/\Delta t)$, which must equal $IR$ by Ohm's law:

$$V - L \frac{\Delta I}{\Delta t} = IR$$

Impressed voltage − induced emf = net voltage

The rate of increase of the current at the moment when the current is $I$ is accordingly

$$\frac{\Delta I}{\Delta t} = \frac{V - IR}{L}$$

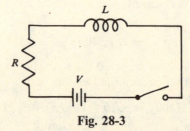

**Fig. 28-3**

The larger the inductance $L$, the more gradually the current increases. When the switch is first closed, $I = 0$ and

$$\frac{\Delta I}{\Delta t} = \frac{V}{L}$$

Eventually the current stops rising and $\Delta I / \Delta t = 0$. From then on

$$I = \frac{V}{R}$$

The final current depends only on $V$ and $R$; the effect of $L$ is to delay the establishment of the final current.

As shown in Fig. 28-4, the current in the circuit of Fig. 28-3 rises gradually in such a manner that after a time $t$ equal to $L/R$ it reaches 63 percent of its final value. The quantity $L/R$ is called the *time constant T* of the circuit; the smaller the time constant, the more rapidly the current changes.

The formula that governs the growth of a current in the circuit of Fig. 28-3 is

$$I = I_0(1 - e^{-t/T})$$

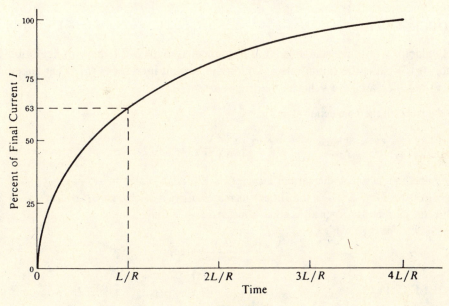

**Fig. 28-4**

In this formula, plotted in Fig. 28-4, $I_0$ is the steady-state current $V/R$, $T$ is the time constant $L/R$, and $I$ is the current at the time $t$ after the switch is closed. (Exponentials such as $e^{-t/T}$ were discussed in Chapter 26.)

If the battery is short-circuited, the current decreases in such a manner that after $t = L/R$ it has fallen to 37 percent of its original value (Fig. 28-5). In this case

$$I = I_0 e^{-t/T}$$

During the establishment of the current of Fig. 28-4, magnetic energy $\frac{1}{2}LI^2$ is being absorbed by the inductance $L$ of the circuit. When the battery is short-circuited so that no voltage is impressed on the circuit, the stored energy is what powers the subsequent decreasing current of Fig. 28-5.

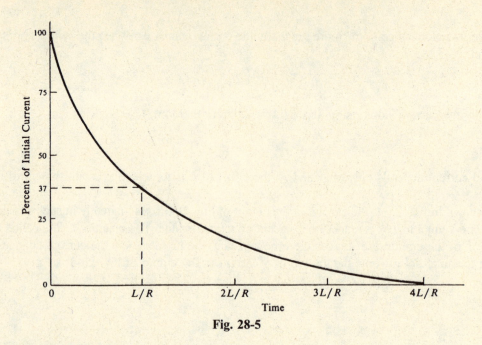

**Fig. 28-5**

## SOLVED PROBLEM 28.14

A 0.1-H inductor whose resistance is 20 Ω is connected to a 12-V battery of negligible internal resistance. (*a*) What is the initial rate at which the current increases? (*b*) What happens to the rate of current increase? (*c*) What is the final current?

(*a*) At the moment the connection is made,

$$\frac{\Delta I}{\Delta t} = \frac{V_{emf}}{L} = \frac{12 \text{ V}}{0.1 \text{ H}} = 120 \text{ A/s}$$

The initial rate at which the current increases is 120 A/s.

(*b*) Since $\Delta I/\Delta t = (V - IR)/L$ as the current increases, its rate of change $\Delta I/\Delta t$ decreases.

(*c*) When the current has reached its final value, $\Delta I/\Delta t = 0$ and

$$I = \frac{V}{R} = \frac{12 \text{ V}}{20 \text{ }\Omega} = 0.6 \text{ A}$$

## SOLVED PROBLEM 28.15

What time is required for the current in the inductor of Prob. 28.14 to reach 63 percent of its final value?

The time constant of the circuit is

$$T = \frac{L}{R} = \frac{0.1 \text{ H}}{20 \text{ }\Omega} = 0.005 \text{ s}$$

The current will reach 63 percent of its final value in this time.

## SOLVED PROBLEM 28.16

A 0.2-H inductor with a resistance of 3 Ω is connected to a 6-V battery whose internal resistance is 1 Ω. (*a*) Find the final current in the circuit. (*b*) Find the current in the circuit 0.01, 0.05, and 0.1 s after the connection is made.

(a)  The total resistance in the circuit is the sum of the 3-$\Omega$ resistance of the inductor and the 1-$\Omega$ internal resistance of the battery, so

$$R = 3\ \Omega + 1\ \Omega = 4\ \Omega$$

The final current in the circuit is therefore

$$I_0 = \frac{V}{R} = \frac{6\ \text{V}}{4\ \Omega} = 1.5\ \text{A}$$

(b)  The time constant of the circuit is

$$T = \frac{L}{R} = \frac{0.2\ \text{H}}{4\ \Omega} = 0.05\ \text{s}$$

At time $t$ after the connection is made, the current is given by

$$I = I_0(1 - e^{-t/T})$$

When $t = 0.01$ s, $t/T = (0.01\ \text{s})/(0.05\ \text{s}) = 0.2$. With the help of a calculator or a table of exponentials, we find that

$$e^{-t/T} = e^{-0.2} = 0.82$$

and so $\qquad I = I_0(1 - e^{-t/T}) = (1.5\ \text{A})(1 - 0.82) = (1.5\ \text{A})(0.18) = 0.27\ \text{A}$

When $t = 0.05$ s, $t = T$ and

$$I = 0.63 I_0 = (0.63)(1.5\ \text{A}) = 0.95\ \text{A}$$

When $t = 0.1$ s, $t/T = (0.1\ \text{s})/(0.05\ \text{s}) = 2$, and

$$e^{-t/T} = e^{-2} = 0.14$$

and so $\qquad I = (I_0)(1 - e^{-t/T}) = (1.5\ \text{A})(1 - 0.14) = (1.5\ \text{A})(0.86) = 1.29\ \text{A}$

## SOLVED PROBLEM 28.17

Find the current in the above inductor 0.01 and 0.1 s after it has been short-circuited, after having been connected to the battery for a long time.

The resistance of the circuit is now just the 3-$\Omega$ resistance of the inductor itself. Hence the time constant is

$$T = \frac{L}{R} = \frac{0.2\ \text{H}}{3\ \Omega} = 0.067\ \text{s}$$

When $t = 0.01$ s, $t/T = (0.01\ \text{s})/(0.067\ \text{s}) = 0.15$, and

$$e^{-t/T} = e^{-0.15} = 0.86$$

Hence the current in the inductor is

$$I = I_0 e^{-t/T} = (1.5\ \text{A})(0.86) = 1.29\ \text{A}$$

When $t = 0.1$ s, $t/T = (0.1\ \text{s})(0.067\ \text{s}) = 1.5$, and

$$e^{-t/T} = e^{-1.5} = 0.22$$

Hence the current in the inductor at this time is

$$I = I_0 e^{-t/T} = (1.5\ \text{A})(0.22) = 0.33\ \text{A}$$

# Multiple-Choice Questions

**28.1.** Upon which one or more of the following does the magnetic flux through a wire loop in a magnetic field depend?

(*a*)   the area of the loop
(*b*)   the shape of the loop
(*c*)   the magnitude of the field
(*d*)   the angle betweeen the plane of the loop and the direction of the field

**28.2.** The cause of the alternating current in the secondary coil of a transformer is an emf produced by

(*a*)   the varying electric field of the primary coil
(*b*)   the varying magnetic field of the primary coil
(*c*)   the varying magnetic field of the secondary coil
(*d*)   the voltage applied to the primary coil

**28.3.** Upon which one or more of the following does the magnetic energy of a coil that carries a current $I$ depend?

(*a*)   $I$
(*b*)   the number of turns in the coil
(*c*)   the resistance of the coil
(*d*)   the presence of an iron core in the coil

**28.4.** The horizontal steel cargo boom of a freighter traveling at 10 m/s is 7.0 m long and is at an angle of $75°$ relative to the direction of the ship's motion. The magnetic field of the earth in that region has a vertical component of $4.0 \times 10^{-5}$ T. The potential difference between the ends of the boom is

(*a*)   0.06 mV          (*c*)   2.7 mV
(*b*)   0.72 mV          (*d*)   2.8 mV

**28.5.** A wire loop that encloses an area of 15 cm$^2$ is perpendicular to a magnetic field of 0.10 T. If the field drops to 0.04 T in 0.2 s, the average emf induced in the loop is

(*a*)   0.3 mV           (*c*)   4 mV
(*b*)   0.45 mV          (*d*)   4.5 V

**28.6.** A 200-turn coil whose resistance is 4 $\Omega$ encloses an area of 20 cm$^2$. A changing magnetic field parallel to the coil axis induces a current of 1.2 A in the coil. How rapidly is the magnetic field changing?

(*a*)   0.75 T/s         (*c*)   14.4 T/s
(*b*)   12 T/s           (*d*)   30 T/s

**28.7.** A transformer has 300 turns in its primary coil and 75 turns in its secondary coil. When the current in the secondary coil is 20 A, the current in the primary coil is

(*a*)   5 A              (*c*)   80 A
(*b*)   25 A             (*d*)   6.4 kA

**28.8.** If the power input to the transformer of Question 28.7 is 40 W, the power output is

(*a*)   2.5 W            (*c*)   40 W
(*b*)   10 W             (*d*)   160 W

**28.9**  While the current in a circuit is falling from 8 to 2 A in 20 ms, the average induced emf in the circuit is 12 V. The inductance of the circuit is

(a)  0.04 H          (c)  100 H
(b)  0.12 H          (d)  3.6 kH

**28.10.**  An 80-mH coil whose resistance is 4 Ω is connected to a 12-V battery of negligible internal resistance. The current in the coil begins to increase at a rate of

(a)  0.15 A/s          (c)  7.5 A/s
(b)  3 A/s             (d)  150 A/s

**28.11.**  The time needed for the current in the coil of Question 28.10 to reach 63 percent of its final value is

(a)  0.02 s          (c)  20 s
(b)  0.05 s          (d)  50 s

**28.12.**  The energy stored in the magnetic field of a 12-mH coil in which the current is 5 A is

(a)  1.8 mJ          (c)  0.15 J
(b)  30 mJ           (d)  0.3 J

# Supplementary Problems

**28.1.**  What would happen if the primary winding of a transformer were connected to a battery?

**28.2.**  What is it whose action on the secondary winding of a transformer causes an alternating potential difference to occur across its ends, even though there is no connection between the primary and secondary windings?

**28.3.**  Verify that $L/R$ has the dimensions of time.

**28.4.**  How fast should a wire 20 cm long be moved through a 0.05-T magnetic field for an emf of 1 V to appear across its ends?

**28.5.**  The magnetic flux through a 50-turn coil increases at the rate of 0.05 Wb/s. (a) What is the induced emf between the ends of the coil? (b) If the coil's resistance is 2 Ω and it is connected to an external circuit whose resistance is 10 Ω, how much current will flow?

**28.6.**  A 30-turn coil 8-cm in diameter is in a magnetic field of 0.1 T that is parallel to its axis. (a) What is the magnetic flux through the coil? (b) In how much time should the field drop to zero to induce an average emf of 0.7 V in the coil?

**28.7.**  A transformer has 50 turns in its primary winding and 100 turns in its secondary. (a) If a 60-Hz, 3-A current passes through the primary winding, what are the nature and magnitude of the current in the secondary? (b) If a 3-A direct current passes through the primary winding, what are the nature and magnitude of the current in the secondary?

**28.8.** The transformer in an electric welding machine draws 3 A from a 240-V ac power line and delivers 400 A. What is the potential difference across the secondary of the transformer?

**28.9.** A 240-V, 400-W electric mixer is connected to a 120-V power line through a transformer. (*a*) What is the ratio of turns in the transformer? (*b*) How much current is drawn from the power line?

**28.10.** A solenoid 2 cm long and 6 mm in diameter has 500 turns of thin wire. What is its inductance?

**28.11.** A 1-mH inductor is to be made by winding wire on a tube 2 cm in diameter and 10 cm long. How many turns are needed?

**28.12.** Find the equivalent inductances of three 2-mH inductors when they are connected in series and in parallel.

**28.13.** Find the emf induced in a 0.1-H coil when the current in it is changing at the rate of 80 A/s.

**28.14.** What rate of change of current is needed to induce an emf of 8 V in a 0.1-H coil?

**28.15.** An average emf of 32 V is induced in a circuit in which the current drops from 10 to 2 A in 0.1 s. What is the inductance of the circuit?

**28.16.** A 5-mH coil carries a current of 2 A. How much energy is stored in it?

**28.17.** How much current must flow through a 40-mH coil in order that it contain 0.1 J of magnetic potential energy?

**28.18.** A potential difference of 100 V is applied to a 50-mH, 40-Ω inductor. (*a*) What is the initial rate at which the current increases? (*b*) What is the rate at which the current is increasing when $I = 1$ A? (*c*) What is the final current?

**28.19.** A potential difference of 50 V is applied across a 12-mH, 8-Ω inductor. (*a*) What is the initial rate at which the current increases? (*b*) What is the current when the rate of change of current is 2000 A/s? (*c*) What is the final current?

**28.20.** What is the time constant of a 50-mH, 3-Ω inductor?

**28.21.** A 60-mH, 5-Ω inductor is connected to a 12-V battery whose internal resistance is 1 Ω. (*a*) Find the time constant of the circuit. (*b*) Find the final current in the circuit. (*c*) Find the current 0.005, 0.01, and 0.05 s after the connection is made.

**28.22.** Find the current in the above inductor 0.005, 0.01, and 0.05 s after it has been short-circuited, after having been connected to the battery for a long time.

**28.23.** A 0.1-H, 4-Ω inductor is connected in series with a 0.2-H, 6-Ω inductor, and the combination is placed across a 24-V battery whose internal resistance is 2 Ω. (*a*) Find the time constant of the circuit. (*b*) Find the final current in the circuit. (*c*) Find the current at 0.01 and 0.1 s after the connection is made.

**28.24.** Find the current in the inductors of Prob. 28.23 at 0.01 and 0.1 s after the battery has been short-circuited, after the inductors have been connected to it for a long time.

## *Answers to Multiple-Choice Questions*

**28.1.**  (*a*), (*c*), (*d*)      **28.7.**    (*a*)

**28.2.**  (*b*)                       **28.8.**    (*c*)

**28.3.**  (*a*), (*b*), (*d*)      **28.9.**    (*a*)

**28.4.**  (*c*)                       **28.10.**   (*d*)

**28.5.**  (*b*)                       **28.11.**   (*a*)

**28.6.**  (*b*)                       **28.12.**   (*c*)

## Answers to Supplementary Problems

**28.1.**  When the connection is made, there will be a momentary current in the secondary winding as the current in the primary builds to its final value. Afterward, since the primary current will be constant and hence its magnetic field will not change, there will be no current in the secondary.

**28.2.**  The changing magnetic field produced by an alternating current in the primary winding

**28.3.**  From their definitions $L = Vt/I$ and $R = V/I$. Hence $L/R = (Vt/I)/(V/I) = t$.

**28.4.**  100 m/s                                      **28.16.**  0.01 J

**28.5.**  (*a*)  2.5 V      (*b*)  0.208 A              **28.17.**  2.24 A

**28.6.**  (*a*)  $5.03 \times 10^{-4}$ Wb      (*b*)  0.0215 s      **28.18.**  (*a*)  2000 A/s      (*b*)  1200 A/s

**28.7.**  (*a*)  60 Hz, 1.5 A      (*b*)  No current              (*c*)  2.5 A

**28.8.**  1.8 V                                        **28.19.**  (*a*)  4167 A/s      (*b*)  3.25 A

**28.9.**  (*a*)  2:1      (*b*)  3.3 A                          (*c*)  6.25 A

**28.10.**  $3.53 \times 10^{-8}$ H                     **28.20.**  0.0167 s

**28.11.**  503 turns                                   **28.21.**  (*a*)  0.01 s      (*b*)  2 A

**28.12.**  6 mH; 0.67 mH                                       (*c*)  0.787 A; 1.264 A; 1.987 A

**28.13.**  8 V                                         **28.22.**  1.318 A; 0.869 A; 0.031 A

**28.14.**  80 A/s                                      **28.23.**  (*a*) 0.025 s  (*b*) 2 A  (*c*) 0.659 A; 1.963 A

**28.15.**  0.4 H                                       **28.24.**  1.433 A; 0.071 A

# Chapter 29

## Alternating-Current Circuits

### ALTERNATING CURRENT

The *frequency* of an alternating current is the number of complete back-and-forth cycles it goes through each second (Fig. 29-1). As in the case of harmonic motion, the unit of frequency is the *hertz* (Hz), where 1 Hz = 1 cycle/s.

An alternating emf of frequency $f$ whose maximum value is $V_{e,\max}$ varies with time according to the formula

$$V_e = V_{e,\max} \sin 2\pi ft = V_{e,\max} \sin \omega t$$

The quantity $\omega = 2\pi f$ is the *angular frequency* of the emf in radians per second. Similarly, an alternating current (ac) of frequency $f$ whose maximum value is $I_{\max}$ varies with time according to the formula

$$I = I_{\max} \sin 2\pi ft = I_{\max} \sin \omega t$$

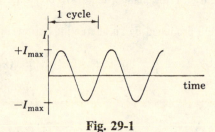

**Fig. 29-1**

### EFFECTIVE VALUES

Because an alternating current changes continuously, its maximum value $\pm I_{\max}$ does not indicate its ability to do work or to produce heat, as does the magnitude of a direct current. Instead it is customary to refer to the *effective current*

$$I_{\text{eff}} = \frac{I_{\max}}{\sqrt{2}} = 0.707 I_{\max}$$

A direct current of this value does as much work or produces as much heat as the alternating current whose maximum value is $\pm I_{\max}$. Similarly, the *effective emf* in an ac circuit is

$$V_{e,\text{eff}} = \frac{V_{e,\max}}{\sqrt{2}} = 0.707 V_{e,\max}$$

Currents and voltages in ac circuits are usually expressed in terms of their effective values. For instance, the potential difference across a 120-V ac power line actually varies between +170 and −170 V since

$$\pm V_{\max} = \frac{\pm V_{\text{eff}}}{0.707} = \frac{\pm 120 \text{ V}}{0.707} = \pm 170 \text{ V}$$

## SOLVED PROBLEM 29.1

The dielectric used in a certain capacitor breaks down at a potential difference of 300 V. Find the maximum effective ac potential difference that can be applied to it.

$$V_{eff} = 0.707 V_{max} = (0.707)(300 \text{ V}) = 212 \text{ V}$$

## SOLVED PROBLEM 29.2

Alternating current with a maximum value of 10 A is passed through a 20-$\Omega$ resistor. At what rate does the resistor dissipate energy?

The effective current is

$$I_{eff} = 0.707 I_{max} = (0.707)(10 \text{ A}) = 7.07 \text{ A}$$

and so the power dissipated is

$$P = I_{eff}^2 R = (7.07 \text{ A})^2 (20 \text{ }\Omega) = 1000 \text{ W}$$

## REACTANCE

The *inductive reactance* of an inductor is a measure of its effectiveness in resisting the flow of an alternating current by virtue of the self-induced back emf that the changing current causes in it. Unlike the case of a resistor, there is no power dissipated in a pure inductor. The inductive reactance $X_L$ of an inductor whose inductance is $L$ (in henries) when the frequency of the current is $f$ (in hertz) is

$$\text{Inductive reactance} = X_L = 2\pi f L$$

When a potential difference $V$ of frequency $f$ is applied across an inductor whose reactance is $X_L$ at frequency $f$, current $I = V/X_L$ will flow. The unit of $X_L$ is the ohm.

The *capacitive reactance* of a capacitor is similarly a measure of its effectiveness in resisting the flow of an alternating current, in this case by virtue of the reverse potential difference across it owing to the accumulation of charge on its plates. No power loss is associated with a capacitor in an ac circuit. The capacitive reactance $X_C$ of a capacitor whose capacitance is $C$ (in farads) when the frequency of the current is $f$ (in hertz) is

$$\text{Capacitive reactance} = X_C = \frac{1}{2\pi f C}$$

When a potential difference of frequency $f$ is applied across a capacitor whose reactance is $X_C$ at frequency $f$, current $I = V/X_C$ will flow. The unit of $X_C$ is the ohm.

## SOLVED PROBLEM 29.3

(a) Sketch graphs that show how $X_L$ and $X_C$ vary with frequency. (b) What happens to $X_L$ and $X_C$ in the limit of $f = 0$? (c) What is the physical meaning of the answer to (b)?

(a)  See Fig. 29-2.

(b)  When $f = 0$, $X_L = 2\pi f L = 0$ and $X_C = 1/(2\pi f C) = \infty$.

(c)  A current with $f = 0$ is a direct current. When a constant current flows in an inductor, there is no self-induced back emf to hamper the current, and the inductive reactance is accordingly zero. A direct current cannot pass through a capacitor because its plates are insulated from each other, so the capacitive reactance is infinite and $I = V/X_C = 0$ when $f = 0$. (An alternating current does not actually pass *through* a capacitor but surges back and forth in the circuit on both sides of it.)

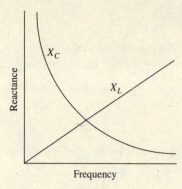

**Fig. 29-2**

## SOLVED PROBLEM 29.4

A 10-$\mu$F capacitor is connected to a 15-V, 5-kHz power source. Find (*a*) the reactance of the capacitor and (*b*) the current that flows.

(*a*)
$$X_C = \frac{1}{2\pi f C} = \frac{1}{(2\pi)(5 \times 10^3 \text{ Hz})(10 \times 10^{-6} \text{ F})} = 3.18 \ \Omega$$

(*b*)
$$I = \frac{V}{X_C} = \frac{15 \text{ V}}{3.18 \ \Omega} = 4.72 \text{ A}$$

## SOLVED PROBLEM 29.5

The reactance of an inductor is 80 $\Omega$ at 500 Hz. Find its inductance.

Since $X_L = 2\pi f L$,

$$L = \frac{X_L}{2\pi f} = \frac{80 \ \Omega}{(2\pi)(500 \text{ Hz})} = 0.0255 \text{ H} = 25.5 \text{ mH}$$

## PHASE ANGLE

A convenient way to represent a quantity that varies sinusoidally (that is, as sin $\theta$ varies with $\theta$) with time is in terms of a rotating vector called a *phasor*. In the case of an ac voltage, the length of the phasor $\mathbf{V}_{\max}$ corresponds to $V_{\max}$, and we imagine it to rotate $f$ times per second in a counterclockwise direction (Fig. 29-3). The vertical component of the phasor at any moment corresponds to the instantaneous voltage *V*. Since the vertical component of $V_{\max}$ is

$$V = V_{\max} \sin \theta = V_{\max} \sin 2\pi f t$$

the result is the same curve as that of Fig. 29-1. In a similar way a phasor $I_{\max}$ can be used to represent an alternating current *I*. Phasors are useful because the voltage and current in an ac circuit or circuit element always have the same frequency *f*, but the peaks in *V* and *I* may occur at different times.

In an ac circuit that contains only resistance, the instantaneous voltage and current are *in phase* with each other; that is, both are zero at the same time, both reach their maximum values in either direction at the same time, and so on, as shown in Fig. 29-4(*a*).

In an ac circuit that contains only inductance, the voltage leads the current by $\frac{1}{4}$ cycle. Since a complete cycle means a change in $2\pi f t$ of 360° and 360°/4 = 90°, it is customary to say that in a pure inductor the voltage leads the current by 90°. The situation is shown in Fig. 29-4(*b*).

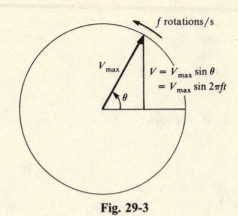

f rotations/s
$V_{max}$
$\theta$
$V = V_{max} \sin \theta$
$= V_{max} \sin 2\pi ft$


Fig. 29-3

R
V
I
t
I = current phasor
V = voltage phasor
I
V
(a)
L
V
I
t
V
I
(b)
C
V
I
t
I
V
(c)


Fig. 29-4

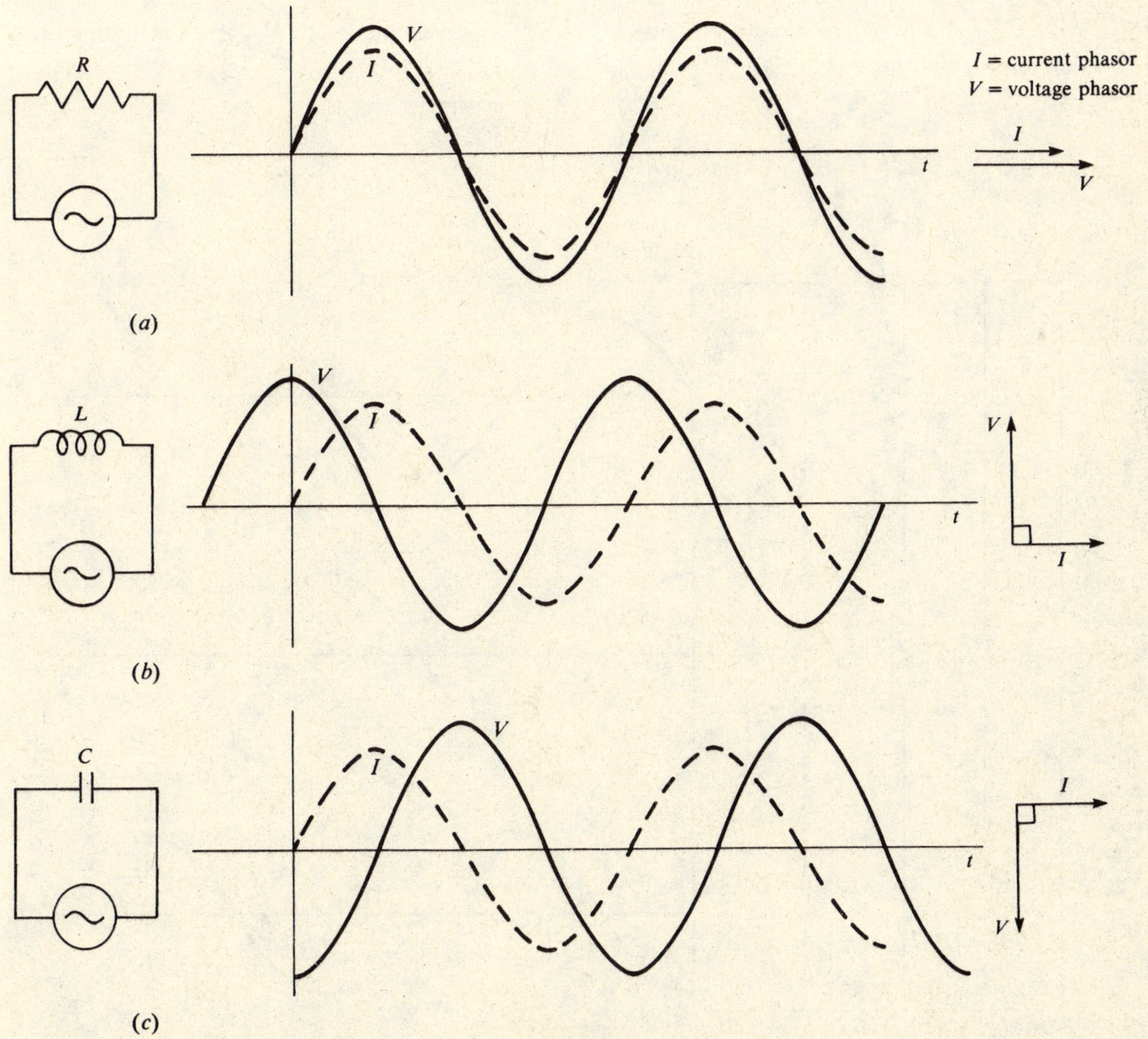

In an ac circuit that contains only capacitance, the voltage lags behind the current by $\frac{1}{4}$ cycle, which is 90°. This situation is shown in Fig. 29-4($c$).

Now we consider an ac circuit that contains resistance, inductance, and capacitance in series, as in Fig. 29-5. The instantaneous voltages across the circuit elements are

$$V_R = IR \qquad V_L = IX_L \qquad V_C = IX_C$$

At any moment the applied voltage $V$ is equal to the sum of the voltage drops $V_R$, $V_L$, and $V_C$:

$$V = V_R + V_L + V_C$$

Because $V_R$, $V_L$, and $V_C$ are out of phase with one another, this formula holds only for the instantaneous voltages, *not* for the effective voltages.

Since we want to work with effective voltages and currents, not instantaneous ones, we must somehow take into account the phase differences. To do this, we can use phasors to represent the various effective quantities. This is done in Fig. 29-6 for the voltages. To find the magnitude $V$ of the sum **V** of the various effective voltages, we proceed in this way:

1.  Find the difference $\mathbf{V}_L - \mathbf{V}_C$. IF $V_L > V_C$, then $\mathbf{V}_L - \mathbf{V}_C$ will be positive and will point upward; if $V_L < V_C$, then $\mathbf{V}_L - \mathbf{V}_C$ will be negative and will point downward.

2.  Add $\mathbf{V}_L - \mathbf{V}_C$ to $\mathbf{V}_R$ to obtain **V**. Since $\mathbf{V}_L - \mathbf{V}_C$ is perpendicular to $\mathbf{V}_R$, use the Pythagorean theorem to find the magnitude $V$:

$$V = \sqrt{V_R^2 + (V_L - V_C)^2}$$

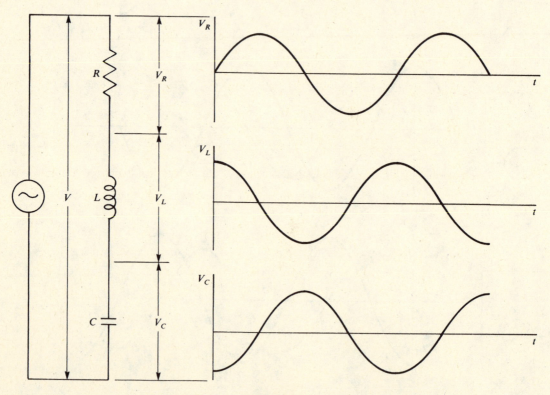

**Fig. 29-5**

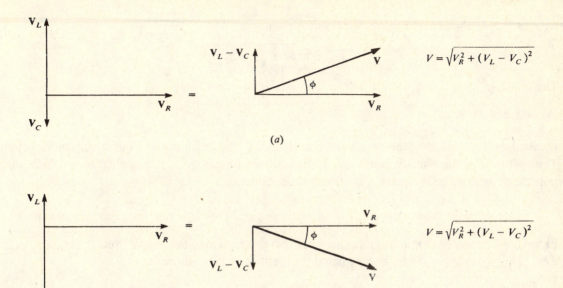

**Fig. 29-6**

The angle $\phi$ between **V** and $\mathbf{V}_R$ is the *phase angle* and can be calculated from the relationships

$$\tan \phi = \frac{V_L - V_C}{V_R} \qquad \text{or} \qquad \cos \phi = \frac{V_R}{V}$$

## SOLVED PROBLEM 29.6

A resistor, a capacitor, and an inductor are connected in series to an ac power source. The effective voltages across the circuit components are $V_R = 5$ V, $V_C = 10$ V, and $V_L = 7$ V. Find (a) the effective voltage of the source and (b) the phase angle in the circuit.

(a)
$$V = \sqrt{V_R^2 + (V_L - V_C)^2} = \sqrt{(5 \text{ V})^2 + (7 \text{ V} - 10 \text{ V})^2}$$
$$= \sqrt{(5 \text{ V})^2 + (-3 \text{ V})^2} = \sqrt{25 \text{ V}^2 + 9 \text{ V}^2} = \sqrt{34 \text{ V}^2} = 5.8 \text{ V}$$

We note that the effective voltages across $C$ and $R$ are greater than the effective applied voltage.

(b)
$$\tan \phi = \frac{V_L - V_C}{V_R} = \frac{7 \text{ V} - 10 \text{ V}}{5 \text{ V}} = -\frac{3}{5} = -0.6$$
$$\phi = -31°$$

The negative phase angle means that the voltage across the resistor is ahead of the applied voltage, as in Fig. 29-6(b). Equivalently we can say that the current in the circuit leads the voltage, as in Fig. 29-4(c). (We recall that phasors rotate counterclockwise.)

## IMPEDANCE

Because

$$V_R = IR \qquad V_L = IX_L \qquad V_C = IX_C$$

we can rewrite the above formula for $V$ in the form

$$V = I\sqrt{R^2 + (X_L - X_C)^2}$$

The quantity

$$Z = \sqrt{R^2 + (X_L - X_C)^2}$$

is called the *impedance* of the circuit and corresponds to the resistance of a direct-current (dc) circuit. The unit of $Z$ is the ohm. When an ac voltage whose frequency is $f$ is applied to a circuit whose impedance is $Z$ at that frequency, the result is the current

$$I = \frac{V}{Z}$$

Figure 29-7 shows phasor impedance diagrams that correspond to the phasor voltage diagrams of Fig. 29-6. The phase angle $\phi$ can be calculated from either of these formulas:

$$\tan\phi = \frac{X_L - X_C}{R} \qquad \text{or} \qquad \cos\phi = \frac{R}{Z}$$

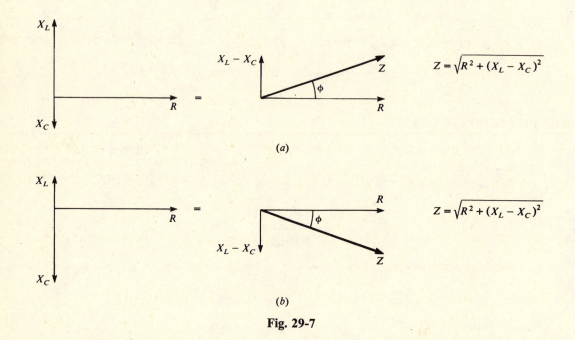

**Fig. 29-7**

## SOLVED PROBLEM 29.7

A 5-$\mu$F capacitor is in series with a 300-$\Omega$ resistor, and a 120-V, 60-Hz voltage is applied to the combination. Find (*a*) the current in the circuit and (*b*) the phase angle.

(*a*)   The reactance of the capacitor at 60 Hz is

$$X_C = \frac{1}{2\pi f C} = \frac{1}{(2\pi)(60\text{ Hz})(5 \times 10^{-6}\text{ F})} = 531\ \Omega$$

$R = 300\ \Omega$
$\phi = -61°$
$X_C = 531\ \Omega$
$Z = 610\ \Omega$


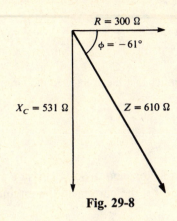

**Fig. 29-8**

Since $X_L = 0$, the impedance of the circuit is (Fig. 29-8)

$$Z = \sqrt{R^2 + (X_L - X_C)^2} = \sqrt{R^2 + (-X_C)^2}$$
$$= \sqrt{R^2 + X_C^2} = \sqrt{(300\ \Omega)^2 + (531\ \Omega)^2} = 610\ \Omega$$

Hence the current is

$$I = \frac{V}{Z} = \frac{120\ \text{V}}{610\ \Omega} = 0.197\ \text{A}$$

(*b*)
$$\tan \phi = \frac{X_L - X_C}{R} = \frac{0 - 531}{300} = -1.77$$
$$\phi = -61°$$

The negative phase angle signifies that the current in the circuit leads the voltage.

## SOLVED PROBLEM 29.8

A 5-mH, 20-Ω inductor is connected to a 28-V, 400-Hz power source. Find (*a*) the current in the inductor and (*b*) the phase angle.

(*a*) The reactance of the inductor is

$$X_L = 2\pi f L = (2\pi)(400\ \text{Hz})(5 \times 10^{-3}\ \text{H}) = 12.6\ \Omega$$

and its impedance is (Fig. 29-9)

$$Z = \sqrt{R^2 + X_L^2} = \sqrt{(20\ \Omega)^2 + (12.6\ \Omega)^2} = 23.6\ \Omega$$

Hence the current is

$$I = \frac{V}{Z} = \frac{28\ \text{V}}{23.6\ \Omega} = 1.18\ \text{A}$$

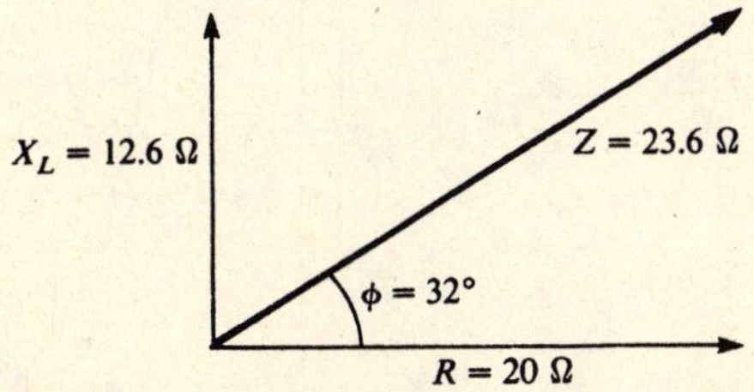

$X_L = 12.6\ \Omega$
$Z = 23.6\ \Omega$
$\phi = 32°$
$R = 20\ \Omega$


**Fig. 29-9**

(b)
$$\tan\phi = \frac{X_L - X_C}{R} = \frac{12.6\ \Omega - 0}{20\ \Omega} = 0.63$$
$$\phi = 32°$$

The positive phase angle signifies that the voltage in the circuit leads the current.

## SOLVED PROBLEM 29.9

A 10-$\mu$F capacitor, a 0.10-H inductor, and a 60-$\Omega$ resistor are connected in series across a 120-V, 60-Hz power source. Find (a) the current in the circuit and (b) the phase angle.

(a)  The reactances are

$$X_L = 2\pi f L = (2\pi)(60\ \text{Hz})(0.10\ \text{H}) = 38\ \Omega$$
$$X_C = \frac{1}{2\pi f C} = \frac{1}{(2\pi)(60\ \text{Hz})(10 \times 10^{-6}\ \text{F})} = 265\ \Omega$$

The impedance is therefore (Fig. 29-10)

$$Z = \sqrt{R^2 + (X_L - X_C)^2} = \sqrt{(60\ \Omega)^2 + (38\ \Omega - 265\ \Omega)^2} = 235\ \Omega$$

Hence the current in the circuit is

$$I = \frac{V}{Z} = \frac{120\ \text{V}}{235\ \Omega} = 0.51\ \text{A}$$

(b)
$$\tan\phi = \frac{X_L - X_C}{R} = \frac{38\ \Omega - 265\ \Omega}{60\ \Omega} = -\frac{227\ \Omega}{60\ \Omega} = -3.78$$
$$\phi = -75°$$

The negative phase angle signifies that the current in the circuit leads the voltage.

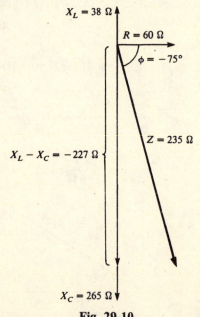

**Fig. 29-10**

## RESONANCE

The impedance in a series ac circuit is a minimum when $X_L = X_C$; under these circumstances $Z = R$ and $I = V/R$. The *resonance frequency* $f_0$ of a circuit is that frequency at which $X_L = X_C$:

$$2\pi f_0 L = \frac{1}{2\pi f_0 C} \qquad f_0 = \frac{1}{2\pi\sqrt{LC}}$$

When the potential difference applied to a circuit has the frequency $f_0$, the current in the circuit is a maximum. This condition is known as *resonance*. At resonance the phase angle is zero since $X_L = X_C$.

### SOLVED PROBLEM 29.10

In the antenna circuit of a radio receiver that is tuned to a particular station, $R = 5\ \Omega$, $L = 5$ mH, and $C = 5$ pF. (*a*) Find the frequency of the station. (*b*) If the potential difference applied to the circuit is $5 \times 10^{-4}$ V, find the current that flows.

(*a*)
$$f_0 = \frac{1}{2\pi\sqrt{LC}} = \frac{1}{2\pi\sqrt{(5 \times 10^{-3}\ \text{H})(5 \times 10^{-12}\ \text{F})}} = 1006\ \text{kHz}$$

(*b*)  At resonance, $X_L = X_C$ and $Z = R$. Hence

$$I = \frac{V}{R} = \frac{5 \times 10^{-4}\ \text{V}}{5\ \Omega} = 10^{-4}\ \text{A} = 0.1\ \text{mA}$$

### SOLVED PROBLEM 29.11

In the antenna circuit of Prob. 29.10, the inductance is fixed but the capacitance can be varied. What should the capacitance be in order to receive an 800-kHz radio signal?

$$C = \frac{1}{(2\pi f)^2 (L)} = \frac{1}{[(2\pi)(800 \times 10^3\ \text{Hz})]^2 (5 \times 10^{-3}\ \text{H})}$$
$$= 7.9 \times 10^{-12}\ \text{F} = 7.9\ \text{pF}$$

### SOLVED PROBLEM 29.12

In a series ac circuit $R = 20\ \Omega$, $X_L = 10\ \Omega$, and $X_C = 25\ \Omega$ when the frequency is 400 Hz. (*a*) Find the impedance of the circuit. (*b*) Find the phase angle. (*c*) Is the resonance frequency of the circuit greater than or less than 400 Hz? (*d*) Find the resonance frequency.

(*a*)
$$Z = \sqrt{R^2 + (X_L - X_C)^2} = \sqrt{(20\ \Omega)^2 + (10\ \Omega - 25\ \Omega)^2} = 25\ \Omega$$

(*b*)
$$\tan\phi = \frac{X_L - X_C}{R} = \frac{10\ \Omega - 25\ \Omega}{20\ \Omega} = -0.75 \qquad \phi = -37°$$

A negative phase angle signifies that the voltage lags behind the current.

(*c*)  At resonance $X_L = X_C$. At 400 Hz, $X_L < X_C$, so the frequency must be changed in such a way as to increase $X_L$ and decrease $X_C$. Since $X_L = 2\pi fL$ and $X_C = 1/(2\pi fC)$, it is clear that increasing the frequency will have this effect. Hence the resonance frequency must be greater than 400 Hz.

(*d*)  Since $X_L = 10\ \Omega$ and $X_C = 25\ \Omega$ when $f = 400$ Hz,

$$L = \frac{X_L}{2\pi f} = \frac{10\ \Omega}{(2\pi)(400\ \text{Hz})} = 4 \times 10^{-3}\ \text{H}$$

$$C = \frac{1}{2\pi f X_C} = \frac{1}{(2\pi)(400\ \text{Hz})(25\ \Omega)} = 1.6 \times 10^{-5}\ \text{F}$$

Hence
$$f_0 = \frac{1}{2\pi\sqrt{LC}} = \frac{1}{2\pi\sqrt{(4 \times 10^{-3}\ \text{H})(1.6 \times 10^{-5}\ \text{F})}} = 629\ \text{Hz}$$

## POWER FACTOR

The power absorbed in an ac circuit is given by

$$P = IV \cos \phi$$

where $\phi$ is the phase angle between voltage and current. The quantity $\cos \phi$ is the *power factor* of the circuit. At resonance, $\phi = 0$, $\cos \phi = 1$, and the power absorbed is a maximum. The power factor in an ac circuit is equal to the ratio between its resistance and its impedance:

$$\text{Power factor} = \cos \phi = \frac{R}{Z} = \frac{R}{\sqrt{R^2 + (X_L + X_C)^2}}$$

Power factors are often expressed as percentages, so a phase angle of, say, $25°$ would give rise to a power factor of $\cos 25° = 0.906 = 90.6$ percent.

Ac power sources are usually specified in terms of *apparent power*, the product of $V_{\text{eff}}$ and $I_{\text{eff}}$, and are measured in *voltamperes* (VA). The *true power* $P = IV \cos \phi$ is not always specified because for a circuit with power factor less than 1, a power greater than the true power $P$ must be supplied. Thus, a power factor of 90.6 percent means that an apparent power of 1 VA must be supplied for each 0.906 W of true power consumed by the circuit. To obtain a true power of 1 W, an apparent power of $1/0.906 = 1.104$ VA must be supplied.

## SOLVED PROBLEM 29.13

A coil of unknown resistance and inductance draws 4 A when it is connected to a 12-V dc power source and 3 A when it is connected to a 12-V, 100-Hz power source. (*a*) Find the values of $R$ and $L$. (*b*) How much power is dissipated when the coil is connected to the dc source? (*c*) When it is connected to the ac source?

(*a*)    There is no inductive reactance when direct current passes through the coil, so its resistance is

$$R = \frac{V_1}{I_1} = \frac{12 \text{ V}}{4 \text{ A}} = 3 \text{ }\Omega$$

At $f = 100$ Hz the impedance of the circuit is

$$Z = \frac{V_2}{I_2} = \frac{12 \text{ V}}{3 \text{ A}} = 4 \text{ }\Omega$$

and so since $Z = \sqrt{R^2 + (X_L^2 - X_C^2)}$ and $X_C = 0$ here,

$$X_L = \sqrt{Z^2 - R^2} = \sqrt{(4 \text{ }\Omega)^2 - (3 \text{ }\Omega)^2} = 2.65 \text{ }\Omega$$

Hence the inductance of the coil is

$$L = \frac{X_L}{2\pi f} = \frac{2.65 \text{ }\Omega}{(2\pi)(100 \text{ Hz})} = 4.22 \text{ mH}$$

(*b*)                     $P_1 = I_1^2 R = (4 \text{ A})^2 (3 \text{ }\Omega) = 48 \text{ W}$

(*c*)                     $P_2 = I_2^2 R = (3 \text{ A})^2 (3 \text{ }\Omega) = 27 \text{ W}$

## SOLVED PROBLEM 29.14

A 50-$\mu F$ capacitor, a 0.3-H inductor, and an 80-$\Omega$ resistor are connected in series with a 120-V, 60-Hz power source (Fig. 29-11). (*a*) What is the impedance of the circuit? (*b*) How much

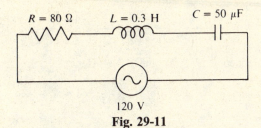

$R = 80\ \Omega$   $L = 0.3\ \text{H}$   $C = 50\ \mu\text{F}$

120 V

**Fig. 29-11**

current flows in it? (*c*) What is the power factor? (*d*) How much power is dissipated by the circuit? (*e*) What must be the minimum rating in volt amperes of the power source?

(*a*)
$$X_L = 2\pi fL = (2\pi)(60\ \text{Hz})(0.3\ \text{H}) = 113\ \Omega$$

$$X_C = \frac{1}{2\pi fC} = \frac{1}{(2\pi)(60\ \text{Hz})(50 \times 10^{-6}\ \text{F})} = 53\ \Omega$$

$$Z = \sqrt{R^2 + (X_L - X_C)^2} = \sqrt{80^2 + (113\ \Omega - 53\ \Omega)^2} = 100\ \Omega$$

(*b*)
$$I = \frac{V}{Z} = \frac{120\ \text{V}}{100\ \Omega} = 1.2\ \text{A}$$

(*c*)
$$\cos\ \phi = \frac{R}{Z} = \frac{80\ \Omega}{100\ \Omega} = 0.8 = 80\%$$

(*d*)
$$\text{True power} = P = IV\ \cos\ \phi = (1.2\ \text{A})(120\ \text{V})(0.8) = 115\ \text{W}$$

Alternatively,
$$P = I^2R = (1.2\ \text{A})^2(80\ \Omega) = 115\ \text{W}$$

(*e*)
$$\text{Apparent power} = IV = (1.2\ \text{A})(120\ \text{V}) = 144\ \text{VA}$$

## SOLVED PROBLEM 29.15

(*a*) Find the potential differences across the resistor, the inductor, and the capacitor in the circuit of Prob. 29.14. (*b*) Are these values in accord with the applied potential difference of 120 V?

(*a*)
$$V_R = IR = (1.2\ \text{A})(80\ \Omega) = 96\ \text{V}$$
$$V_L = IX_L = (1.2\ \text{A})(113\ \Omega) = 136\ \text{V}$$
$$V_C = IX_C = (1.2\ \text{A})(53\ \Omega) = 64\ \text{V}$$

(*b*)  The sum of these potential differences is 296 V, more than twice the 120 V applied to the circuit. However, this is a meaningless way to combine the potential differences since they are not in phase with one another: $V_L$ is 90° ahead of $V_R$ and $V_C$ is 90° behind $V_R$. The correct way to find the total potential difference across the circuit is as follows:

$$V = \sqrt{V_R^2 + (V_L - V_C)^2} = \sqrt{(96\ \text{V})^2 + (136\ \text{V} - 64\ \text{V})^2} = 120\ \text{V}$$

This result is in agreement with the applied potential difference of 120 V. We note that the voltage across an inductor or capacitor in an ac circuit can be greater than the voltage applied to the circuit.

## SOLVED PROBLEM 29.16

(*a*) Find the resonance frequency $f_0$ of the circuit of Prob. 29.14. (*b*) What current will flow in the circuit if it is connected to a 120-V power source whose frequency is $f_0$? (*c*) What will be the power factor in this case? (*d*) How much power will be dissipated by the circuit? (*e*) What must be the minimum rating in voltamperes of the power source now?

(a)
$$f_0 = \frac{1}{2\pi\sqrt{LC}} = \frac{1}{2\pi\sqrt{(0.3\text{ H})(50 \times 10^{-6}\text{ F})}} = 41\text{ Hz}$$

(b)  At the resonance frequency, $X_L = X_C$ and $Z = R$. Hence

$$I = \frac{V}{R} = \frac{120\text{ V}}{80\ \Omega} = 1.5\text{ A}$$

(c)
$$\cos\phi = \frac{R}{Z} = \frac{R}{R} = 1 = 100\%$$

(d)  True power $= IV \cos\phi = (1.5\text{ A})(120\text{ V})(1) = 180\text{ W}$

(e)  Apparent power $= IV = (1.5\text{ A})(120\text{ V}) = 180\text{ VA}$

## SOLVED PROBLEM 29.17

(a) Find the potential difference across the resistor, the inductor, and the capacitor in the circuit of Prob. 29.14 when it is connected to a 120-V ac source whose frequency is equal to the circuit's resonance frequency of 41 Hz. (b) Are these values in accord with the applied potential difference of 120 V?

(a)  At the resonance frequency of $f_0 = 41$ Hz, the inductive and capacitive reactances are, respectively,

$$X_L = 2\pi f_0 L = (2\pi)(41\text{ Hz})(0.3\text{ H}) = 77\ \Omega$$

$$X_C = \frac{1}{2\pi f_0 C} = \frac{1}{(2\pi)(41\text{ Hz})(50 \times 10^{-6}\text{ F})} = 77\ \Omega$$

The various potential differences are therefore

$$V_R = IR = (1.5\text{ A})(80\ \Omega) = 120\text{ V}$$
$$V_L = IX_L = (1.5\text{ A})(77\ \Omega) = 116\text{ V}$$
$$V_C = IX_C = (1.5\text{ A})(77\ \Omega) = 116\text{ V}$$

(b)  The total potential difference across the circuit is

$$V = \sqrt{V_R^2 + (V_L - V_C)^2} = \sqrt{(120\text{ V})^2 + (0)^2} = 120\text{ V}$$

which is equal to the applied potential difference.

## SOLVED PROBLEM 29.18

A 5-hp electric motor is 80 percent efficient and has an inductive power factor of 75 percent. (a) What minimum rating in kilovoltamperes must its power source have? (b) A capacitor is connected in series with the motor to raise the power factor to 100 percent. What minimum rating in kilovoltamperes must the power source now have?

(a)  The power required by the motor is

$$P = \frac{(5\text{ hp})(0.746\text{ kW/hp})}{0.8} = 4.66\text{ kW}$$

Since $P = IV \cos\phi$, the power source must have the minimum rating

$$IV = \frac{P}{\cos\phi} = \frac{4.66\text{ kW}}{0.75} = 6.22\text{ kVA}$$

(b)  When $\cos\phi = 1$, $IV = P = 4.66$ kVA.

**SOLVED PROBLEM 29.19**

A coil connected to a 120-V, 25-Hz power line draws a current of 0.5 A and dissipates 50 W. (*a*) What is its power factor? (*b*) What capacitance should be connected in series with the coil to increase the power factor to 100 percent? (*c*) What would the current in the circuit be then? (*d*) How much power would the circuit then dissipate?

(*a*)   Since $P = IV \cos \phi$,

$$\cos \phi = \frac{P}{IV} = \frac{50 \text{ W}}{(0.5 \text{ A})(120 \text{ V})} = 0.833 = 83.3\%$$

(*b*)   The power factor will be 100 percent at resonance, when $X_L = X_C$. The first step is to find $X_L$, which can be done from the formula $\tan \phi = (X_L - X_C)/R$. Here $X_C = 0$ and, since $\cos \phi = 0.833$, $\phi = 34°$, and $\tan \phi = 0.663$. Since $P = I^2 R$,

$$R = \frac{P}{I^2} = \frac{50 \text{ W}}{(0.5 \text{ A})^2} = 200 \ \Omega$$

Hence          $X_L = R \tan \phi + X_C = (200 \ \Omega)(0.663) + 0 = 133 \ \Omega$

This must also be the value of $X_C$ when $f = 25$ Hz, and so

$$C = \frac{1}{2\pi f X_C} = \frac{1}{(2\pi)(25 \text{ Hz})(133 \ \Omega)} = 4.8 \times 10^{-5} \text{ F} = 48 \ \mu\text{F}$$

(*c*)   At resonance $Z = R$, so

$$I = \frac{V}{Z} = \frac{V}{R} = \frac{120 \text{ V}}{200 \ \Omega} = 0.6 \text{ A}$$

(*d*)          $P = I^2 R = (0.6 \text{ A})^2 (200 \ \Omega) = 72 \text{ W}$

## PARALLEL AC CIRCUITS

When a resistor, an inductor, and a capacitor are connected in parallel across an ac source, as in Fig. 29-12(*a*), the voltage is the same across each circuit element:

$$V = V_R = V_L = V_C$$

The total instantaneous current is the sum of the instantaneous currents in each branch, as in a parallel dc circuit, but this is not true of the total effective current $I$ because the branch currents are not in phase. Although the current $I_R$ in the resistor is in phase with $V$, the current $I_C$ in the capacitor leads $V$ by 90° and the current $I_L$ in the inductor lags $V$ by 90°. To find the total current $I$, the phasors that represent $I_R$, $I_C$, and $I_L$ must be added vectorially, as in Fig. 29-12(*b*).

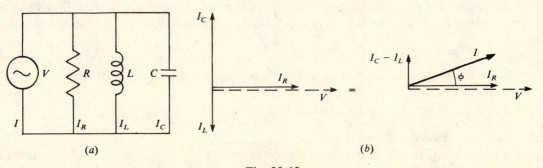

(*a*)                    (*b*)

**Fig. 29-12**

The branch currents in the parallel circuit of Fig. 29-12($a$) are given by

$$I_R = \frac{V}{R} \qquad I_C = \frac{V}{X_C} \qquad I_L = \frac{V}{X_L}$$

Adding these currents vectorially with the help of the Pythagorean theorem gives

$$I = \sqrt{I_R^2 + (I_C - I_L)^2}$$

The phase angle $\phi$ between current and voltage is specified by

$$\cos \phi = \frac{I_R}{I}$$

If $I_C$ is greater than $I_L$, the current leads the voltage and the phase angle is considered positive; if $I_L$ is greater than $I_C$, the current lags the voltage and the phase angle is considered negative. The power dissipated in a parallel ac circuit is given by the same formula as in a series circuit, namely,

$$P = IV \cos \phi$$

## RESONANCE IN PARALLEL CIRCUITS

Figure 29-13($a$) shows an inductor and a capacitor connected in parallel to a power source. The currents in the inductor and capacitor are 180° apart in phase, as the phasor diagram shows, so the total current $I$ in the circuit is the *difference* between the currents in $L$ and $C$:

$$I = I_C - I_L$$

The current that circulates between the inductor and the capacitor without contributing to $I$ is called the *tank current* and may be greater than $I$.

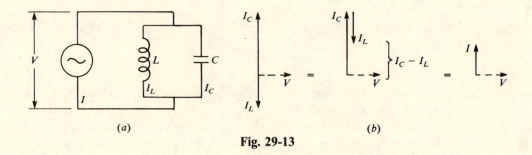

(a)                                                                 (b)

**Fig. 29-13**

In the event that $X_C = X_L$, currents $I_C$ and $I_L$ are also equal. Since $I_C$ and $I_L$ are 180° out of phase, the total current $I = 0$: The currents in the inductor and capacitor cancel. This situation is called *resonance*.

In a series $RLC$ circuit, as discussed earlier, the impedance has its minimum value $Z = R$ when $X_C = X_L$, a situation also called resonance. The frequency for which $X_C = X_L$ is

$$f_0 = \frac{1}{2\pi\sqrt{LC}}$$

and is called the resonance frequency.

In a parallel $RLC$ circuit, resonance again corresponds to $X_C = X_L$, but here the impedance is a *maximum* at $f_0$. At $f_0$, the currents in the inductor and capacitor are equal in magnitude but 180° out of phase, so no current passes through the combination. Thus $I = I_R$ and $Z = R$. At frequencies higher and lower than $f_0$, $I_C$ is not equal to $I_L$ and some current can pass through the inductor-capacitor part of the circuit, which reduces the impedance $Z$ to less than $R$. Thus a series circuit can be used as a selector to favor a particular frequency, and a parallel circuit with the same $L$ and $C$ can be used as a selector to discriminate against the same frequency.

**SOLVED PROBLEM 29.20**

The reactances of a coil and a capacitor connected in parallel and supplied by a 15-V, 1000-Hz power source are, respectively, $X_L = 20\ \Omega$ and $X_C = 30\ \Omega$. Find (a) the currents in each component, (b) the total current, (c) the impedance of the circuit, and (d) the phase angle and total power dissipated in the circuit.

(a)
$$I_L = \frac{V}{X_L} = \frac{15\ \text{V}}{20\ \Omega} = 0.75\ \text{A} \qquad I_C = \frac{V}{X_C} = \frac{15\ \text{V}}{30\ \Omega} = 0.5\ \text{A}$$

(b)
$$I = I_C - I_L = 0.5\ \text{A} - 0.75\ \text{A} = -0.25\ \text{A}$$

The minus sign means that the total current lags 90° behind the voltage [the opposite of the situation shown in Fig. 29-13(b)] and can be disregarded in the other calculations.

(c)
$$Z = \frac{V}{I} = \frac{15\ \text{V}}{0.25\ \text{A}} = 60\ \Omega$$

The impedance not only is greater than $X_L$ or $X_C$ but also is greater than their arithmetical sum.

(d) Because the phase angle here is 90°, $\cos \phi = \cos 90° = 0$ and the power drawn by the circuit is

$$P = IV \cos \phi = 0$$

This conclusion follows from the absence of resistance in the circuit.

**SOLVED PROBLEM 29.21**

A 10-$\Omega$ resistor, an 8-$\mu$F capacitor, and a 2-mH inductor are connected in parallel across a 10-V, 1000-Hz power source, as in Fig. 29-14(a). Find (a) the current in each component, (b) the total current in the circuit, (c) the impedance of the circuit, and (d) the phase angle and the total power dissipation of the circuit.

(a)
$$X_C = \frac{1}{2\pi f C} = \frac{1}{(2\pi)(10^3\ \text{Hz})(8 \times 10^{-6}\ \text{F})} = 20\ \Omega$$

$$X_L = 2\pi f L = (2\pi)(10^3\ \text{Hz})(2 \times 10^{-3}\ \text{H}) = 12.6\ \Omega$$

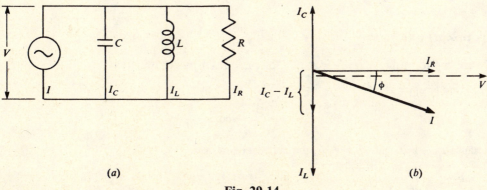

(a)                                                                           (b)

**Fig. 29-14**

Hence

$$I_C = \frac{10\text{ V}}{20\ \Omega} = 0.5\text{ A} \qquad I_L = \frac{10\text{ V}}{12.6\ \Omega} = 0.8\text{ A} \qquad I_R = \frac{10\text{ V}}{10\ \Omega} = 1.0\text{ A}$$

(b)   The phasor diagram of Fig. 29-14(b) shows how the various currents are to be added. We have

$$I = \sqrt{I_A^2 + (I_C - I_L)^2} = \sqrt{(1.0\text{ A})^2 + (0.5\text{ A} - 0.8\text{ A})^2}$$

$$= \sqrt{(1.0\text{ A})^2 + (-0.3\text{ A})^2} = \sqrt{1.00\text{ A}^2 + 0.09\text{ A}^2} = \sqrt{1.09\text{ A}^2} = 1.04\text{ A}$$

(c)

$$Z = \frac{V}{I} = \frac{10\text{ V}}{1.04\text{ A}} = 9.6\ \Omega$$

(d)

$$\cos \frac{I_R}{I} = \frac{1.0\text{ A}}{1.04\text{ A}} = 0.962$$

$$\phi = -16°$$

The current lags behind the voltage by 16°, as in Fig. 29-14(b):

$$P = IV \cos \phi = (10\text{ V})(1.04\text{ A})(0.962) = 10\text{ W}$$

We can also find $P$ in this way:

$$P = I_R^2 R = (1.0\text{ A})^2 (10\ \Omega) = 10\text{ W}$$

## IMPEDANCE MATCHING

In Chapter 25 we saw that power transfer between two dc circuits is a maximum when their resistances are equal, an effect called impedance matching. The same effect occurs in an ac circuit with the impedance of the circuits instead of their resistances as the quantities that must be the same for maximum power transfer. An additional consideration is that an impedance mismatch between ac circuits may result in a distorted signal by altering the properties of the circuits, for instance, by changing their resonance frequencies.

A transformer enables impedances between ac circuits to be matched. An example is an audio system in which the impedance of the amplifier circuit might be several thousand ohms whereas the impedance of the voice coil of the loudspeaker might be only a few ohms. Connecting the voice coil directly to the amplifier would be extremely inefficient; using a transformer is the answer. The ratio of turns $N_1/N_2$ between the windings of the transformer for maximum power transfer from a circuit of impedance $Z_1$ to a circuit of impedance $Z_2$ is

$$\frac{N_1}{N_2} = \sqrt{\frac{Z_1}{Z_2}}$$

## SOLVED PROBLEM 29.22

Derive the formula $N_1/N_2 = \sqrt{Z_1/Z_2}$ for maximum power transfer.

From Chapter 28

$$\frac{V_1}{V_2} = \frac{N_1}{N_2} \qquad \text{and} \qquad \frac{I_2}{I_1} = \frac{N_1}{N_2}$$

Because $Z_1 = V_1/I_1$ and $Z_2 = V_2/I_2$, the impedance ratio is

$$\frac{Z_1}{Z_2} = \frac{V_1 I_2}{V_2 I_1} = \left(\frac{N_1}{N_2}\right)^2$$

The ratio of turns needed to match the impedances is therefore

$$\frac{N_1}{N_2} = \sqrt{\frac{Z_1}{Z_2}}$$

**SOLVED PROBLEM 29.23**

A loudspeaker whose voice coil has an impedance of 6 $\Omega$ is to be used with an amplifier whose load impedance is 7350 $\Omega$. Find the ratio of turns in a transformer that will match these impedances.

Let us use the subscripts 1 for the amplifier and 2 for the voice coil. From the formula for maximum power transfer

$$\frac{N_1}{N_2} = \sqrt{\frac{Z_1}{Z_2}} = \sqrt{\frac{7350\ \Omega}{6\ \Omega}} = \sqrt{1225} = 35$$

Thus the ratio of turns is 35:1, and the transformer winding connected to the amplifier should have 35 times as many turns as the winding connected to the voice coil.


# *Multiple-Choice Questions*

29.1.    The current in an ac circuit

    (*a*)   leads the voltage         (*c*)   is in phase with the voltage
    (*b*)   lags behind the voltage    (*d*)   is any of the above, depending on the circuit


29.2.    The current and voltage cannot be exactly in phase in a circuit that contains

    (*a*)   only resistance     (*c*)   inductance and capacitance
    (*b*)   only inductance     (*d*)   resistance, inductance, and capacitance


29.3.    A capacitor, in inductor, and a resistor are connected in series. If the capacitance is increased, the impedance of the circuit

    (*a*)   decreases    (*c*)   decreases or increases
    (*b*)   increases    (*d*)   decreases, increases, or stays the same


29.4.    In a certain circuit $X_C = X_L$. The power factor of the circuit

    (*a*)   is 0      (*c*)   is 1
    (*b*)   is 0.5    (*d*)   depends on the resistance in the circuit


29.5.    A coil of negligible resistance is connected to a 60-V, 400-Hz ac source. If the current in the coil is 1.2 A, the inductance of the coil is

    (*a*)   8 $\mu$H     (*c*)   20 mH
    (*b*)   14 mH    (*d*)   50 H

**29.6.** The current that flows when a 5.0-$\mu$F capacitor is connected to a 120-V, 60-Hz ac source is

(a) 0.16 A     (c) 4.4 A
(b) 0.23 A     (d) 64 kA

**29.7.** When a capacitor, an inductor, and a resistor are connected in series to an ac source, the effective voltages across the circuit components are $V_C = 12.0$ V, $V_L = 8.0$ V, and $V_R = 6.0$ V. The effective voltage of the source is

(a) 4.5 V     (c) 21 V
(b) 7.2 V     (d) 26 V

**29.8.** The phase angle in the circuit of Question 29.7 is

(a) $-24°$     (c) $+24°$
(b) $-34°$     (d) $+34°$

**29.9.** A circuit that contains a 0.50-$\mu$F capacitor, a 20-mH inductor, and an 8.0-$\Omega$ resistor has a resonance frequency of

(a) 1.0 kHz     (c) 6.3 kHz
(b) 1.6 kHz     (d) 16 MHz

**29.10.** A coil draws a current of 0.250 A and dissipates 10.0 W when it is connected to a 45.0-V ac source. A capacitor is then connected in series to bring the power factor to 1. The circuit now dissipates

(a) 10.0 W     (c) 12.7 W
(b) 11.3 W     (d) 14.1 W

**29.11.** When a 24.0-V, 400-Hz ac source is connected to a series combination of a 5.00-$\mu$F capacitor, a 20.0-mH inductor, and a 40.0-$\Omega$ resistor, the current in the circuit is

(a) 0.484 A     (c) 0.882 A
(b) 0.600 A     (d) 2.25 A

**29.12.** The phase angle in the circuit of Question 29.11 is

(a) $-36°$     (c) $+36°$
(b) $-51°$     (d) $+51°$

**29.13.** The power dissipated in the circuit of Question 29.11 is

(a) 2.5 W     (c) 11.6 W
(b) 9.4 W     (d) 14.4 W

**29.14.** The ac source of the circuit of Question 29.11 should have a minimum rating in voltamperes of

(a) 2.5 VA     (c) 11.6 VA
(b) 9.4 VA     (d) 14.4 VA

**29.15.** The circuit elements of Question 29.11 are now connected in parallel. When the same ac source is connected to the combination, the total current the source provides is

(a) 0.424 A     (c) 0.983 A
(b) 0.625 A     (d) 1.38 A

# Supplementary Problems

**29.1.** The frequency of the alternating emf applied to a series circuit containing resistance, inductance, and capacitance is doubled. What happens to $R$, $X_L$, and $X_C$?

**29.2.** (*a*) What is the minimum value the power factor of a circuit can have? Under what circumstances can this occur? (*b*) What is the maximum value the power factor can have? Under what circumstances can this occur?

**29.3.** A voltmeter across an ac circuit reads 40 V, and an ammeter in series with the circuit reads 6 A. (*a*) What is the maximum potential difference across the circuit? (*b*) What is the maximum current in the circuit? (*c*) What other information is needed to determine the power consumed in the circuit?

**29.4.** Find the reactance of a 10-mH coil when it is in a 500-Hz circuit.

**29.5.** The reactance of an inductor is 1000 $\Omega$ at 20 Hz. Find its inductance.

**29.6.** A current of 0.20 A flows through a 0.15-H inductor of negligible resistance that is connected to an 80-V source of alternating current. What is the frequency of the source?

**29.7.** Find the reactance of a 5-$\mu$F capacitor at 10 Hz and at 10 kHz.

**29.8.** A capacitor has a reactance of 200 $\Omega$ at 1000 Hz. What is its capacitance?

**29.9.** A 5-$\mu$F capacitor is connected to a 6-kHz alternating emf, and a current of 2 A flows. Find the effective magnitude of the emf.

**29.10.** A 5-$\mu$F capacitor draws 1 A when it is connected to a 60-V ac source. What is the frequency of the source?

**29.11.** A 25-$\mu$F capacitor is connected in series with a 50-$\Omega$ resistor, and a 12-V, 60-Hz potential difference is applied. Find the current in the circuit and the power dissipated in it.

**29.12.** A capacitor is connected in series with an 8-$\Omega$ resistor, and the combination is placed across a 24-V, 1000-Hz power source. A current of 2 A flows. Find (*a*) the capacitance of the capacitor, (*b*) the phase angle, and (*c*) the power dissipated by the circuit.

**29.13.** A 0.1-H, 30-$\Omega$ inductor is connected to a 50-V, 100-Hz power source. Find the current in the inductor and the power dissipated in it.

**29.14.** The current in a resistor is 1 A when it is connected to a 50-V, 100-Hz power source. (*a*) How much inductive reactance is required to reduce the current to 0.5 A? What value of $L$ will accomplish this? (*b*) How much capacitive reactance is required to reduce the current to 0.5 A? What value of $C$ will accomplish this? (*c*) What will the current be if the above inductance and capacitance are both placed in series with the resistor?

**29.15.** A pure resistor, a pure capacitor, and a pure inductor are connected in series across an ac power source. An ac voltmeter placed across each of these circuit elements in turn reads 10, 20, and 30 V. What is the potential difference of the source?

**29.16.** An inductive circuit with a power factor of 80 percent consumes 750 W of power. What minimum rating in voltamperes must its power source have?

**29.17.** An inductive load dissipates 75 W of power when it draws 1.0 A from a 120-V, 60-Hz power line. (*a*) What is the power factor? (*b*) What capacitance should be connected in series in the circuit to increase the power factor to 100 percent? (*c*) What would the current then be? (*d*) How much power would the circuit then dissipate?

**29.18.** A circuit that consists of a capacitor in series with a 50-$\Omega$ resistor draws 4 A from a 250-V, 200-Hz power source. (*a*) How much power is dissipated? (*b*) What is the power factor? (*c*) What inductance should be connected in series in the circuit to increase the power factor to 100 percent? (*d*) What would the current then be? (*e*) How much power would the circuit then dissipate?

**29.19.** In a series circuit $R = 100\ \Omega$, $X_L = 120\ \Omega$, and $X_C = 60\ \Omega$ when it is connected to an 80-V ac power source. Find (*a*) the current in the circuit, (*b*) the phase angle, and (*c*) the current if the frequency of the power source were changed to the resonance frequency of the circuit.

**29.20.** A 10-$\mu$F capacitor, a 10-mH inductor, and a 10-$\Omega$ resistor are connected in series with a 45-V, 400-Hz power source. Find (*a*) the impedance of the circuit, (*b*) the current in it, (*c*) the power it dissipates, and (*d*) the minimum rating in voltamperes of the power source.

**29.21.** (*a*) Find the resonance frequency $f_0$ of the circuit of Prob. 29.20. (*b*) What current will flow in the circuit if it is connected to a 45-V power source whose frequency is $f_0$? (*c*) How much power will be dissipated by the circuit? (*d*) What must be the minimum rating of the power source in voltamperes?

**29.22.** A 60-$\mu$F capacitor, a 0.3-H inductor, and a 50-$\Omega$ resistor are connected in series with a 120-V, 60-Hz power source. Find (*a*) the impedance of the circuit, (*b*) the current in it, (*c*) the power it dissipates, and (*d*) the minimum rating in voltamperes of the power source.

**29.23.** (*a*) Find the resonance frequency $f_0$ of the circuit of Prob. 29.22. (*b*) What current will flow in the circuit if it is connected to a 120-V power source whose frequency is $f_0$? (*c*) How much power will be dissipated by the circuit? (*d*) What must be the minimum rating of the power source in voltamperes?

**29.24.** A 10-$\Omega$ resistor and an 8-$\mu$F capacitor are connected in parallel across a 10-V, 1000-Hz power source. Find (*a*) the current in each component, (*b*) the total current, (*c*) the impedance of the circuit, and (*d*) the phase angle and power dissipation of the circuit.

**29.25.** A 10-$\Omega$ resistor and a 2-mH inductor are connected in parallel across a 10-V, 1000-Hz power source. Find (*a*) the current in each component, (*b*) the total current, (*c*) the impedance of the circuit, and (*d*) the phase angle and the total power dissipation of the circuit.

**29.26.** A 25-$\Omega$ resistor, a 40-$\mu$F capacitor, and a 40-mH coil are connected in parallel to a 24-V, 100-Hz source. Find (*a*) the current in each component, (*b*) the total current in the circuit, (*c*) the impedance of the circuit, (*d*) the phase angle between current and voltage, and (*e*) the total power dissipated by the circuit.

**29.27.** The circuit of Prob. 29.26 is connected to a 24-V, 200-Hz source. Answer the same questions for this case.

**29.28.** A microphone whose impedance is 20 $\Omega$ is to be used with an amplifier whose input impedance is 50,000 $\Omega$. Find the ratio of turns in the transformer needed to match these two impedances.

## Answers to Multiple-Choice Questions

**29.1.** (d)    **29.9.** (b)

**29.2.** (b)    **29.10.** (c)

**29.3.** (c)    **29.11.** (a)

**29.4.** (c)    **29.12.** (a)

**29.5.** (c)    **29.13.** (b)

**29.6.** (b)    **29.14.** (c)

**29.7.** (b)    **29.15.** (b)

**29.8.** (b)

## Answers to Supplementary Problems

**29.1.** $R$ is unchanged, $X_L$ is doubled, and $X_C$ is halved.

**29.2.** (a)   The minimum value is 0, which can occur only if $R = 0$ in a circuit in which $X_L$ is not equal to $X_C$.

    (b)   The maximum value is 1, which occurs at resonance when $X_L = X_C$.

**29.3.** (a)   56.6 V    (b)   8.5 A    (c)   Phase angle

**29.4.** 31.4 $\Omega$

**29.5.** 0.796 H

**29.6.** 424 Hz

**29.7.** 3183 $\Omega$; 3.183 $\Omega$

**29.8.** 0.796 $\mu$F

**29.9.** 10.6 V

**29.10.** 531 Hz

**29.11.** 0.102 A; 0.524 W

**29.12.** (a)   17.8 $\mu$F    (b)   $-48°$    (c)   32 W

**29.13.** 0.72 A; 15.6 W

**29.14.** (a)  86.6 $\Omega$; 137.8 mH      (b)  86.6 $\Omega$; 18.4 $\mu$F      (c)  1 A

**29.15.**  14 V

**29.16.**  938 VA

**29.17.** (a)  0.625      (b)  28.6 $\mu$F      (c)  1.6 A      (d)  192 W

**29.18.** (a)  800 W      (b)  80 percent      (c)  30 mH      (d)  5 A      (e)  1250 W

**29.19.** (a)  0.69 A      (b)  31°      (c)  0.80 A

**29.20.** (a)  18 $\Omega$      (b)  2.5 A      (c)  62.5 W      (d)  112.5 VA

**29.21.** (a)  503 Hz      (b)  4.5 A      (c)  202.5 W      (d)  202.5 VA

**29.22.** (a)  85 $\Omega$      (b)  1.41 A      (c)  99.4 W      (d)  169 VA

**29.23.** (a)  37.5 Hz      (b)  2.4 A      (c)  288 W      (d)  288 VA

**29.24.** (a)  $I_C = 0.5$ A, $I_R = 10$ A      (b)  1.12 A      (c)  8.9 $\Omega$      (d)  27°; 10 W

**29.25.** (a)  $I_L = 0.8$ A, $I_R = 1.0$ A      (b)  1.3 A      (c)  7.7 $\Omega$      (d)  −40°; 10 W

**29.26.** (a)  0.960 A; 0.603 A; 0.955 A      (b)  1.022 A      (c)  23.48 $\Omega$
(d)  *I* lags *V* by 20°      (e)  23 W

**29.27.** (a)  0.960 A; 1.206 A; 0.477 A      (b)  1.205 A      (c)  19.91 $\Omega$
(d)  *I* leads *V* by 37°      (e)  23 W

**29.28.**  The ratio of turns is 50:1; the winding with the greater number of turns should be connected to the amplifier since it has the greater impedance.

# Chapter 30

# Light

## ELECTROMAGNETIC WAVES

*Electromagnetic waves* consist of coupled electric and magnetic fields that vary periodically as they move through space. The electric and magnetic fields are perpendicular to each other and to the direction in which the waves travel (Fig. 30-1), so the waves are transverse, and the variations in **E** and **B** occur simultaneously. Electromagnetic waves transport energy and require no material medium for their passage. Radio waves, light waves, X-rays, and gamma rays are examples of electromagnetic waves, and they differ only in frequency. The color sensation produced by light waves depends on their frequency, with red light having the lowest visible frequencies and violet light the highest. White light contains light waves of all frequencies.

Electromagnetic waves are generated by accelerated electric charges, usually electrons. Electrons oscillating back and forth in an antenna give off radio waves, for instance, and accelerated electrons in atoms give off light waves.

In free space all electromagnetic waves have the *velocity of light* which is

$$\text{Velocity of light} = c = 3.00 \times 10^8 \text{ m/s} = 186{,}000 \text{ mi/s}$$

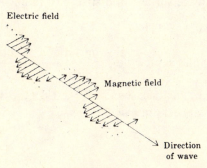

Electric field

Magnetic field

Direction
of wave

**Fig. 30-1**

## SOLVED PROBLEM 30.1

Why can light waves travel through a vacuum whereas sound waves cannot?

Light waves consist of coupled fluctuations in electric and magnetic fields and hence require no material medium for their passage. Sound waves, however, are pressure fluctuations and cannot occur without a material medium to transmit them.

## SOLVED PROBLEM 30.2

A marine radar operates at a wavelength of 3.2 cm. What is the frequency of the radar waves?

Radar waves are electromagnetic and hence travel with the velocity of light $c$. Therefore

$$f = \frac{c}{\lambda} = \frac{3 \times 10^8 \text{ m/s}}{3.2 \times 10^{-2} \text{ m}} = 9.4 \times 10^9 \text{ Hz}$$

## LUMINOUS INTENSITY AND FLUX

The eye responds to only part of the electromagnetic radiation emitted by most light sources (about 10 percent in the case of an ordinary light bulb) and is not equally sensitive to light of different colors (the greatest sensitivity is to yellow-green light). For these reasons the watt is not a useful unit for comparing light sources and the illumination they provide, and other units, more closely based on the visual response of the eye, are necessary.

The brightness of a light source is called its *luminous intensity I*, whose unit is the *candela* (cd). The candela is defined in terms of the light emitted by a blackbody (see Chapter 22) at the freezing temperature of platinum, 1773°C. The intensity of a light source is sometimes referred to as its *candlepower*.

The amount of visible light that falls on a given surface is called *luminous flux F*, whose unit is the *lumen* (lm). One lumen is equal to the luminous flux which falls on each 1 m$^2$ of a sphere 1 m in radius when a 1-cd isotrophic light source (one that radiates equally in all directions) is at the center of the sphere. Since the surface area of a sphere of radius $r$ is $4\pi r^2$, a sphere whose radius is 1 m has $4\pi$ m$^2$ of area, and the total luminous flux emitted by a 1-cd source is therefore $4\pi$ lm. Thus the luminous flux emitted by an isotropic light source of intensity $I$ is given by

$$F = 4\pi I$$

Luminous flux $= (4\pi)$(luminous intensity)

The above formula does not apply to a light source that radiates different fluxes in different directions. In such a situation the concept of *solid angle* is needed. A solid angle is the counterpart in three dimensions of an ordinary angle in two dimensions. The solid angle $\Omega$ (Greek capital letter *omega*) subtended by area $A$ on the surface of a sphere of radius $r$ is given by

$$\Omega = \frac{A}{r^2}$$

Solid angle $= \dfrac{\text{area on surface of sphere}}{(\text{radius of sphere})^2}$

The unit of solid angle is the *steradian* (sr); see Fig. 30-2. Like the degree and the radian, the steradian is a dimensionless ratio that disappears in calculations.

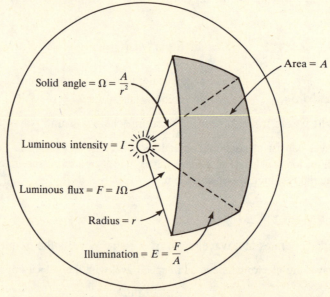

**Fig. 30-2**

The general definition of luminous flux is

$$F = I\Omega$$

Luminous flux = (luminous intensity)(solid angle)

Since the total area of a sphere is $4\pi r^2$, the total solid angle it subtends is $4\pi r^2/r^2$ sr = $4\pi$ sr. This definition of $F$ thus gives $F = 4\pi I$ for the total flux emitted by an isotropic source, as stated earlier. The luminous flux a 1-cd source gives off per steradian therefore equals 1 lm, and 1 cd equals 1 lm/sr.

The *luminous efficiency* of a light source is the amount of luminous flux it radiates per watt of input power. The luminous efficiency of ordinary tungsten-filament lamps increases with their power, because the higher the power of such a lamp, the greater its temperature and the more of its radiation is in the visible part of the spectrum. The efficiencies of such lamps range from about 8 lm/W for a 10-W lamp to 22 lm/W for a 1000-W lamp. Fluorescent lamps have efficiencies from 40 to 75 lm/W.

## ILLUMINATION

The *illumination* (or *illuminance*) $E$ of a surface is the luminous flux per unit area that reaches the surface:

$$E = \frac{F}{A}$$

Illumination = $\dfrac{\text{luminous flux}}{\text{area}}$

In SI, the unit of illumination is the lumen per square meter, or *lux* (lx); in the British system it is the lumen per square foot, or *footcandle* (fc) (Table 30-1).

**Table 30-1**

| Quantity | Symbol | Meaning | Formula | Unit |
|---|---|---|---|---|
| Luminous intensity | $I$ | Brightness of light source | | Candela (cd) |
| Solid angle | $\Omega$ | Three-dimensional equivalent of a plane angle | $\Omega = A/r^2$ | Steradian (sr) |
| Luminous flux | $F$ | Amount of visible light | $F = I\Omega$ | Lumen (lm) |
| Luminous efficiency | | Ratio of luminous flux to input power of light source | $F/P$ | Lumen/watt (lm/W) |
| Illumination | $E$ | Luminous flux per unit area | $E = F/A$ | Lux (lx) (= $lm/m^2$) Footcandle (fc) (= $lm/ft^2$) |

The illumination on a surface a distance $R$ away from an isotropic source of light of intensity $I$ is

$$E = \frac{I \cos \theta}{R^2}$$

where $\theta$ is the angle between the direction of the light and the normal to the surface (Fig. 30-3). Thus the illumination from such a source varies inversely as $R^2$, just as in the case of sound waves; doubling

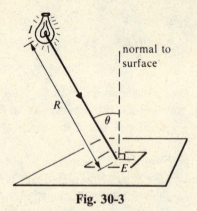

**Fig. 30-3**

the distance means, reducing the illumination to $(\frac{1}{2})^2 = \frac{1}{4}$ its former value. For light perpendicularly incident on a surface, $\theta = 0$ and $\cos \theta = 1$, so in this situation $E = I/R^2$.

## SOLVED PROBLEM 30.3

A 10-W fluorescent lamp has a luminous intensity of 35 cd. Find (a) the luminous flux it emits and (b) its luminous efficiency.

(a) $$F = 4\pi I = (4\pi)(35 \text{ cd}) = 440 \text{ lm}$$

(b) $$\text{Luminous efficiency} = \frac{F}{P} = \frac{440 \text{ lm}}{10 \text{ W}} = 44 \text{ lm/W}$$

## SOLVED PROBLEM 30.4

A spotlight concentrates all the light from a 100-cd bulb in a circle 1.3 m in radius on a wall. If the spotlight beam is perpendicular to the wall, find the illumination it produces.

The luminous flux emitted by the bulb is

$$F = 4\pi I = (4\pi)(100 \text{ cd}) = 400\pi \text{ lm}$$

The area of a circle 1.3 m in radius is $A = \pi r^2 = (\pi)(1.3 \text{ m})^2 = 1.69\pi \text{ m}^2$. Hence the illumination is

$$E = \frac{F}{A} = \frac{400\pi \text{ lm}}{1.69\pi \text{ m}^2} = 237 \text{ lm/m}^2 = 237 \text{ lx}$$

## SOLVED PROBLEM 30.5

What area on the surface of a sphere of radius 60 cm is cut by a solid angle of 0.2 sr from its center?

Since $\Omega = A/r^2$, here

$$A = \Omega r^2 = (0.2 \text{ sr})(60 \text{ cm})^2 = 720 \text{ cm}^2$$

We note that the steradian, which is a dimensionless ratio, does not appear in the result.

## SOLVED PROBLEM 30.6

A spotlight concentrates the light from a 150-cd bulb into a circle 0.8 m in radius a distance 25 m away. Find the luminous intensity of the source looking into the beam. This is the luminous intensity of an isotropic source that provides the same luminous flux on the illuminated circle.

The bulb emits the total luminous flux

$$F = 4\pi I = (4\pi)(150 \text{ cd}) = 1885 \text{ lm}$$

Because the radius $R$ of the illuminated circle is small compared with the radius $r$ of a sphere whose center is the spotlight, we can neglect the difference between the circle's plane area of $R^2$ and its area measured on the surface of the sphere. Hence the solid angle of the spotlight beam is

$$\Omega = \frac{A}{r^2} = \frac{\pi R^2}{r^2} = \frac{(\pi)(0.8 \text{ m})^2}{(25 \text{ m})^2} = 0.0032 \text{ sr}$$

The luminous intensity $I'$ of the spotlight looking into its beam is therefore

$$I' = \frac{F}{\Omega} = \frac{1885 \text{ lm}}{0.0032 \text{ sr}} = 5.89 \times 10^5 \text{ cd}$$

The source brightness has been increased by a factor of $I'/I = 3927$ by the optical system of the spotlight.

## SOLVED PROBLEM 30.7

An illumination of about 800 lx is recommended for reading. How far away from a book should a 75-W lamp of intensity 90 cd be located if the angle between the light rays and the plane of the opened book is 60°?

The angle between the light rays and the normal to the book is $\theta = 30°$. Since $E = (I \cos \theta)/R^2$,

$$R = \sqrt{\frac{I \cos \theta}{E}} = \sqrt{\frac{(90 \text{ cd})(\cos 30°)}{800 \text{ lm/m}^2}} = 0.31 \text{ m} = 31 \text{ cm}$$

## SOLVED PROBLEM 30.8

A 60-W light bulb whose luminious efficiency is 14 lm/W is suspended 2 m over a table. (*a*) What is the illumination on the table directly under the bulb? (*b*) How high over the table should the bulb be in order to double that illumination?

(*a*)   The luminous flux emitted by the bulb is

$$F = (60 \text{ W})(14 \text{ lm/W}) = 840 \text{ lm}$$

Since $F = 4\pi I$, the intensity of the bulb is

$$I = \frac{F}{4\pi} = \frac{840 \text{ lm}}{4\pi} = 66.8 \text{ cd}$$

The illumination at a distance of $R = 2$ m is

$$E = \frac{I}{R^2} = \frac{66.8 \text{ cd}}{(2 \text{ m})^2} = 16.7 \text{ lm/m}^2 = 16.7 \text{ lx}$$

(*b*)   Since $E = I/R^2$,

$$\frac{E_2}{E_1} = \frac{I_2 R_1^2}{I_1 R_2^2}$$

Here $I_1 = I_2$ and $E_2/E_1 = 2$, so

$$R_2 = R_1 \sqrt{\frac{E_1}{E_2}} = (2 \text{ m})\sqrt{\frac{1}{2}} = 1.41 \text{ m}$$

**SOLVED PROBLEM 30.9**

A lamp is suspended 6 ft over a large table. How much greater is the illumination directly under the lamp as compared with the illumination at a point on the table 4 ft to one side?

Directly under the lamp the illumination is

$$E_1 = \frac{I}{R_1^2}$$

and at a distance $d$ away on the table it is

$$E_2 = \frac{I \cos \theta}{R_2^2}$$

With the help of Fig. 30-4 we have

$$R_2 = \sqrt{R_1^2 + d^2} = \sqrt{(6 \text{ ft})^2 + (4 \text{ ft})^2} = 7.2 \text{ ft}$$

and

$$\cos \theta = \frac{R_1}{R_2} = \frac{6 \text{ ft}}{7.2 \text{ ft}} = 0.83$$

Hence

$$\frac{E_1}{E_2} = \frac{I/R_1^2}{(I \cos \theta)/R_2^2} = \frac{R_2^2}{R_1^2 \cos \theta} = \frac{(7.2 \text{ ft})^2}{(6 \text{ ft})^2 (0.83)} = 1.7$$

The illumination is 1.7 times greater directly under the lamp.

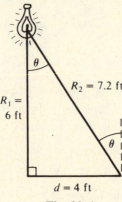

**Fig. 30-4**

## REFLECTION OF LIGHT

When a beam of light is reflected from a smooth, plane surface, the angle of reflection equals the angle of incidence (Fig. 30-5). The image of an object in a plane mirror has the same size and shape as the object but with left and right reversed; the image is the same distance behind the mirror as the object is in front of it.

**SOLVED PROBLEM 30.10**

A woman 170 cm tall wishes to buy a mirror in which she can see herself at full length. What is the minimum height of such a mirror? How far from the mirror should she stand?

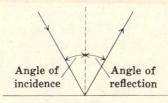

**Fig. 30-5**

Since the angle of reflection equals the angle of incidence, the mirror should be half her height (85 cm) and placed so its top is level with the middle of her forehead (Fig. 30-6). The distance between the mirror and the woman does not matter.

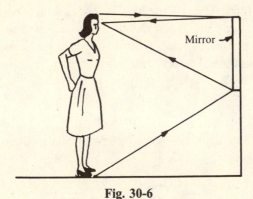

**Fig. 30-6**

## REFRACTION OF LIGHT

When a beam of light passes obliquely from one medium to another in which its velocity is different, its direction changes (Fig. 30-7). The greater the ratio between the two velocities, the greater the deflection. If the light goes from the medium of high velocity to the one of low velocity, it is bent toward the normal to the surface; if the light goes the other way, it is bent away from the normal. Light moving along the normal is not deflected.

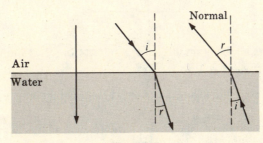

**Fig. 30-7**

The *index of refraction* of a transparent medium is the ratio between the velocity of light in free space $c$ and its velocity in the medium $v$:

$$\text{Index of refraction} = n = \frac{c}{v}$$

The greater its index of refraction, the more a beam of light is deflected on entering a medium from air. The index of refraction of air is about 1.0003, so for most purposes it can be considered equal to 1.

According to *Snell's law*, the angles of incidence $i$ and refraction $r$ shown in Fig. 30-7 are related by the formula

$$\frac{\sin i}{\sin r} = \frac{v_1}{v_2} = \frac{n_2}{n_1}$$

where $v_1$ and $n_1$ are, respectively, the velocity of light and index of refraction of the first medium and $v_2$ and $n_2$ are the corresponding quantities in the second medium. Snell's law is often written

$$n_1 \sin i = n_2 \sin r$$

In general, the index of refraction of a medium increases with increasing frequency of the light. For this reason a beam of white light is separated into its component frequencies, each of which produces the sensation of a particular color, when it passes through an object whose sides are not parallel, for instance, a glass prism. The resulting band of color is called a *spectrum*.

## SOLVED PROBLEM 30.11

Why is a beam of white light that passes perpendicularly through a flat pane of glass not dispersed into a spectrum?

Light incident perpendicular to a surface is not deflected, so light of the various frequencies in white light stays together despite the different velocities in the glass.

## SOLVED PROBLEM 30.12

The index of refraction of diamond is 2.42. What is the velocity of light in diamond?

Since $n = c/v$, here

$$v = \frac{c}{n} = \frac{3 \times 10^8 \text{ m/s}}{2.42} = 1.24 \times 10^8 \text{ m/s}$$

## SOLVED PROBLEM 30.13

A beam of light enters a lake at an angle of incidence of 40°. Find the angle of refraction. The index of refraction of water is 1.33.

Here medium 1 is air and medium 2 is water. From Snell's law,

$$\sin r = \frac{n_1}{n_2} \sin i = \frac{1.00}{1.33} \sin 40° = 0.483 \qquad r = 29°$$

The angle of refraction is less than the angle of incidence because $n_2 > n_1$.

## SOLVED PROBLEM 30.14

A lantern held by a submerged skin diver directs a beam of light at the surface of a lake at an angle of incidence of 40°. Find the angle of refraction.

Now medium 1 is water and medium 2 is air. Hence

$$\sin r = \frac{n_1}{n_2} \sin i = \frac{1.33}{1.00} \sin 40° = 0.855$$
$$r = 59°$$

The angle of refraction is greater than the angle of incidence because $n_2 < n_1$.

## SOLVED PROBLEM 30.15

A beam of light strikes a pane of glass at an angle of incidence of 50°. If the angle of refraction is 30°, find the index of refraction of the glass.

According to Snell's law, $n_1 \sin i = n_2 \sin r$. Here air is medium 1, so $n_1 = 1.00$ and

$$n_2 = (n_1)\left(\frac{\sin i}{\sin r}\right) = (1.00)\left(\frac{\sin 50°}{\sin 30°}\right) = 1.53$$

## SOLVED PROBLEM 30.16

A ray of light enters a glass plate whose index of refraction is 1.5 at an angle of incidence of 50° (Fig. 30-8). At what angle does the ray leave the other side of the plate?

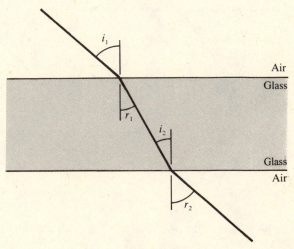

**Fig. 30-8**

We start by finding the angle of refraction $r_1$ at the first side of the plate:

$$\sin r_1 = \left(\frac{n_{air}}{n_{glass}}\right)(\sin i_1) = \left(\frac{1.0}{1.5}\right)(\sin 50°)$$
$$r_1 = \sin^{-1}\left(\frac{1.0}{1.5}\sin 50°\right) = \sin^{-1} 0.51 = 30.7°$$

The angle of incidence $i_2$ at the other side of the plate is equal to $r_1$. Hence

$$\sin r_2 = \left(\frac{n_{glass}}{n_{air}}\right)(\sin i_2) = \left(\frac{n_{glass}}{n_{air}}\right)(\sin r_1) = \left(\frac{1.5}{1.0}\right)(\sin 30.7°)$$
$$r_2 = \sin^{-1}\left(\frac{1.5}{1.0}\sin 30.7°\right) = 50°$$

The ray leaves the glass plate parallel to its original direction but shifted to one side. This result would be true regardless of the index of refraction of the glass.

## TOTAL INTERNAL REFLECTION

The phenomenon of *total internal reflection* can occur when light goes from a medium of high index of refraction to one of low index of refraction, for example, from glass or water to air. The angle of refraction in this situation is greater than the angle of incidence, and a light ray is bent away from the normal, as in Fig. 30-9(a), at the interface between the two media. At the *critical angle* of incidence, the angle of refraction is 90° [Fig. 30-9(b)], and at angles of incidence greater than this the refracted rays are reflected back into the original medium [Fig. 30-9(c)]. If the critical angle is $i_c$,

$$\sin i_c = \frac{n_2}{n_1}$$

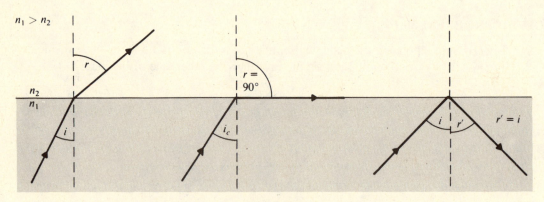

**Fig. 30-9**

## SOLVED PROBLEM 30.17

Derive the formula $\sin i_c = n_2/n_1$.

At the critical angle $i_c$ the angle of refraction $r$ is 90°, so $\sin r = \sin 90° = 1$ when $i = i_c$. Substituting $\sin r = 1$ into Snell's law

$$n_1 \sin i = n_2 \sin r$$

yields                              $n_1 \sin i_c = n_2 \qquad \sin i_c = \frac{n_2}{n_1}$

## SOLVED PROBLEM 30.18

Find the critical angle for light going from crown glass ($n = 1.52$) to air ($n = 1.00$) and for light going from crown glass to water ($n = 1.33$).

For light going from glass into air,

$$\sin i_c = \frac{1.00}{1.52} = 0.658 \qquad i_c = 41°$$

For light going from glass into water,

$$\sin i_c = \frac{1.33}{1.52} = 0.875 \qquad i_c = 61°$$

## SOLVED PROBLEM 30.19

Prisms are used in optical instruments instead of mirrors to change the direction of light beams by 90° because total internal reflection better preserves the sharpness and brightness of light beams. What is the minimum index of refraction of the glass used in such prisms?

As Fig. 30-10 shows, the angle of incidence must be 45°, so the critical angle must be at least 45°. Hence the minimum value of $n$ is

$$n = \frac{1}{\sin i_c} = \frac{1}{\sin 45°} = 1.41$$

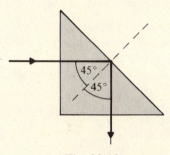

**Fig. 30-10**

## APPARENT DEPTH

An object submerged in water or other transparent liquid appears closer to the surface than it actually is. As Fig. 30-11 shows, light leaving the object is bent away from the normal to the water-air surface as it leaves the water. Since an observer interprets what she or he sees in terms of the straight-line propagation of light, the object seems at a shallower depth than its true one. The ratio between apparent and true depths is

$$\frac{\text{Apparent depth}}{\text{True depth}} = \frac{h'}{h} = \frac{n_2}{n_1}$$

where $n_1$ is the index of refraction of the liquid and $n_2$ is the index of refraction of air.

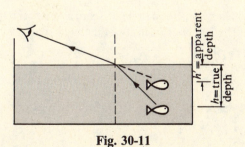

**Fig. 30-11**

**SOLVED PROBLEM 30.20**

The water in a swimming pool is 2 m deep. How deep does it appear to be to someone looking down into it?

Here $n_1 = 1.33$ and $n_2 = 1.00$, so

$$h' = h\,\frac{n_2}{n_1} = (2\text{ m})\left(\frac{1.00}{1.33}\right) = 1.5\text{ m}$$

## *Multiple-Choice Questions*

**30.1.** The magnetic field of an electromagnetic wave is

(a) parallel to the electric field and to the wave direction
(b) parallel to the electric field and perpendicular to the wave direction
(c) perpendicular to the electric field and to the wave direction
(d) perpendicular to the electric field and parallel to the wave direction

**30.2.** A transparent medium has an index of refraction that is

(a) less than 1     (c) greater than 1
(b) 1               (d) any of the above

**30.3.** For total internal reflection to occur when light passes from one medium to another, the second medium must have an index of refraction relative to that of the first medium that is

(a) lower     (c) higher
(b) the same  (d) any of the above, depending on the angle of incidence

**30.4.** The unit of the luminous intensity (brightness) of a light source is the

(a) lumen   (c) footcandle
(b) lux     (d) candela

**30.5.** A hemisphere subtends a solid angle about its center of

(a) $\pi/2$ sr   (c) $2\pi$ sr
(b) $\pi$ sr     (d) $4\pi$ sr

**30.6.** The luminous flux emitted by an isotropic light source is concentrated on an area of 0.5 m$^2$ whose illumination is 750 lx. The luminous intensity of the light source is

(a) 30 cd    (c) 375 cd
(b) 60 cd    (d) 4712 cd

**30.7.** The velocity of light in diamond is $1.24 \times 10^8$ m/s. The index of refraction of diamond is

(a) 1.24   (c) 2.42
(b) 2.31   (d) 3.72

**30.8.** Light enters a glass plate at an angle of incidence of 25°. If the index of refraction of the glass is 1.6, the angle of refraction is

(a)   15°        (c)   40°
(b)   16°        (d)   43°

**30.9.** A diver with a flashlight shines a light beam upward from the bottom of a pool at an angle of incidence of 40°. The light leaves the surface of the pool at an angle of refraction of 60°. The index of refraction of water is

(a)   0.67       (c)   1.3
(b)   0.74       (d)   1.5

**30.10.** A medal in a plastic cube appears to be 12 mm below its upper surface. If $n = 1.5$ for the plastic, the depth of the medal is actually

(a)   6 mm       (c)   18 mm
(b)   8 mm       (d)   24 mm

**30.11.** Light going from glass of $n = 1.5$ to ethanol has a critical angle of incidence of 63°. The index of refraction of ethanol is

(a)   0.59       (c)   1.3
(b)   0.75       (d)   2.4

# Supplementary Problems

**30.1.** When a light beam enters one medium from another, which (if any) of the following quantities never changes—the direction of the beam, its velocity, its frequency, its wavelength?

**30.2.** What is the relationship between the direction of an electromagnetic wave and the directions of its electric and magnetic fields?

**30.3.** What is the frequency of radio waves whose wavelength is 20 m?

**30.4.** A certain radio station transmits at a frequency of 1050 kHz. What is the wavelength of these waves?

**30.5.** Is it possible for the index of refraction of a substance to be less than 1?

**30.6.** A 100-W tungsten-filament lamp has a luminous efficiency of 16 lm/W. Find its intensity in candelas.

**30.7.** A newspaper is held 20 ft from a 2000-cd street lamp at midnight. What is the maximum illumination of the newspaper?

**30.8.** A 150-cd lamp is to be substituted for a 100-cd lamp over a workbench. If the original lamp was 2 m above the workbench, how much higher can the new lamp be and still provide the same illumination?

**30.9.** Find the solid angle subtended at the center of a sphere of radius 2.0 m by an area of 50 cm$^2$ on its surface.

**30.10.** An illumination of $10^4$ lx is produced on a screen by a projector 6 m away. Find the luminous intensity of the projector looking into its beam.

**30.11.** What intensity should a lamp have if it is to provide an illumination of 400 lx at a distance of 3 m on a surface whose normal is 20° from the direction of the light rays?

**30.12.** A playing field 250 ft by 400 ft is to have an average illumination at night of 30 ft. How many 4000-cd lamps are needed if reflectors are used that enable 40 percent of their luminous flux to reach the field?

**30.13.** The index of refraction of benzene is 1.50. Find the velocity of light in benzene.

**30.14.** The velocity of light in ice is $2.3 \times 10^8$ m/s. What is its index of refraction?

**30.15.** A beam of light enters a plate of flint glass ($n = 1.63$) at an angle of incidence of 40°. Find the angle of refraction.

**30.16.** A beam of light enters a tank of glycerin at an angle of incidence of 45°. The angle of refraction is 29°. Find the index of refraction of glycerin.

**30.17.** Find the critical angle for total internal reflection for light going from ice ($n = 1.31$) into air.

**30.18.** A stick frozen into a pond in winter appears to be 3 cm below the surface. What is its actual depth in the ice?

## *Answers to Multiple-Choice Questions*

**30.1.**  *(c)*      **30.7.**  *(c)*

**30.2.**  *(c)*      **30.8.**  *(a)*

**30.3.**  *(a)*      **30.9.**  *(c)*

**30.4.**  *(d)*      **30.10.** *(c)*

**30.5.**  *(c)*      **30.11.** *(c)*

**30.6.**  *(a)*

## Answers to Supplementary Problems

**30.1.**     Only the frequency never changes.

**30.2.**     The electric and magnetic fields are perpendicular to the direction of the wave and to each other.

**30.3.**    $1.5 \times 10^7$ Hz = 15 MHz

**30.4.**    286 m

**30.5.**    No, because this would mean a velocity of light greater than its velocity in free space, which is not possible.

**30.6.**    127 cd

**30.7.**    5 fc

**30.8.**    0.45 m higher for a total height of 2.45 m

**30.9.**    0.00125 sr

**30.10.**   $3.6 \times 10^5$ cd

**30.11.**   3831 cd

**30.12.**   149 lamps

**30.13.**   $2 \times 10^8$ m/s

**30.14.**   1.3

**30.15.**   23°

**30.16.**   1.46

**30.17.**   50°

**30.18.**   3.93 cm

# Chapter 31

# Spherical Mirrors

## FOCAL LENGTH

Figure 31-1 shows how a concave mirror converges a parallel beam of light to a real focal point $F$, and Fig. 31-2 shows how a convex mirror diverges a parallel beam of light so that the reflected rays appear to come from a virtual focal point $F$ behind the mirror. In either case, if the radius of curvature of the mirror is $R$, the *focal length f* is $R/2$. for a concave mirror, $f$ is positive, and for a convex mirror, $f$ is negative. Thus

$$\text{Concave mirror}: \quad f = +\frac{R}{2}$$

$$\text{Convex mirror}: \quad f = -\frac{R}{2}$$

The *axis* of a mirror of either kind is the straight line that passes through $C$ and $F$.

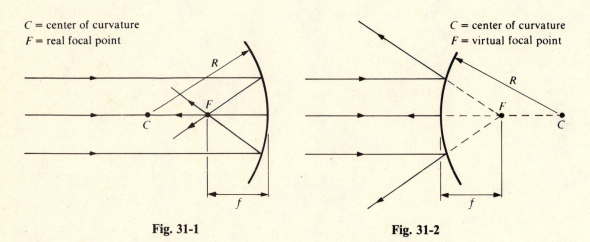

| |
|---|
| $C$ = center of curvature |
| $F$ = real focal point |

| |
|---|
| $C$ = center of curvature |
| $F$ = virtual focal point |

**Fig. 31-1**          **Fig. 31-2**

## RAY TRACING

The position and size of the image formed by a spherical mirror of an object in front of it can be found by constructing a scale drawing. What is done is to trace two different light rays from each point of interest in the object to where they (or their extensions, in the case of a virtual image) intersect after being reflected by the mirror. Three rays especially useful for this purpose are shown in Fig. 31-3; any two are sufficient:

1.  A ray that leaves the object parallel to the axis of the mirror. After reflection, this ray passes through the focal point of a concave mirror or seems to come from the focal point of a convex mirror.

2.  A ray that passes through the focal point of a concave mirror or is directed toward the focal point of a convex mirror. After reflection, this ray travels parallel to the axis of the mirror.

3.  A ray that leaves the object along a radius of the mirror. After reflection, this ray returns along the same radius.

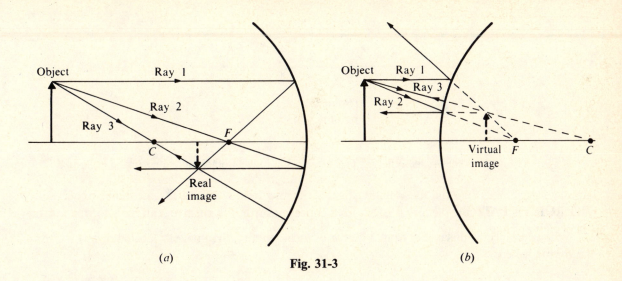

(a)                                    **Fig. 31-3**                                    (b)

## SOLVED PROBLEM 31.1

What is *spherical aberration?*

Spherical aberration refers to the fact that light rays from a point on an object that are reflected at different distances from the axis of a spherical mirror do not converge to (or appear to diverge from) a single point. This effect is shown in Fig. 31-4 for parallel rays reaching a concave mirror: Rays reflected from the outer parts of the mirror converge at focal points closer to the mirror than those reflected near the mirror's axis. As a result, a spherical mirror, which has a circular cross section, produces sharp images only close to the mirror axis. Concave mirrors whose cross sections are parabolic do not suffer from spherical aberration and are used when high-quality images are required, for instance, in telescopes.

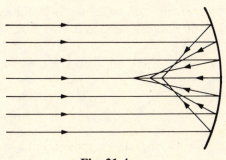

**Fig. 31-4**

## SOLVED PROBLEM 31.2

What is the nature of the image of a real object formed by a convex mirror?

The image is virtual, erect, and smaller in size than the object, as in Fig. 31-3(*b*).

## SOLVED PROBLEM 31.3

Describe the image formed by a concave mirror of an object placed at the focal point of the mirror.

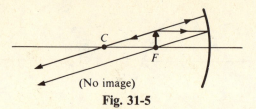

(No image)

**Fig. 31-5**

As Fig. 31-5 shows, the reflected rays are parallel to each other and so no image is formed.

## MIRROR EQUATION

When an object is distance $p$ from a mirror of focal length $f$, the image is located distance $q$ from the mirror, where

$$\frac{1}{p} + \frac{1}{q} = \frac{1}{f}$$

$$\frac{1}{\text{Object distance}} + \frac{1}{\text{image distance}} = \frac{1}{\text{focal length}}$$

This equation holds for both concave and convex mirrors (see Fig. 31-6). The mirror equation is readily solved for $p$, $q$, or $f$:

$$p = \frac{qf}{q-f} \qquad q = \frac{pf}{p-f} \qquad f = \frac{pq}{p+q}$$

A positive value of $p$ or $q$ denotes a real object or image, and a negative value denotes a virtual object or image. A *real object* is in front of a mirror; a *virtual object* appears to be located behind the mirror and must itself be an image produced by another mirror or lens. A *real image* is formed by light rays that actually pass through the image, so a real image will appear on a screen placed at the position of the image. But a *virtual image* can be seen only by the eye since the light rays that appear to come from the image actually do not pass through it. Real images are located in front of a mirror, virtual images behind it.

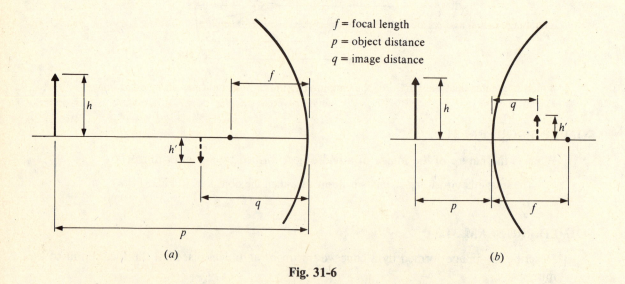

$f$ = focal length
$p$ = object distance
$q$ = image distance

**Fig. 31-6**

## MAGNIFICATION

The *linear magnification m* of any optical system is the ratio between the size (height or width or other transverse linear dimension) of the image and the size of the object. In the case of a mirror,

$$m = \frac{h'}{h} = -\frac{q}{p}$$

$$\text{Linear magnification} = \frac{\text{image height}}{\text{object height}} = -\frac{\text{image distance}}{\text{object distance}}$$

A positive magnification signifies an erect image, as in Fig. 31-6(*b*); a negative one signifies an inverted image, as in Fig. 31-6(*a*). Table 31.1 is a summary of the sign conventions used in connection with spherical mirrors.

**Table 31-1**

| Quantity | Positive | Negative |
|---|---|---|
| Focal length $f$ | Concave mirror | Convex mirror |
| Object distance $p$ | Real object | Virtual object |
| Image distance $q$ | Real image | Virtual image |
| Magnification $m$ | Erect image | Inverted image |

## SOLVED PROBLEM 31.4

A candle 5 cm high is placed 20 cm in front of a concave mirror whose focal length is 15 cm. Find the location, size, and nature of the image.

Here $p = 20$ cm and $f = 15$ cm, so the image distance is

$$q = \frac{pf}{p - f} = \frac{(20 \text{ cm})(15 \text{ cm})}{20 \text{ cm} - 15 \text{ cm}} = 60 \text{ cm}$$

The image is real and on the same side of the mirror as the candle (Fig. 31-7). The height of the image is

$$h' = -h\,\frac{q}{p} = -(5 \text{ cm})\left(\frac{60 \text{ cm}}{20 \text{ cm}}\right) = -15 \text{ cm}$$

which is greater than the height of the candle. The minus sign indicates an inverted image.

In general, an object placed between the focal point and the center of curvature $C$ of a concave mirror (that is, with $p$ greater than $f$ but less than $2f$ ) will have a real, inverted image that is larger than the object.

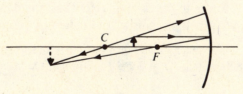

**Fig. 31-7**

**SOLVED PROBLEM 31.5**

A pencil 12 cm long is placed at the center of curvature of a concave mirror whose focal length is 40 cm. Find the location, size, and nature of the image.

Since $f = R/2$ for a concave mirror, $p = R = 2f = 80$ cm here. The image distance is therefore

$$q = \frac{pf}{p-f} = \frac{(80 \text{ cm})(40 \text{ cm})}{80 \text{ cm} - 40 \text{ cm}} = 80 \text{ cm}$$

The image is real and at the same distance from the mirror as the object (Fig. 31-8). The height of the image is

$$h' = -h\,\frac{q}{p} = -(12 \text{ cm})\left(\frac{80 \text{ cm}}{80 \text{ cm}}\right) = -12 \text{ cm}$$

which is the same as the height of the object. The minus sign indicates an inverted image.

In general, an object placed at the center of curvature of a concave mirror has a real, inverted image the same size as the object and at the same distance from the mirror.

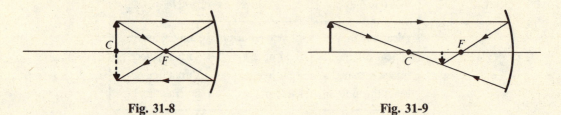

Fig. 31-8                                                        Fig. 31-9

**SOLVED PROBLEM 31.6**

A cigar 15 cm long is placed 75 cm in front of a concave mirror whose focal length is 30 cm. Find the location, size, and nature of the image.

Here $p = 75$ cm and $f = 30$ cm, so the image distance is

$$q = \frac{pf}{p-f} = \frac{(75 \text{ cm})(30 \text{ cm})}{75 \text{ cm} - 30 \text{ cm}} = 50 \text{ cm}$$

The image is real and on the same side of the mirror as the cigar (Fig. 31-9). The length of the image is

$$h' = -h\,\frac{q}{p} = -(15 \text{ cm})\left(\frac{50 \text{ cm}}{75 \text{ cm}}\right) = -10 \text{ cm}$$

which is smaller than the length of the cigar. The minus sign indicates an inverted image.

**SOLVED PROBLEM 31.7**

A *reflecting telescope* that uses a concave mirror whose radius of curvature is 4 m produces an image of the moon on a photographic plate. The moon's diameter is approximately 3500 km, and it is about 384,000 km from the earth. (*a*) How far should the photographic plate be placed from the mirror? (*b*) What will be the size and nature of the moon's image?

(*a*)   The object distance is so much greater than the mirror's focal length of $f = R/2 = 2$ m that we can let $p = \infty$ here. Substituting $1/p = 0$ and $f = 2$ m into the mirror equation yields

$$\frac{1}{p} + \frac{1}{q} = \frac{1}{f} \qquad 0 + \frac{1}{q} = \frac{1}{2 \text{ m}} \qquad q = 2 \text{ m}$$

The image is real and on the same side of the mirror as the moon.

(*b*)   Since the moon's diameter is $h = 3500$ km $= 3.5 \times 10^6$ m and object distance is $p = 384,000$ km $= 3.84 \times 10^8$ m, the diameter of the moon's image is

$$h' = -h\,\frac{q}{p} = -(3.5 \times 10^6 \text{ m})\left(\frac{2 \text{ m}}{3.84 \times 10^8 \text{ m}}\right) = -0.018 \text{ m} = -18 \text{ mm}$$

The minus sign indicates an inverted image (Fig. 31-10).

In general, an object very far away from a concave mirror relative to its focal length will have a real, inverted image smaller than the object and located very nearly at the focal point of the mirror.

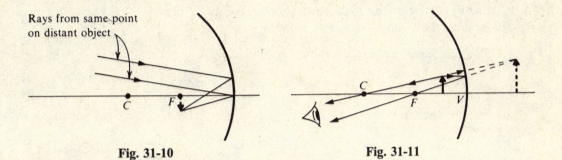

**Fig. 31-10**                                                   **Fig. 31-11**

## SOLVED PROBLEM 31.8

A concave mirror has a radius of curvature of 120 cm. How far from the mirror should one's face be so that the image is erect and twice the size of the actual face? Is the image real or virtual?

The focal length of the mirror is $f = R/2 = 60$ cm and is positive since the mirror is concave. Because the image is to be erect and twice the size of the object, $m = +2$. We know $m$ and $f$ and are to find the object distance $p$. There are various ways to do this, one of which is as follows. Since $m = -q/p$, we have $q = -mp$. The negative image distance signifies a virtual image (Fig. 31-11). Substituting $q = -mp$ in the mirror equation enables us to solve for $p$:

$$\frac{1}{p} + \frac{1}{q} = \frac{1}{f} \qquad \frac{1}{p} - \frac{1}{mp} = \frac{1}{f} \qquad \frac{m-1}{mp} = \frac{1}{f} \qquad \frac{mp}{m-1} = f$$

$$p = \frac{f}{m}\,(m - 1) = \left(\frac{60 \text{ cm}}{2}\right)(2 - 1) = 30 \text{ cm}$$

In general, an object placed closer to a concave mirror than its focal point (that is, closer than $f$) will have a virtual, erect image that is larger than the object.

## SOLVED PROBLEM 31.9

How far away from a concave mirror of 40-cm focal length should a real object 30 mm long be located in order that its image be 8 mm long?

When a concave mirror forms an image that is smaller than the object, the image is always inverted (see Fig. 31-9). An inverted image is considered to have a negative height, so here the magnification is

$$m = \frac{h'}{h} = \frac{-8 \text{ mm}}{30 \text{ mm}} = -0.267$$

From the solution to Prob. 31.8,

$$p = \frac{f}{m}\,(m-1) = \left(\frac{40 \text{ cm}}{-0.267}\right)(-0.267 - 1) = 190 \text{ cm}$$

If we had not known that the image was inverted and had used a magnification of $+0.267$, the result would have been an object distance of $-110$ cm. But a negative object distance signifies a virtual object, whereas here we are given a real object, so it would be clear that the wrong sign has been used for the magnification.

## SOLVED PROBLEM 31.10

A grasshopper 5 cm long is 25 cm in front of a convex mirror whose radius of curvature is 80 cm. Find the location, size, and nature of the image.

The focal length of the mirror is

$$f = -\frac{R}{2} = -\frac{80 \text{ cm}}{2} = -40 \text{ cm}$$

The image distance is

$$q = \frac{pf}{p-f} = \frac{(25 \text{ cm})(-40 \text{ cm})}{25 \text{ cm} - (40 \text{ cm})} = -\frac{(25 \text{ cm})(40 \text{ cm})}{25 \text{ cm} + 40 \text{ cm}} = -15.4 \text{ cm}$$

The minus sign indicates a virtual image located behind the mirror. The length of the image is

$$h' = -h\,\frac{q}{p} = (-5 \text{ cm})\left(\frac{-15.4 \text{ cm}}{25 \text{ cm}}\right) = 3.1 \text{ cm}$$

A positive value for $h'$ signifies an erect image.

# *Multiple-Choice Questions*

**31.1.**      When a spherical mirror forms a real image, the image relative to its object is always

     (a)   smaller      (c)   erect
     (b)   larger       (d)   inverted

**31.2.**      An image distance that is negative means that the image is

     (a)   erect        (c)   real
     (b)   inverted     (d)   virtual

**31.3.**      A magnification that is negative means that the image is

     (a)   erect        (c)   smaller than the object
     (b)   inverted     (d)   larger than the object

**31.4.**      An erect image is formed by a concave mirror when the object distance is

     (a)   less than $f$      (c)   between $f$ and $2f$
     (b)   $f$             (d)   greater than $2f$

**31.5.** A concave mirror forms an enlarged image of an object

    (*a*)   for no values of *p*      (*c*)   when *p* is more than 2*f*

    (*b*)   when *p* is less than 2*f*    (*d*)   for all values of *p*

**31.6.** A convex mirror forms an enlarged image of an object

    (*a*)   for no values of *p*      (*c*)   when *p* is more than 2*f*

    (*b*)   when *p* is less than 2*f*    (*d*)   for all values of *p*

**31.7.** The focal length of a convex mirror whose radius of curvature is 24 cm is

    (*a*)   $-12$ cm    (*c*)   $+12$ cm

    (*b*)   $-48$ cm    (*d*)   $+48$ cm

**31.8.** A candle 6 cm high is 40 cm in front of a mirror whose focal length is $+60$ cm. The image is

    (*a*)   2 cm high, erect    (*c*)   2 cm high, inverted

    (*b*)   18 cm high, erect    (*d*)   18 cm high, inverted

**31.9.** The candle of Question 31.8 is 100 cm in front of the same mirror. The image is now

    (*a*)   4 cm long, erect    (*c*)   4 cm long, inverted

    (*b*)   9 cm long, erect    (*d*)   9 cm long, inverted

**31.10.** A candle 6 cm high is 40 cm in front of a mirror whose focal length is $-60$ cm. The image is

    (*a*)   3.6 cm long, erect    (*c*)   3.6 cm long, inverted

    (*b*)   10 cm long, erect    (*d*)   10 cm long, inverted

# Supplementary Problems

**31.1.** What does a negative magnification signify? A magnification that is less than 1?

**31.2.** Under what circumstances does a concave mirror produce a real image of a real object? A virtual image?

**31.3.** A moth flies toward a convex mirror. Does its image become larger or smaller as it approaches the mirror's focal point? What kind of image is it?

**31.4.** (*a*) A concave mirror has a radius of curvature of 30 cm. What is its focal length? (*b*) A convex mirror has a radius of curvature of 30 cm. What is its focal length?

**31.5.** A match 6 cm long is placed 30 cm in front of a concave mirror whose focal length is 50 cm. Find the location, size, and nature of the image.

**31.6.** The match of Prob. 31.5 is placed 50 cm in front of the same mirror. Find the location, size, and nature of the image.

**31.7.** The match of Prob. 31.5 is placed 80 cm in front of the same mirror. Find the location, size, and nature of the image.

**31.8.** The match of Prob. 31.5 is placed 100 cm in front of the same mirror. Find the location, size, and nature of the image.

**31.9.** The match of Prob. 31.5 is placed 3 m in front of the same mirror. Find the location, size, and nature of the image.

**31.10.** A pencil 6 in. long is placed 8 in. in front of a mirror whose focal length is −15 in. Find the location, size, and nature of the image.

**31.11.** How far away from a concave mirror of a 25-cm focal length should a real object be located so that its image is one-third its actual size?

**31.12.** A shaving mirror is intended to produce an erect image of a man's face 2.5 times its actual size when the face is 30 cm in front of it. (*a*) Should the mirror be concave or convex? (*b*) What should its radius of curvature be?

## *Answers to Multiple-Choice Questions*

**31.1.**  (*d*)       **31.6.**   (*a*)

**31.2.**  (*d*)       **31.7.**   (*a*)

**31.3.**  (*b*)       **31.8.**   (*b*)

**31.4.**  (*a*)       **31.9.**   (*d*)

**31.5.**  (*b*)       **31.10.**  (*a*)

## Answers to Supplementary Problems

**31.1.**   An inverted image; an image smaller than the object.

**31.2.**   When the object distance is greater than the focal length of the mirror; when the object distance is less than the focal length.

**31.3.**   Larger, virtual, and erect.

**31.4.**   (*a*)  15 cm      (*b*)   −15 cm

**31.5.**   The image is 75 cm behind the mirror, 15 cm long, virtual, and erect.

**31.6.**   No image is formed.

**31.7.**   The image is 133 cm in front of the mirror, 10 cm long, real, and inverted.

**31.8.**    The image is 100 cm in front of the mirror, 6 cm long, real, and inverted.

**31.9.**    The image is 60 cm in front of the mirror, 1.2 cm long, real, and inverted.

**31.10.**    The image is 5.22 in. behind the mirror, 3.91 in. long, virtual, and erect.

**31.11.**    100 cm

**31.12.**    (*a*)  Concave       (*b*)   200 cm

# Chapter 32

## Lenses

### FOCAL LENGTH

Figure 32-1 shows how a converging lens brings a parallel beam of light to a real focal point $F$, and Fig. 32-2 shows how a diverging lens spreads out a parallel beam of light so that the refracted rays appear to come from a virtual focal point $F$. In this chapter we consider only *thin lenses*, whose thickness can be neglected as far as optical effects are concerned. The focal length $f$ of a thin lens is given by the *lensmaker's equation*:

$$\frac{1}{f} = (n - 1)\left(\frac{1}{R_1} + \frac{1}{R_2}\right)$$

In this equation $n$ is the index of refraction of the lens material relative to the medium it is in, and $R_1$ and $R_2$ are the radii of curvature of the two surfaces of the lens. Both $R_1$ and $R_2$ are considered as plus for a convex (curved outward) surface and as minus for a concave (curved inward) surface; obviously it does not matter which surface is labeled as 1 and which as 2.

A positive focal length corresponds to a converging lens and a negative focal length to a diverging lens.

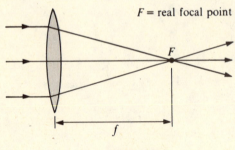

$F$ = real focal point

**Fig. 32-1**

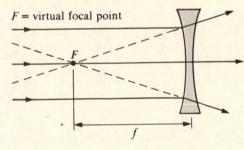

$F$ = virtual focal point

**Fig. 32-2**

### SOLVED PROBLEM 32.1

A *planoconvex lens* has one plane surface and one convex surface. If a planoconvex lens of focal length 12 cm is to be ground from glass of index of refraction 1.60, find the radius of curvature of the convex surface.

The radius of curvature of the plane surface is infinity, so if we call this surface 1, then $1/R_1 = 0$. From the lensmaker's equation

$$\frac{1}{f} = (n - 1)\left(\frac{1}{R_1} + \frac{1}{R_2}\right) = (n - 1)\left(\frac{1}{R_2}\right)$$

and so

$$R_2 = (n - 1)f = (1.60 - 1.00)(12 \text{ cm}) = 7.2 \text{ cm}$$

### SOLVED PROBLEM 32.2

A *meniscus lens* has one concave and one convex surface. The concave surface of a particular meniscus lens has a radius of curvature of 30 cm, and its convex surface has a radius of curvature of 50 cm. The index of refraction of the glass used is 1.50. (a) Find the focal length of the lens. (b) Is it a converging lens or a diverging lens?

(a)   Here $R_1 = -30$ cm (since the first surface is concave, its radius is considered negative), and $R_2 = +50$ cm. Hence

$$\frac{1}{f} = (n-1)\left(\frac{1}{R_1} + \frac{1}{R_2}\right) = (1.50 - 1.00)\left(-\frac{1}{30 \text{ cm}} + \frac{1}{50 \text{ cm}}\right) = -0.00667 \text{ cm}^{-1}$$

$$f = -\frac{1}{0.00667 \text{ cm}^{-1}} = -150 \text{ cm}$$

(b)   A negative focal length signifies a diverging lens.

## SOLVED PROBLEM 32.3

A lens made of glass whose index of refraction is 1.60 has a focal length of +20 cm in air. Find its focal length in water, whose index of refraction is 1.33.

Let $f$ be the focal length of the lens in air and $f'$ be its focal length in water. The index of refraction of the glass relative to air is $n = 1.60$ since the index of refraction of air is very nearly equal to 1. The index of refraction of the glass relative to water is

$$n' = \frac{\text{index of refraction of glass}}{\text{index of refraction of water}} = \frac{1.60}{1.33} = 1.20$$

From the lensmaker's equation, since $R_1$ and $R_2$ are the same in both air and water,

$$\frac{f'}{f} = \frac{n-1}{n'-1} = \frac{1.60 - 1.00}{1.20 - 1.00} = 3$$

and so                                    $$f' = 3f = (3)(+20 \text{ cm}) = +60 \text{ cm}$$

The focal length of any lens made of this glass is three times longer in water than in air.

## SOLVED PROBLEM 32.4

Both surfaces of a double-concave lens whose focal length is −9 cm/cm have a radii of 10 in. Find the index of refraction of the glass.

From the lensmaker's equation

$$n - 1 = \frac{1}{f(1/R_1 + 1/R_2)} = \frac{1}{(-9 \text{ cm})[1/(-10 \text{ cm}) + 1/(-10 \text{ cm})]} = 0.56$$

$$n = 0.56 + 1 = 1.56$$

## SOLVED PROBLEM 32.5

The focal length $f$ of a combination of two thin lenses in contact whose individual focal lengths are $f_1$ and $f_2$ is given by

$$\frac{1}{f} = \frac{1}{f_1} + \frac{1}{f_2}$$

Use this formula to find the focal length of a combination of a converging lens of $f = +10$ cm and a diverging lens of $f = -20$ cm that are in contact.

The above formula can be rewritten in the more convenient form

$$f = \frac{f_1 f_2}{f_1 + f_2}$$

Since $f_1 = 10$ cm and $f_2 = -20$ cm,

$$f = \frac{(10 \text{ cm})(-20 \text{ cm})}{10 \text{ cm} - 20 \text{ cm}} = +20 \text{ cm}$$

The combination acts as a converging lens of focal length +20 cm.

## RAY TRACING

As with a spherical mirror, the position and size of the image of an object formed by a lens can be found by constructing a scale drawing. Again, what is done is to trace two different light rays from a point of interest in the object to where they (or their extensions, in the case of a virtual image) intersect after being refracted by the lens. Three rays especially useful for this purpose are shown in Fig. 32-3; any two of these are sufficient.

1.  A ray that leaves the object parallel to the axis of the lens. After refraction, this ray passes through the far focal point of a converging lens or seems to come from the near focal point of a diverging lens.

2.  A ray that passes through the focal point of a converging lens or is directed toward the far focal point of a diverging lens. After refraction, this ray travels parallel to this axis of the lens.

3.  A ray that leaves the object and proceeds toward the center of the lens. This ray is not deviated by refraction.

## SOLVED PROBLEM 32.6

What is the nature of the image of a real object formed by a diverging lens?

It is virtual, erect, and smaller than the object, as in Fig. 32-3(b).

## SOLVED PROBLEM 32.7

Describe the image formed by a converging lens of an object located at the focal point of the lens.

As Fig. 32-4 shows, the refracted rays are parallel to each other, and so no image is formed.

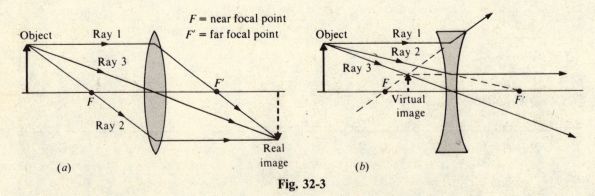

Fig. 32-3

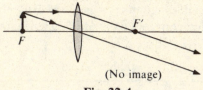

(No image)

**Fig. 32-4**

## LENS EQUATION

The object distance $p$, image distance $q$, and focal length $f$ of a lens (Fig. 32-5) are related by the lens equation:

$$\frac{1}{p} + \frac{1}{q} = \frac{1}{f}$$

$$\frac{1}{\text{Object distance}} + \frac{1}{\text{image distance}} = \frac{1}{\text{focal length}}$$

This equation holds for both converging and diverging lenses. The lens equation is readily solved for $p$, $q$, or $f$:

$$p = \frac{qf}{q-f} \qquad q = \frac{pf}{p-f} \qquad f = \frac{pq}{p+q}$$

As in the case of mirrors, a positive value of $p$ or $q$ denotes a real object or image, and a negative value denotes a virtual object or image. A real image of a real object is always on the opposite side of the lens from the object, and a virtual image is on the same side. Thus if a real object is on the left of a lens, a positive image distance $q$ signifies a real image to the right of the lens, whereas a negative image distance $q$ denotes a virtual image to the left of the lens.

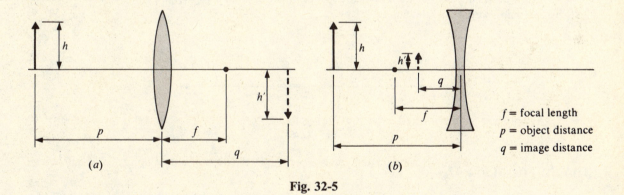

**Fig. 32-5**

## MAGNIFICATION

The linear magnification $m$ produced by a lens is given by the same formula that applies for mirrors:

$$m = \frac{h'}{h} = -\frac{q}{p}$$

$$\text{Linear magnification} = \frac{\text{image height}}{\text{object height}} = -\frac{\text{image distance}}{\text{object distance}}$$

Again, a positive magnification signifies an erect image, a negative one signifies an inverted image. Table 32.1 is a summary of the sign conventions used in connnection with lenses.

**Table 32-1**

| Quantity | Positive | Negative |
|---|---|---|
| Focal length $f$ | Concave lens | Diverging lens |
| Object distance $p$ | Real object | Virtual object |
| Image distance $q$ | Real image | Virtual image |
| Magnification $m$ | Erect image | Inverted image |

## SOLVED PROBLEM 32.8

A coin 3 cm in diameter is placed 24 cm from a converging lens whose focal length is 16 cm. Find the location, size, and nature of the image.

Here $p = 24$ cm and $f = +16$ cm, so the image distance is

$$q = \frac{pf}{p-f} = \frac{(24 \text{ cm})(16 \text{ cm})}{24 \text{ cm} - 16 \text{ cm}} = 48 \text{ cm}$$

The image is real since $q$ is positive (Fig. 32-6). The diameter of the coin's image is, since $m = h'/h = -q/p$.

$$h' = -h\frac{q}{p} = -(3 \text{ cm})\left(\frac{48 \text{ cm}}{24 \text{ cm}}\right) = -6 \text{ cm}$$

The image is inverted (since $h'$ is negative) and twice as large as the object.

In general, an object that is a distance between $f$ and $2f$ from a converging lens has a real, inverted image that is larger than the object.

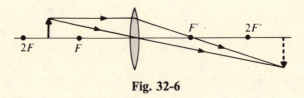

**Fig. 32-6**

## SOLVED PROBLEM 32.9

A sardine 8 cm long is 30 cm from a converging lens whose focal length is 15 cm. Find the location, size, and nature of the image.

Here $p = 30$ cm and $f = +15$ cm, so the image distance is

$$q = \frac{pf}{p-f} = \frac{(30 \text{ cm})(15 \text{ cm})}{30 \text{ cm} - 15 \text{ cm}} = 30 \text{ cm}$$

The image is real since $q$ is positive (Fig. 32-7). The length of the sardine's image is

$$h' = -h\frac{q}{p} = -(8 \text{ cm})\left(\frac{30 \text{ cm}}{30 \text{ cm}}\right) = -8 \text{ cm}$$

The image is inverted (since $h'$ is negative) and is the same size as the object.

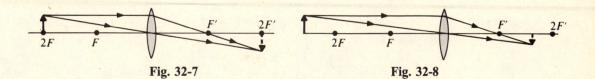

Fig. 32-7                                      Fig. 32-8

In general, an object that is the distance $2f$ from a converging lens has a real, inverted image the same size as the object with an image distance equal to $2f$.

## SOLVED PROBLEM 32.10

A key 6 cm long is 100 cm from a converging lens whose focal length is 40 cm. Find the location, size, and nature of the image.

Here $p = 100$ cm and $f = +40$ cm, so the image distance is

$$q = \frac{pf}{p-f} = \frac{(100 \text{ cm})(40 \text{ cm})}{100 \text{ cm} - 40 \text{ cm}} = 66.7 \text{ cm}$$

The image is real since $q$ is positive (Fig. 32-8). The length of the key's image is

$$h' = -h\,\frac{q}{p} = (-6 \text{ cm})\left(\frac{66.7 \text{ cm}}{100 \text{ cm}}\right) = -4 \text{ cm}$$

The image is inverted (since $h'$ is negative) and is smaller than the object.

In general, an object that is farther than $2f$ from a converging lens has a real, inverted image smaller than the object with an image distance between $f$ and $2f$.

## SOLVED PROBLEM 32.11

A diverging lens has a focal length of $-2$ ft. What are the location, size, and nature of the image formed by the lens when it is used to look at an object 12 ft away?

Here $p = 12$ ft and $f = -2$ ft, so the image distance is

$$q = \frac{pf}{p-f} = \frac{(12 \text{ ft})(-2 \text{ ft})}{12 \text{ ft} - (-2 \text{ ft})} = -1.71 \text{ ft}$$

A negative image distance signifies a virtual image. The magnification is

$$m = -\frac{q}{p} = -\frac{(-1.71 \text{ ft})}{12 \text{ ft}} = 0.143 = \frac{1}{7}$$

The image is erect (since $m$ is positive) and one-seventh the size of the object.

## SOLVED PROBLEM 32.12

A double-convex lens has a focal length of 6 cm. (*a*) How far from an insect 2 mm long should the lens be held in order to produce an erect image 5 mm long? (*b*) What is the image distance?

(*a*)  A double-convex lens is always converging, so the focal length of the lens is $+6$ cm. An erect image means a positive magnification, which is

$$m = \frac{h'}{h} = \frac{5 \text{ mm}}{2 \text{ mm}} = 2.5$$

Since $m = -q/p$, the image distance is $q = -mp$. We proceed by substituting $q = -mp$ in the lens equation and solving for $p$:

$$\frac{1}{p} = \frac{1}{q} = \frac{1}{f} \qquad \frac{1}{p} - \frac{1}{mp} = \frac{1}{f} \qquad \frac{m-1}{mp} = \frac{1}{f}$$

$$p = f\left(\frac{m-1}{m}\right) = (+6 \text{ cm})\left(\frac{2.5-1}{2.5}\right) = 3.6 \text{ cm}$$

(b)
$$q = -mp = -(2.5)(3.6 \text{ cm}) = -9 \text{ cm}$$

The negative image distance signifies a virtual image (Fig. 32-9).

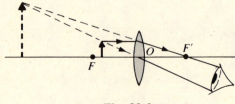

**Fig. 32-9**

## SOLVED PROBLEM 32.13

Find the focal length of a magnifying glass that produces an erect image magnified three times of an object 1.5 in. away.

An erect image magnified three times means a magnification of $+3$. Since the object distance is $p = 1.5$ in. and $m = -q/p$,

$$q = -mp = (-3)(1.5 \text{ in.}) = -4.5 \text{ in.}$$

The negative image distance signifies a virtual image. The required focal length is

$$f = \frac{pq}{p+q} = \frac{(1.5 \text{ in.})(-4.5 \text{ in.})}{1.5 \text{ in.} - 4.5 \text{ in.}} = +2.25 \text{ in.}$$

## SOLVED PROBLEM 32.14

A 35-mm camera has a telephoto lens whose focal length is 150 mm. What range of adjustments should the lens have in order to be able to bring to a sharp focus objects as close as 1.5 m from the camera?

The image distance that corresponds to $f = 0.15$ m and $p = 1.5$ m is

$$q = \frac{pf}{p-f} = \frac{(1.5 \text{ m})(0.15 \text{ m})}{1.5 \text{ m} - 0.15 \text{ m}} = 0.167 \text{ m} = 167 \text{ mm}$$

An object at $p = \infty$ is brought to a focus at $q = f = 150$ mm (see Fig. 32-10). Hence a range of adjustment of 167 mm $-$ 150 mm $=$ 17 mm will permit objects to be photographed at distances from 1.5 m to infinity.

## SOLVED PROBLEM 32.15

A slide projector uses a converging lens to produce a real, inverted image of a transparent slide on a screen. What should be the focal length of the lens of a slide projector if a slide 5 cm by 5 cm is to appear 90 cm by 90 cm on a screen 5 m away?

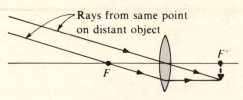

**Fig. 32-10**

Here $h = 5$ cm and $h' = -90$ cm ($h'$ is negative because the image is inverted), so the magnification is

$$m = \frac{h'}{h} = \frac{-90 \text{ cm}}{5 \text{ cm}} = -18$$

Since $q = 5$ m $= 500$ cm and $m = -q/p$, the object distance is

$$p = -\frac{q}{m} = -\frac{500 \text{ cm}}{-18} = 27.8 \text{ cm}$$

The required focal length is therefore

$$f = \frac{pq}{p+q} = \frac{(27.8 \text{ cm})(500 \text{ cm})}{27.8 \text{ cm} + 500 \text{ cm}} = 26.3 \text{ cm} = 263 \text{ mm}$$

**SOLVED PROBLEM 32.16**

For the most distinct vision with a normal eye, an object should be about 25 cm (10 in.) away. (a) Find a formula for the magnification of a converging lens of focal length $f$ when it is used as a magnifying glass with an image distance of $-25$ cm. (b) Use this formula to find the magnification of a lens of focal length $+5$ cm.

(a)  From $m = -q/p$ the object distance that corresponds to an image distance of $q = -25$ cm is

$$p = -\frac{q}{m} = \frac{25 \text{ cm}}{m}$$

Substituting $p = (25 \text{ cm})/m$ and $q = -25$ cm in the lens equation yields

$$\frac{1}{f} = \frac{1}{p} + \frac{1}{q} = \frac{m}{25 \text{ cm}} - \frac{1}{25 \text{ cm}} = \frac{m-1}{25 \text{ cm}}$$

$$m = \frac{25 \text{ cm}}{f} + 1$$

(b)                               $$m = \frac{25 \text{ cm}}{5 \text{ cm}} + 1 = 5 + 1 = 6$$

## LENS SYSTEMS

When a system of lenses is used to produce an image of an object, for instance, in a telescope or microscope, the procedure for finding the position and nature of the final image is to let the image formed by each lens in turn be the object for the next lens in the system. Thus to find the image produced by a system of two lenses, the first step is to determine the image formed by the lens nearest the object. This image then serves as the object for the second lens, with the usual sign convention: If the image is on the front side of the second lens, the object distance is considered positive, whereas if the image is on the back side, the object distance is considered negative.

The total magnification produced by a system of lenses is equal to the product of the magnifications of the individual lenses. Thus if the magnification of the objective lens of a microscope or telescope is $m_1$ and that of the eyepiece is $m_2$, the total magnification is $m = m_1 m_2$.

## SOLVED PROBLEM 32.17

A microscope has an objective of focal length 6 mm and an eyepiece of focal length 25 mm. If the image distance of the objective is 160 mm and that of the eyepiece is 250 mm (both of which are typical figures), find the magnification produced by (a) the objective, (b) the eyepiece, (c) the entire microscope.

(a)   The object distance of the objective is

$$p = \frac{qf}{q-f} = \frac{(160 \text{ mm})(6 \text{ mm})}{160 \text{ mm} - 6 \text{ mm}} = 6.23 \text{ mm}$$

and so its magnification is

$$m_1 = -\frac{q}{p} = \frac{-160 \text{ mm}}{6.23 \text{ mm}} = -25.7$$

The minus sign means in inverted image.

(b)   From the result of Prob. 32.16,

$$m_2 = \frac{250 \text{ mm}}{f} + 1 = \frac{250 \text{ mm}}{25 \text{ mm}} + 1 = 11$$

Alternatively the same figure can be obtained by the procedure of part (a).

(c)   The total magnification is the product of $m_1$ and $m_2$:

$$m = m_1 m_2 = (-25.7)(11) = -283$$

## SOLVED PROBLEM 32.18

The *angular magnification* of a telescope when used to view a distant object is

$$m_{\text{ang}} = \frac{f_o}{f_e} = \frac{\text{focal length of objective}}{\text{focal length of eyepiece}}$$

(a)   The objective of a telescope has a focal length of 120 cm. What should the focal length of the eyepiece be to produce an angular magnification of 40? (b) The telescope is used to examine a boat 20 m long which is 600 m away. If the image distance of the telescope eyepiece is 25 cm, what is the apparent length of the boat?

(a)                             $$f_e = \frac{f_o}{m_{\text{ang}}} = \frac{120 \text{ cm}}{40} = 3 \text{ cm}$$

(b)   The angle subtended by the boat from the location of the telescope is

$$\theta = \frac{\text{object length}}{\text{object distance}} = \frac{20 \text{ m}}{600 \text{ m}} = 0.033 \text{ rad}$$

If $L$ is the length of the boat's image as seen through the telescope, the angle that image subtends is

$$\theta' = \frac{\text{image length}}{\text{image distance}} = \frac{L}{25 \text{ cm}}$$

Since the angular magnification of the telescope is 40 and $m_{\text{ang}} = \theta'/\theta$, $\theta' = m_{\text{ang}}\theta$, so

$$\frac{L}{25 \text{ cm}} = (40)(0.033 \text{ rad}) = 1.3 \text{ rad}$$

$$L = (1.3 \text{ rad})(25 \text{ cm}) = 33 \text{ cm}$$

The boat seems to be 33 cm long and to be located 25 cm from the viewer's eye.

## *Multiple-Choice Questions*

**32.1.** The image of a real object farther from a converging lens than $f$ is always which one or more of the following?

    (*a*)  smaller than the object      (*c*)  virtual
    (*b*)  the same size as the object    (*d*)  inverted

**32.2.** The image of a real object closer to a converging lens than $f$ is always which one or more of the following?

    (*a*)  smaller than the object      (*c*)  virtual
    (*b*)  the same size as the object    (*d*)  inverted

**32.3.** The image of a real object the distance $f$ from a converging lens

    (*a*)  does not exist
    (*b*)  is virtual, erect, and larger than the object
    (*c*)  is real, inverted, and larger than the object
    (*d*)  is real, inverted, and the same size as the object

**32.4.** A real image formed by a lens is always which one or more of the following?

    (*a*)  smaller than the object    (*c*)  erect
    (*b*)  larger than the object     (*d*)  inverted

**32.5.** The image formed by a diverging lens of a real object is never which one or more of the following?

    (*a*)  real      (*c*)  erect
    (*b*)  virtual   (*d*)  smaller than the object

**32.6.** The object distance of a converging lens of focal length $f$ used as a magnifying glass must be

    (*a*)  less than $f$    (*c*)  between $f$ and $2f$
    (*b*)  $f$           (*d*)  more than $2f$

**32.7.** The image distance of an object located 12 cm from a converging lens of focal length 16 cm is

    (*a*)  $-4$ cm    (*c*)  $+4$ cm
    (*b*)  $-48$ cm   (*d*)  $+48$ cm

**32.8.** The image distance of an object located 16 cm from a converging lens of focal length 12 cm is

    (*a*)  $-4$ cm    (*c*)  $+4$ cm
    (*b*)  $-48$ cm   (*d*)  $+48$ cm

**32.9.** If the image of an object 6 cm from a lens is 6 cm behind the object, the lens has a focal length of

(*a*)  −12 cm      (*c*)  +4 cm
(*b*)  +3 cm      (*d*)  +12 cm

**32.10.**  A candle 6 cm high is 80 cm in front of a lens whose focal length is +60 cm. The image is

(*a*)  2 cm high, erect      (*c*)  18 cm high, erect
(*b*)  2 cm high, inverted      (*d*)  18 cm high, inverted

**32.11.**  A candle 6 cm high is 120 cm in front of a lens whose focal length is +60 cm. The image is

(*a*)  3 cm high, erect      (*c*)  6 cm high, erect
(*b*)  3 cm high, inverted      (*d*)  6 cm high, inverted

**32.12.**  A candle 6 cm high is 150 cm in front of a lens whose focal length is +60 cm. The image is

(*a*)  4 cm long, erect      (*c*)  9 cm long, erect
(*b*)  4 cm long, inverted      (*d*)  9 cm long, inverted

**32.13.**  The focal length of a magnifying glass that produces an image six times larger than an object 10 mm away is

(*a*)  +1.4 mm      (*c*)  +8.6 mm
(*b*)  +2.0 mm      (*d*)  +12 mm

**32.14.**  A projector whose lens has a focal length of +12 cm forms an image 90 cm high of a slide whose picture area is 30 mm high. How far is the lens from the screen?

(*a*)  348 cm      (*c*)  372 cm
(*b*)  360 cm      (*d*)  384 cm

# Supplementary Problems

**32.1.**  Can a diverging lens ever form an inverted image of a real object? Can a converging lens?

**32.2.**  If the screen is moved closer to a movie projector, how should the projector's lens be moved to restore the image to a sharp focus?

**32.3.**  A double-convex lens has surfaces whose radii are both 50 cm. The index of refraction of the glass is 1.52. Find the focal length of the lens.

**32.4.**  A meniscus lens has a concave surface of radius 30 cm and a convex surface of radius 25 cm. The index of refraction of the glass is 1.50. (*a*) Find the focal length of the lens. (*b*) Is it a converging or a diverging lens?

**32.5.**  A *planoconcave lens* has one plane surface and one concave surface. If a planoconcave lens of focal length −10 cm is to be ground from optical glass of index of refraction 1.50, find the radius of curvature of the concave surface.

**32.6.**  Glycerin has an index of refraction of 1.47. Find the focal length in glycerin of a lens made of flint glass ($n = 1.63$) whose focal length in air is +10 cm.

**32.7.** Find the index of refraction of the glass used in a planoconvex lens of focal length 12 cm whose convex surface has a radius of 7 cm.

**32.8.** A button 1 cm in diameter is held 10 cm from a converging lens whose focal length is +25 cm. Find the location, size, and nature of the image.

**32.9.** The button of Prob. 32.8 is held 25 cm from the same lens. Find the location, size, and nature of the image.

**32.10.** The button of Prob. 32.8 is held 40 cm from the same lens. Find the location, size, and nature of the image.

**32.11.** The button of Prob. 32.8 is held 50 cm from the same lens. Find the location, size, and nature of the image.

**32.12.** The button of Prob. 32.8 is held 100 cm from the same lens. Find the location, size, and nature of the image.

**32.13.** A diverging lens with a focal length of −1 m is used to examine a man 1.8 m tall who is standing 4 m away. How tall does the man appear to be?

**32.14.** A magnifying glass of focal length 75 mm is held 25 mm from a postage stamp. What magnification does it produce? What is the nature of the image?

**32.15.** A converging lens has a focal length of 10 cm. (*a*) How far away from an object should the lens be held to produce an erect, virtual image magnified three times? (*b*) What is the image distance?

**32.16.** A slide projector has a lens whose focal length is 8 in. How far from the lens should the screen be located if a slide 2 in. by 2 in. is to appear 40 in. by 40 in.?

**32.17.** What is the focal length of a combination of two thin lenses in contact both of whose focal lengths are +8 in.?

**32.18.** A microscope has an objective of focal length 4 mm which is used with a $15\times$ eyepiece. If the image distance for the objective is 160 mm, find the magnification of the instrument.

**32.19.** A telescope has an objective lens whose focal length is 120 cm and an eyepiece whose focal length is 4 cm. (*a*) What is the angular magnification of this telescope? (*b*) The moon has an angular diameter of about 0.5° as seen from the earth. What is its angular diameter through the telescope? (*c*) If the image distance is 25 cm, what is the apparent diameter of the moon as seen through the telescope?

## *Answers to Multiple-Choice Questions*

**32.1.** (*d*)     **32.3.** (*a*)

**32.2.** (*c*)     **32.4.** (*d*)

**32.5.**  (*a*)      **32.10.**  (*d*)

**32.6.**  (*a*)      **32.11.**  (*d*)

**32.7.**  (*b*)      **32.12.**  (*b*)

**32.8.**  (*d*)      **32.13.**  (*d*)

**32.9.**  (*d*)      **32.14.**  (*a*)

# Answers to Supplementary Problems

**32.1.**   No; yes

**32.2.**   The lens should be moved farther away from the film.

**32.3.**   48 cm

**32.4**   (*a*)  300 cm       (*b*)  Converging

**32.5.**   −5 cm

**32.6.**   +58 cm

**32.7.**   1.58

**32.8.**   The image is 16.7 cm in front of the lens, 1.67 cm in diameter, virtual, and erect.

**32.9.**   No image is formed.

**32.10.**   The image is 66.7 cm behind the lens, 1.67 cm in diameter, real, and inverted.

**32.11.**   The image is 50 cm behind the lens, 1 cm in diameter, real, and inverted.

**32.12.**   The image is 33.3 cm behind the lens, 3.3 mm in diameter, real, and inverted.

**32.13.**   36 cm

**32.14.**   1.5; erect and virtual

**32.15.**   (*a*)  6.67 cm       (*b*)  20 cm

**32.16.**   14 ft

**32.17.**   +4 cm

**32.18.**   585

**32.19.**   (*a*)  30       (*b*)  15°       (*c*)  6.54 cm

# Chapter 33

# Physical and Quantum Optics

## INTERFERENCE

In examining the reflection and refraction of light, it is sufficient to consider light as though it consisted of rays that travel in straight lines in a uniform medium. The study of such phenomena, therefore, is called *geometrical optics*. Other phenomena, notably interference, diffraction, and polarization, can be understood only in terms of the wave nature of light, and the study of these phenomena is called *physical optics*.

*Interference* occurs when waves of the same nature from different sources meet at the same place. In *constructive interference* the waves are in phase ("in step") and reinforce each other; in *destructive interference* the waves are out of phase and partially or completely cancel (Fig. 33-1). All types of waves exhibit interference under appropriate circumstances. Thus water waves interfere to produce the irregular surface of the sea, sound waves close in frequency interfere to produce beats, and light waves interfere to produce the fringes seen around the images formed by optical instruments and the bright colors of soap bubbles and thin films of oil on water.

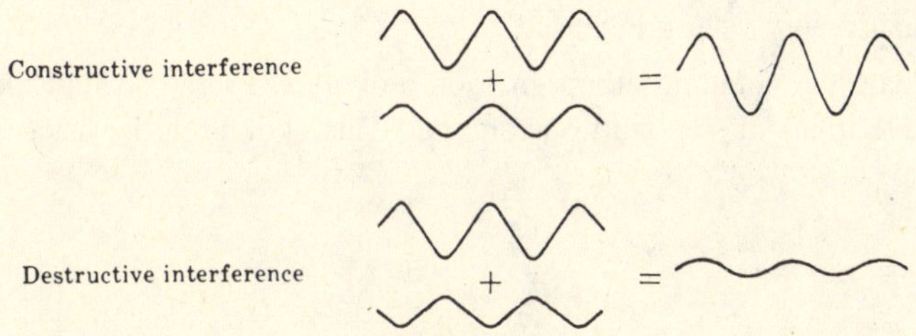

Constructive interference

Destructive interference

**Fig. 33-1**

## SOLVED PROBLEM 33.1

When is it appropriate to think of light as consisting of waves and when as consisting of rays?

When paths or path differences are involved whose lengths are comparable with the wavelengths found in light, the wave nature is significant and must be taken into account. Thus diffraction and interference can be understood only on a wave basis. When paths are involved that are many wavelengths long and neither diffraction nor interference occurs, as in reflection and refraction, it is more convenient to consider light as consisting of rays.

## SOLVED PROBLEM 33.2

When two light beams of the same wavelength interfere, the result is a pattern of bright and dark lines. What becomes of the energy of the light waves whose destructive interference leads to the dark lines?

The missing energy is found in the bright lines, whose brightness is greater than the simple addition of the two light beams would produce in the absence of interference. The total energy remains the same.

411

**SOLVED PROBLEM 33.3**

When two trains of waves meet on the surface of a body of water, the resulting interference pattern is obvious. However, when the light beams from two flashlights overlap on a screen, there is no evidence of an interference pattern. Why not? Is there any way in which the interference of light can be demonstrated?

There are two reasons why such an experiment does not yield a conspicuous interference pattern. First, the wavelengths found in light are so short that such a pattern would be on an extremely small scale. Second (and more important), all sources of light (except lasers) emit light waves as short trains of random phase and not as continuous trains. The interference that occurs between light beams from two independent sources is therefore averaged during all but the briefest of observation times and cannot be seen by eye or recorded on photographic film. Such light sources are said to be *incoherent*.

To exhibit an interference pattern in light, sources must be used whose waves have fixed phase relationships during the observation period. The waves from one source can be in step with those from the other when they are produced, or out of step, or something in between, but the essential thing is that the relationship be constant. Such sources are *coherent*. Three ways to construct coherent sources are as follows:

1.  Pass light from a single source (such as an illuminated slit or a narrow filament) through two or more other slits. The waves that emerge from the latter slits are necessarily coordinated and can interfere to produce a visible pattern.

2.  Combine a direct light beam from a source with an indirect beam from the same source produced by refraction or reflection. This is how the interference patterns produced by thin oil films floating on water are caused.

3.  Coordinate the radiating atoms in each individual source so that the radiating atoms always emit wave trains in step with one another. This is done in the laser.

## DIFFRACTION

The ability of a wave to bend around the edge of an obstacle is called *diffraction*. Owing to the combined effects of diffraction and interference, the image of a point source of light is always a small disk with bright and dark fringes around it. The smaller the lens or mirror used to form the image, the larger the disk. The angular width in radians of the image disk of a point source is about

$$\theta_0 = (1.22)\left(\frac{\lambda}{D}\right)$$

where $\lambda$ is the wavelength of the light and $D$ is the lens or mirror diameter. The images of objects closer than $\theta_0$ will overlap and hence cannot be resolved no matter how great the magnification produced by the lens or mirror. In the case of a telescope or microscope, $D$ refers to the diameter of the objective lens. If two objects $d_0$ apart that can just be resolved are the distance $L$ from the observer, the angle in radians between them is $\theta_0 = d_0/L$, so the above formula can be rewritten in the form

$$\text{Resolving power} = d_0 = (1.22)\left(\frac{\lambda L}{D}\right)$$

## POLARIZATION

A *polarized* beam of light is one in which the electric fields of the waves are all in the same direction. If the electric fields are in random directions (though, of course, always in a plane perpendicular to the direction of propagation), the beam is *unpolarized*. Various substances affect differently light with different directions of polarization, and these substances can be used to prepare devices that permit only light polarized in a certain direction to pass through them.

### SOLVED PROBLEM 33.4

What is a *diffraction grating*?

A diffraction grating consists of a series of closely spaced parallel slits that diffract light passing through them. The diffracted wave trains of a particular wavelength interfere constructively in certain directions only (Fig. 33-2). When light is directed at a grating, each wavelength present undergoes constructive interference in difference directions from those of other wavelengths, and the result is a series of spectra. A grating is better able than a prism to separate nearby wavelengths, and so gratings rather than prisms are used in nearly all spectrographs.

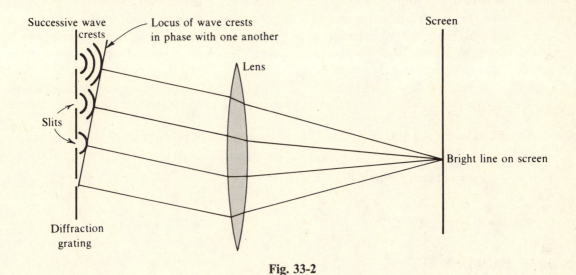

**Fig. 33-2**

### SOLVED PROBLEM 33.5

A pair of "7 × 50" binoculars has a magnification of 7 and objective lens diameters of 50 mm. Find the length of the smallest detail that can possibly be resolved by such binoculars when something 1 km away is examined. Consider the wavelength of the light to be $5 \times 10^{-7}$ m, which is near the middle of the visible spectrum and corresponds to green.

Since $D = 50$ mm $= 5 \times 10^{-2}$ m and $L = 1$ km $= 10^3$ m,

$$d_0 = (1.22)\left(\frac{\lambda L}{D}\right) = \frac{(1.22)(5 \times 10^{-7} \text{ m})(10^3 \text{ m})}{5 \times 10^{-2} \text{ m}} = 1.22 \times 10^{-2} \text{ m} = 1.22 \text{ cm}$$

### SOLVED PROBLEM 33.6

A radar operating at a wavelength of 3 cm is to have a resolving power of 30 m at a range of 1 km. Find the minimum width its antenna must have.

The width of a radar antenna corresponds to the diameter of the objective lens of an optical system. Here $L = 1$ km $= 1000$ m, $d_0 = 30$ m, and $\lambda = 0.03$ m, so

$$D = (1.22)\left(\frac{\lambda L}{d_0}\right) = \frac{(1.22)(0.03 \text{ m})(1000 \text{ m})}{30 \text{ m}} = 1.22 \text{ m}$$

## QUANTUM THEORY OF LIGHT

Certain features of the behavior of light can be explained only on the basis that light consists of individual *quanta*, or *photons*. The energy of a photon of light whose frequency is $f$ is

$$\text{Quantum energy} = E = hf$$

where $h$ is *Planck's constant*:

$$\text{Planck's constant} = h = 6.63 \times 10^{-34} \text{ J·s}$$

A photon has most of the properties associated with particles—it is localized in space and possesses energy and momentum—but it has no mass. Photons travel with the velocity of light.

The electromagnetic and quantum theories of light complement each other: Under some circumstances light exhibits a wave character, under other circumstances it exhibits a particle character. Both are aspects of the same basic phenomenon.

## SOLVED PROBLEM 33.7

The human eye can respond to as few as three photons of light. If the light is yellow ($f = 5 \times 10^{14}$ Hz), how much energy does this represent?

The energy of each photon is

$$E = hf = (6.63 \times 10^{-34} \text{ J·s})(5 \times 10^{14} \text{ Hz}) = 3.3 \times 10^{-19} \text{ J}$$

The total energy is $3E = 10^{-18}$ J.

## SOLVED PROBLEM 33.8

The average wavelength of the light emitted by a certain 100-W light bulb is $5.5 \times 10^{-7}$ m. How many photons per second does the light bulb emit?

The frequency of the light is

$$f = \frac{c}{\lambda} = \frac{3 \times 10^8 \text{ m/s}}{5.5 \times 10^{-7} \text{ m}} = 5.5 \times 10^{14} \text{ Hz}$$

and the energy of each photon is

$$E = hf = (6.63 \times 10^{-34} \text{ J·s})(5.5 \times 10^{14} \text{ Hz}) = 3.6 \times 10^{-19} \text{ J}$$

Since 100 W $= 100$ J/s, the number of photons emitted per second is

$$\frac{100 \text{ J/s}}{3.6 \times 10^{-19} \text{ J/photon}} = 2.8 \times 10^{20} \text{ photons/s}$$

## X-RAYS

X-rays are high-frequency electromagnetic waves produced when fast electrons impinge on a target. If the electrons are accelerated through a potential difference of $V$, each electron has the energy

KE = $eV$. If all this energy goes into creating an X-ray photon, then

$$eV = hf$$

Electron kinetic energy = X-ray photon energy

and the frequency of the X-rays is $f = eV/h$.

## SOLVED PROBLEM 33.9

In a certain television picture tube, electrons are accelerated through a potential difference of 10,000 V. Find the frequency of the X-rays emitted when these electrons strike the screen.

Since $hf = eV$, here we have

$$f = \frac{eV}{h} = \frac{(1.6 \times 10^{-19} \text{ C})(10^4 \text{ V})}{6.63 \times 10^{-34} \text{ J·s}} = 2.4 \times 10^{18} \text{ Hz}$$

## SOLVED PROBLEM 33.10

An X-ray tube emits X-rays whose wavelength is $2 \times 10^{-11}$ m. What is the operating voltage of the tube?

The frequency of the X-rays is

$$f = \frac{c}{\lambda} = \frac{3 \times 10^8 \text{ m/s}}{2 \times 10^{-11} \text{ m}} = 1.5 \times 10^{19} \text{ Hz}$$

Since $hf = eV$,

$$V = \frac{hf}{e} = \frac{(6.63 \times 10^{-34} \text{ J·s})(1.5 \times 10^{19} \text{ Hz})}{1.6 \times 10^{-19} \text{ C}} = 6.2 \times 10^4 \text{ V} = 62 \text{ keV}$$

## ELECTRONVOLT

A common energy unit in atomic and quantum physics is the *electronvolt* (eV), defined as the energy an electron gains when it moves through a potential difference of 1 V. Hence

$$1 \text{ eV} = 1.60 \times 10^{-19} \text{ J}$$

Multiples of the electronvolt are the kiloelectronvolt (keV), megaelectronvolt (MeV), and gigaelectronvolt (GeV), where

$$1 \text{ keV} = 10^3 \text{ eV} \qquad 1 \text{ MeV} = 10^6 \text{ eV} \qquad 1 \text{ GeV} = 10^9 \text{ eV}$$

## SOLVED PROBLEM 33.11

In the *photoelectric effect*, light directed at the surface of certain metals causes electrons to be emitted. In the case of potassium, 2 eV of work must be done to remove an electron from the surface. (*a*) If light of wavelength $5 \times 10^{-7}$ m falls on a potassium surface, what is the maximum energy of the photoelectrons that emerge? (*b*) If light of wavelength $4 \times 10^{-7}$ m falls on the same surface, will the photoelectrons have more or less energy?

(*a*)   Since $c = f\lambda$, the frequency of the light is

$$f = \frac{c}{\lambda} = \frac{3 \times 10^8 \text{ m/s}}{5 \times 10^{-7} \text{ m}} = 6 \times 10^{14} \text{ Hz}$$

The energy of each photon is therefore

$$E = hf = (6.63 \times 10^{-34} \text{ J·s})(6 \times 10^{14} \text{ Hz}) = 3.98 \times 10^{-19} \text{ J}$$

Since 1 eV = $1.6 \times 10^{-19}$ J,

$$E = \frac{3.98 \times 10^{-19} \text{ J}}{1.6 \times 10^{-19} \text{ J/eV}} = 2.49 \text{ eV}$$

This is the maximum energy that can be given to an electron by a photon of this light. Because 2 eV is needed to remove an electron, the maximum energy of the photoelectrons in this situation is 0.49 eV.

(b)   A shorter wavelength means a higher frequency and hence more energy to be imparted to the photoelectrons.

## SOLVED PROBLEM 33.12

What is the kinetic energy in electronvolts of an electron whose velocity is $10^7$ m/s?

$$\text{KE} = \tfrac{1}{2}mv^2 = (\tfrac{1}{2})(9.1 \times 10^{-31} \text{ kg})(10^7 \text{ m/s})^2 = 4.55 \times 10^{-17} \text{ J}$$

Since 1 eV = $1.6 \times 10^{-19}$ J,

$$\text{KE} = \frac{4.55 \times 10^{-17} \text{ J}}{1.6 \times 10^{-19} \text{ J/eV}} = 284 \text{ eV}$$

## SOLVED PROBLEM 33.13

A proton ($m = 1.67 \times 10^{-27}$ kg) is accelerated through a potential difference of 200 V. (a) What is its kinetic energy in electronvolts? (b) What is its velocity?

(a)   Since the charge on the proton is $+e$, its kinetic energy is 200 eV.

(b)

$$\text{KE} = (200 \text{ eV})(1.6 \times 10^{-19} \text{ J/eV}) = 3.2 \times 10^{-17} \text{ J}$$

Since KE = $\tfrac{1}{2}mv^2$,

$$v = \sqrt{\frac{2(\text{KE})}{m}} = \sqrt{\frac{(2)(3.2 \times 10^{-17} \text{ J})}{1.67 \times 10^{-27} \text{ kg}}} = 2 \times 10^5 \text{ m/s}$$

## *Multiple-Choice Questions*

**33.1.**   Which one or more of the following produce coherent electromagnetic waves?

(a)   two lasers of the same frequency
(b)   two antennas fed by the same radio transmitter
(c)   a pinhole in a cover over a monochromatic light source and its reflection in a mirror
(d)   two pinholes in a cover over a monochromatic light source

**33.2.**   The resolving power of a lens can be improved by increasing which one or more of the following?

(a)   the diameter of the lens      (c)   the wavelength of the light
(b)   the object distance           (d)   the brightness of the light

**33.3.** A beam of transverse waves whose vibrations occur in all directions perpendicular to their direction of motion is

    (*a*)   polarized     (*c*)   resolved

    (*b*)   unpolarized   (*d*)   diffracted

**33.4.** Which one or more of the following cannot be polarized?

    (*a*)   sound waves    (*c*)   radio waves

    (*b*)   white light     (*d*)   X-rays

**33.5.** Photons in a vacuum have the same

    (*a*)   velocity    (*c*)   frequency

    (*b*)   energy     (*d*)   wavelength

**33.6.** When the voltage applied to an X-ray tube is increased, the X-rays have a greater

    (*a*)   number per second    (*c*)   energy

    (*b*)   velocity            (*d*)   wavelength

**33.7.** The photons in red light whose wavelength is 650 nm have an energy of

    (*a*)   $4.3 \times 10^{-40}$ J    (*c*)   $1.0 \times 10^{-27}$ J

    (*b*)   $1.3 \times 10^{-31}$ J    (*d*)   $3.1 \times 10^{-19}$ J

**33.8.** An X-ray tube produces a 0.50-W beam of $1.0 \times 10^{19}$ Hz X-rays. The tube emits

    (*a*)   $7.5 \times 10^{13}$ photons/s    (*c*)   $2.2 \times 10^{15}$ photons/s

    (*b*)   $3.0 \times 10^{14}$ photons/s    (*d*)   $3.0 \times 10^{18}$ photons/s

**33.9.** The kinetic energy of an electron whose velocity is $1.5 \times 10^7$ m/s is

    (*a*)   $1.64 \times 10^{-35}$ eV    (*c*)   0.64 keV

    (*b*)   $1.56 \times 10^{-3}$ eV    (*d*)   1.64 keV

**33.10.** The velocity of an electron whose kinetic energy is 1.5 keV is

    (*a*)   $7.3 \times 10^5$ m/s    (*c*)   $2.3 \times 10^7$ m/s

    (*b*)   $5.1 \times 10^6$ m/s    (*d*)   $7.3 \times 10^{12}$ m/s

# Supplementary Problems

**33.1.** Which of the following phenomena occur only in transverse waves (such as light) and not in longitudinal waves (such as sound)—reflection, refraction, interference, diffraction, polarization?

**33.2.** Which of the following optical phenomena, if any, are independent of the wavelength of the light involved—interference, diffraction, resolving power, polarization?

**33.3.** Why does the electric field of an electromagnetic wave rather than its magnetic field determine its direction of polarization?

**33.4.** State two advantages in having the objective lens or mirror of a telescope be of larger diameter.

**33.5.** Why do radio waves readily diffract around buildings whereas light waves, which are also electromagnetic in nature, do not?

**33.6.** The energy of a light beam is carried by separate photons, yet we do not perceive light as a series of tiny flashes. Why not?

**33.7.** If Planck's constant were equal to 6.63 J·s instead of $6.63 \times 10^{-34}$ J·s, would quantum phenomena be more or less conspicuous in everyday life than they are now?

**33.8.** When light is directed at a metal surface, upon what property of the light does the maximum energy of the emitted electrons depend?

**33.9.** A radar whose operating frequency is 9000 mHz has an antenna 1.5 m wide. What is its resolving power at a range of 5 km?

**33.10.** A telescope has an objective lens 10 cm in diameter. What is the maximum distance at which the telescope can resolve two objects 1 cm apart? Assume that $\lambda = 5 \times 10^{-7}$ m.

**33.11.** Light from the sun arrives at the earth at the rate of about 1400 W/m$^2$ of area perpendicular to the direction of the light. Assuming that sunlight consists exclusively of light of wavelength $6 \times 10^{-7}$ m, find the number of photons per second that fall on each square meter of the earth's surface directly facing the sun.

**33.12.** How many photons per second are emitted by a 50-kW radio transmitter that operates at a frequency of 1200 kHz?

**33.13.** What is the operating voltage of an X-ray tube that produces X-rays of frequency $10^{19}$ Hz?

**33.14.** Find the wavelength of the X-rays produced by a 50,000-V X-ray machine.

**33.15.** The work needed to remove an electron from the surface of sodium is 2.3 eV. Find the maximum wavelength of light that will cause photoelectrons to be emitted from sodium.

**33.16.** Photoelectrons are emitted by a copper surface only when light whose frequency is $1.1 \times 10^{15}$ Hz or more is directed at it. What is the maximum energy of the photoelectrons when light of frequency $1.5 \times 10^{15}$ Hz is directed at the surface?

**33.17.** What is the kinetic energy in electronvolts of a proton ($m = 1.67 \times 10^{-27}$ kg) whose velocity is $5 \times 10^6$ m/s?

**33.18.** Find the velocity of a 50-eV electron.

# Answers to Multiple-Choice Questions

**33.1.**  $(a)$, $(b)$, $(c)$      **33.3.**     $(b)$

**33.2.**  $(a)$                    **33.4.**     $(a)$

**33.5.**  *(a)*        **33.8.**  *(a)*

**33.6.**  *(c)*        **33.9.**  *(c)*

**33.7.**  *(d)*        **33.10.**  *(c)*

# Answers to Supplementary Problems

**33.1.**  Polarization

**33.2.**  Polarization

**33.3.**  The interaction between the electric field of an electromagnetic wave and the matter it passes through is responsible for nearly all optical effects; hence this field is used to specify the direction of polarization.

**33.4.**  (1) The larger the diameter, the greater the ability to resolve objects close together. (2) A large diameter means that more light reaches the eye from a given object; hence it can be seen even if poorly illuminated.

**33.5.**  The wavelengths of visible light are very short relative to the size of a building, so their diffraction is imperceptible. The wavelengths of radio waves are more nearly comparable with the size of a building.

**33.6.**  Even a weak light involves many photons per second. Visual responses persist for a short time, so successive photons give the impression of a continuous transfer of energy.

**33.7.**  More conspicuous

**33.8.**  Frequency

**33.9.**  136 m

**33.10.**  1.64 km

**33.11.**  $4.2 \times 10^{21}$ photons/(m$^2 \cdot$s)

**33.12.**  $6.3 \times 10^{31}$ photons/s

**33.13.**  41,400 V

**33.14.**  $2.5 \times 10^{-11}$ m

**33.15.**  $5.4 \times 10^{-7}$ m

**33.16.**  1.7 eV

**33.17.**  $1.3 \times 10^5$ eV $= 0.13$ MeV

**33.18.**  $4.2 \times 10^6$ m/s

# Chapter 34

# Atomic Physics

## MATTER WAVES

Under certain conditions moving bodies exhibit wave properties. The quantity whose variations constitute the *matter waves* (or *de Broglie waves*) of a moving body is known as its *wave function* $\psi$ (Greek letter psi). The likelihood of finding the body at a particular time and place is proportional to the value of $\psi^2$ at that time and place. A large value of $\psi^2$ signifies a high probability of finding the body; a small value of $\psi^2$ signifies a low probability of finding the body.

The matter waves associated with a moving body are in the form of a group of waves that travels with the same velocity as the body.

The wavelength of the matter waves of a body of mass $m$ and velocity $v$ is

$$\text{de Broglie wavelength} = \lambda = \frac{h}{mv}$$

## UNCERTAINTY PRINCIPLE

A consequence of the wave nature of moving bodies is the *uncertainty principle*: It is impossible to determine both the exact position and the exact momentum of a body at the same time. If $\Delta x$ is the uncertainty in position and $\Delta(mv)$ is the uncertainty in momentum, then

$$\Delta x \, \Delta(mv) \geq \frac{h}{2\pi}$$

where the symbol $\geq$ means "equal to or greater than." Since Planck's constant $h$ is so small, the uncertainty principle is significant only for very small bodies such as elementary particles.

## SOLVED PROBLEM 34.1

What is the de Broglie wavelength of a 1000-kg car whose velocity is 20 m/s? Would you expect the wave properties of such a car to be noticeable?

$$\lambda = \frac{h}{mv} = \frac{6.63 \times 10^{-34} \text{ J·s}}{(10^3 \text{ kg})(20 \text{ m/s})} = 3.3 \times 10^{-38} \text{ m}$$

The wave properties of such a car would be impossible to observe.

## SOLVED PROBLEM 34.2

An electron microscope uses a beam of fast electrons that are focused by electric and magnetic fields to produce an enlarged image of a thin specimen on a screen or photographic plate. Find the resolving power of an electron microscope which uses 15-keV electrons by assuming that this is equal to the electron wavelength.

The kinetic energy of the electrons is

$$KE = (1.5 \times 10^4 \text{ eV})(1.6 \times 10^{-19} \text{ J/eV}) = 2.4 \times 10^{-15} \text{ J}$$

Since $KE = \frac{1}{2}mv^2$, the velocity of the electrons is

$$v = \sqrt{\frac{2(KE)}{m}} = \sqrt{\frac{(2)(2.4 \times 10^{-15} \text{ J})}{9.1 \times 10^{-31} \text{ kg}}} = 7.26 \times 10^7 \text{ m/s}$$

The electron wavelength is therefore

$$\lambda = \frac{h}{mv} = \frac{6.63 \times 10^{-34} \text{ J·s}}{(9.1 \times 10^{-31} \text{ kg})(7.26 \times 10^7 \text{ m/s})} = 1.0 \times 10^{-11} \text{ m}$$

## SOLVED PROBLEM 34.3

The position of a certain electron is determined at a certain time with an uncertainty of $10^{-9}$ m. (a) Find the uncertainty in the electron's momentum. (b) Find the uncertainty in the electron's velocity and, from this, the uncertainty in its position 1 s after the original measurement was made.

(a)
$$\Delta(mv) = \frac{h}{2\pi \, \Delta x} = \frac{6.63 \times 10^{-34} \text{ J·s}}{(2\pi)(10^{-9} \text{ m})} = 1.06 \times 10^{-25} \text{ kg·m/s}$$

(b)
$$\Delta v = \frac{\Delta(mv)}{m} = \frac{1.06 \times 10^{-25} \text{ kg·m/s}}{9.1 \times 10^{-31} \text{ kg}} = 1.2 \times 10^5 \text{ m/s}$$

Hence the uncertainty in the electron's position 1 s later is

$$\Delta x' = t \, \Delta v = (1 \text{ s})(1.2 \times 10^5 \text{ m/s}) = 1.2 \times 10^5 \text{ m}$$

which is 120 km!

## SOLVED PROBLEM 34.4

Most atoms are a little over $10^{-10}$ m in radius. (a) Find the uncertainty in the momentum of an electron whose position is known to within $10^{-10}$ m. (b) Find the corresponding uncertainty in the electron's kinetic energy. What is the significance of this figure?

(a)   The momentum uncertainty is

$$\Delta(mv) = \frac{h}{2\pi \, \Delta x} = \frac{6.63 \times 10^{-34} \text{ J·s}}{(2\pi)(10^{-10} \text{ m})} = 1.06 \times 10^{-24} \text{ kg·m/s}$$

(b)   Since $KE = \frac{1}{2}mv^2 = 1/(mv)^2/2m$,

$$\Delta KE = \frac{1}{2m}(\Delta mv)^2 = \frac{(1.06 \times 10^{-24} \text{ kg·m/s})^2}{(2)(9.11 \times 10^{-31} \text{ kg})} = 6.1 \times 10^{-19} \text{ J} = 3.8 \text{ eV}$$

Electrons in atoms must have greater kinetic energies than this, as in fact they do.

## BOHR MODEL OF THE HYDROGEN ATOM

The hydrogen atom consists of a single electron and a single proton, which is the nucleus. In the classical model, the electron is imagined to circle the proton in an orbit such that the electric attaction of the proton provides the required centripetal force. The flaw in this model is that, according to electromagnetic theory, because the electron is accelerated, it must radiate electromagnetic waves and thus will lose energy until it spirals into the proton.

In the Bohr model, it is postulated that stable orbits exist in which the angular momentum of the electron is a multiple of $h/(2\pi)$, that is, that the angular momentum is $nh/(2\pi)$, where $n = 1, 2, 3, \ldots$. This postulate is equivalent to requiring that each orbit be a whole number of de Broglie wavelengths in circumference. If $n$ is the *quantum number* of an orbit, then the orbit radius is

$$r_n = n^2 r_1 \qquad n = 1, 2, 3, \ldots$$

where $r_1$, the radius of the smallest orbit, is $5.3 \times 10^{-11}$ m.

## ENERGY LEVELS

The total energy (kinetic energy plus electric potential energy) of a hydrogen atom whose electron is in the *n*th orbit is

$$E_n = \frac{E_1}{n^2} \qquad n = 1, 2, 3, \ldots$$

where $E_1 = -13.6$ eV $= -2.18 \times 10^{-18}$ J. The permitted energies of an atom are called its *energy levels*. The energy levels are all negative, which means that the electron does not have enough energy to escape from the proton. The lowest level, corresponding to $n = 1$, is the *ground state*; higher levels are *excited states*. As $n$ increases, $E_n$ approaches zero; when $E_n = 0$, the electron is no longer bound to the proton and the atom breaks up. The work needed to remove an electron from an atom in its ground state is called the *ionization energy*; it is 13.6 eV in the case of hydrogen.

## SOLVED PROBLEM 34.5

Find the energies of the first three excited states of the hydrogen atom.

Since $E_n = E_1/n^2$ and $E_1 = -13.6$ eV, we have

$$E_2 = \frac{E_1}{2^2} = \frac{-13.6 \text{ eV}}{4} = -3.4 \text{ eV}$$

$$E_3 = \frac{E_1}{3^2} = \frac{-13.6 \text{ eV}}{9} = -1.51 \text{ eV}$$

$$E_4 = \frac{E_1}{4^2} = \frac{-13.6 \text{ eV}}{16} = -0.85 \text{ eV}$$

## SOLVED PROBLEM 34.6

Find the velocity of the electron in a ground-state hydrogen atom according to the Bohr model.

In the Bohr model, the de Broglie wavelength $\lambda = h/(mv)$ of the electron in the $n = 1$ state is equal to the orbit circumference of $2\pi r_1$. Hence

$$\lambda = \frac{h}{mv} = 2\pi r_1$$

$$v = \frac{h}{2\pi m r_1} = \frac{6.63 \times 10^{-34} \text{ J·s}}{(2\pi)(9.1 \times 10^{-31} \text{ kg})(5.3 \times 10^{-11} \text{ m})} = 2.2 \times 10^6 \text{ m/s}$$

**SOLVED PROBLEM 34.7**

To what temperature must a hydrogen sample be heated so that the average molecular energy equals the binding energy of the hydrogen atom?

The binding (or ionization) energy of the hydrogen atom is 13.6 eV = $2.18 \times 10^{-18}$ J. Since the average molecular energy is a gas whose absolute temperature $T$ is equal to $\frac{3}{2}kT$, here

$$\frac{3}{2}\,kT = E$$

$$T = \frac{2E}{3k} = \frac{(2)(2.18 \times 10^{-18}\ \text{J})}{(3)(1.38 \times 10^{-23}\ \text{J/K})} = 1.05 \times 10^5\ \text{K}$$

## ATOMIC SPECTRA

When a gas or vapor is excited by the passage of an electric current, light is given off which consists of certain specific wavelengths. Every element has a characteristic *emission line spectrum*. The wavelengths in this spectrum fall into definite series whose member wavelengths are related by simple formulas.

When white light is passed through a cool gas or vapor, light of certain specific wavelengths is absorbed. The wavelengths in the resulting *absorption line spectrum* correspond to a number of the wavelengths in the emission spectrum of that element.

Line spectra owe their origin to the presence of energy levels in atoms. An atom in an excited state can remain there only a brief time (normally about $10^{-8}$ s) before dropping to a lower state. The difference in energy appears as a photon of frequency $f$, where

$$E_{\text{initial}} - E_{\text{final}} = hf$$

Figure 34-1 is an energy-level diagram which shows the possible transitions in the hydrogen atom that are responsible for its emission line spectrum. The larger the energy difference between initial and final energy states, as indicated by the lengths of the arrows, the higher the frequency of the photon that is emitted. The names of the various spectral series in hydrogen are indicated.

An absorption spectrum is produced by transitions in the opposite direction, from the ground state to excited states. Consider atoms illuminated by a beam of light whose spectrum is continuous (that is, a spectrum which contains all frequencies). These atoms will absorb only light of frequencies that correspond to particular energy differences; these are the differences that correspond to transitions from the ground state to excited states. The excited atoms then fall back to their ground states, reradiating light as they do so. However, since the reradiation occurs randomly in all directions, only a small fraction of the reradiated light is in the direction of the original beam.

## THE LASER

A laser is a device that produces an intense beam of monochromatic, coherent light from the cooperative radiation of excited atoms. The light waves in a coherent beam are all in phase with one another, which greatly increases their effectiveness. A laser beam is virtually nondivergent and hence remains as a narrow pencil of light even after traveling a large distance.

The word *laser* stands for *l*ight *a*mplification by *s*timulated *e*mission of *r*adiation. In a laser, atoms of a particular kind are raised to *metastable* (temporarily stable) states of energy $hf$, which are excited

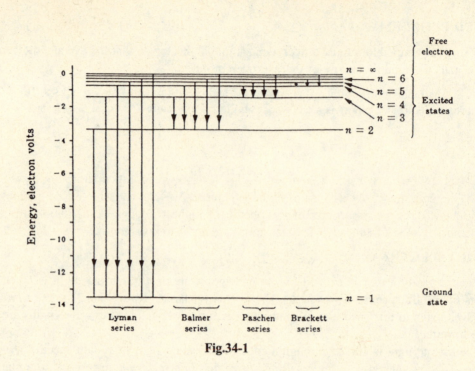

**Fig.34-1**

states of relatively long lifetimes. Radiation of frequency $f$ induces the excited atoms to emit photons of the same frequency and thereby to return to their ground (normal) states so that a small amount of initial radiation can be greatly amplified.

**SOLVED PROBLEM 34.8**

Describe two mechanisms by which the atoms of a gas can be excited so that they emit light whose frequencies make up the characteristic line spectrum of the element involved.

(1)   One mechanism is a collision with another atom, as a result of which some of their kinetic energy becomes excitation energy within one or both atoms. The excited atoms then lose this energy by emitting one or more photons. In an electric discharge in a gas, for instance in a neon sign or a sodium-vapor highway lamp, an electric field accelerates electrons and ions to velocities sufficient for atomic excitation.

(2)   Another mechanism is the absorption by an atom of light for which $hf$ is just right to raise the atom from its ground state to one of its excited states. If white light is directed at the gas, photons of those energies that correspond to such transitions are absorbed. When the energy is reradiated, all the possible transitions from the highest excited state reached show up in the emitted light.

**SOLVED PROBLEM 34.9**

A proton and an electron come together to form a hydrogen atom in its ground state. Under the assumption that a single photon is emitted in this process, what is its frequency?

The energy of a hydrogen atom in its ground state is $-13.6$ eV $= -2.13 \times 10^{-18}$ J. Hence $2.18 \times 10^{-18}$ J of energy must be given off when the atom is being formed. If a single photon is emitted, its frequency is found from $E = hf$ to be

$$f = \frac{E}{h} = \frac{2.18 \times 10^{-18} \text{ J}}{6.63 \times 10^{-34} \text{ J} \cdot \text{s}} = 3.3 \times 10^{15} \text{ Hz}$$

## SOLVED PROBLEM 34.10

Of the following transitions between energy levels in a hydrogen atom, which one involves (a) the emission of the photon of highest frequency, (b) the emission of the photon of lowest frequency, (c) the absorption of the photon of highest frequency, (d) the absorption of the photon of lowest frequency? The transitions are $n = 1$ to $n = 2$; $n = 2$ to $n = 1$; $n = 2$ to $n = 6$; and $n = 6$ to $n = 2$.

In general, photon emission occurs during a transition to a state of lower $n$, and photon absorption occurs during a transition to a state of higher $n$. By inspecting the energy level diagram of hydrogen, we see that the energy difference between the $n = 1$ and $n = 2$ levels is greater than that between the $n = 2$ and $n = 6$ levels. Hence the answers are as follows: (a) $n = 2$ to $n = 1$; (b) $n = 6$ to $n = 2$; (c) $n = 1$ to $n = 2$; (d) $n = 2$ to $n = 6$.

## SOLVED PROBLEM 34.11

A sample of hydrogen gas is bombarded by a beam of electrons. How much energy must the electrons have if the first line of the Balmer spectral series, corresponding to a transition from the $n = 3$ state to the $n = 2$ state, is to be radiated?

The energy of the $n = 3$ state in hydrogen is

$$E_3 = \frac{E_1}{3^2} = \frac{-13.6 \text{ eV}}{9} = -1.5 \text{ eV}$$

The difference in energy between the ground ($n = 1$) state and the $n = 3$ state is $13.6$ eV $- 1.5$ eV $= 12.1$ eV, so the energy needed by the bombarding electrons is $12.1$ eV. The reason the energy difference between the $n = 1$ and $n = 3$ states is involved is that the hydrogen atoms are initially in the $n = 1$ state.

## QUANTUM THEORY OF THE ATOM

In the quantum theory of the atom, no compromise is made with mechanical analogies; instead, an entirely probabilistic concept is developed. This theory holds for many-electron atoms as well as for the hydrogen atom. Four quantum numbers are needed to describe the physical state of an atomic electron, in place of the single quantum number of the Bohr model. These are shown in Table 34.1.

The possible energies of the electron are chiefly determined by $n$ and only to a smaller extent by $l$ and $m_l$. For the hydrogen atom, $E_n = E_1/n^2$ in the quantum theory as in the Bohr theory.

Every electron behaves in certain respects as though it is a spinning charged sphere. The amount of spin is the same for every electron, but there are two possible directions in which the angular momentum vector can point in a magnetic field: "up" ($m_s = +\frac{1}{2}$) and "down" ($m_s = -\frac{1}{2}$).

**Table 34.1**

| Name | Symbol | Possible Values | Quantity Determined |
|------|--------|-----------------|---------------------|
| Principal | $n$ | 1, 2, 3, . . . | Electron energy |
| Orbital | $l$ | $0, 1, 2, \ldots, n-1$ | Magnitude of angular momentum |
| Magnetic | $m_l$ | $-l, \ldots, 0, \ldots, +l$ | Direction of angular momentum |
| Spin magnetic | $m_s$ | $-\frac{1}{2}, +\frac{1}{2}$ | Direction of electron spin |

## ATOMIC ORBITALS

The distribution in space of the *probability density* $\psi^2$ of an atomic electron depends on its quantum numbers $n$, $l$, and $m_l$. As mentioned above, the larger the value of $\psi^2$ in a certain place at a certain time, the greater the likelihood of finding the electron there. The quantum theory of the atom enables $\psi^2$ to be calculated for any combination of $n$, $l$, and $m_l$ values for any atom. The region in space where $\psi^2$ is appreciable for a given quantum state is called an *orbital*.

## ATOMIC STRUCTURE

The two basic rules that govern the electron structures of many-electron atoms are:

1. An atom is stable when all its electrons are in quantum states of the lowest energy possible.

2. Only one electron can occupy each quantum state of an atom; this is the *exclusion principle*. Thus each electron must be described by a different set of quantum numbers $n$, $l$, $m_l$, $m_s$.

Electrons with the same principal quantum number $n$ in an atom are usually about the same distance from the nucleus and have similar energies. Such electrons are said to occupy the same *shell*. The higher the value of $n$, the greater the energy. Electrons in a given shell that have different orbital quantum numbers $l$ have different probability density distributions and therefore somewhat different energies; those with the same value of $l$ are said to occupy the same *subshell* and have very nearly the same energy. The higher the value of $l$ in a given subshell, the greater the energy.

## SOLVED PROBLEM 34.12

In the Bohr model of the hydrogen atom, the radius of the electron's orbit is $5.3 \times 10^{-11}$ m in the ground state. What aspect of the quantum theoretical model might correspond to this relationship?

The maximum probability of finding the electron occurs at this distance from the nucleus.

## SOLVED PROBLEM 34.13

How many electrons are able to share an orbital in an atom?

Since an orbital is characterized by a given $n$, $l$, and $m_l$, the exclusion principle permits two electrons to occupy each orbital in an atom, one with $m_s = +\frac{1}{2}$ and the other with $m_s = -\frac{1}{2}$.

**SOLVED PROBLEM 34.14**

What quantum numbers characterize the atomic state in which an electron has the lowest energy?

$$n = 1 \qquad l = 0 \qquad m_l = 0 \qquad m_s = \pm\tfrac{1}{2}$$

**SOLVED PROBLEM 34.15**

What are the possible values of the orbital and magnetic quantum numbers of an atomic electron whose principal quantum number is $n = 3$?

Since $l$ ranges from 0 up to $n - 1$ for an electron of principal quantum number $n$ and since $m_l$ ranges from $-l$ through 0 to $+l$ for an electron of orbital quantum number $l$, we have

$$l = 0 \qquad m_l = 0$$

$$l = 1 \quad \begin{cases} m_l = +1 \\ m_l = 0 \\ m_l = -1 \end{cases}$$

$$l = 2 \quad \begin{cases} m_l = +2 \\ m_l = +1 \\ m_l = 0 \\ m_l = -1 \\ m_l = -2 \end{cases}$$

**SOLVED PROBLEM 34.16**

What is the effective nuclear charge that acts on each electron in the outer shell of the chlorine ($Z = 17$) atom?

The chlorine atom has closed $n = 1$ and $n = 2$ shells with, respectively, two and eight electrons in each. Hence the $n = 3$ shell contains seven electrons that are shielded by the 10 inner ones, and the effective charge acting on these $n = 3$ electrons is accordingly $+7e$ (Fig. 34-2).

**SOLVED PROBLEM 34.17**

Why is a chlorine atom able to pick up only a single electron when it reacts chemically whereas an oxygen atom can pick up two electrons?

A chlorine atom lacks one electron of a closed outer shell. When it picks up an electron, which it can do easily because the other electrons in the outer shell do not shield the nuclear charge, this shell is

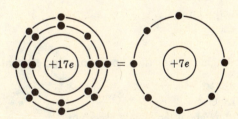

**Fig. 34-2**

closed, and any further electrons would have to go into the next shell. However, an electron in the next shell would find a net charge of $-e$ inside its orbital, not a net charge of $+7e$ as the electrons in the previous shell do, and hence would be repelled rather than held to the atom. Since an oxygen atom lacks two electrons of a closed outer shell, it can accommodate two additional electrons in this shell; further electrons would be repelled as in the case of a $Cl^-$ ion.

## SOLVED PROBLEM 34.18

Why does a closed $l = 2$ subshell contain 10 electrons?

The possible values of the magnetic quantum number $m_l$ when the orbital quantum number is $l = 2$ are $m_l = -2, -1, 0, +1, +2$, which is a total of 5. An electron with a given value of $m_l$ can have $m_s = \pm\frac{1}{2}$, so the total number of permitted quantum states in a $d$ subshell is twice 5, or 10.

# *Multiple-Choice Questions*

**34.1.** Wave behavior is exhibited by

    (*a*)   only particles at rest    (*c*)   only charged particles
    (*b*)   only moving particles    (*d*)   all particles

**34.2.** The velocity of the wave packet that corresponds to a moving particle is

    (*a*)   lower than the particle's velocity
    (*b*)   equal to the particle's velocity
    (*c*)   higher than the particle's velocity
    (*d*)   may be any of the above

**34.3.** According to the uncertainty principle, it is impossible to precisely determine at the same time a particle's

    (*a*)   position and charge    (*c*)   momentum and energy
    (*b*)   position and momentum    (*d*)   charge and mass

**34.4.** A hydrogen atom is in its ground state when its electron is

    (*a*)   at rest    (*c*)   in its lowest energy level
    (*b*)   inside the nucleus    (*d*)   in its highest energy level

**34.5.** A photon is emitted by an atom when one of the atom's electrons

    (*a*)   leaves the atom
    (*b*)   collides with another of its electrons
    (*c*)   shifts to a lower energy level
    (*d*)   shifts to a higher energy level

**34.6.** The wavelengths in the bright-line emission spectrum of an element are

    (*a*)   characteristic of the particular element
    (*b*)   the same for all elements
    (*c*)   evenly distributed throughout the visible spectrum
    (*d*)   different from the wavelengths in its dark-line absorption spectrum

**34.7.** Of the following transitions in a hydrogen atom, the one that results in the emission of the photon of lowest frequency is

(a)  $n = 1 \rightarrow n = 2$     (c)  $n = 2 \rightarrow n = 6$
(b)  $n = 2 \rightarrow n = 1$     (d)  $n = 6 \rightarrow n = 2$

**34.8.** Of the following transitions in a hydrogen atom, the one that results in the absorption of the photon of highest frequency is

(a)  $n = 1 \rightarrow n = 2$     (c)  $n = 2 \rightarrow n = 6$
(b)  $n = 2 \rightarrow n = 1$     (d)  $n = 6 \rightarrow n = 2$

**34.9.** The exclusion principle states that no two electrons in an atom can have the same

(a)  velocity      (c)  spin
(b)  orbit         (d)  set of quantum numbers

**34.10.** Which one or more of the following events cannot raise an atom from its ground state to an excited state?

(a)  spontaneous emission of a photon
(b)  induced emission of a photon
(c)  absorption of a photon
(d)  a collision with another atom

**34.11.** The operation of the laser is based on which one or more of the following?

(a)  the uncertainty principle
(b)  the exclusion principle
(c)  induced emission of radiation
(d)  interference of matter waves

**34.12.** The waves emitted by a laser do not

(a)  all have the same wavelength
(b)  emerge in step with one another
(c)  form a narrow beam
(d)  have higher photon energies than waves of the same frequency from an ordinary source

**34.13.** An electron with a velocity of $1.5 \times 10^7$ m/s has a de Broglie wavelength of

(a)  $9.1 \times 10^{-57}$ m     (c)  $4.9 \times 10^{-11}$ m
(b)  $6.5 \times 10^{-18}$ m     (d)  $4.9 \times 10^{-10}$ m

**34.14.** The kinetic energy of a neutron whose de Broglie wavelength is $2.0 \times 10^{-14}$ m is

(a)  $3.3 \times 10^{-13}$ eV     (c)  2.0 MeV
(b)  0.21 eV                      (d)  0.37 GeV

# Supplementary Problems

**34.1.** Why is the uncertainty principle only significant for such extremely small particles as electrons and protons even though it applies to objects of all sizes?

**34.2.** What relationship would you expect between the chemical activity of a metal and its ionization energy?

**34.3.** What is the nature of the spectrum found in (*a*) light from the hot filament of a light bulb, (*b*) light from a neon sign, and (*c*) light originating as in (*a*) that has passed through cool neon gas?

**34.4.** How are the compositions of the sun and stars determined?

**34.5.** Why does the spectrum of hydrogen consist of many lines even though a hydrogen atom has only a single electron?

**34.6.** In the Bohr model of the hydrogen atom, what condition must be obeyed by the electron if it is to move in its orbit indefinitely without radiating energy?

**34.7.** Find the de Broglie wavelength of a 10-g rifle bullet traveling at the velocity of sound, 331 m/s.

**34.8.** The mass of an electron moving at 75 percent of the velocity of light is 1.5 times its rest mass. Find the de Broglie wavelength of such an electron.

**34.9.** An experiment is planned in which the momentum of an electron is to be measured to within $\pm 10^{-20}$ kg·m/s while its position is known to within $\pm 10^{-10}$ m. Do you think the experiment will succeed?

**34.10.** A proton is confined to a region $10^{-9}$ m across. (*a*) Find the uncertainty in its momentum. (*b*) Find the minimum energy it can have.

**34.11.** The earth's mass is $6 \times 10^{24}$ kg, the radius of its orbit around the sun is $1.5 \times 10^{11}$ m, and its orbital velocity is $3 \times 10^4$ m/s. (*a*) Find the de Broglie wavelength of the earth. (*b*) Find the quantum number of the earth's orbit. (*c*) Do you think quantum considerations play an important part in the earth's orbital motion?

**34.12.** How much energy is needed to remove the electron from a hydrogen atom when it is in the $n = 4$ state?

**34.13.** How does the average energy per molecule in a gas at 20°C compare with the energy needed to raise a hydrogen atom from its ground state to its lowest excited state?

**34.14.** Find the velocity of the electron in the $n = 5$ state of a hydrogen atom according to the Bohr model. Is this more or less than the velocity of the electron when it is in the $n = 1$ state?

**34.15.** What $m_l$ values are possible for an electron in a state of $n = 4$, $l = 3$?

**34.16.** How many electrons are there in a closed $l = 3$ subshell?

**34.17.** What is the effective nuclear charge that acts on each electron in the outer shell of the calcium ($Z = 20$) atom? Would you think that such an electron is relatively easy or relatively hard to detach from the atom?

**34.18.** What is the effective nuclear charge that acts on each electron in the outer shell of the sulfur ($Z = 16$) atom? Would you think that such an electron is relatively easy or relatively hard to detach from the atom?

# Answers to Multiple-Choice Questions

**34.1.** (b)　　**34.8.** (a)

**34.2.** (b)　　**34.9.** (d)

**34.3.** (b)　　**34.10.** (a), (b)

**34.4.** (c)　　**34.11.** (c)

**34.5.** (c)　　**34.12.** (d)

**34.6.** (a)　　**34.13.** (c)

**34.7.** (d)　　**34.14.** (c)

# Answers to Supplementary Problems

**34.1.**　The uncertainties in the position and momentum of an object much larger than an elementary particle are so small compared with its dimensions and momentum as to be undetectable.

**34.2.**　The lower the ionization energy of a metal, the more easily one of its electrons can be detached in a chemical reaction and hence the more active it is. Thus potassium, whose ionization energy is 4.3 eV, is more active chemically than zinc, whose ionization energy is 9.4 eV.

**34.3.**　(a) Continuous emission spectrum as described in Chapter 22　　(b) Emission line spectrum　(c) Absorption line spectrum

**34.4.**　The presence of the spectral lines of a particular element in the spectrum of the sun or a star means that this element must be present there.

**34.5.**　A hydrogen sample contains a great many atoms, each of which can undergo a variety of transitions between energy levels.

**34.6.**　The orbit must be one de Broglie wavelength in circumference; or equivalently, the orbital angular momentum of the electron must be equal to $h/(2\pi)$.

**34.7.**　$2 \times 10^{-34}$ m

**34.8.**　$2.16 \times 10^{-12}$ m

**34.9.**　No, because the uncertainty principle limits $\Delta(mv)\Delta x$ to a minimum of $1.05 \times 10^{-34}$ J·s.

**34.10.**　(a) $1.05 \times 10^{-25}$ kg·m/s　　(b) $3.3 \times 10^{-24}$ J = 0.21 MeV

**34.11.**　(a) $3.68 \times 10^{-63}$ m　(b) $2.56 \times 10^{74}$　(c) No

**34.12.**   0.85 eV

**34.13.**   $KE_{av} = 6.07 \times 10^{-21}$ J; the energy difference between the $n = 1$ and $n = 2$ states is $1.64 \times 10^{-18}$ J, which is 270 times greater.

**34.14.**   $8.8 \times 10^4$ m/s; less

**34.15.**   $m_l = 0, +1, -1, +2, -2, +3, -3$

**34.16.**   14

**34.17.**   $+2e$; relatively easy

**34.18.**   $+6e$; relatively hard

# Chapter 35

## The Solid State

### CHEMICAL BONDS

When a compound is formed, atoms of the elements present are linked by *chemical bonds*. It is customary to classify chemical bonds as *ionic* or *covalent*, although actual bonds are often intermediate between the two extremes. In an ionic bond, one or more electrons from one atom are transferred to another atom, and the resulting positive and negative ions then attract each other. In a covalent bond, one or more pairs of electrons are shared by two adjacent atoms. As these electrons move about, they spend more time between the atoms than elsewhere, which results in an attractive electric force that holds the atoms together.

A *molecule* is a group of atoms that are held together tightly enough by covalent bonds to behave as a single particle. A molecule always has a definite composition and structure and has little tendency to gain or lose atoms. Ionic bonds usually result in crystalline solids, not in molecules; such solids consist of aggregates of positive and negative ions in a stable arrangement characteristic of the compound involved. Some crystalline solids are covalent rather than ionic, as discussed below.

### CRYSTALS

Most solids are crystalline, with the ions, atoms, or molecules of which they consist arranged in a regular pattern. Four kinds of bonds are found in crystals: ionic, covalent, metallic, and van der Waals.

A crystal of ordinary salt, NaCl, is an example of an ionic solid, with $Na^+$ and $Cl^-$ ions in alternate positions in a simple lattice (Fig. 35-1).

An example of a covalent solid is diamond, each of whose carbon atoms is joined by covalent bonds to four other carbon atoms in a structure that is repeated throughout the crystal (Fig. 35-2). Both ionic and covalent solids are hard and have high melting points, which are reflections of the strength of the bonds. Ionic solids are much more common than covalent ones.

In a metal, the outermost electrons of each atom are shared by the entire assembly, so that a "gas" or "sea" of electrons moves relatively freely throughout. The interaction between this electron sea and

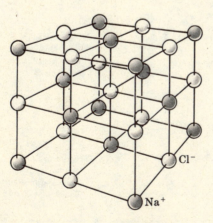

**Fig. 35-1**

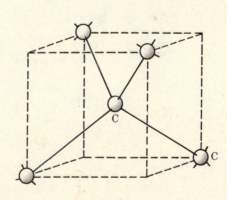

**Fig. 35-2**

the positive metal ions leads to a cohesive force, much as in the case of the shared electrons in a covalent bond but on a larger scale. The presence of the free electrons accounts for such typical properties of metals as their opacity, surface luster, and high electric and heat conductivities.

All molecules, and even inert-gas atoms such as those of helium, exhibit weak, short-range attractions for one another due to *van der Waals forces*. These forces are responsible for the condensation of gases into liquids and the freezing of liquids into solids even in the absence of ionic, covalent or metallic bonds between the atoms or molecules involved. Such familiar aspects of the behavior of matter as friction, viscosity, and adhesion are due to van der Waals forces. Van der Waals forces arise from the lack of symmetry in the momentary distributions of the electrons in a molecule. When two molecules are close together, these momentary charge asymmetries tend to shift together, with the positive part of one molecule always near the negative part of the other even though the locations of these parts are always changing. Van der Waals forces are quite weak, and substances composed of whole molecules, such as water, usually have low melting and boiling points and little mechanical strength in the solid state. Table 35.1 shows the four types of crystalline solids.

### Table 35.1.   Crystal Types*

| Type | | Bond | Example | Properties |
|---|---|---|---|---|
| Ionic | negative ion<br>positive ion | Electric attraction | Sodium chloride NaCl | Hard; high melting points; may be soluble in polar liquids such as water |
| Covalent | shared electrons | Shared electrons | Diamond C | Very hard; high melting points; insoluble in nearly all solvents |
| Metallic | metal ion<br>electron gas | Electron gas | Sodium Na | Ductile; metallic luster, high electric and thermal conductivity |
| Molecular | instantaneous charge separation in molecule | Van der Waals forces | Methane CH$_4$ | Soft; low melting and boiling points; soluble in covalent liquids |

* From A. Beiser, *Concepts of Modern Physics*, McGraw-Hill.

## SOLVED PROBLEM 35.1

Are all solids crystalline?

No. The atoms, ions, or molecules of which a crystalline solid is composed fall into regular, repeated patterns. The presence of such long-range order is the defining property of crystals. Other solids lack long-range order in their structures and may be regarded as supercooled liquids whose stiffness is due to exceptionally high viscosity. Glass, pitch, and many plastics are examples of such *amorphous* ("without form") solids.

## SOLVED PROBLEM 35.2

How are the crystal structures of solids usually determined?

The structure of a crystal is usually determined by the interference patterns produced when an X-ray beam passes through it. A crystal consists of a regular array of atoms, each of which is able to scatter an electromagnetic wave that happens to strike it. A beam of X-rays, all with the same wavelength, that falls upon a crystal will be scattered in all directions within it; but, owing to the regular arrangement of the atoms, in certain directions the scattered waves will constructively interfere with one another while in others they will destructively interfere. The phenomenon is known as *X-ray diffraction*. The resulting pattern of high and low X-ray intensities can be analyzed to yield the arrangement in space of the scattering centers, which are the atoms of the crystal.

## ENERGY BANDS

The atoms in almost all crystalline solids, whether metals or not, are so close together that their outer electrons constitute a single system of electrons common to the entire crystal. In place of each precisely defined characteristic energy level of an individual atom, the entire crystal possesses an *allowed energy band* which spans a range of possible energies. The allowed energy bands in a solid thus correspond to the energy levels in an atom, and an electron in a solid can have only those energies that fall within these energy bands. If adjacent allowed energy bands do not overlap, the intervals between them represent energies which their electrons cannot have. Such intervals are called *forbidden bands*. The electrical behavior of a crystalline solid is determined both by its energy-band structure and by how these bands are normally filled with electrons. The highest energy band that normally may contain electrons is known as the *upper energy band*.

## SOLVED PROBLEM 35.3

The upper energy band of a metal is only partly filled with electrons (Fig. 35-3); that is, the band does not contain the maximum number of electrons that can have energies in its range. How does this fact account for the ability of metals to conduct electric current?

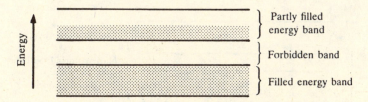

**Fig. 35-3**   Energy bands of a metal.

When an electric field is established in a metal, electrons easily acquire additional energy while remaining in their original energy band. The additional energy is in the form of kinetic energy, and the moving electrons constitute an electric current.

## SOLVED PROBLEM 35.4

The upper energy band of an insulator is completely filled with electrons and is separated from the next higher energy band by a forbidden band several electronvolts wide (Fig. 35-4). How does this fact account for the inability of insulators to conduct electric current?

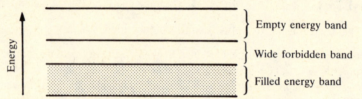

**Fig. 35-4**    Energy bands of an insulator.

An electron in an insulator must acquire at least as much energy as the width of the forbidden band if it is to have the kinetic energy required to move through the crystal. An energy increment of several electronvolts cannot be readily given to an electron in a solid by an electric field because of the presence of so many other electrons and nuclei, so insulators are very poor conductors of electric current.

## SOLVED PROBLEM 35.5

Why are metals opaque to visible light, whereas insulators are transparent when in the form of regular crystals?

Photons of visible light have energies of between about 1 and 3 eV. Such amounts of energy are readily absorbed by a "free electron" in a metal, since its allowed energy band is only partly filled, and metals are accordingly opaque. The electrons in an insulator, however, need more than 3 eV of energy to jump across the forbidden band to the next allowed band. Insulators therefore cannot absorb photons of visible light and so are transparent. Of course, most samples of insulating materials do not appear transparent, but that is because of such other factors as the scattering of light by irregularities in their structures or an amorphous character.

## SOLVED PROBLEM 35.6

What is the energy-band structure of a semiconductor?

In semiconductors, a very narrow forbidden band separates a filled upper energy band from the next empty one (Fig. 35-5), and some electrons have enough kinetic energy of thermal origin to jump this gap. Such a substance can conduct electric current to a limited extent. Some semiconductors contain small amounts of impurities which provide energy levels in the forbidden band, thus reducing the width of the energy gap electrons must overcome in order to move freely.

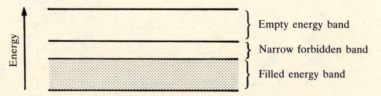

**Fig. 35-5**    Energy bands of a semiconductor.

## SOLVED PROBLEM 35.7

What is an *n*-type semiconductor?

An *n*-type semiconductor is one in which current is carried by negative charges. These are excess electrons from impurity atoms whose outermost shells contain too many electrons to fit into the crystal's electron structure. For example, arsenic atoms have five electrons in their outer shells, whereas silicon atoms have four. When an arsenic atom replaces a silicon atom in a silicon crystal, four of its electrons are incorporated in covalent bonds with its nearest neighbors. The fifth electron needs little energy to be detached and move about in the crystal (Fig. 35-6). In an energy-band diagram, such as that of Fig. 35-7, the effect of arsenic as an impurity is to provide occupied energy levels, called *donor levels*, just below an empty energy band.

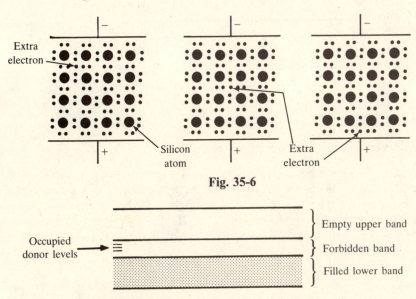

**Fig. 35-6**

**Fig. 35-7**   Energy bands of an *n*-type semiconductor.

## SOLVED PROBLEM 35.8

What is a *p*-type semiconductor?

A *p*-type semiconductor is one in which current is carried by the motion of *holes*, which are vacancies in the crystal's electron structure that behave as positive charges. An electron needs relatively little energy to enter a hole; but as it does so, it leaves a new hole in its former location. When an electric field is applied across a crystal that contains holes, electrons move toward the positive electrode by successively filling holes. The flow of current in this situation is most conveniently described with reference to the holes, whose behavior is like that of positive charges since they move toward the negative electrode (Fig. 35-8).

An example of a *p*-type semiconductor is a silicon crystal with gallium as an impurity. Gallium atoms have only three electrons in their outer shells, and their presence leaves holes in the electron strucutre of the crystal. In an energy-band diagram, such as that of Fig. 35-9, the effect of gallium as an impurity is to provide unoccupied energy levels, called *acceptor levels*, just above the highest filled energy band.

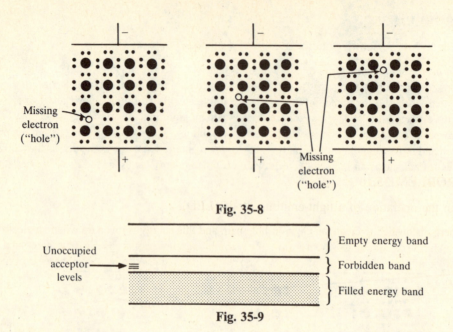

**Fig. 35-8**

Unoccupied
acceptor → ≡
levels

{ Empty energy band

{ Forbidden band

{ Filled energy band

**Fig. 35-9**

## SOLVED PROBLEM 35.9

Explain the operation of a semiconductor diode.

The operation of a semiconductor diode is based on the properties of a junction between $p$- and $n$-type materials. In the diode shown in Fig. 35-10, the left-hand end is a $p$-type region in which conduction occurs by the motion of holes, and the right-hand end is an $n$-type region in which conduction occurs by the motion of electrons. In Fig. 35-10($b$), a voltage is applied across the crystal so that the $p$ end is negative and the $n$ end is positive. This situation is called *reverse bias*. The holes in the

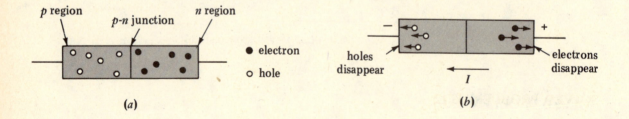

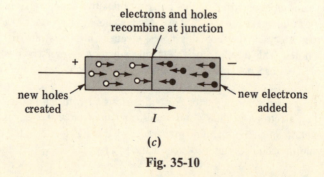

**Fig. 35-10**

*p* region migrate to the left and disappear at the negative terminal, while the electrons in the *n* end migrate to the right and disappear at the positive terminal. New electron-hole pairs are spontaneously created by thermal excitation, but they are few in number and the resulting current is extremely small.

Figure 35-10(*c*) shows the same crystal with a *forward bias*: The *p* end is now positive, and the *n* end is now negative. In this case new holes are created continuously by the removal of electrons at the positive terminal, and new electrons are added at the negative terminal. The holes migrate to the right and the electrons to the left under the influence of the applied voltage. The holes and electrons meet at the *p-n* junction and recombine there. Thus current can flow readily in one direction through a *p-n* junction but hardly at all in the other, which makes a diode of this kind an ideal rectifier.

### SOLVED PROBLEM 35.10

Explain the operation of a light-emitting diode (LED).

Energy is needed to create an electron-hole pair, and this energy is given up when an electron and a hole recombine. In silicon and germanium, the recombination energy is absorbed by the crystal as heat, but in certain other semiconductors, such as gallium arsenide, a photon is emitted when recombination occurs. This is the basis of the light-emitting diode.

### SOLVED PROBLEM 35.11

Explain the operation of a simple junction transistor.

Figure 35-11 shows an *n-p-n* junction transistor, which consists of a thin *p*-type region, called the *base*, that is sandwiched between two *n*-type regions, called the *emitter* and the *collector*. The transistor is given a forward bias across the emitter-base junction and a reverse bias across the base-collector junction. The emitter is more heavily "doped" with impurity atoms than the base, so nearly all the current across the emitter-base junction consists of electrons moving from left to right. Because the base is made very thin and the concentration of holes there is low, most of the electrons that enter the base diffuse through it to the base-collector junction where the high positive potential attracts them to the collector. Changes in the input circuit current are thus mirrored by changes in the output circuit current. The ability of the transistor of Fig. 35-11 to produce amplification comes from the reverse bias across the base-collector junction, which permits a much higher voltage in the output circuit than that in the input circuit. Since electric power = (current)(voltage), the power of the output signal can greatly exceed the power of the input signal. By using different circuits, a transistor can also be used as a current or voltage amplifier.

### SOLVED PROBLEM 35.12

Explain the operation of a field-effect transistor.

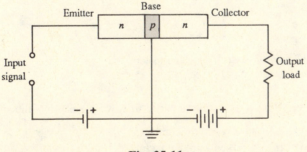

**Fig. 35-11**

Figure 35-12 shows an *n*-channel field-effect transistor that consists of a strip of *n*-type material with contacts at both ends together with a strip of *p*-type material, called the *gate*, on one side. When connected as shown, electrons move from the *source* terminal to the *drain* terminal through the *n*-type channel. The *p-n* junction is given a reverse bias, and as a result both the *n* and *p* materials near the junction are depleted of charge carriers. The higher the reverse potential on the gate, the larger the depleted region in the channel, and the fewer the electrons available to carry the current. Thus the gate voltage controls the channel current.

In a metal oxide semiconductor field-effect transistor (MOSFET), the semiconductor gate is replaced by a metal film separated from the channel by an insulating layer of silicon dioxide. The metal film is thus capacitively coupled to the channel, and its potential controls the drain current through the number of induced charges in the channel. A MOSFET is easier to make than the field-effect transistor of Fig. 35-12 and occupies only a few percent of the area needed for a junction transistor.

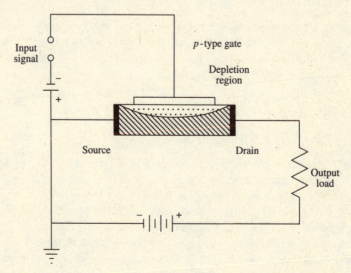

**Fig. 35-12**

# Multiple-Choice Questions

**35.1.** Solids held together by van der Waals bonds usually

   (*a*)   have low melting points
   (*b*)   have high melting points
   (*c*)   are good conductors of electricity
   (*d*)   are extremely hard

**35.2.** The electron "gas" in a metal is not directly responsible for its

   (*a*)   electrical conductivity    (*c*)   surface luster
   (*b*)   thermal conductivity     (*d*)   strength

**35.3.** When the upper energy band of a solid is partly filled with electrons, the solid is

   (*a*)   an insulator     (*c*)   an *n*-type semiconductor
   (*b*)   a conductor     (*d*)   a *p*-type semiconductor

**35.4.**  When a wide forbidden band in a solid separates an empty upper energy band from a filled lower band, the solid is

(*a*)  an insulator      (*c*)  an *n*-type semiconductor
(*b*)  a conductor       (*d*)  a *p*-type semiconductor

**35.5.**  When an empty energy band in a solid has occupied donor levels below it, the solid is

(*a*)  an insulator      (*c*)  an *n*-type semiconductor
(*b*)  a conductor       (*d*)  a *p*-type semiconductor

**35.6.**  A hole in a *p*-type semiconductor is which one or more of the following?

(*a*)  a missing atom
(*b*)  a missing electron
(*c*)  equivalent to a positive charge
(*d*)  equivalent to a negative charge

**35.7.**  A current in a *p*-type semiconductor involves the motion of

(*a*)  electrons       (*c*)  positive atomic ions
(*b*)  holes           (*d*)  negative atomic ions

**35.8.**  A current in an *n*-type semiconductor involves the motion of

(*a*)  electrons       (*c*)  positive atomic ions
(*b*)  holes           (*d*)  negative atomic ions

**35.9.**  A junction between *n*- and *p*-type semiconductors conducts electric current best when

(*a*)  both ends are positive
(*b*)  both ends are negative
(*c*)  the *p* end is positive and the *n* end is negative
(*d*)  the *p* end is negative and the *n* end is positive

**35.10.**  A photon is given off by a light-emitting diode when

(*a*)  a hole is created
(*b*)  a hole is filled
(*c*)  two holes collide
(*d*)  two electrons collide

# Supplementary Problems

**35.1.**  You are given two solids of almost identical appearance, one of which is held together by ionic bonds and the other by van der Waals bonds. How could you tell them apart?

**35.2.**  Why can metals be deformed with relative ease whereas covalent and ionic solids are quite brittle?

**35.3.**   Van der Waals forces hold inert-gas atoms together to form solids, but they cannot hold such atoms together to form molecules in the gaseous state. Why not?

**35.4.**   How good a conductor is (*a*) a crystal whose upper energy band is partly filled with electrons? (*b*) A crystal that has a wide forbidden band between a filled lower band and an empty upper band?

**35.5.**   Does the "gas" of freely moving electrons in a metal include all the electrons of the metal atoms?

**35.6.**   The forbidden band in silicon is 1.1 eV wide, and in diamond it is 6 eV wide. What bearing do these figures have on the transparency to visible light of silicon and diamond?

**35.7.**   An indium atom has three electrons in its outer shell, and a germanium atom has four. Does the addition of a small amount of indium to germanium produced an *n*- or a *p*-type semiconductor?

**35.8.**   At a *p-n* junction, (*a*) which direction of current represents forward bias and which reverse bias? (*b*) In which direction does current flow more rapidly?

## *Answers to Multiple-Choice Questions*

**35.1.** (*a*)      **35.6.** (*b*), (*c*)

**35.2.** (*d*)      **35.7.** (*b*)

**35.3.** (*b*)      **35.8.** (*a*)

**35.4.** (*a*)      **35.9.** (*c*)

**35.5.** (*c*)      **35.10.** (*b*)

## Answers to Supplementary Problems

**35.1.**   The van der Waals solid will be softer and will melt at a much lower temperature.

**35.2.**   Atoms in a metal can be readily rearranged in position because the bonding occurs by means of a sea of freely moving electrons. In a covalent crystal, the bonds are localized between adjacent atoms and must be ruptured to deform the crystal. In an ionic crystal, the bonding process requires a configuration of alternate positive and negative ions whose relative positions cannot be altered without breaking the crystal apart.

**35.3.**   Such forces are too weak to hold inert-gas molecules together against the forces exerted during collisions in the gaseous state.

**35.4.**   (*a*) Excellent   (*b*) Very poor

**35.5.** Only the outermost electrons in each metal atom become part of the electron "gas."

**35.6.** Silicon is opaque because its electrons readily absorb photons of visible light and enter the upper energy band. Diamond is transparent because photons of visible light do not have enough energy for electrons to absorb them and enter the upper band.

**35.7.** $p$-type

**35.8.** (a) $p \rightarrow n$, forward bias; $n \rightarrow p$, reverse bias   (b) Forward bias

# Chapter 36

## Nuclear Physics

### NUCLEAR STRUCTURE

The nucleus of an atom is composed of protons and neutrons whose masses are, respectively,

$$m_p = 1.673 \times 10^{-27} \text{ kg} = 1.007277 \text{ u}$$
$$m_n = 1.675 \times 10^{-27} \text{ kg} = 1.008665 \text{ u}$$

The proton has a charge of $+e$, and the neutron is uncharged. The *atomic number* of an element is the number of protons in the nucleus of one of its atoms. Protons and neutrons are jointly called *nucleons*.

Although all the atoms of an element have the same number of protons in their nuclei, the number of neutrons may be different. Each variety of nucleus found in a given element is called an *isotope* of the element. Symbols for isotopes follow the pattern

$$^A_Z X$$

where $X$ = chemical symbol of element
$Z$ = atomic number of element = number of protons in nucleus
$A$ = mass number of isotope = number of protons + neutrons in nucleus

### FUNDAMENTAL FORCES

The force between nucleons that holds an atomic nucleus together despite the repulsive electric forces its protons exert on each other is the result of what is known as the *strong interaction*. This is a fundamental interaction in the same sense as the gravitational and electromagnetic interactions are: None can be explained in terms of any of the others. The strong interaction has only a very short range, unlike the gravitational and electromagnetic interactions, and is effective only within nuclei.

There is another interaction involving nuclei called the *weak interaction* which is responsible for beta decays. The weak and electromagnetic interactions are closely related.

### SOLVED PROBLEM 36.1

The largest stable nucleus is that of the bismuth isotope $^{209}_{83}\text{Bi}$. Why are larger nuclei unstable?

The range of the strong interaction, which provides the attractive forces that hold nucleons together, is quite short, whereas the electric repulsive forces that act between protons have unlimited range. Hence beyond a certain size the repulsive forces become comparable with the attractive ones, and such nuclei are unstable.

### SOLVED PROBLEM 36.2

State the number of protons and neutrons in the following nuclei:

$$^6_3\text{Li} \qquad ^{12}_6\text{C} \qquad ^{36}_{16}\text{S} \qquad ^{137}_{56}\text{Ba}$$

A nucleus designated $^A_Z X$ contains $Z$ protons and $A - Z$ neutrons. Accordingly the numbers of protons and neutrons in the given nuclei are as follows:

$$^6_3\text{Li}: \qquad 3 \text{ protons, } 3 \text{ neutrons}$$

$$^{12}_6\text{C}: \qquad 6 \text{ protons, } 6 \text{ neutrons}$$

$$^{36}_{16}\text{S}: \qquad 16 \text{ protons, } 20 \text{ neutrons}$$

$$^{137}_{56}\text{Ba}: \qquad 56 \text{ protons, } 81 \text{ neutrons}$$

## SOLVED PROBLEM 36.3

Ordinary chlorine is a mixture of 75.53 percent of the $^{35}_{17}\text{Cl}$ isotope and 24.47 percent of the $^{37}_{17}\text{Cl}$ isotope. The atomic masses of these isotopes are, respectively, 34.969 and 36.966 u. Find the atomic mass of ordinary chlorine.

The procedure is to multiply the mass of each isotope by the proportion of the whole it represents and then to add the results. Thus we obtain

$$(0.7553)(34.969 \text{ u}) + (0.2447)(36.966 \text{ u}) = 35.458 \text{ u}$$

which is the atomic mass of ordinary chlorine.

## BINDING ENERGY

The mass of an atom is always less than the sum of the masses of the neutrons, protons, and electrons of which it is composed. The energy equivalent of the missing mass is called the *binding energy* of the nucleus; the greater its binding energy, the more stable the nucleus. The mass defect $\Delta m$ of a nucleus with $Z$ protons and $N$ neutrons may be found from its atomic mass $m$ by using the formula

$$\Delta m = Z m_\text{H} + N m_n - m$$

where $m_\text{H}$, the mass of the hydrogen atom (which consists of a proton and an electron), is

$$m_\text{H} = 1.007825 \text{ u}$$

To find the binding energy in megaelectronvolts, the usual unit, $\Delta m$ can be multiplied by the conversion factor 931 MeV/u.

## SOLVED PROBLEM 36.4

The atomic mass of $^{16}_8\text{O}$ is 15.9949 u. (*a*) What is its binding energy? (*b*) What is its binding energy per nucleon?

(*a*)   The $^{16}_8\text{O}$ contains 8 protons and 8 neutrons in its nucleus. The mass of 8 H atoms is $8 m_\text{H} = (8)(1.007825) u = 8.0626 \text{ u}$, and the mass of 8 neutrons is $8 m_n = (8)(1.008665) u = 8.0693 \text{ u}$. Hence the mass deficit is $^{16}_8\text{O}$ is

$$\Delta m = (8.0626 + 8.0693) \text{ u} - 15.9949 \text{ u} = 0.1370 \text{ u}$$

and since 1 u = 931 MeV, the binding energy is

$$\Delta E = (0.1370 \text{ u})(931 \text{ MeV/u}) = 127.5 \text{ MeV}$$

(*b*)   There are 16 nucleons in $^{16}_8\text{O}$, so the binding energy per nucleon is 127.6 MeV/16 nucleons = 7.97 MeV per nucleon.

## SOLVED PROBLEM 36.5

The binding energy of $^{20}_{10}$Ne is 160.6 MeV. Find its atomic mass.

The $^{20}_{10}$Ne contains 10 protons and 10 neutrons in its nucleus. The mass of 10 H atoms and 10 neutrons is

$$m_0 = 10.07825 \text{ u} + 10.08665 \text{ u} = 20.1649 \text{ u}$$

The mass equivalent of 160.6 MeV is

$$\Delta m = \frac{160.6 \text{ MeV}}{931 \text{ MeV/u}} = 0.1725 \text{ u}$$

and so the mass of the $^{20}_{10}$Ne atom is

$$m = m_0 - \Delta m = 20.1649 \text{ u} - 0.1725 \text{ u} = 19.9924 \text{ u}$$

## NUCLEAR REACTIONS

Nuclei can be transformed into others of a different kind by interaction with each other. Since nuclei are all positively charged, a high-energy collision is necessary between two nuclei if they are to get close enough together to react. Because it has no charge, a neutron can initiate a nuclear reaction even if it is moving slowly. In any nuclear reaction, the total number of neutrons and the total number of protons in the products must be equal to the corresponding total numbers in the reactants.

## FISSION AND FUSION

Nuclei of intermediate size have the highest binding energies per nucleon and therefore are more stable than lighter and heavier nuclei. If a heavy nucleus is split into two smaller ones, the greater binding energy of the latter means that energy will be liberated. This process is called *nuclear fission*. Certain very large nuclei, such as $^{235}_{92}$U, undergo fission when they absorb a neutron. Since the products of the fission include several neutrons as well as two daughter nuclei, a *chain reaction* can be established in an assembly of a suitable fissionable isotope. If it is uncontrolled, the result is an atomic bomb. If it is controlled so that the rate at which fission events occur is constant, the result is a nuclear reactor which can serve as an energy source for generating electricity or for ship propulsion.

In *nuclear fusion*, two light nuclei combine to form a heavier one whose binding energy per nucleon is greater. The difference in binding energies is liberated in the process. To bring about a fusion reaction, the initial nuclei must be moving rapidly when they collide to overcome their electric repulsion. Nuclear fusion is the source of energy in the sun and stars, where the high temperatures in the interiors mean that nuclei there have sufficiently high velocities and the high pressures mean that nuclear collisions occur frequently. In the operation of a hydrogen bomb, a fission bomb is first detonated to produce the high temperature and pressure necessary for fusion reactions to occur. The problem in constructing a fusion reactor for controlled energy production is to contain a sufficiently hot and dense mixture of suitable isotopes for long enough to yield a net energy output.

## SOLVED PROBLEM 36.6

Complete the following nuclear reactions:

$$^{6}_{3}\text{Li} + {}^{2}_{1}\text{H} \rightarrow {}^{4}_{2}\text{He} + \ ?$$
$$^{35}_{17}\text{Cl} + \ ? \rightarrow {}^{32}_{16}\text{S} + {}^{4}_{2}\text{He}$$
$$^{9}_{4}\text{Be} + {}^{4}_{2}\text{He} \rightarrow {}^{1}_{0}n + \ ?$$

In each of these reactions, the number of protons and the number of neutrons must be the same on both sides of the equation. Hence the complete reactions must be as follows:

$$_3^6\text{Li} + {}_1^2\text{H} \rightarrow {}_2^4\text{He} + {}_2^4\text{He}$$
$$_{17}^{35}\text{Cl} + {}_1^1\text{H} \rightarrow {}_{16}^{32}\text{S} + {}_2^4\text{He}$$
$$_4^9\text{Be} + {}_2^4\text{He} \rightarrow {}_0^1 n + {}_6^{12}\text{C}$$

## SOLVED PROBLEM 36.7

In a typical fission reaction, a $_{92}^{235}\text{U}$ nucleus absorbs a neutron and splits into a $_{54}^{140}\text{Xe}$ nucleus and a $_{38}^{94}\text{Sr}$ nucleus. How many neutrons are liberated in this process?

In order that the total numbers of protons and neutrons be the same before and after the fission reaction, two neutrons must be liberated. Hence the reaction is

$$_{92}^{235}\text{U} + {}_0^1 n \rightarrow {}_{54}^{140}\text{Xe} + {}_{38}^{94}\text{Sr} + {}_0^1 n + {}_0^1 n + \Delta E$$

In this case $\Delta E$ is about 200 MeV.

## SOLVED PROBLEM 36.8

When $_{92}^{235}\text{U}$ undergoes fission, about 0.1 percent of the original mass is released as energy. (a) How much energy is released when 1 kg of $_{92}^{235}\text{U}$ undergoes fission? (b) How much $_{92}^{235}\text{U}$ must undergo fission per day in a nuclear reactor that provides energy to a 100-MW ($10^8$-W) electric power plant? Assume perfect efficiency. (c) When coal is burned, about 32.6 MJ/kg of heat is liberated. How many kilograms of coal would be consumed per day by a conventional coal-fired 100-MW electric power plant?

(a)             $$E = mc^2 = (0.001 \text{ kg})(3 \times 10^8 \text{ m/s})^2 = 9 \times 10^{13} \text{ J}$$

(b)   Energy = power × time, and so here

$$E = Pt = (10^8 \text{ W})(3600 \text{ s/h})(24 \text{ h/day}) = 8.64 \times 10^{12} \text{ J/day}$$

Hence the mass of $_{92}^{235}\text{U}$ required is

$$\frac{8.64 \times 10^{12} \text{ J/day}}{9 \times 10^{13} \text{ J/kg}} = 9.6 \times 10^{-2} \text{ kg/day} = 96 \text{ g/day}$$

(c)   The energy liberated per kilogram of coal burned is $3.26 \times 10^7$ J. Hence the mass of coal required is

$$\frac{8.64 \times 10^{12} \text{ J/day}}{3.26 \times 10^7 \text{ J/kg}} = 2.65 \times 10^5 \text{ kg/day}$$

which is 265 metric tons.

## SOLVED PROBLEM 36.9

In the sun and most other stars the principal energy-liberating process is the conversion of hydrogen to helium in a series of nuclear fusion reactions in the course of which *positrons* (positively charged electrons) are emitted. (a) Write the equation for the overall process in which four protons form a helium nucleus (alpha particle). (b) How much energy is liberated in each such process? The masses of $_1^1\text{H}$, $_2^4\text{He}$, and the electron are, respectively, 1.007825, 4.002603, and 0.000549 u.

(*a*) Two positrons must be given off so that charge will be conserved. Hence the overall process is

$$^1_1\text{H} + ^1_1\text{H} + ^1_1\text{H} + ^1_1\text{H} \rightarrow ^4_2\text{He} + e^+ + e^+$$

(*b*) Since a helium atom has only two electrons around its nucleus, two electrons as well as two positrons are lost when each helium atom is formed. The mass change is therefore

$$\Delta m = 4m_\text{H} - (m_\text{He} + 4m_e) = (4)(1.007825 \text{ u}) - [4.002603 \text{ u} + (4)(0.000549 \text{ u})]$$
$$= 0.026501 \text{ u}$$

and the energy liberated is $(0.026501 \text{ u})(931 \text{ MeV/u}) = 24.7 \text{ MeV}$.

# RADIOACTIVITY

Certain nuclei are unstable and undergo *radioactive decay* into more stable ones. The five types of radioactive decay are shown in Fig. 36-1. A *positron* is a positive electron and an *alpha particle* is a nucleus of the helium atom $^4_2\text{He}$.

| Type of Decay | Parent nucleus | Decay event | Daughter nucleus | Reason for instability |
|---|---|---|---|---|
| Gamma decay | | Emission of gamma ray reduces energy of nucleus | | Nucleus has excess energy |
| Alpha decay | | Emission of alpha particle reduces size of nucleus | | Nucleus is too large |
| Beta decay | | Emission of electron by nuclear neutron changes it to a proton | | Nucleus has too many neutrons relative to number of protons |
| Electron capture | | Capture of electron by nuclear proton changes it to a neutron | | Nucleus has too many protons relative to number of neutrons |
| Positron emission | | Emission of positron by nuclear proton changes it to a neutron | | Nucleus has too many protons relative to number of neutrons |

● Proton (charge $= +e$)　　　　　● Electron (charge $= -e$)
○ Neutron (charge $= 0$)　　　　　○ Positron (charge $= +e$)

**Fig. 36-1**　Radioactive decay.

## HALF-LIFE

A nucleus subject to radioactive decay always has a certain definite probability of decay during any time interval. The *half-life* of a radioactive isotope is the time required for half of any initial quantity to decay. If an isotope has a half-life of, say 5 h, and we start with 100 g of it, 50 g will be left undecayed after 5 h; 25 g will be left undecayed after 10 h; 12.5 g will be left undecayed after 15 h; and so on.

## SOLVED PROBLEM 36.10

What happens to the atomic number and mass number of a nucleus that (*a*) emits an electron, (*b*) undergoes electron capture, (*c*) emits an alpha particle?

(*a*) $Z$ increases by 1, $A$ is unchanged. (*b*) $Z$ decreases by 1, $A$ is unchanged. (*c*) $Z$ decreases by 2, $A$ decreases by 4.

## SOLVED PROBLEM 36.11

How many successive alpha decays occur in the decay of the thorium isotope $^{228}_{90}$Th into the lead isotope $^{212}_{82}$Pb?

Each alpha decay means a reduction of 2 in atomic number and of 4 in mass number. Here $Z$ decreases by 8 and $A$ by 16, which means that 4 alpha particles are emitted.

## SOLVED PROBLEM 36.12

Tritium is the hydrogen isotope $^3_1$H whose nucleus contains two neutrons and a proton. Tritium is beta-radioactive and emits an electron. (*a*) What does tritium become after beta decay? (*b*) The half-life of tritium is 12.5 yr. How much of a 1-g sample will remain undecayed after 25 yr?

(*a*)   In the beta decay of a nucleus, one of its neutrons becomes a proton. Since the atomic number 2 corresponds to helium, the beta decay of $^3_1$H is given by

$$^3_1\text{H} \rightarrow \, ^3_2\text{He} + e^-$$

and the new atom is $^3_2$He.

(*b*)   Twenty-five years is two half-lives of tritium, and so $\frac{1}{2} \times \frac{1}{2} \times 1 \text{ g} = \frac{1}{4} \text{ g}$ of tritium remains undecayed.

## SOLVED PROBLEM 36.13

The half-life of the sodium isotope $^{24}_{11}$Na against beta decay is 15 h. How long does it take for seven-eighths of a sample of this isotope to decay?

After seven-eighths has decayed, one-eighth is left, and $\frac{1}{8} = \frac{1}{2} \times \frac{1}{2} \times \frac{1}{2}$ which is three half-lives. Hence the answer is (3)(15) h = 45 h.

## SOLVED PROBLEM 36.14

The carbon isotope $^{14}_6$C (called *radiocarbon*) is beta-radioactive with a half-life of 5600 yr. Radiocarbon is produced in the earth's atmosphere by the action of cosmic rays on nitrogen atoms, and the carbon dioxide of the atmosphere contains a small proportion of radiocarbon as a result. All plants and animals therefore contain a certain amount of radiocarbon along with the stable isotope $^{12}_{6}C$. When a living thing dies, it stops taking in radiocarbon and the radiocarbon it already contains decays steadily. By measuring the ratio between the $^{14}_{6}C$ and $^{12}_{6}C$ contents of the remains of an animal or plant and comparing it with the ratio of these isotopes in living organisms, the time that has passed since the death of the animal or plant can be found. (*a*) How old is a piece of wood from an ancient dwelling if its relative radiocarbon content is one-fourth that of a modern specimen? (*b*) If it is one-sixteenth that of a modern specimen?

(*a*)   Since $\frac{1}{4} = \frac{1}{2} \times \frac{1}{2}$, the specimen is two half-lives old, which is 11,200 yr old.

(*b*)   Since $\frac{1}{16} = \frac{1}{2} \times \frac{1}{2} \times \frac{1}{2} \times \frac{1}{2}$, the specimen is four half-lives old, which is 22,400 yr old.

## SOLVED PROBLEM 36.15

The rate at which a sample of radioactive substance decays is called its *activity*. A unit of activity is the *curie* (Ci), where 1 curie $= 3.7 \times 10^{10}$ decays/s. If a luminous watch dial contains 5 $\mu$Ci of the radium isotope $^{226}_{88}Ra$, how many decays per second occur in it? (This isotope emits alpha particles which cause flashes of light when they strike a special material the isotope is mixed with.)

Since 1 $\mu$Ci is $10^{-6}$ Ci, the activity of the watch dial is

$$\left(3.7 \times 10^{10} \ \frac{\text{decays/s}}{\text{Ci}}\right)(5 \times 10^{-6} \ \text{Ci}) = 1.85 \times 10^{5} \ \text{decays/s}$$

# *Multiple-Choice Questions*

**36.1.**   The isotopes of an element all have the same

   (*a*)   atomic number
   (*b*)   mass number
   (*c*)   binding energy
   (*d*)   half-life

**36.2.**   Each nucleus of the nitrogen isotope $^{16}_{7}N$ contains

   (*a*)   7 neutrons
   (*b*)   9 neutrons
   (*c*)   16 neutrons
   (*d*)   23 neutrons

**36.3.**   Nuclear fusion and fission reactions give off energy because

   (*a*)   the binding energy per nucleon is least for nuclei of intermediate size
   (*b*)   the binding energy per nucleon is most for nuclei of intermediate size
   (*c*)   they liberate neutrons
   (*d*)   they liberate protons

**36.4.**   The half-life of a radionuclide equals

   (*a*)   half the time needed for a sample to completely decay
   (*b*)   half the time a sample can be kept before it starts to decay
   (*c*)   the time needed for half a sample to decay
   (*d*)   the time needed for the rest of a sample to decay once half of it has already decayed

**36.5.** During the decay of a radionuclide, its half-life

(a) decreases        (c) increases
(b) does not change  (d) any of the above, depending on the nuclide

**36.6.** In a chain reaction

(a) protons and neutrons join to form atomic nuclei
(b) light nuclei join to form heavy ones
(c) neutrons emitted during the fission of heavy nuclei induce fussions in other nuclei
(d) uranium is burned in a type of furnace called a reactor

**36.7.** The energy that heats the sun has its origin in

(a) radioactivity
(b) nuclear fission
(c) the production of helium from hydrogen
(d) the production of hydrogen from helium

**36.8.** The sum of the masses of 10 protons and 10 neutrons is 0.172 u more than the mass of a $^{20}_{10}$Ne nucleus. The binding energy per nucleon in this nucleus is

(a) $8.6 \times 10^{-3}$ eV   (c) 16.0 MeV
(b) 8.0 MeV                    (d) $7.7 \times 10^{14}$ eV

**36.9.** When the uranium isotope $^{234}_{92}$U undergoes alpha decay, the result is the nuclide

(a) $^{230}_{90}$Th   (c) $^{232}_{88}$Ra
(b) $^{230}_{92}$U    (d) $^{230}_{88}$Ra

**36.10.** When the strontium isotope $^{87}_{38}$Sr undergoes gamma decay, the result is the nuclide

(a) $^{87}_{37}$Rb   (c) $^{87}_{39}$Y
(b) $^{87}_{38}$Sr   (d) $^{83}_{36}$Kr

**36.11.** The copper isotope $^{64}_{29}$Cu decays into the nickel isotope $^{64}_{28}$Ni by emitting

(a) an electron       (c) an alpha particle
(b) a positron        (d) a gamma ray

**36.12.** A certain radionuclide has a half-life of 12 h. Starting from 1.00 g of the nuclide, the amount left after 2 d will be

(a) 0          (c) 0.16 g
(b) 0.0625 g   (d) 0.25 g

# Supplementary Problems

**36.1.** (a) Which of the fundamental interactions has the least significance in nuclear physics? (b) Which two are closely related?

**36.2.** In experiments involving nuclear fusion, magnetic fields rather than solid containers are used to confine atomic nuclei that are to react. Why?

**36.3.** What are the similarities and differences between nuclear fission and nuclear fusion?

**36.4.** What parts of its structure are chiefly responsible for an atom's mass and for its chemical behavior?

**36.5.** State the numbers of protons and neutrons in each of the following nuclei: $^{15}_{7}$N, $^{35}_{17}$Cl, $^{64}_{30}$Zn, $^{200}_{80}$Hg.

**36.6.** Ordinary boron is a mixture of 20 percent of the $^{10}_{5}$B isotope and 80 percent of the $^{11}_{5}$B isotope. The atomic masses of these isotopes are, respectively, 10.013 and 11.009 u. Find the atomic mass of ordinary boron.

**36.7.** The atomic mass of $^{3}_{2}$He is 3.01603. (a) What is its binding energy? (b) What is its binding energy per nucleon?

**36.8.** The atomic mass of $^{35}_{17}$Cl is 34.96885 u. (a) What is its binding energy? (b) What is its binding energy per nucleon?

**36.9.** The binding energy of $^{42}_{20}$Ca is 361.7 MeV. Find its atomic mass.

**36.10.** Complete the following nuclear reactions:

$$^{14}_{7}\text{N} + ^{4}_{2}\text{He} \rightarrow ^{1}_{1}\text{H} + ?$$
$$^{11}_{5}\text{B} + ^{1}_{1}\text{H} \rightarrow ^{11}_{6}\text{C} + ?$$
$$^{6}_{3}\text{Li} + ? \rightarrow ^{7}_{4}\text{Be} + ^{1}_{0}n$$

**36.11.** When $^{235}_{92}$U undergoes fission, about 0.1 percent of the original mass is released as energy. (a) How much energy is released by an atomic bomb that contains 10 kg of $^{235}_{92}$U? (b) When 1 ton of TNT is exploded, about $4 \times 10^{9}$ J is released. How many tons of TNT are equivalent in destructive power to the above bomb?

**36.12.** In some stars three $^{4}_{2}$He nuclei fuse in sequence to form a $^{12}_{6}$C nucleus ($m = 12.000000$ u). How much energy is liberated each time this happens?

**36.13.** Radium spontaneously decays into the elements helium and radon. Why is radium itself considered an element and not simply a chemical compound of helium and radon?

**36.14.** What happens to the atomic number and mass number of a nucleus that emits a gamma-ray photon? What happens to its mass?

**36.15.** The uranium isotope $^{238}_{92}$U decays into a stable lead isotope through the successive emission of eight alpha particles and six electrons. What is the symbol of the lead isotope?

**36.16.** The half-life of $^{238}_{92}$U against alpha decay is $4.5 \times 10^{9}$ yr. (a) How long does it take for seven-eighths of a sample of this isotope to decay? (b) For fifteen-sixteenths to decay?

**36.17.** The half-life against beta decay of the strontium isotope $^{90}_{38}$Sr is 28 yr. (a) What does $^{90}_{38}$Sr become after beta decay? (b) What percentage of a sample of $^{90}_{38}$Sr will remain undecayed after 112 yr?

## Answers to Multiple-Choice Questions

**36.1.**  (a)       **36.7.**   (c)

**36.2.**  (b)       **36.8.**   (b)

**36.3.**  (b)       **36.9.**   (a)

**36.4.**  (c)       **36.10.**  (b)

**36.5.**  (b)       **36.11.**  (b)

**36.6.**  (c)       **36.12.**  (b)

## Answers to Supplementary Problems

**36.1.**   (a)  The gravitational interaction      (b)  The weak and electromagnetic interactions

**36.2.**   In such experiments, the nuclei form a gas at very high temperature that is called a *plasma*. A plasma would be cooled upon contact with a solid container, and atoms of the container would also be dislodged and enter the plasma, where they might affect the reaction unfavorably. It is not likely that the container would actually melt, since the total internal energy of the plasma, as distinguished from its temperature, is not very great.

**36.3.**   In fission, a large nucleus splits into smaller ones; in fusion, two small nuclei join to form a larger one. In both processes, the products of the reaction have less mass than the original nucleus or nuclei, with the missing mass being released as energy.

**36.4.**   The number of protons and neutrons in its nucleus determines the mass of an atom, and the number of electrons in the electron cloud surrounding the nucleus governs its chemical behavior.

**36.5.**   $7p, 8n$; $17p, 18n$; $30p, 34n$; $80p, 120n$

**36.6.**   10.81 u

**36.7.**   (a)  7.71 MeV      (b)  2.57 MeV

**36.8.**   (a)  298 MeV      (b)  8.5 MeV

**36.9.**   41.9586 u

**36.10.**  $^{17}_{8}O$; $^{1}_{0}n$; $^{2}_{1}H$

**36.11.**  (a)  $9 \times 10^{14}$ J      (b)  $2.25 \times 10^5$ tons

**36.12.**  7.27 MeV

**36.13.** Helium and radon cannot be combined to form radium, nor can radium be broken down into helium and radon by chemical means.

**36.14.** $Z$ and $A$ are unchanged, but the actual mass decreases in proportion to the energy lost.

**36.15.** $^{206}_{82}\text{Pb}$

**36.16.** (*a*)  $1.35 \times 10^{10}$ yr      (*b*)  $1.8 \times 10^{10}$ yr

**36.17.** (*a*)  $^{90}_{39}\text{Y}$      (*b*)  6.25 percent

# Appendix A

## Physical Constants and Quantities

| Quantity | Symbol | Value |
|---|---|---|
| Absolute zero | 0 K, 0°R | $-273°C = -460°F$ |
| Acceleration of gravity of earth's surface | $g$ | $9.81$ m/s$^2$ = $32.2$ ft/s$^2$ |
| Avogadro's number | $N$ | $6.023 \times 10^{23}$ atoms or molecules per mole |
| Boltzmann's constant | $k$ | $1.38 \times 10^{-23}$ J/K |
| Coulomb constant | $k$ | $8.99 \times 10^{9}$ N·m$^2$/C$^2$ |
| Electron charge | $e$ | $1.60 \times 10^{-19}$ C |
| Electron mass | $m_e$ | $9.1 \times 10^{-31}$ kg |
| Gravitational constant | $G$ | $6.67 \times 10^{-11}$ N·m$^2$/kg$^2$ <br> $= 3.44 \times 10^{-8}$ lb·ft$^2$/slug$^2$ |
| Molar volume at STP | $V_0$ | $22.4$ L/mol |
| Permeability of free space | $\mu_0$ | $4\pi \times 10^{-7}$ T·m/A |
| Permittivity of free space | $\varepsilon_0$ | $8.85 \times 10^{-12}$ C$^2$/(N·m$^2$) |
| Planck's constant | $h$ | $6.626 \times 10^{-34}$ J·s |
| Stefan-Boltzmann constant | $\sigma$ | $5.67 \times 10^{-8}$ W/(m$^2$·K$^4$) |
| Universal gas constant | $R$ | $8.31 \times 10^{3}$ J/(mol·K) <br> $= 0.0821$ atm·L/(mol·K) |
| Velocity of light in free space | $c$ | $3.00 \times 10^{8}$ m/s |

# Appendix B

## Conversion Factors

**Time**

1 day = 1.44 $10^3$ min = 8.64 $10^4$ s

1 yr = 8.76 $10^3$ h = 5.26 $10^5$ min = 3.15 $10^7$ s

**Length**

1 meter (m) = 100 cm = 39.4 in. = 3.28 ft

1 centimeter (cm) = 10 millimeters (mm) = 0.394 in.

1 kilometer (km) = $10^3$ m = 0.621 mi

1 foot (ft) = 12 in. = 0.305 m = 30.5 cm

1 inch (in.) = 0.0833 ft = 2.54 cm = 0.0254 m

1 mile (mi) = 5280 ft = 1.61 km

**Area**

1 $m^2$ = $10^4$ $cm^2$ = 1.55 $10^3$ $in.^2$ = 10.76 $ft^2$

1 $cm^2$ = $10^{-4}$ $m^2$ = 0.155 $in.^2$

1 $ft^2$ = 144 $in.^2$ = 9.29 $10^{-2}$ $m^2$ = 929 $cm^2$

**Volume**

1 $m^3$ = $10^3$ L = $10^6$ $cm^3$ = 35.3 $ft^3$ = 6.10 $10^4$ $in.^3$

1 $ft^3$ = 1728 $in.^2$ = 2.83 $10^{-2}$ $m^3$ = 28.3 L

**Velocity**

1 m/s = 3.28 ft/s = 2.24 mi/h = 3.60 km/h

1 ft/s = 0.305 m/s = 0.682 mi/h = 1.10 km/h

(*Note*:  It is often convenient to remember that 88 ft/s = 60 mi/h.)

1 km/h = 0.278 m/s = 0.913 ft/s = 0.621 mi/h

1 mi/h = 1.47 ft/s = 0.447 m/s = 1.61 km/h

**Mass**

1 kilogram (kg)  = $10^3$ grams (g) = 0.0685 slug

(*Note*:  1 kg corresponds to 2.21 lb in the sense that the *weight*
of 1 kg at the earth's surface is 2.21 lb.)

1 slug = 14.6 kg

(*Note*:  1 slug corresponds to 32.2 lb in the sense that the *weight*
of 1 slug at the earth's surface is 32.2 lb.)

1 atomic mass unit (u) = 1.66 $10^{-27}$ kg = 1.49 $10^{-10}$ J = 931 MeV

**Force**

1 newton (N) = 0.225 lb = 3.60 oz

1 pound (lb) = 16 ounces (oz) = 4.45 N

(*Note*:  1 lb corresponds to 0.454 kg = 454 g in the sense that the *mass*
of something that weighs 1 lb at the earth's surface is 0.454 kg.)

**Pressure**

1 Pa = 1 $N/m^2$ = 2.09 $10^{-2}$ $lb/ft^2$ = 1.45 $10^{-4}$ $lb/in.^2$

1 $lb/in.^2$ = 144 $lb/ft^2$ = 6.90 $10^3$ Pa

1 atm = 1.013 $10^5$ Pa = 14.7 $lb/in.^2$

**Energy**

1 joule (J) = 0.738 ft·lb = 2.39　$10^{-4}$ kcal = 6.24　$10^{18}$ eV

1 foot-pound (ft·lb) = 1.36 J = 1.29　$10^{-3}$ Btu = 3.25　$10^{-4}$ kcal

1 kilocalorie (kcal) = 4185 J = 3.97 Btu = 3077 ft·lb

1 Btu = 0.252 kcal = 788 ft·lb

1 electronvolt (eV) = $10^{-6}$ MeV = $10^{-9}$ GeV = 1.60　$10^{-19}$ J

**Power**

1 watt (W) = 1 J/s = 0.738 ft·lb/s

1 kilowatt (kW) = $10^3$ W = 1.34 hp

1 horsepower (hp) = 550 ft·lb/s = 746 W

1 refrigeration ton = 12,000 Btu/h

**Temperature**

$T_C = (T_F - 32°)$

$T_F = T_C + 32°$

$T_K = T_C + 273°$

$T_R = T_F + 460°$

# Appendix C

## Natural Trigonometric Functions

| Angle | | | | | Angle | | | | |
|-------|------|------|------|------|-------|------|------|------|------|
| Deg. | Rad. | Sin | Cos | Tan | Deg. | Rad. | Sin | Cos | Tan |
| 0° | 0.000 | 0.000 | 1.000 | 0.000 | | | | | |
| 1° | 0.017 | 0.018 | 1.000 | 0.018 | 46° | 0.803 | 0.719 | 0.695 | 1.036 |
| 2° | 0.035 | 0.035 | 0.999 | 0.035 | 47° | 0.820 | 0.731 | 0.682 | 1.072 |
| 3° | 0.052 | 0.052 | 0.999 | 0.052 | 48° | 0.838 | 0.743 | 0.669 | 1.111 |
| 4° | 0.070 | 0.070 | 0.998 | 0.070 | 49° | 0.855 | 0.755 | 0.656 | 1.150 |
| 5° | 0.087 | 0.087 | 0.996 | 0.088 | 50° | 0.873 | 0.766 | 0.643 | 1.192 |
| 6° | 0.105 | 0.105 | 0.995 | 0.105 | 51° | 0.890 | 0.777 | 0.629 | 1.235 |
| 7° | 0.122 | 0.122 | 0.993 | 0.123 | 52° | 0.908 | 0.788 | 0.616 | 1.280 |
| 8° | 0.140 | 0.139 | 0.990 | 0.141 | 53° | 0.925 | 0.799 | 0.602 | 1.327 |
| 9° | 0.157 | 0.156 | 0.988 | 0.158 | 54° | 0.942 | 0.809 | 0.588 | 1.376 |
| 10° | 0.175 | 0.174 | 0.985 | 0.176 | 55° | 0.960 | 0.819 | 0.574 | 1.428 |
| 11° | 0.192 | 0.191 | 0.982 | 0.194 | 56° | 0.977 | 0.829 | 0.559 | 1.483 |
| 12° | 0.209 | 0.208 | 0.978 | 0.213 | 57° | 0.995 | 0.839 | 0.545 | 1.540 |
| 13° | 0.227 | 0.225 | 0.974 | 0.231 | 58° | 1.012 | 0.848 | 0.530 | 1.600 |
| 14° | 0.244 | 0.242 | 0.970 | 0.249 | 59° | 1.030 | 0.857 | 0.515 | 1.664 |
| 15° | 0.262 | 0.259 | 0.966 | 0.268 | 60° | 1.047 | 0.866 | 0.500 | 1.732 |
| 16° | 0.279 | 0.276 | 0.961 | 0.287 | 61° | 1.065 | 0.875 | 0.485 | 1.804 |
| 17° | 0.297 | 0.292 | 0.956 | 0.306 | 62° | 1.082 | 0.883 | 0.470 | 1.881 |
| 18° | 0.314 | 0.309 | 0.951 | 0.325 | 63° | 1.100 | 0.891 | 0.454 | 1.963 |
| 19° | 0.332 | 0.326 | 0.946 | 0.344 | 64° | 1.117 | 0.899 | 0.438 | 2.050 |
| 20° | 0.349 | 0.342 | 0.940 | 0.364 | 65° | 1.134 | 0.906 | 0.423 | 2.145 |
| 21° | 0.367 | 0.358 | 0.934 | 0.384 | 66° | 1.152 | 0.914 | 0.407 | 2.246 |
| 22° | 0.384 | 0.375 | 0.927 | 0.404 | 67° | 1.169 | 0.921 | 0.391 | 2.356 |
| 23° | 0.401 | 0.391 | 0.921 | 0.425 | 68° | 1.187 | 0.927 | 0.375 | 2.475 |
| 24° | 0.419 | 0.407 | 0.914 | 0.445 | 69° | 1.204 | 0.934 | 0.358 | 2.605 |
| 25° | 0.436 | 0.423 | 0.906 | 0.466 | 70° | 1.222 | 0.940 | 0.342 | 2.747 |
| 26° | 0.454 | 0.438 | 0.899 | 0.488 | 71° | 1.239 | 0.946 | 0.326 | 2.904 |
| 27° | 0.471 | 0.454 | 0.891 | 0.510 | 72° | 1.257 | 0.951 | 0.309 | 3.078 |
| 28° | 0.489 | 0.740 | 0.883 | 0.532 | 73° | 1.274 | 0.956 | 0.292 | 3.271 |
| 29° | 0.506 | 0.485 | 0.875 | 0.554 | 74° | 1.292 | 0.961 | 0.276 | 3.487 |
| 30° | 0.524 | 0.500 | 0.866 | 0.577 | 75° | 1.309 | 0.966 | 0.259 | 3.732 |
| 31° | 0.541 | 0.515 | 0.857 | 0.601 | 76° | 1.326 | 0.970 | 0.242 | 4.011 |
| 32° | 0.559 | 0.530 | 0.848 | 0.625 | 77° | 1.344 | 0.974 | 0.225 | 4.331 |
| 33° | 0.576 | 0.545 | 0.839 | 0.649 | 78° | 1.361 | 0.978 | 0.208 | 4.705 |
| 34° | 0.593 | 0.559 | 0.829 | 0.675 | 79° | 1.379 | 0.982 | 0.191 | 5.145 |
| 35° | 0.611 | 0.574 | 0.819 | 0.700 | 80° | 1.396 | 0.985 | 0.174 | 5.671 |
| 36° | 0.628 | 0.588 | 0.809 | 0.727 | 81° | 1.414 | 0.988 | 0.156 | 6.314 |
| 37° | 0.646 | 0.602 | 0.799 | 0.754 | 82° | 1.431 | 0.990 | 0.139 | 7.115 |
| 38° | 0.663 | 0.616 | 0.788 | 0.781 | 83° | 1.449 | 0.993 | 0.122 | 8.144 |
| 39° | 0.681 | 0.629 | 0.777 | 0.810 | 84° | 1.466 | 0.995 | 0.105 | 9.514 |
| 40° | 0.698 | 0.643 | 0.766 | 0.839 | 85° | 1.484 | 0.996 | 0.087 | 11.43 |
| 41° | 0.716 | 0.658 | 0.755 | 0.869 | 86° | 1.501 | 0.998 | 0.070 | 14.30 |
| 42° | 0.733 | 0.669 | 0.743 | 0.900 | 87° | 1.518 | 0.999 | 0.052 | 19.08 |
| 43° | 0.751 | 0.682 | 0.731 | 0.933 | 88° | 1.536 | 0.999 | 0.035 | 28.64 |
| 44° | 0.768 | 0.695 | 0.719 | 0.966 | 89° | 1.553 | 1.000 | 0.018 | 57.29 |
| 45° | 0.785 | 0.707 | 0.707 | 0.000 | 90° | 1.571 | 1.000 | 0.000 | |

# Index